Non-relativistic

Quantum Theory

Dynamics, Symmetry, and Geometry

Kai S Lam

California State Polytechnic University, Pomona, USA

Non-relativistic Quantum Theory

Dynamics, Symmetry, and Geometry

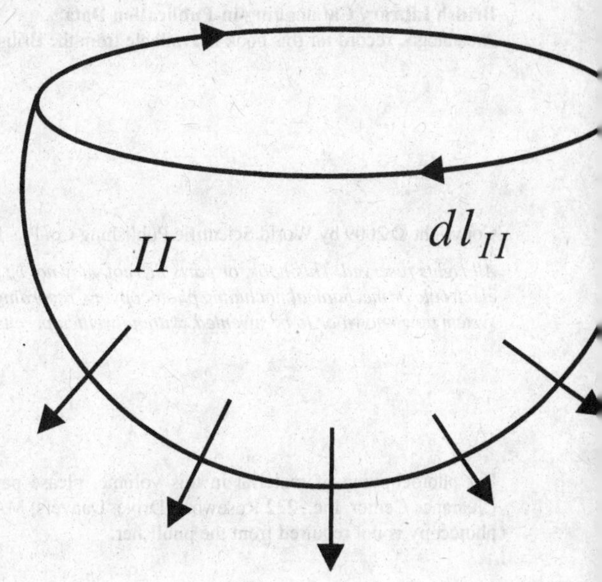

NEW JERSEY • LONDON • SINGAPORE • BEIJING • SHANGHAI • HONG KONG • TAIPEI • CHENNAI

Published by

World Scientific Publishing Co. Pte. Ltd.

5 Toh Tuck Link, Singapore 596224

USA office: 27 Warren Street, Suite 401-402, Hackensack, NJ 07601

UK office: 57 Shelton Street, Covent Garden, London WC2H 9HE

Library of Congress Cataloging-in-Publication Data

Lam, K. S. (Kai Shue), 1949–

Non-relativistic quantum theory : dynamics, symmetry, and geometry / Kai S Lam.

p. cm.

Includes bibliographical references and index.

ISBN-13: 978-981-4271-79-0 (hardcover : alk. paper)

ISBN-10: 981-4271-79-9 (hardcover : alk. paper)

1. Nonrelativistic quantum mechanics. 2. Quantum theory--Mathematics. I. Title.

QC174.24.N64L36 2009

530.12--dc22

2009016162

British Library Cataloguing-in-Publication Data

A catalogue record for this book is available from the British Library.

Printed in Singapore.

to Bonnie,
Nathan, Reuben, Aaron,
and
my Parents

Preface

The present book is intended mainly to be a textbook for physics students at the advanced undergraduate and beginning graduate levels, especially those with a theoretical inclination. Its chief purpose is to give a systematic introduction to the main ingredients of the fundamentals of quantum theory, with special emphasis on those aspects of *quantum dynamics*, *group theory* and *differential geometry* that are relevant in modern developments of the subject. Of these three areas, the first is without doubt of primary importance from the viewpoint of physics, and is usually given prominent place in most texts. Its treatment from first principles is also accorded the same status in the present text, and indeed occupies a good forty per cent of the material. But it is really the second and the third areas that the author wishes to highlight, in relation to the first. While the importance of group theory in quantum mechanics is almost universally recognized, and there indeed exist numerous excellent monographs and texts on this specialized topic at the graduate level, its inclusion at the undergraduate level is still a matter of controversy. The recognition of the importance of differential geometry and topology in quantum theory, for example, in the topics of *geometric phases* and *topological quantum numbers*, is of even more recent vintage. Although they are beginning to be mentioned in quantum mechanics texts, their mathematical underpinnings (in terms of the geometric notions of *fiber bundles*) are usually well hidden. The present book is an attempt to inject *both* group theory and differential geometry into the traditional framework of non-relativistic quantum dynamics in an integral fashion, beginning at the undergraduate level. In doing so it also seeks to present the fundamental physical ideas and mathematical framework in a unified manner, with the underlying purpose of providing the student with an overview of the key elements of the theory, as well as a solid preparation in calculational techniques which are universally applicable to diverse disciplines of physics where quantum mechanics plays a role. The focus of the book is on fundamental theory as well as practical calculations. Experimental aspects of the subject, though indispensable epistemologically in an overall appreciation, will not be emphasized for the sake of stylistic and thematic unity.

Perhaps a brief explanation of the title of the book is in order here. It is meant to convey the thematic unity of what the author sees as the three interlocking elements of modern quantum theory. While *dynamics* – the understanding of the space-time evolution of a physical system based on the relevant equations of motion – is the immediate objective of the theory, its effective use, and deep understanding, cannot be achieved without an appreciation of the constraints at a deep level due to the presence of various kinds of symmetries (both space-time and internal, the latter also referred to as *gauge symmetries*) in a physical system. These constraints manifest themselves as *conservation principles*, independent of many details of the system. Group theory *is* the mathematical language of symmetries in general; and *group representation theory*, in particular, provides the most natural framework for the description and indeed, the classification, of the quantum states of a system possessing various symmetries. Finally, the geometric notions of *complex vector and tensor bundles*, with identified *gauge groups*, provide the exact mathematical framework for the description of the dynamics of quantum states (represented by vectors in complex vector spaces) "living" in space-time and constrained by *gauge forces*. In many interesting cases, as in that of "conserved" *topological quantum numbers*, the physical constraints are manifested as a result of topological ones, at an even deeper level than geometrical ones. All of what is said in this paragraph so far is also relevant in relativistic quantum theory, namely, *quantum field theory*, although this subject, one that is much more involved technically than its non-relativistic counterpart, is not covered in the present text.

Some of the distinguishing features of the book are as follows. The topics are mostly introductory; yet the presentation is more advanced and in-depth than customary. Unlike many traditional texts, more care is devoted to the motivation and explanation of the relevant mathematical ideas than usual, and an attempt has been made to demonstrate their thematic unity beyond seemingly unrelated applications. On occasion, proofs of certain key mathematical theorems are presented in some detail; but only when they are relatively simple and of importance in applications. Calculational techniques are also stressed throughout. For this purpose, a mode of exposition with a degree of mathematical rigor somewhere between a physics text and mathematics text is adopted.

Another quite unique feature of the book is the inclusion in the first few chapters of the historical genesis of quantum theory derived from the canonical works of Schrödinger, Heisenberg, and Dirac. Again, this fascinating material is not usually included in modern texts at either the undergraduate or graduate levels. We hope that it will allow students to appreciate better the significant continuity of theoretical themes (if not the interpretational ones) in the passage from classical to quantum mechanics. On the other hand, there are numerous topics in non-relativistic quantum theory that are of vital importance but not treated in this text. These include, among others, many-particle systems, quantum statistics and non-relativistic quantum field theory, and topics of more current interests such as quantum decoherence, quantum entanglement, and quantum computation. It is hoped, however, that the present text will lay the requisite groundwork for these more specialized topics.

I have deliberately tried to arrange the material of this book into a large number of topically well-defined chapters, with each more or less focused on a single topic that is clearly discernable from the title. The chapters, however, are far from logically independent. Almost without exception, relatively unfamiliar concepts and results found anywhere in the text can be traced to a point of origin in an earlier part, and it is hoped that the organic connections between many topics will become apparent as they are discussed within different contexts. To facilitate this, some effort has been made to provide for clear cross references within the text by specific chapter numbers, equation numbers, or other means, as well as a comprehensive index. To promote further study, I have also not shied away from including more advanced literature as well as references to original work in the Bibliography. Problems of various degrees of difficulty, sometimes accompanied by generous hints, are interspersed within the text at strategic points. These are meant to complement the text in crucial ways; and the reader is encouraged to attempt as many as possible in order to derive maximal benefit from the book.

Organizationally, the flow of the book is as follows. The first three chapters give a portrait of the "pre-history" of quantum mechanics. Chapters 4 to 16 treat the fundamental principles of quantum dynamics, embodied in the Schrödinger equation, the Heisenberg equation, and the uncertainty relation. In these chapters, the basic mathematical framework of quantum mechanics employing the theory of self-adjoint operators in Hilbert spaces is also introduced, and discussed by means of several specific applications. Beginning with Chapter 17, and running through Chapter 30, the relevance of the notion of symmetries in quantum theory and its mathematical description by means of group theory are discussed and illustrated by various examples. The main groups discussed are the Lie groups $SU(2)$ and $SO(3)$ (for their relevance in rotational invariance), and the discrete symmetric groups (for their relevance in systems of identical particles). Chapters 31 through 37 resume with more conventional topics within the purview of dynamics, where the focus is on specific perturbative and non-perturbative techniques of the solution of the Schrödinger equation. The final part of the book, from Chapters 38 to 44, deals with the applications of topology and geometry in quantum mechanics. The main point of entry into this fascinating area is the physics of geometric phases. In this part, we also provide a rather unconventional introduction of sorts to the mathematical machinery of homotopy, homology and de Rham cohomology, differential forms, connections on vector bundles, Chern classes, and Chern-Simons classes. There is probably more than sufficient material in this book for an entire academic year's worth of instruction. It can be used in a sequential fashion; but the arrangement of topics as described above may also allow for the selection of groups of chapters to focus on any of the three main themes of the book.

In terms of the physics background required of the reader, she/he is expected to be familiar with the fundamentals of classical mechanics and electromagnetism at the advanced undergraduate level, including an appreciation of Lagrangian and Hamiltonian mechanics, and the concept of gauge transformations of the electromagnetic vector potential. A course in introductory modern physics at the sophomore level, including some exposure to the beginnings of quantum theory, is not absolutely required but will be useful. The requisite mathematical background is perhaps not so easily spelled out. In general, a working knowledge of those topics usually covered in a mathematical physics sequence at the junior to senior level is also assumed. This include, roughly speaking, vector calculus, some ordinary and partial differential equations, some complex analysis and special functions, and especially, a good dose of linear algebra. Within the text, however, a body of mathematical knowledge much beyond these items will be presented. The rudiments of functional analysis (Hilbert space theory) are developed, as far as needed, in the text itself, as are those of the more non-traditional topics of group theory and topology/differential geometry. The presentation of these latter mathematical topics in a physics text presents special challenges. As far as possible, and in general, I aimed at providing some physical motivation before delving into the exposition of the mathematics; and abstract mathematical concepts and facts (of which there are unavoidably many in this book) are introduced in light of hopefully more intuitive and concrete physical examples. To this end, I have, somewhat ambivalently, refrained from the more usual practice of relegating the collection of useful mathematical definitions and theorems to appendices, and have, instead, tried to "weave in" the mathematics with the physics. I realize that this is done at the price of mathematical rigor, completeness, and at times even accuracy, and only hope that the physics reader will be more engaged, by being assured at each point that the mathematics will be put to good use, and by not having to skip back and forth between the main text and the appendices. My modest wish is that the physics student, after a serious perusal of the text, will be persuaded of the essential importance of group theory and differential geometry in physics, and motivated to engage in a more systematic study of these subjects. On the other hand, I will also be thrilled if an occasional mathematics reader will be enticed to look within these pages for some non-trivial use of abstract mathematical formalisms in fundamental physics problems.

The first half or so of the book had been developed from an evolving set of class notes for an upper-division quantum mechanics sequence that I have taught at Cal Poly Pomona, on and off, since the early 1990's. These notes were made available to students, and many of them have provided valuable feedback and encouragement over the years. In a very real sense, they were the motivating force for me to write the book. Motivation and even a strong desire to write would not have been sufficient for progress, had not another key requirement – time – been generously fulfilled by the Faculty Sabbatical Program of the California State University. The late, preeminent geometer, Shiing-Shen Chern, through his generous and patient guidance on a collaborative textbook on differential geometry (Chern, Chen and Lam 1999), taught me most of what is found in these pages on that subject, and imparted to me a heightened appreciation of the often tortured and nuanced relationship between physics and mathematics. Dr. Soumya Chakravarti, my colleague in the Physics Department at Cal Poly Pomona, has kindly and with good humor acted as a sounding board for many of my half-baked ideas on how to present mathematical concepts convincingly in a physics text, and provided many valuable suggestions. Dr. Ertan Salik, another Physics colleague of mine, and Mr. Hector Maciel, also from our Physics Department, have both provided useful advice on the preparation of the figures. Ms. E. H. Chionh, my able editor at World Scientific, has from the start guided me through the intricacies of the publication aspects of this project expertly, efficiently, and professionally. My wife Dr. Bonnie Buratti, and my three sons, Nathan, Reuben, and Aaron, who have transitioned from boys to fine young adults over the years that the writing of this book was in progress, have all given me indispensable sustenance, both intellectual and emotional, through the ups and downs of this project. Finally, my mother steered me towards the path of confidence in school geometry at a critical time in my early years when my interests and competence in elementary mathematics were fast declining. To each and all of these parties, and many other unsung heroes who have impacted me on my professional life, I owe a heartfelt debt of gratitude and appreciation.

Kai S. Lam
California State Polytechnic University, Pomona *May 2009*

Contents

Chapter 1

How Did Schrödinger Get His Equation?

The short answer to the question posed in the title of this chapter is: from Hamilton's formulation of classical mechanics.

In one of his seminal papers, "*Quantization as a Problem of Proper Values (Part II)*" (Schrödinger 1926, an English translation of which can be found in Schrödinger 1982), Schrödinger recognized and stressed the importance of the following pair of correspondences:

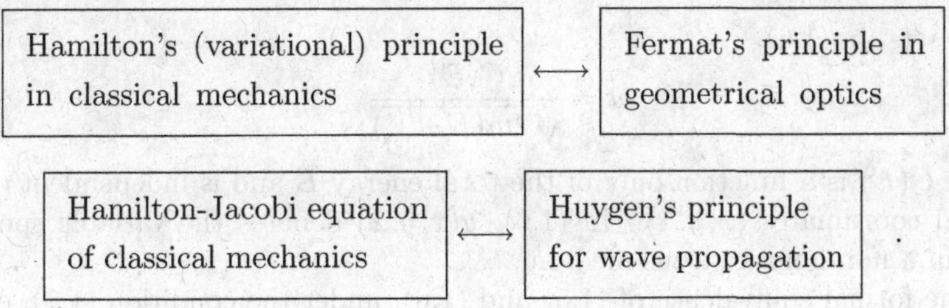

In this paper Schrödinger exploited fully the analogy between the Hamiltonian formulation of classical mechanics and optics. This is one of the most fascinating episodes of theory building in the history of physics. In this chapter we will give a brief and heuristic account of the main ideas behind it.

Consider the classical motion of a point particle from point A to point B in three-dimensional Euclidean space under the influence of a given potential energy function $V(x, y, z)$. The total energy $E = T + V$, where T is the kinetic energy of the particle, is conserved. Out of the totality of classical paths beginning at point A at time t_1 and ending at point B at time t_2, **Hamilton's variational principle** picks out the correct one. This principle states that

$$\boxed{\delta \int_{t_1}^{t_2} L(q, \dot{q})\, dt = 0} \quad , \tag{1.1}$$

where $L = T - V$, called the **Lagrangian** of the system, is a function of the (collective) spatial coordinates q and their time derivatives $\dot{q}$. For $E =$ constant,

$$L = T - V = T - (E - T) = 2T - E\,. \tag{1.2}$$

Then Hamilton's principle translates to

$$\delta \int_{t_1}^{t_2} 2T\,dt = 0\,, \tag{1.3}$$

which is the form of Hamilton's principle given by Maupertuis. Now, since $T = p^2/2m$,

$$2T = m\left(\frac{ds}{dt}\right)^2 = p\frac{ds}{dt} = \sqrt{2mT}\left(\frac{ds}{dt}\right) = \sqrt{2m(E-V)}\,\frac{ds}{dt}\,. \tag{1.4}$$

Hence (1.3) implies

$$\delta \int_A^B \sqrt{2m(E-V)}\,ds = 0\,. \tag{1.5}$$

Hamilton noticed that (1.5) is formally the same as **Fermat's principle of least time** in the geometrical optics of non-homogeneous media:

$$\delta \int_A^B \frac{ds}{u} = 0\,, \tag{1.6}$$

if one sets, in (1.6),

$$u = \frac{C(E)}{\sqrt{2m(E-V)}}\,, \tag{1.7}$$

where $C(E)$ is a function only of the total energy E and is independent of the spatial coordinates (x, y, z). In (1.6), $u(x, y, z)$ denotes the variable speed of light in a non-homogeneous medium.

The formal equivalence of (1.5) and (1.6), under the condition (1.7), establishes the correspondence between Hamilton's principle for the classical mechanical motion of a particle on the one hand, and Fermat's principle for the motion of light rays in geometrical optics, on the other.

It is important to distinguish between the two velocities $u(E;\, x, y, z)$ [given by (1.7)] and

$$v = \frac{1}{m}\sqrt{2m(E-V)}\,. \tag{1.8}$$

While u is the velocity of a light ray, v is the velocity of a mass point. The fact that u is a function of E suggests that we are also dealing with a **dispersive medium**, namely, one in which the velocity of light depends on the frequency of the light ray, in addition to a non-homogeneous one. This is implied by Einstein's formula for the energy of photons:

$$E = hf\,, \tag{1.9}$$

where h is Planck's constant.

The two velocities $u(E;x,y,z)$ [(1.7)] and $v(E;x,y,z)$ [(1.8)] cannot be the same for arbitrary potential energy function $V(x,y,z)$ since $C(E)$ is not a function of (x,y,z). To establish a connection between u and v, we first note that u, as the velocity of a light ray in geometrical optics, is also the **phase velocity** of a light wave in a non-homogeneous and dispersive medium. Consider a **localized wave packet** constructed from a superposition of plane waves with different frequencies. Corresponding to a wave packet one has the **group velocity** v_g given by

$$v_g = \frac{d\omega}{dk} = \frac{df}{d\left(\dfrac{1}{\lambda}\right)} = \frac{df}{d\left(\dfrac{f}{u}\right)} \,. \tag{1.10}$$

Thus

$$\frac{1}{v_g} = \frac{d}{df}\left(\frac{f}{u}\right) = \frac{d}{dE}\left(\frac{E}{u}\right) , \tag{1.11}$$

where in the second equality of the above equation we have used Einstein's formula (1.9).

At this point we impose the condition

$$v = v_g \,. \tag{1.12}$$

Then, using (1.7) and (1.8) in (1.11), we have

$$\frac{d}{dE}\left(\frac{E\sqrt{2m(E-V)}}{C(E)}\right) = \frac{m}{\sqrt{2m(E-V)}} = \frac{d}{dE}\sqrt{2m(E-V)} \,, \tag{1.13}$$

which implies

$$\frac{E\sqrt{2m(E-V)}}{C(E)} = \sqrt{2m(E-V)} + \text{a quantity independent of } E \,, \tag{1.14}$$

or

$$\left(\frac{E}{C(E)} - 1\right)\sqrt{2m(E-V)} = \text{a quantity independent of } E \,. \tag{1.15}$$

In order for the last equation to hold, it must be true that

$$C(E) = E \,. \tag{1.16}$$

It then follows from (1.7) that

$$u(E;x,y,z) = \frac{E}{\sqrt{2m(E-V)}} \,. \tag{1.17}$$

Problem 1.1 Consider a wave packet

$$\psi(\boldsymbol{x}) = \int d^3k\, a(\boldsymbol{k}) e^{i(\boldsymbol{k}\cdot\boldsymbol{x}-\omega t)} \quad ,$$

where $\omega(\boldsymbol{k})$ is a function of $\boldsymbol{k}$ (given by a dispersion relation). Assume that $a(\boldsymbol{k})$ is significantly different from zero only in some small region around $\boldsymbol{k}_0$. The position of the center of the packet is then given by

$$\langle \boldsymbol{x}(t) \rangle = \frac{\int d^3x\, \boldsymbol{x} |\psi|^2}{\int d^3x\, |\psi|^2} \,.$$

Show that

$$\langle \boldsymbol{x}(t) \rangle = \boldsymbol{x}_0 + \boldsymbol{v}_g t \quad ,$$

where

$$\boldsymbol{x}_0 \equiv \frac{i \int d^3k\, a^*(\boldsymbol{k}) \nabla_{\boldsymbol{k}} a(\boldsymbol{k})}{\int d^3k\, a^* a} \,,$$

and the group velocity of the packet is given by

$$\boldsymbol{v}_g = \langle \nabla_{\boldsymbol{k}} \omega \rangle \equiv \frac{\int d^3k\, |\, a(\boldsymbol{k})\,|^2 \nabla_{\boldsymbol{k}} \omega(\boldsymbol{k})}{\int d^3k\, |\, a(\boldsymbol{k})\,|^2} \,.$$

Now let us take a more formal look at how Hamilton's formulation of classical mechanics gives rise to Schrödinger's wave mechanics, this time via the so-called **Hamilton-Jacobi equation** :

$$\boxed{\frac{\partial W}{\partial t} + T\left(q^i, \frac{\partial W}{\partial q^i}\right) + V(q^i) = 0} \quad , \tag{1.18}$$

where $W = W(q^i, P_i, t)$ is the **generating function** of the **canonical transformation**

$$(q^i, p_i) \longrightarrow (Q^i, P_i) \,, \qquad H \longrightarrow H + \frac{\partial W}{\partial t} \,, \tag{1.19}$$

with

$$p_i = \frac{\partial W}{\partial q^i} \,, \qquad Q^i = \frac{\partial W}{\partial P_i} \,. \tag{1.20}$$

The new variables P_i are usually taken to be the constants of motion if the Hamiltonian is **integrable**. One of these is the total energy E. Let

$$W = -Et + S(q^i) \tag{1.21}$$

and $T = p^2/(2m)$. Then the Hamilton-Jacobi equation yields

$$-E + \frac{(\nabla S)^2}{2m} + V = 0 \,, \tag{1.22}$$

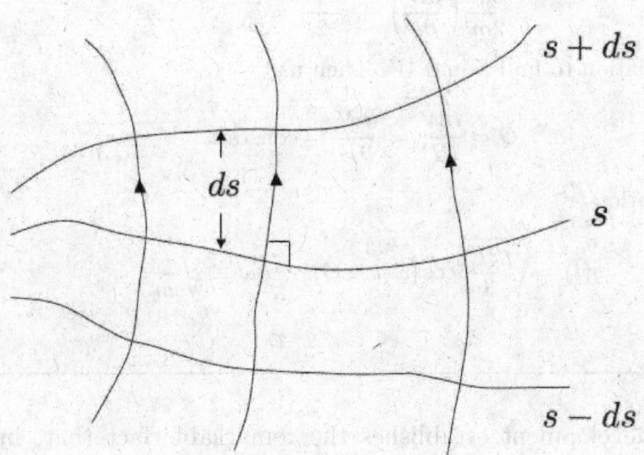

Fig. 1.1

or

$$ds = \frac{dS}{\sqrt{2m(E-V)}} \ , \tag{1.23}$$

where ds denotes the direction normal to the surfaces of constant "phase" S. (See Fig. 1.1.) The picture presented by the Hamilton-Jacobi equation is thus the propagation of surfaces of constant phase (constant value of W) in time. Thus, from (1.21), the constant phase condition $dW = 0$ implies $dS = E\,dt$, which, together with (1.23), gives

$$u = \frac{ds}{dt} = \frac{E}{\sqrt{2m(E-V)}} \ . \tag{1.24}$$

This, of course, is none other than (1.17)!

Problem 1.2 Solve the simple harmonic oscillator problem the hard way: using the Hamilton-Jacobi equation. Let

$$H = \frac{p^2}{2m} + \frac{kq^2}{2} \ .$$

Set

$$W(q,E,t) = S(q) - Et \ , \qquad (P = E) \ .$$

Then the Hamilton-Jacobi equation becomes

$$\frac{1}{2m}\left(\frac{\partial W}{\partial q}\right)^2 + \frac{kq^2}{2} + \frac{\partial W}{\partial t} = 0 \ ,$$

or

$$\frac{1}{2m}\left(\frac{\partial S}{\partial q}\right)^2 + \frac{kq^2}{2} = E \quad .$$

Integrate this equation to find S and W. Then use

$$Q = \frac{\partial W}{\partial P} = \frac{\partial W}{\partial E} = \text{constant}$$

to obtain the solution

$$q(t) = \sqrt{\frac{2E}{k}}\cos\{\omega(t+Q)\}\,, \quad \omega \equiv \sqrt{\frac{k}{m}} \quad .$$

The above development establishes the remarkable fact that, inherent in the Hamiltonian formulation of classical mechanics, there is a picture of wave propagation. Now, whereas Hamilton explicitly recognized the analogy between classical mechanics and geometrical optics, an insight that Schrödinger called "powerful and momentous", Schrödinger himself went one step further and established a relationship between the wave mechanics of particles and physical (wave) optics. The following diagram illustrates the amazing interplay between the relevant theories.

Hamiltonian mechanics	$\overset{\text{(Hamilton)}}{\longleftrightarrow}$	geometrical optics
↓ (Schrödinger)		↓ (classical wave theory)
wave mechanics	$\longleftrightarrow$	physical optics .

Let us now see how Schrödinger provided the crucial steps linking Hamiltonian mechanics to wave mechanics.

The classical **wave equation** in physical optics for a wave function $\Psi(x, y, z; t)$ is given by

$$\boxed{\nabla^2\Psi = \frac{1}{u^2}\frac{\partial^2\Psi}{\partial t^2}} \quad . \tag{1.25}$$

The solution of this equation for a monochromatic wave of frequency f must be of the form

$$\Psi(x, y, z; t) = \psi(x, y, z)\, e^{-2\pi i f t} \quad . \tag{1.26}$$

Plugging this into (1.25), we obtain the time-independent equation

$$\nabla^2\psi + \frac{4\pi^2 f^2}{u^2}\psi = 0 \quad , \tag{1.27}$$

where u is the phase velocity of the wave. Now use (1.17) for u in (1.27) to get

$$\nabla^2\psi + \frac{4\pi^2 f^2 \cdot 2m(E-V)}{E^2}\,\psi = 0 \quad . \tag{1.28}$$

Finally, with the help of Einstein's equation $E = hf$, one arrives at the so-called **time-independent Schrödinger equation** :

$$\nabla^2\psi + \frac{8\pi^2 m}{h^2}(E-V)\psi = 0 \quad . \tag{1.29}$$

To restore the time dependence, we multiply (1.29) by $\exp(-2\pi iEt/h)$:

$$\nabla^2(\psi e^{-2\pi iEt/h}) + \frac{8\pi^2 m}{h^2}E\psi e^{-2\pi iEt/h} - \frac{8\pi^2 mV}{h^2}e^{-2\pi iEt/h} = 0 \quad . \tag{1.30}$$

Recognizing that the second term on the LHS is equal to $\dfrac{4\pi im}{h}\dfrac{\partial}{\partial t}(\psi e^{-2\pi iEt/h})$ we can clean up the above equation by rewriting $\psi e^{-2\pi iEt/h}$ as ψ and obtain

$$\nabla^2\psi - \frac{8\pi^2 m}{h^2}V\psi = -4\pi i\frac{m}{h}\frac{\partial\psi}{\partial t} \quad . \tag{1.31}$$

The standard form of the so-called **time-dependent Schrödinger equation** is finally obtained by multiplying both sides of (1.31) by $-\hbar^2/(2m)$, where $\hbar \equiv h/(2\pi)$:

$$\boxed{-\frac{\hbar^2}{2m}\nabla^2\psi + V\psi = i\hbar\frac{\partial\psi}{\partial t}} \quad . \tag{1.32}$$

This time-dependent equation is a trivial generalization of the time-independent equation (1.29) *if* the potential function V is time-independent. But Schrödinger took the bold step of postulating that (1.32) is a completely general equation, even though V may be a function of time [in which case the time-dependence of ψ will no longer be of the simple form $\exp(-2\pi iEt/h)$]. He offered no logical justification and only claimed that "success will justify our procedure". The breath-taking development of quantum mechanics would soon vindicate his leap of faith, at least as far as non-relativistic quantum mechanics was concerned.

It is interesting to note that if we make the formal correspondence

$$\boxed{p_i \longleftrightarrow \frac{\hbar}{i}\frac{\partial}{\partial q^i}} \quad , \tag{1.33}$$

where p_i is the **canonical momentum** conjugate to the **canonical coordinate** q^i, we obtain the "operator" equation

$$H \equiv \frac{p^2}{2m} + V = -\frac{\hbar^2}{2m}\nabla^2 + V \quad . \tag{1.34}$$

H is called the **Hamiltonian operator** (operating on the function ψ). The time-independent Schrödinger equation then appears as

$$\boxed{H\psi \equiv \left(-\frac{\hbar^2}{2m}\nabla^2 + V\right)\psi = E\psi} \quad . \tag{1.35}$$

The function $\psi(x, y, z; t)$ is called the **wave function** of a non-relativistic quantum mechanical system, whose dynamics is governed by the time-dependent Schrödinger equation (1.32), in which the potential function V may depend explicitly on time.

The big question is: What does ψ mean, physically?

We cannot answer this question definitively at this time, except to say that knowledge of ψ allows one to know everything possible about the system, in the sense that one can use ψ to maximally predict all observational results obtained by performing experiments on the system. The question, in fact, has never been answered definitively, and has been the subject of heated debates since the inception of quantum mechanics. This is a big topic, and goes under the name of "**the interpretation of quantum mechanics**". Suffice it to mention at this time that, under the most commonly accepted interpretation (the so-called **Copenhagen interpretation**), the quantity $\psi^*\psi$ is interpreted as the spatial probability density of a quantum mechanical object described by the wave function at the time t. We will consider this question in more detail in Chapters 10 and 11.

Problem 1.3 Consider a free particle (not subjected to any force) of mass m moving in 3-dimensional Euclidean space, and its classical trajectories starting at time t_0 at positions $\boldsymbol{x}_0$ and ending at time t at (one) position $\boldsymbol{x}$. The **Green's function (propagator)** relating two space-time points is defined to be the function $G(\boldsymbol{x} - \boldsymbol{x}_0; t - t_0)$ satisfying the following relationship between the Schrödinger wave functions at different space-time points:

$$\psi(\boldsymbol{x}, t) = \int d^3x_0\, G(\boldsymbol{x} - \boldsymbol{x}_0; t - t_0)\psi(\boldsymbol{x}_0, t_0) \quad ,$$

where the integral is over all space. Show that the free-particle Green's function is given by

$$G(\boldsymbol{x} - \boldsymbol{x}_0; t - t_0) = \left(\frac{m}{2\pi i\hbar(t - t_0)}\right)^{3/2} \exp\left\{\frac{im|\,\boldsymbol{x} - \boldsymbol{x}_0\,|^2}{2\hbar(t - t_0)}\right\} \quad ,$$

by plugging this expression for G into the above defining equation for the Green's function and showing that the resulting expression for the wave function $\psi(\boldsymbol{x}, t)$ satisfies the time-dependent Schrödinger equation.

Problem 1.4 Consider a free particle of mass m which, at $t = 0$, is prepared in the state

$$\begin{aligned} \psi(\boldsymbol{x}, 0) &= 1/\sqrt{V}\,, & (x \le R)\,, \\ &= 0\,, & (x > R)\,, \end{aligned}$$

where $V = 4\pi R^3/3$. Show that for $x \gg R$, and for all t,

$$\psi(\boldsymbol{x}, t) = \left(\frac{2}{\pi V}\right)^{1/2} \left(\frac{aR}{ix^2}\right)^{3/2} \left[\sin\left(\frac{x}{a}\right) - \frac{x}{a}\cos\left(\frac{x}{a}\right)\right] \exp\{ix^2 m/(2\hbar t)\}\,,$$

where $a \equiv \hbar t/(mR)$. What is the significance of the length a? Determine the behavior of the probability distribution in the limits $x \gg a$ and $x \ll a$. [*Hint*: Use the free particle Green's function to compute $\psi(\boldsymbol{x}, t)$ from $\psi(\boldsymbol{x}_0, 0)$.]

Problem 1.5 Consider the one-particle Schrödinger equation

$$i\hbar\frac{\partial}{\partial t}\psi(\boldsymbol{x}, t) = -\frac{\hbar^2}{2m}\nabla^2\psi(\boldsymbol{x}, t) + [\,V_1(\boldsymbol{x}) + iV_2(\boldsymbol{x})\,]\,\psi(\boldsymbol{x}, t)\quad,$$

where V_1 and V_2 are real functions. If we interpret $\rho = \psi^*\psi$ as the spatial probability density at the time t, show that the probability is not conserved, and give an expression for the rate at which probability is "lost" or "gained" in a spatial volume Ω. *Hint*: Compute

$$\frac{\partial}{\partial t}\int d^3x\,\psi^*(\boldsymbol{x}, t)\psi(\boldsymbol{x}, t)\,,$$

and show that the **equation of continuity**

$$\nabla\cdot\boldsymbol{j} + \frac{\partial\rho}{\partial t} = 0$$

is *not* satisfied, where

$$\boldsymbol{j} \equiv -\frac{i\hbar}{2m}\left(\psi^*\nabla\psi - \psi\nabla\psi^*\right)$$

can be interpreted as the quantum mechanical **current density**.

We close this chapter with some technical matters of the utmost importance. Define the **commutator** of two operators A and B, denoted by $[\,A, B\,]$, by

$$[\,A, B\,] \equiv AB - BA\quad. \tag{1.36}$$

Then (1.33) implies the following commutation relations for q^i and p_i, regarded as operators:

$$[\,q^i, p_j\,] = i\hbar\,\delta^i_j, \quad [\,q^i, q^j\,] = [\,p_i, p_j\,] = 0\quad. \tag{1.37}$$

These are also called the **quantum conditions** for the canonical coordinate and canonical momentum operators. As will be seen later (in Chapter 11), they constitute the mathematical basis of **Heisenberg's Uncertainty Principle**.

In the so-called Schrödinger picture, q^i and $p_j = \frac{\hbar}{i}\frac{\partial}{\partial q^j}$ are all operators on **square-integrable functions** $\psi(q^i)$. These are defined to be functions satisfying the condition

$$\int dq^1 \dots dq^{3N}\, \psi^*\psi < \infty \quad , \tag{1.38}$$

where N is the total number of particles in the system. Were this condition to be violated, the Copenhagen interpretation of the wave function mentioned above would not make sense.

Finally, we mention that if ψ is a solution of the time-dependent Schrödinger equation (1.32), then, for systems of spinless particles, ψ^* (the complex conjugate of ψ) is a solution of the so-called **time-reversed** Schrödinger equation, which is obtained by replacing t in (1.32) by $-t$. This can be shown easily by taking the complex conjugate of (1.32):

$$-\frac{\hbar^2}{2m}\nabla^2\psi^* + V\psi^* = -i\hbar\frac{\partial\psi^*}{\partial t} = i\hbar\frac{\partial\psi^*}{\partial(-t)} \; , \tag{1.39}$$

where we have assumed that V is real.

Problem 1.6 Verify the quantum conditions (1.37) by using the correspondence (1.33).

Problem 1.7 In special relativity the relationship between the energy E and the momentum $\boldsymbol{p}$ of a free particle of rest mass m is given by

$$E^2 = p^2c^2 + m^2c^4 \; ,$$

where c is the speed of light. Use the quantum translation keys

$$\boldsymbol{p} \longrightarrow -i\hbar\nabla \quad \text{and} \quad E \longrightarrow i\hbar\frac{\partial}{\partial t}$$

to construct a relativistic wave equation for the particle. (The energy correspondence will be discussed in more detail in Chapter 17.) The resulting equation

$$\left(\nabla^2 - \frac{1}{c^2}\frac{\partial^2}{\partial t^2}\right)\psi = \frac{m^2c^2}{\hbar^2}\psi$$

is called the **Klein-Gordon equation** for a spin-zero particle. For a massless particle ($m = 0$, as for the photon), the **wave equation** results. Note that if we had used the non-relativistic relationship $E = p^2/(2m)$ we would have ended up with the non-relativistic Schrödinger equation instead.

Chapter 2

Heisenberg's Matrix Mechanics and Dirac's Re-creation of it

Matrix mechanics was actually developed somewhat before Schrödinger's work as described in the last chapter and was built upon Heisenberg's early positivistic outlook. Heisenberg reasoned as follows. Physical theory ought to focus on quantities immediately related to observables. Since the Bohr orbits used in the Early Quantum Theory are never directly observable, they should not play a fundamental role in the theory. Now all observable quantities, such as emission frequencies, transition rates, etc., are always related to a pair of Bohr orbits, but never just a single one. Hence Heisenberg sought to build a quantum mechanics based on theoretical constructs linking the totality of all pairs of orbits of a system. Since one usually has an infinity of orbits (stationary states), the most natural mathematical object to use is an infinite matrix:

$$\begin{pmatrix} \times & \times & \times & \dots & \dots \\ \times & \times & \times & \dots & \dots \\ \times & \times & \times & \dots & \dots \\ & \vdots & \vdots & \vdots & \\ & \vdots & \vdots & \vdots & \end{pmatrix}.$$

Heisenberg conjectured that to each Newtonian observable, such as position q or momentum p, there corresponds an infinite matrix, and that the elements of these matrices would actually yield measurable (observable) quantities.

But there was a big problem. Newtonian observables commute; matrices do not in general. This fact caused Heisenberg great anxiety at first. But later it turned out to be the cornerstone of the theory. In fact, as pointed out earlier, it is the mathematical basis of the Uncertainty Principle.

We will not go into the details of exactly how Heisenberg and others (Born and Jordan) constructed matrix mechanics. This was laid out in a series of three seminal papers (Heisenberg 1925; Born and Jordan 1925; Born, Heisenberg and Jordan 1925, English translations of which can be found in Van der Waerden 1968). Instead we will present briefly an elegant account given by Dirac (Dirac 1925). In his paper Dirac observed: "*Heisenberg puts forward a new theory which suggests that it is not the equations of classical mechanics that are in any way at fault, but that the mathematical operations by which physical results are deduced from them require modification. All the information supplied by the classical theory can thus be made use of in the new theory*".

We go back to Hamilton's reformulation of Newtonian mechanics. In the Hamiltonian formulation, one encounters the so-called **Poisson brackets** of the functions of the canonical coordinates and momenta (q, p):

$$\boxed{\{u, v\} \equiv \sum_i \left(\frac{\partial u}{\partial q^i} \frac{\partial v}{\partial p_i} - \frac{\partial u}{\partial p_i} \frac{\partial v}{\partial q^i} \right)} \quad . \tag{2.1}$$

This definition immediately implies that

$$\{q^i, q^j\} = \{p_i, p_j\} = 0 \quad ; \quad \{q^i, p_j\} = \delta^i_j \quad . \tag{2.2}$$

The Poisson brackets are also invariant with respect to **canonical transformations** [cf. (1.19) and (1.20)]:

$$\{u, v\}_{q,p} = \{u, v\}_{Q,P} \quad . \tag{2.3}$$

Two other important properties of the Poisson brackets are

$$\{u, q^i\} = -\frac{\partial u}{\partial p_i} \, , \tag{2.4}$$

$$\{u, p_i\} = \frac{\partial u}{\partial q^i} \, , \tag{2.5}$$

which imply that

$$\{q^i, H\} = \frac{\partial H}{\partial p_i} \, , \tag{2.6}$$

$$\{p_i, H\} = -\frac{\partial H}{\partial q^i} \, , \tag{2.7}$$

where H is the Hamiltonian function.

By Hamilton's equations of motion

$$\frac{dq^i}{dt} = \frac{\partial H}{\partial p_i} \quad , \quad \frac{dp_i}{dt} = -\frac{\partial H}{\partial q^i} \quad , \tag{2.8}$$

(2.6) and (2.7) give

$$\frac{dq^i}{dt} = \{q^i, H\} \quad , \quad \frac{dp_i}{dt} = \{p_i, H\} \quad . \tag{2.9}$$

The time derivative of a general function $u(p_i, q^i)$ of the canonical coordinates and momenta is then obtained from Hamilton's equations of motion as follows:

$$\frac{du}{dt} = \frac{\partial u}{\partial q^i}\frac{dq^i}{dt} + \frac{\partial u}{\partial p_i}\frac{dp_i}{dt} + \frac{\partial u}{\partial t} = \frac{\partial u}{\partial q^i}\frac{\partial H}{\partial p_i} - \frac{\partial u}{\partial p_i}\frac{\partial H}{\partial q^i} + \frac{\partial u}{\partial t} \quad , \tag{2.10}$$

that is,

$$\boxed{\frac{du}{dt} = \{u, H\} + \frac{\partial u}{\partial t}} \quad . \tag{2.11}$$

Problem 2.1 Using the definition of the Poisson bracket given by (2.1), prove (2.2), (2.4), and (2.5), and the fact that Poisson brackets are invariant under canonical transformations [(2.3)].

Dirac gave the following recipe to convert (2.11) into a quantum mechanical equation of motion for dynamical quantities:

$$\boxed{\{\,a\,,b\,\} \longrightarrow \frac{[\,a\,,\,b\,]}{i\hbar}} \quad , \tag{2.12}$$

where $[a, b] \equiv ab - ba$ is the commutator of the operators (or matrices) a and b. Thus the correct quantum mechanical equation of motion is

$$\boxed{\frac{du}{dt} = \frac{[u, H]}{i\hbar} + \frac{\partial u}{\partial t}} \quad , \tag{2.13}$$

where u and H are in general infinite matrices. This is known as **Heisenberg's equation of motion.**

The recipe (2.12), together with (2.2), immediately yields the quantum conditions

$$[\,q^i\,,\,p_j\,] = i\hbar\,\delta^i_j \quad , \quad [\,q^i\,,\,q^j\,] = [\,p_i\,,\,p_j\,] = 0 \quad , \tag{2.14}$$

which are the same as (1.37), derived from quite a different route.

Dirac justifies recipe (2.12) by the following arguments (see Dirac 1967, pp. 86, 87). First observe that the classical Poisson brackets $\{\ ,\ \}$ satisfy the following properties:

$$\{\,u\,,v\,\} = -\{\,v\,,u\,\} \quad , \tag{2.15}$$

$$\begin{aligned} \{ u_1 + u_2 , v \} &= \{ u_1 , v \} + \{ u_2 , v \} , \\ \{ u , v_1 + v_2 \} &= \{ u , v_1 \} + \{ u , v_2 \} , \end{aligned} \tag{2.16}$$

$$\begin{aligned} \{ u_1 u_2 , v \} &= \{ u_1 , v \} u_2 + u_1 \{ u_2 , v \} , \\ \{ u , v_1 v_2 \} &= \{ u , v_1 \} v_2 + v_1 \{ u , v_2 \} , \end{aligned} \tag{2.17}$$

as well as the **Jacobi identity**

$$\boxed{\{ u , \{ v , w \}\} + \{ v , \{ w , u \}\} + \{ w , \{ u , v \}\} = 0} \quad . \tag{2.18}$$

Problem 2.2 Using the definition (2.1), verify the properties of the Poisson bracket as given by Eqs. (2.15) to (2.18).

Now evaluate $\{ u_1 u_2 , v_1 v_2 \}$ in two different ways, using one and then the other of the two formulas in (2.17) first:

$$\begin{aligned} &\{ u_1 u_2 , v_1 v_2 \} = \{ u_1 , v_1 v_2 \} u_2 + u_1 \{ u_2 , v_1 v_2 \} \\ &= (\{ u_1 , v_1 \} v_2 + v_1 \{ u_1 , v_2 \}) u_2 + u_1 (\{ u_2 , v_1 \} v_2 + v_1 \{ u_2 , v_2 \}) \\ &= \{ u_1 , v_1 \}, v_2 u_2 + v_1 \{ u_1 , v_2 \} u_2 + u_1 \{ u_2 , v_1 \} v_2 + u_1 v_1 \{ u_2 , v_2 \} \quad ; \end{aligned}$$

$$\begin{aligned} &\{ u_1 u_2 , v_1 v_2 \} = \{ u_1 u_2 , v_1 \} v_2 + v_1 \{ u_1 u_2 , v_2 \} \\ &= (\{ u_1 , v_1 \} u_2 + u_1 \{ u_2 , v_1 \}) v_2 + v_1 (\{ u_1 , v_2 \} u_2 + u_1 \{ u_2 , v_2 \}) \\ &= \{ u_1 , v_1 \}, u_2 v_2 + u_1 \{ u_2 , v_1 \} v_2 + v_1 \{ u_1 , v_2 \} u_2 + v_1 u_1 \{ u_2 , v_2 \} \quad . \end{aligned}$$

Equating these two results, we obtain

$$\{ u_1 , v_1 \} [u_2 , v_2] = [u_1 , v_1] \{ u_2 , v_2 \} \quad .$$

Since u_1 , u_2 , v_1 , v_2 are all independent of each other, the above condition implies

$$[u_1 , v_1] = i\hbar \{ u_1 , v_1 \} \quad , \quad [u_2 , v_2] = i\hbar \{ u_2 , v_2 \} \quad , \tag{2.19}$$

where $\hbar$ is a real number independent of the u's and v's, and has the dimensions of action (energy $\times$ time). The appearance of the imaginary unit i in (2.19) is of the utmost significance. Its introduction is dictated by the fact that the u's and v's must be **hermitian matrices** if they are to represent **dynamical variables (observables)** in quantum theory, where the mathematical property of **hermiticity** of a matrix is defined by the condition

$$u = u^\dagger \equiv (u^*)^T \quad .$$

In the above equation, $*$ denotes complex conjugation and T denotes the transpose of a matrix. $u^\dagger$ is called the **hermitian adjoint** of u.

The important mathematical requirement that observables be represented by hermitian matrices (operators) in the formalism of matrix mechanics is due to the physical requirement that measured values of observable quantities must be real. According to the formalism (discussed in Chapter 8), these are given by the eigenvalues of the matrices (or the **spectrum** of the corresponding operators). The property of hermiticity of a matrix guarantees that its eigenvalues must be real. (This will also be shown in Chapter 8.)

Comparing (2.11) and (2.13) we see that in the transition from classical to quantum mechanics, classical Poisson brackets, which are classical observables and are real, must be replaced by "quantum Poisson brackets", which are required to be hermitian. Note that $[u\,,\,v]$ is not necessarily hermitian even if u and v are. On the other hand, it is straightforward to see that, if u and v are hermitian, then $\pm i[u\,,\,v]$ are both hermitian:

$$\begin{aligned}(\pm\, i[u\,,\,v])^\dagger &= \mp\, i(uv - vu)^\dagger = \mp\, i(v^\dagger u^\dagger - u^\dagger v^\dagger) \\ &= \mp\, i(vu - uv) = \pm\, i(uv - vu) = \pm\, i[u\,,\,v] \quad .\end{aligned}$$

In (2.19) the choice $+i$ has been picked by convention.

Suppose u does not depend explicitly on time in (2.13), so that $\partial u/\partial t = 0$. Then the Heisenberg equation of motion reads

$$\frac{dq}{dt} = \frac{[\,q\,,\,H\,]}{i\hbar} \tag{2.20}$$

for a system with one degree of freedom. Equation (2.20) is actually a matrix equation, and thus represents (since q is an infinite matrix in general) an infinite set of coupled equations:

$$\frac{dq_{nm}}{dt} = \frac{(qH)_{nm} - (Hq)_{nm}}{i\hbar} = \sum_k \frac{(q_{nk}H_{km} - H_{nk}q_{km})}{i\hbar} \quad . \tag{2.21}$$

If we can find a **representation** in which the Hamiltonian H is diagonal, that is, $H_{nk} = \delta_{nk}H_{nn}$, then the equations become decoupled:

$$\begin{aligned}\frac{dq_{nm}}{dt} &= \sum_k (q_{nk}\delta_{km}H_{mm} - H_{nn}\delta_{nk}q_{km}) \,/\, (i\hbar) \\ &= q_{nm}(H_{mm} - H_{nn}) \,/\, (i\hbar) \\ &= -i\omega_{mn}q_{nm} = i\omega_{nm}q_{nm} \quad ,\end{aligned} \tag{2.22}$$

where

$$\omega_{mn} \equiv \frac{(H_{mm} - H_{nn})}{\hbar} \quad . \tag{2.23}$$

Equation (22) can be solved easily to obtain

$$q_{nm}(t) = q_{nm}(0)\, e^{i\omega_{nm}t} \quad . \tag{2.24}$$

Hence the fundamental problem of Heisenberg's matrix mechanics is to find infinite matrices q^i and p_i such that the quantum conditions (2.14) hold, and such that the Hamiltonian $H(q^1, \ldots, q^N; p_1, \ldots, p_N)$ becomes a diagonal matrix.

On the face of it, this seems to have nothing to do with Schrödinger's approach discussed in the last chapter. It is indeed one of the most remarkable results in the development of the quantum theory that the two approaches, Schrödinger's wave mechanics and Heisenberg's matrix mechanics, are in fact equivalent. This will be further discussed in Chapters 4 and 5.

Chapter 3

Dirac's Derivation of the Quantum Conditions

Before we go on to show the equivalence of Heisenberg's matrix mechanics and Schrödinger's wave mechanics, we pause to look at a different approach, again due to Dirac, to derive the quantum conditions (2.14). This approach more closely captures the actual historical development, since it shows clearly the genesis of the quantum conditions from the equations of classical mechancs, via the **correspondence principle**. [The details are given in (Dirac 1925).]

For a multiply-periodic nondegenerate classical dynamical system of N degrees of freedom, a dynamical variable, say q, can be expanded as a multiple Fourier series in the time t:

$$q(t) = \sum_{\alpha_1,\ldots,\alpha_N} q_{\alpha_1\ldots\alpha_N}\, e^{i(\alpha_1\omega_1+\cdots+\alpha_N\omega_N)t} = \sum_{\alpha} q_\alpha\, e^{i\boldsymbol{\alpha}\cdot\boldsymbol{\omega}t} \quad , \tag{3.1}$$

if the problem is **integrable**, that is, if there exist N constants of motion, $\kappa_1,\ldots,\kappa_N$, such that q_α and $\boldsymbol{\omega}$ are functions of $(\kappa_1,\ldots,\kappa_N)$. Then we can write

$$q_\kappa(t) = \sum_\alpha q_{\alpha\kappa}\, e^{i\boldsymbol{\alpha}\cdot\boldsymbol{\omega}_\kappa t} \quad . \tag{3.2}$$

Obviously, the following linear relationship is satisfied classically:

$$\boldsymbol{\alpha}\cdot\boldsymbol{\omega}_\kappa + \boldsymbol{\beta}\cdot\boldsymbol{\omega}_\kappa = (\boldsymbol{\alpha}+\boldsymbol{\beta})\cdot\boldsymbol{\omega}_\kappa \quad . \tag{3.3}$$

Heisenberg took the crucial step to replace (3.3) by the following equation:

$$\omega(n, n-\alpha) + \omega(n-\alpha, n-\alpha-\beta) = \omega(n, n-\alpha-\beta) \tag{3.4}$$

in the quantum theory; or

$$\omega(n, m) + \omega(m, k) = \omega(n, k) \quad , \tag{3.5}$$

where there is a one-to-one correspondence

$$n \longrightarrow (\kappa_1, \ldots, \kappa_N) \quad . \tag{3.6}$$

Classically, when two quantities a and b each of the form (3.2) are multiplied, we have

$$a_\kappa(t) b_\kappa(t) = \sum_\alpha \sum_\beta a_{\alpha\kappa} b_{\beta\kappa} \, e^{i(\boldsymbol{\alpha}+\boldsymbol{\beta})\cdot\boldsymbol{\omega}_\kappa t} = \sum_{\alpha,\beta} (ab)_{\alpha+\beta,\kappa} \, e^{i(\boldsymbol{\alpha}+\boldsymbol{\beta})\cdot\boldsymbol{\omega}_\kappa t} \quad , \tag{3.7}$$

where

$$(ab)_{\alpha+\beta,\kappa} \equiv a_{\alpha\kappa} b_{\beta\kappa} \quad . \tag{3.8}$$

Quantum mechanically, one makes the following replacements:

$$a_\kappa(t) \longrightarrow a_{nm}(t) \quad ; \quad a_{\alpha\kappa} \longrightarrow a_{nl} \quad , \tag{3.9}$$

so that the classical expression (3.7), with the leftmost side set equal to $(ab)_\kappa(t)$, is replaced by

$$\begin{aligned} (ab)_{nm}(t) &= \sum_l a_{nl} \, {}^{i\omega_{nl} t} \, b_{lm} \, e^{i\omega_{lm} t} \\ &= \left(\sum_l a_{nl} b_{lm} \right) e^{i\omega_{nm} t} = (ab)_{nm} \, e^{i\omega_{nm} t} \quad . \end{aligned} \tag{3.10}$$

We see that the quantities a_{nm} satisfy the matrix multiplication rule. This important point was first observed by M. Born.

Dirac asked the question: To what classical expression will the matrix element

$$(ab)_{nm}(t) - (ba)_{nm}(t)$$

correspond when n and m are large (i.e., for large quantum numbers)? In Fig. 3.1 we represent the non-degenerate energy levels of a quantum system labelled by the quantum numbers $(1, 2, 3, 4, \ldots, m, \ldots, n, \ldots)$. In the spirit of the correspondence principle, we assume that n, m are large, and $\alpha, \beta \ll n, m$. The latter assumption on the smallness of α, β is made so that differences can be replaced by derivatives multiplied by small changes in α and β.

Let

$$(\kappa_1, \ldots, \kappa_N) = \hbar(n_1, \ldots, n_N) \quad . \tag{3.11}$$

This is suggested by the **Bohr-Sommerfeld quantization rule**, where $\kappa_1, \ldots, \kappa_N$ may be the constant values assumed by the action variables, and $(n_1, \ldots, n_N)$ are integers. Suppose

$$a_{n,n-\alpha} \equiv a(n, n-\alpha) \longrightarrow a_{\alpha\kappa} \quad . \tag{3.12}$$

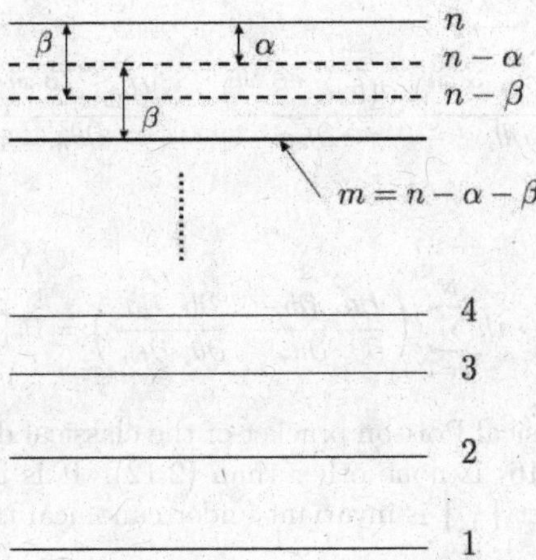

Fig. 3.1

Then

$$\begin{aligned}
&a(n, n-\alpha)b(n-\alpha, n-\alpha-\beta) - b(n, n-\beta)a(n-\beta, n-\alpha-\beta) \\
&= \{a(n, n-\alpha) - a(n-\beta, n-\alpha-\beta)\}b(n-\alpha, n-\alpha-\beta) \\
&- \{b(n, n-\beta) - b(n-\alpha, n-\alpha-\beta)\}a(n-\beta, n-\alpha-\beta) \\
&\longrightarrow \hbar \sum_{r=1}^{N} \left\{ \beta_r \frac{\partial a_{\alpha\kappa}}{\partial \kappa_r} b_{\beta\kappa} - \alpha_r \frac{\partial b_{\beta\kappa}}{\partial \kappa_r} a_{\alpha\kappa} \right\} \quad .
\end{aligned} \tag{3.13}$$

Let the **angle variables** be

$$\theta_r = \omega_r t \quad , \quad r = 1, \ldots, N \quad . \tag{3.14}$$

Then

$$\frac{\partial}{\partial \theta_r} \{ b_{\beta\kappa} \, e^{i\boldsymbol{\beta}\cdot\boldsymbol{\omega}t} \} = i\beta_r b_{\beta\kappa} e^{i\boldsymbol{\beta}\cdot\boldsymbol{\omega}t} \quad . \tag{3.15}$$

Multiplying the LHS of the $\longrightarrow$ relation in (3.13) by $\exp(i\omega_{n,n-\alpha-\beta}t)$ and the RHS by $\exp(i\boldsymbol{\alpha}\cdot\boldsymbol{\omega}t)\exp(i\boldsymbol{\beta}\cdot\boldsymbol{\omega}t)$ we have

$$\begin{aligned}
&(a_{n,n-\alpha}\, b_{n-\alpha,n-\alpha-\beta} - b_{n,n-\beta}\, a_{n-\beta,n-\alpha-\beta})\, e^{i\omega_{n,n-\alpha-\beta}t} \\
&\longrightarrow -i\hbar \sum_{r=1}^{N} \left\{ \frac{\partial}{\partial \kappa_r} (a_{\alpha\kappa}\, e^{i\boldsymbol{\alpha}\cdot\boldsymbol{\omega}t}) \frac{\partial}{\partial \theta_r} (b_{\beta\kappa}\, e^{i\boldsymbol{\beta}\cdot\boldsymbol{\omega}t}) - \frac{\partial}{\partial \kappa_r} (b_{\beta\kappa}\, e^{i\boldsymbol{\beta}\cdot\boldsymbol{\omega}t}) \frac{\partial}{\partial \theta_r} (a_{\alpha\kappa}\, e^{i\boldsymbol{\alpha}\cdot\boldsymbol{\omega}t}) \right\} \quad .
\end{aligned}$$

Hence

$$[a, b]_{nm} \equiv (ab - ba)_{nm} \longrightarrow$$

$$i\hbar \sum_{\alpha+\beta=n-m} \sum_{r=1}^{N} \left(\frac{\partial(a_{\alpha\kappa}\, e^{i\boldsymbol{\alpha}\cdot\boldsymbol{\omega}t})}{\partial\theta_r} \frac{\partial(b_{\beta\kappa}\, e^{i\boldsymbol{\beta}\cdot\boldsymbol{\omega}t})}{\partial\kappa_r} - \frac{\partial(b_{\beta\kappa}\, e^{i\boldsymbol{\beta}\cdot\boldsymbol{\omega}t})}{\partial\theta_r} \frac{\partial(a_{\alpha\kappa}\, e^{i\boldsymbol{\alpha}\cdot\boldsymbol{\omega}t})}{\partial\kappa_r} \right) .$$

Finally, as matrices,

$$[a, b] \equiv ab - ba \longrightarrow i\hbar \sum_{r=1}^{N} \left(\frac{\partial a}{\partial\theta_r} \frac{\partial b}{\partial\kappa_r} - \frac{\partial b}{\partial\theta_r} \frac{\partial a}{\partial\kappa_r} \right) = i\hbar\{ a, b \} \quad , \qquad (3.16)$$

where $\{a, b\}$ is the classical Poisson bracket of the classical dynamical variables a and b. Equation (3.16) is none other than (2.12). It is important to recall that the Poisson bracket $\{\ ,\ \}$ is invariant under canonical transformations.

Chapter 4

The Equivalence between Matrix Mechanics and Wave Mechanics

The mathematical equivalence between matrix mechanics and wave mechanics was first established by Schrödinger (Schrödinger 1926) and later by von Neumann (von Neumann 1955). The present chapter and the next will be devoted to addressing this problem. We will follow the approach given by von Neumann.

Recall that in Heisenberg's matrx mechanics [see the discussion following (2.24)], the fundamental problem is to find infinite matrices q^i and p_j such that the quantum conditions (2.14) hold, and that the Hamiltonian $H(q^i, p_j)$ is diagonal. As an example, we will consider the one-dimensional simple harmonic oscillator, which is probably the most important and fundamental example in quantum theory. (A full solution of this problem will be presented in Chapter 14.) The Hamiltonian is given by

$$H = \frac{p^2}{2m} + \frac{1}{2} m\omega^2 q^2 \quad , \tag{4.1}$$

where m is the mass of the oscillator and ω its natural frequency. Define the dimensionless coordinate and momentum as follows:

$$Q \equiv \sqrt{\frac{m\omega}{\hbar}}\, q \ , \quad P \equiv \frac{1}{\sqrt{m\hbar\omega}}\, p \quad . \tag{4.2}$$

Then

$$H = \frac{\hbar\omega}{2}\,(P^2 + Q^2) \quad . \tag{4.3}$$

Further define the (matrix) operator

$$a \equiv \frac{1}{\sqrt{2}}\,(Q + iP) \tag{4.4}$$

with its **hermitian conjugate** given by

$$a^\dagger = \frac{1}{\sqrt{2}}\,(Q - iP) \quad . \tag{4.5}$$

Then

$$Q = \frac{1}{\sqrt{2}}\,(a + a^\dagger) \quad , \quad P = -\frac{i}{\sqrt{2}}\,(a - a^\dagger) \quad , \tag{4.6}$$

from which it follows that

$$q = \sqrt{\frac{\hbar}{m\omega}}\,\frac{1}{\sqrt{2}}\,(a + a^\dagger) \quad , \quad p = \sqrt{m\hbar\omega}\left(-\frac{i}{\sqrt{2}}\right)(a - a^\dagger) \quad . \tag{4.7}$$

We claim that the infinite matrices

$$a = \begin{pmatrix} 0 & \sqrt{1} & 0 & 0 & \dots & \dots \\ 0 & 0 & \sqrt{2} & 0 & \dots & \dots \\ 0 & 0 & 0 & \sqrt{3} & 0 & \dots \\ & & \vdots & & \vdots & \\ & & \vdots & & \vdots & \end{pmatrix} , \tag{4.8}$$

and

$$a^\dagger = \begin{pmatrix} 0 & 0 & 0 & 0 & \dots & \dots \\ \sqrt{1} & 0 & 0 & 0 & \dots & \dots \\ 0 & \sqrt{2} & 0 & 0 & \dots & \dots \\ 0 & 0 & \sqrt{3} & 0 & \dots & \dots \\ & & \vdots & & \vdots & \\ & & \vdots & & \vdots & \end{pmatrix} \tag{4.9}$$

will do the trick, i.e., make the infinite matrices q and p [given by (4.7)] satisfy the quantum condition

$$[\,q,\,p\,] = i\hbar$$

and the Hamiltonian H diagonal. The latter fact can be verified easily by direct calculation on using (4.3) and (4.6):

$$Q = \frac{1}{\sqrt{2}}(a + a^\dagger) = \frac{1}{\sqrt{2}} \begin{pmatrix} 0 & \sqrt{1} & 0 & 0 & \cdots \\ \sqrt{1} & 0 & \sqrt{2} & 0 & \cdots \\ 0 & \sqrt{2} & 0 & \sqrt{3} & \cdots \\ 0 & 0 & \sqrt{3} & 0 & \cdots \\ & & \vdots & & \end{pmatrix} \quad , \tag{4.10}$$

$$P = -\frac{i}{\sqrt{2}}(a - a^\dagger) = -\frac{i}{\sqrt{2}} \begin{pmatrix} 0 & \sqrt{1} & 0 & 0 & \cdots \\ -\sqrt{1} & 0 & \sqrt{2} & 0 & \cdots \\ 0 & -\sqrt{2} & 0 & \sqrt{3} & \cdots \\ 0 & 0 & -\sqrt{3} & 0 & \cdots \\ & & \vdots & & \end{pmatrix} \quad ; \tag{4.11}$$

and thus

$$H = \frac{\hbar\omega}{2}(P^2 + Q^2) = \frac{\hbar\omega}{2} \begin{pmatrix} 1 & 0 & 0 & 0 & \cdots \\ 0 & 3 & 0 & 0 & \cdots \\ 0 & 0 & 5 & 0 & \cdots \\ 0 & 0 & 0 & 7 & \cdots \\ & & \vdots & & \end{pmatrix} \quad . \tag{4.12}$$

The diagonal elements of H are the energy eigenvalues of the simple harmonic oscillator, given by

$$E = \hbar\omega\,(n + \frac{1}{2}) \quad ; \quad n = 0, 1, 2, \ldots \quad . \tag{4.13}$$

We will study the simple harmonic oscillator in more detail in Chapter 14.

Problem 4.1 Show explicitly that the matrix expression for a given by (4.8) when plugged into the expressions for q and p given by (4.7) will make q and p satisfy the quantum condition $[\,q,\, p\,] = i\hbar$.

Returning to the general problem, we first assume that infinite matrices $\overline{q}^1,\ldots,\overline{q}^N;\overline{p}_1,\ldots,\overline{p}_N$ have been found such that the quantum conditions are satisfied, but such that

$$\overline{H} = \overline{H}(\overline{q}^1,\ldots,\overline{q}^N;\overline{p}_1,\ldots,\overline{p}_N)$$

is not diagonal. The desired matrices $(q^1,\ldots,q^N;p_1,\ldots,p_N)$ which render the Hamiltonian diagonal can be obtained by **similarity transformations** of the original matrices:

$$q^1 = S^{-1}\overline{q}^1 S\,,\ldots\,,\; q^N = S^{-1}\overline{q}^N S\;;\; p_1 = S^{-1}\overline{p}_1 S\,,\ldots\,,\; p_N = S^{-1}\overline{p}_N S\;, \tag{4.14}$$

where S is an invertible matrix. The quantum conditions are preserved under these transformations. Indeed

$$\begin{aligned}[q^i\,,\,p_j] &= q^i p_j - p_j q^i = S^{-1}\overline{q}^i S S^{-1}\overline{p}_j S - S^{-1}\overline{p}_j S S^{-1}\overline{q}^i S \\ &= S^{-1}\,[\overline{q}^i\,,\,\overline{p}_j\,]S = i\hbar\,\delta^i_j S^{-1}S = i\hbar\,\delta^i_j \quad .\end{aligned} \tag{4.15}$$

We will consider Hamiltonian functions that can be expressed as power series in the (q^i, p_j). Then

$$\begin{aligned}H &= H(q^1,\ldots,q^N\,;\,p_1,\ldots,p_N) \\ &= H(S^{-1}\overline{q}^1 S,\ldots,S^{-1}\overline{q}^N S\,;\,S^{-1}\overline{p}_1 S,\ldots,S^{-1}\overline{p}_N S) \\ &= S^{-1}\overline{H}(\overline{q}^1,\ldots,\overline{q}^N\,;\,\overline{p}_1,\ldots,\overline{p}_N)S \quad .\end{aligned} \tag{4.16}$$

Our problem is to diagonalize $\overline{H}$, that is, to find an invertible matrix S such that

$$H = S^{-1}\overline{H}S$$

is diagonal, that is,

$$H_{ij} = \delta_{ij}W_j \quad , \tag{4.17}$$

where W_i are the diagonal elements of H. We thus require

$$(SH)_{ij} = (\overline{H}S)_{ij} = \sum_k S_{ik}\delta_{kj}W_j \quad , \tag{4.18}$$

or

$$\sum_k \overline{H}_{ik}S_{kj} = W_j S_{ij} \quad . \tag{4.19}$$

The individual columns of the matrix S

$$\begin{pmatrix} S_{11} \\ S_{21} \\ \vdots \\ \vdots \end{pmatrix}\;,\quad \begin{pmatrix} S_{12} \\ S_{22} \\ \vdots \\ \vdots \end{pmatrix}\;,\quad \ldots\;,\quad \begin{pmatrix} S_{1j} \\ S_{2j} \\ \vdots \\ \vdots \end{pmatrix}\;,\ldots$$

and the corresponding diagonal elements W_j of the matrix H are therefore solutions of the **eigenvalue problem**

$$\sum_k \overline{H}_{ik} x_k = \lambda x_i \quad , \qquad (i = 1, 2, \ldots) \; , \tag{4.20}$$

or in matrix form

$$\overline{H} x = \lambda x \quad , \tag{4.21}$$

where x is a column matrix.

The trivial solution (all $x_i = 0$) is excluded since it would introduce a column consisting entirely of zeros in S and hence make it non-invertible. We will now show that all solutions of (4.21) are generated by the columns of S. Indeed, let x be a solution of (4.21) with eigenvalue λ. Then

$$S^{-1}\overline{H} x = \lambda S^{-1} x \quad .$$

But from the fact that $H = S^{-1}\overline{H} S$, we have $S^{-1}\overline{H} = H S^{-1}$. Hence

$$H S^{-1} x = \lambda S^{-1} x \quad . \tag{4.22}$$

Let $y \equiv S^{-1}x$. Then $Hy = \lambda y$. But H is diagonal by assumption. Writing its matrix elements as $H_{ij} = W_i \delta_{ij}$, we have

$$\sum_j W_i \delta_{ij} \, y_j = \lambda y_i \quad ,$$

or

$$(W_i - \lambda) y_i = 0 \quad . \tag{4.23}$$

Thus if $\lambda \neq W_i$, then $y_i = 0$. Suppose λ is equal to none of the W's, then, as a column matrix, $y = 0$. It follows that in this case, $x = Sy = 0$ also. Suppose λ is equal to only one of the W's, say W_i. Then (4.23) implies that y has the form

$$y = a\,(0, 0, \ldots, 1, \ldots, 0, \ldots)^T$$

where a is an arbitrary complex constant, T denotes transpose, and the only non-vanishing matrix element is the 1 occurring in the i-th row. In this case

$$x = Sy = a \begin{pmatrix} S_{1i} \\ S_{2i} \\ S_{3i} \\ \vdots \end{pmatrix} \quad .$$

Finally suppose λ is equal to a finite number of the W's, say

$$\lambda = W_1 = W_2 = \cdots = W_n \quad .$$

Then

$$y = a_1 \begin{pmatrix} 1 \\ 0 \\ 0 \\ \vdots \\ \vdots \end{pmatrix} + a_2 \begin{pmatrix} 0 \\ 1 \\ 0 \\ \vdots \\ \vdots \end{pmatrix} + \cdots + a_n \begin{pmatrix} 0 \\ \vdots \\ 1 \\ 0 \\ \vdots \end{pmatrix},$$

where $a_1, \ldots, a_n$ are arbitrary complex constants, and in the last column matrix the 1 occurs in the n-th row. It follows that

$$x = Sy = a_1 \begin{pmatrix} S_{11} \\ S_{21} \\ \vdots \\ \vdots \end{pmatrix} + a_2 \begin{pmatrix} S_{12} \\ S_{22} \\ \vdots \\ \vdots \end{pmatrix} + \cdots + a_n \begin{pmatrix} S_{1n} \\ S_{2n} \\ \vdots \\ \vdots \end{pmatrix},$$

that is, x is a linear combination of those columns of S which are **eigenvectors** of H with the same eigenvalue.

We are now in a position to compare the fundamental problem of Heisenberg's matrix mechanics with that of Schrödinger's wave mechanics. In the former, the problem is the solution of an eigenvalue problem for an infinite matrix, while in the latter, the problem is the solution of a partial differential equation (Schrödinger's equation). These are recalled explicitly in the following table.

Heisenberg's matrix mechanics	Schrödinger's wave mechanics
$\sum_j H_{ij}x_j = \lambda x_i$ $(i, j = 1, 2, \ldots, \infty)$	$H\psi(q^1, \ldots, q^N) = \lambda\psi(q^1, \ldots, q^N)$

In what sense are the two problems mathematically equivalent? We will explore this question further in the next chapter.

Chapter 5

The Dirac Delta Function

One approach to answer the question posed at the end of the previous chapter is through a mathematical device known as the Dirac delta function. This function is one of the most useful tools in theoretical physics. In this chapter we will discuss the motivation for its use in the present context and give a brief account of some of its properties.

In Heisenberg's matrix mechanics we work with infinite-dimensional complex vectors. Consider one such vector, written $x = (x_1, x_2, \dots)$, as a complex-valued function on the set of all positive integers $\mathbb{Z}_+ = \{1, 2, 3, \dots\}$, that is, $x : \mathbb{Z}_+ \longrightarrow \mathbb{C}$, $i \in \mathbb{Z}_+ \mapsto x_i \in \mathbb{C}$. On the other hand, in Schrödinger's wave mechanics, the complex-valued wave function $\psi(q^1, \dots, q^N)$ is a function $\psi : \Omega \longrightarrow \mathbb{C}$ whose domain is the N-dimensional configuration space $\Omega \subset \mathbb{R}^N$, that is, $(q^1, \dots, q^N) \in \Omega \mapsto \psi(q^1, \dots, q^N) \in \mathbb{C}$.

Now the linear transformations $\sum_j H_{ij} x_j$ and $H\psi$ bear little similarity to each other superficially. To establish a correspondence, the sum $\sum_j$ has to go over into an integral as follows:

$$\sum_j \longrightarrow \int_\Omega dq^1 \dots dq^N \quad ; \tag{5.1}$$

and so the algebraic transformation in matrix mechanics

$$x_i \longrightarrow \sum_j H_{ij} x_j \tag{5.2}$$

becomes an integral transformation

$$\psi(q^1, \dots, q^N) \longrightarrow \int_\Omega dq'^1 \dots dq'^N \, h(q^1, \dots, q^N ; q'^1, \dots, q'^N) \psi(q'^1, \dots, q'^N) \tag{5.3}$$

in wave mechanics. The eigenvalue problem

$$\sum_j H_{ij} x_j = \lambda x_i \tag{5.4}$$

thus becomes the following problem:

$$\int_\Omega dq'^1 \dots dq'^N \, h(q^1, \dots, q^N ; q'^1, \dots, q'^N)\psi(q'^1, \dots, q'^N) = \lambda\psi(q^1, \dots, q^N) \,. \tag{5.5}$$

This type of eigenvalue problem is known as an **integral equation**, and has been investigated extensively by the mathematicians Fredholm and Hilbert. Unfortunately, problem (5.5) cannot be identified with problem (5.4), unless a function

$$h(q^1, \dots, q^N ; q'^1, \dots, q'^N)$$

can be found for the differential operator

$$H\left(q^1, \dots, q^N ; \frac{\hbar}{i}\frac{\partial}{\partial q^1}, \dots, \frac{\hbar}{i}\frac{\partial}{\partial q^N}\right)$$

such that the equation

$$H\psi(q^1, \dots, q^N) = \int_\Omega dq'^1 \dots dq'^N \, h(q^1, \dots, q^N ; q'^1, \dots, q'^N)\psi(q'^1, \dots, q'^N) \tag{5.6}$$

is satisfied identically, that is, for all $\psi(^1, \dots, q^N)$. The function

$$h(q^1, \dots, q^N ; q'^1, \dots, q'^N) \,,$$

if it exists, is called the **kernel** of the functional operator H, and H is itself then called an **integral operator**.

The bad news is that this is in general impossible: a differential operator like H is never an integral operator, that is, it does not possess a kernel.

Let us convince ourselves of this. Let H be the identity operator, i.e., $H = 1$. Consider the case $N = 1$. Then for H to have a kernel, we require the existence of a function $h(q; q')$ such that

$$\psi(q) = \int_{-\infty}^{\infty} dq' \, h(q; q')\psi(q') \quad . \tag{5.7}$$

We will first show that $h(q; q')$ only depends on the difference $q' - q$. Replace q by 0 in (5.7) to get

$$\psi(0) = \int_{-\infty}^{\infty} dq' \, h(0; q')\psi(q') \quad . \tag{5.8}$$

Then change the variable of integration to $q'' = q' + q_0$ to obtain

$$\psi(0) = \int_{-\infty}^{\infty} dq'' \, h(0; q'' - q_0)\psi(q'' - q_0) \quad . \tag{5.9}$$

Now let $\psi(q'' - q_0) \equiv \psi'(q'')$. Then

$$\psi'(q_0) = \int_{-\infty}^{\infty} dq'' \, h(0; q'' - q_0)\psi'(q'') \quad . \tag{5.10}$$

On renaming symbols

$$q_0 \longrightarrow q\,, \quad q'' \longrightarrow q'\,, \quad \psi' \longrightarrow \psi\,,$$

Eq. (5.10) becomes

$$\psi(q) = \int_{-\infty}^{\infty} dq'\, h(0; q' - q)\psi(q') \quad . \tag{5.11}$$

Comparison of the above equation with (5.7) finally yields

$$h(q; q') = h(0; q' - q) = h(q' - q) \quad , \tag{5.12}$$

as asserted earlier.

In view of (5.12) we can write the requirement (5.7) as

$$\psi(q) = \int_{-\infty}^{\infty} dq'\, h(q' - q)\psi(q') \quad , \tag{5.13}$$

which implies

$$\psi(0) = \int_{-\infty}^{\infty} dq\, h(q)\psi(q) \quad . \tag{5.14}$$

Now change the variable of integration again, to $q' = -q$; then

$$\psi(0) = -\int_{\infty}^{-\infty} dq'\, h(-q')\psi(-q') = \int_{-\infty}^{\infty} dq'\, h(-q')\psi(-q') \quad . \tag{5.15}$$

Let $\psi'(q') \equiv \psi(-q')$. Equation (5.15) can then be written

$$\psi'(0) = \int_{-\infty}^{\infty} dq'\, h(-q')\psi'(q') \quad . \tag{5.16}$$

Rewrite the above equation by the renaming of variables $q' \to q$ and $\psi' \to \psi$:

$$\psi(0) = \int_{-\infty}^{\infty} dq\, h(-q)\psi(q) \quad . \tag{5.17}$$

Thus, if $h(q)$ is a solution of (5.14), so is $h(-q)$. It follows that

$$h_1(q) \equiv \frac{1}{2}[h(q) + h(-q)]$$

is also a solution. Indeed, we may regard $h(q)$ as an even function of q in general, i.e., $h(q) = h(-q)$.

Remember that in (5.14) ψ is supposed to be an arbitrary function such that the integral in that equation makes sense. Choose $\psi(q)$ to be specified by the following conditions:

$$\psi(0) = 0\,, \quad \psi(q) = h(q) \quad \text{for } q \neq 0\,.$$

Then it must follow from (5.14) that $h(q) = 0$ for $q \neq 0$, and so

$$0 = \int_{-\infty}^{\infty} h(q)dq \quad . \tag{5.18}$$

If, on the other hand, we choose $\psi(q) \equiv 1$, then (5.14) also implies

$$1 = \int_{-\infty}^{\infty} h(q)dq \quad . \tag{5.19}$$

Thus we arrive at a contradiction! So $h(q)$ cannot be a well-defined function.

Dirac was nevertheless able to "get away with it" and assumed the existence of, and then used very effectively, such a "function". We will discuss this in more detail in Chapter 9, where it will be seen that this "function" arises naturally as eigenfunctions corresponding to a continuous spectrum. The so-called **Dirac delta-function** is postulated to satisfy the following properties:

$$\delta(q) = 0 \quad (\text{when } q \neq 0)\,; \quad \delta(q) = \delta(-q)\,; \quad \int_{-\infty}^{\infty} \delta(q)dq = 1\,. \tag{5.20}$$

By virtue of these properties, it is easily seen that the Dirac delta-function satisfies (5.14):

$$\begin{aligned}\int_{-\infty}^{\infty} \delta(q)\psi(q)dq &= \psi(0)\int_{-\infty}^{\infty} \delta(q)dq + \int_{-\infty}^{\infty} \delta(q)(\psi(q) - \psi(0))dq \\ &= \psi(0)\cdot 1 + \int_{-\infty}^{\infty} 0 \cdot dq = \psi(0) \quad ,\end{aligned} \tag{5.21}$$

and hence (5.7) and (5.13) also.

We will give a few formal representations of the Dirac delta-function as follows:

$$\delta(q) = \lim_{a\to\infty} \sqrt{\frac{a}{\pi}}\, e^{-aq^2} \quad , \tag{5.22}$$

$$\delta(q) = \frac{1}{2\pi}\int_{-\infty}^{\infty} dk\, e^{ikq} \quad , \tag{5.23}$$

$$\delta(q - q_0) = \frac{1}{\pi}\lim_{\epsilon\to 0^+} \frac{\epsilon}{(q - q_0)^2 + \epsilon^2} \quad , \tag{5.24}$$

$$\delta(q - q_0) = \lim_{\eta\to 0} \frac{\theta(q - q_0 + \eta) - \theta(q - q_0)}{\eta} \quad . \tag{5.25}$$

In the last equation $\theta(q - q_0)$ is the step function shown in Fig. 5.1.
The integral representation (5.23) will be particularly useful in later applications.

Besides (5.20), we list below some more important properties of the Dirac delta-function:

$$x\delta(x) = 0 \quad , \tag{5.26}$$

$$\delta(ax) = \frac{1}{|a|}\,\delta(x) \quad , \tag{5.27}$$

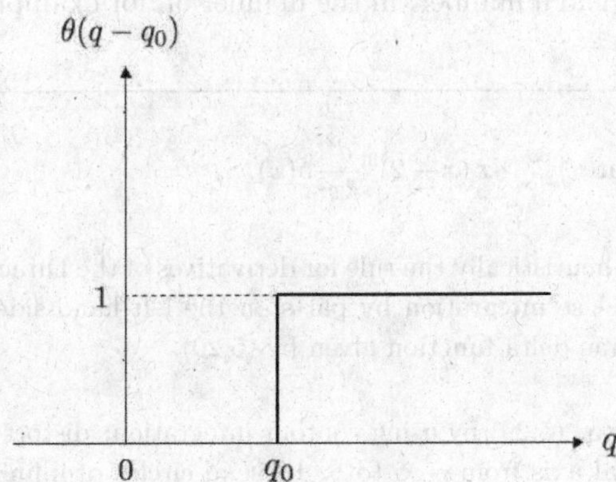

Fig. 5.1

$$\delta(g(x)) = \sum_n \frac{1}{|\,g'(x_n)\,|}\,\delta(x-x_n)\ ; \quad g(x_n)=0\ ,\ g'(x_n) = \left.\frac{dg}{dx}\right|_{x=x_n} \neq 0\quad , \tag{5.28}$$

$$\lim_{\epsilon\to 0}\frac{1}{x-x_0\pm i\epsilon} = P\left(\frac{1}{x-x_0}\right) \mp i\pi\delta(x-x_0)\quad . \tag{5.29}$$

In the last equation P stands for the principal value and it is understood that each term occurs inside an integral sign.

Using the Dirac delta function we can represent differential operators by integral operators, and hence solve (5.6) (which establishes the equivalence between Heisenberg's matrix mechanics and Schrödinger's wave mechanics) if we introduce the following rule for the derivatives of $\delta(q)$:

$$\int_{-\infty}^{\infty} dq\, f(q)\delta^{(m)}(q) = (-1)^m\, f^{(m)}(0)\quad . \tag{5.30}$$

The operators $\frac{d^n}{dq^n}$ and q^n then have the kernels $\delta^{(n)}(q-q')$ and $q^n\delta(q-q')$, respectively, since

$$\begin{aligned}\frac{d^n}{dq^n}\psi(q) &= \frac{d^n}{dq^n}\int_{-\infty}^{\infty}\delta(q-q')\psi(q')dq' = \int_{-\infty}^{\infty}\frac{\partial^n}{\partial q^n}\delta(q-q')\psi(q')dq' \\ &= \int_{-\infty}^{\infty}\delta^{(n)}(q-q')\psi(q')dq'\quad ;\end{aligned} \tag{5.31}$$

and

$$q^n\psi(q) = \int_{-\infty}^{\infty}\delta(q-q')q^n\psi(q')dq'\quad . \tag{5.32}$$

In more correct mathematical terminology, the Dirac delta function is an example of a so-called **distribution**, which is a **functional** acting on well-behaved functions to yield a number, in the manner of, for example, Eq. (5.21).

Problem 5.1 Calculate $\int_{-\infty}^{\infty} dx\,(x-2)^3 \frac{d^2}{dx^2}\delta(x)$.

Problem 5.2 Justify heuristically the rule for derivatives of the Dirac delta function given by (5.30). *Hints*: Use integration by parts on the left-hand-side of (5.30) and the properties of the Dirac delta function given by (5.20).

Problem 5.3 Prove Eq. (5.29) by using contour integration: distort the contour of integration along the real axis from $-\infty$ to ∞ by semi-circles of infinitesimal radii ε centered at x_0.

Chapter 6

Why Do We Need Hilbert Space?

In Schrödinger's wave mechanics we are interested in the so-called L^2 (square-integrable) functions $\psi : \Omega \to \mathbb{C}$, that is, functions such that

$$\int_{\Omega} dq^1 \ldots dq^N \mid \psi \mid^2 < \infty \quad ;$$

while in Heisenberg's matrix mechanics we are interested in functions $x : \mathbb{Z}_+ \to \mathbb{C}$ such that

$$\sum_{\nu} \mid x_\nu \mid^2 < \infty \quad .$$

We say that the functions are **normalizable** in each case, or have finite norms. The spaces $\mathbb{Z}_+$ and Ω, as domains of the functions, are very different; but it turns out that the spaces of our normalizable functions on them have the same mathematical structure as vector spaces. This is the content of the **Fischer-Riesz theorem** in **functional analysis**. We will state this theorem more precisely as follows. (For a proof, see Reed and Simon 1972.)

Theorem 6.1. *Let*

$$F_Z = \{ x \in f_Z \mid \sum_{\nu} \mid x_\nu \mid^2 < \infty \} \tag{6.1}$$

$$\text{and} \qquad F_\Omega = \{ \psi \in f_\Omega \mid \int_\Omega dq^1 \ldots dq^N \mid \psi \mid^2 < \infty \} \, , \tag{6.2}$$

*where f_Z is the set of all functions $x : \mathbb{Z}_+ \to \mathbb{C}$ and f_Ω is the set of all functions $\psi : \Omega \to \mathbb{C}$. Then F_Z is isomorphic to F_Ω, and the **isomorphism** is linear and isometric. By **linearity** we mean that if $x \longleftrightarrow \phi$ and $y \longleftrightarrow \psi$ then*

$$ax + by \longleftrightarrow a\phi + b\psi \ , \quad a, b, \in \mathbb{C} \ .$$

By ***isometry*** *we mean that under the same conditions*

$$\sum_{\nu} x_{\nu} y_{\nu}^{*} = \int_{\Omega} dq^1 \ldots dq^N \, \phi\psi^* \quad . \tag{6.3}$$

Equation (6.3) immediately implies that, if $x \longleftrightarrow \phi$, then

$$\sum_{\nu} | x_{\nu} |^2 = \int_{\Omega} dq^1 \ldots dq^N \, | \phi |^2 \quad . \tag{6.4}$$

Thus the norms (lengths) of vectos are preserved under the isomorphism.

Each of F_Z and F_{Ω} is an infinite-dimensional linear vector space called a **Hilbert space.**

Definition 6.1. *A Hilbert space is a complete, separable infinite-dimensional linear vector space over the field of complex numbers* $\mathbb{C}$ *with a hermitian inner product.*

Remark. The property of separability is sometimes not included in the mathematics literature; but in physics applications, this property is generally required.

We will now proceed to explain the technical terms in the above definition.

A **hermitian inner product** in a Hilbert space $\mathcal{H}$ over $\mathbb{C}$ is an inner product $(\, , \,)$ with the following properties: for any $\psi, \psi_1, \psi_2, \phi \in \mathcal{H}$,

i) $(\psi, \phi) \in \mathbb{C}$,

ii) $(\psi, \phi) = (\phi, \psi)^*$,

iii) $(\phi, c_1\psi_1 + c_2\psi_2) = c_1(\phi, \psi_1) + c_2(\phi, \psi_2) \quad , \quad c_1, c_2 \in \mathbb{C}$,

iv) $(\phi, \phi) \geq 0$, the equality holds if and only if $\phi = 0$ (the null vector).

From ii) and iii) it follows that

$$(c_1\psi_1 + c_2\psi_2 \, , \, \phi) = (\phi \, , \, c_1\psi_1 + c_2\psi_2)^* = c_1^*(\phi, \psi_1)^* + c_2^*(\phi, \psi_2)^* \quad .$$

Thus

$$(c_1\psi_1 + c_2\psi_2, \phi) = c_1^*(\psi_1, \phi) + c_2^*(\psi_2, \phi) \quad . \tag{6.5}$$

We say that the hermitian product is linear in its second argument and antilinear in its first argument.

The hermitian product permits the definition of the **norm** (or length) of a vector $\phi \in \mathcal{H}$:

$$\|\phi\| \equiv \sqrt{(\phi \, , \, \phi)} \quad . \tag{6.6}$$

The hermitian product also satisfies the following two important inequalities:

$$| (\psi, \phi) | \leq \| \psi \| \, \| \phi \| \qquad \textbf{(Schwarz inequality)} \quad , \tag{6.7}$$

$$\| \psi + \phi \| \leq \| \psi \| + \| \phi \| \qquad \textbf{(triangle inequality)} \quad . \tag{6.8}$$

Problem 6.1 Discuss the geometrical meaning of the Schwarz inequality by analogy with the scalar product of 3-dimensional real vectors.

Definition 6.2. *A sequence* ψ_i, $i = 1, 2, \ldots$ *in* $\mathcal{H}$ *is said to converge (in the norm) to* $\psi \in \mathcal{H}$ *if*

$$\lim_{i\to\infty} \| \psi - \psi_i \| = 0 \quad . \tag{6.9}$$

We simply write $\lim_{i\to\infty} \psi_i = \psi$.

Definition 6.3. *An infinite-dimensional vector space* $\mathcal{H}$ *with a hermitian inner product is* ***complete*** *if every* ***Cauchy sequence*** *in* $\mathcal{H}$ *converges. A sequence* ψ_i *is Cauchy if for each* $\epsilon > 0$, *there exists an integer* $N(\epsilon)$ *such that* $\| \psi_n - \psi_m \| < \epsilon$ *for all* $n, m \geq N(\epsilon)$.

An example of an infinite-dimensional linear vector space that fails to be complete is the space of polynomials $\mathcal{P}$ in t, $0 \leq t \leq 1$, with inner product defined by

$$(x, y) \equiv \int_0^1 x^*(t)y(t)dt \quad .$$

Definition 6.4. *An infinite-dimensional vector space* $\mathcal{H}$ *with a hermitian inner product is* ***separable*** *if there exists a sequence* $\{ f_i \}$ *in* $\mathcal{H}$ *which is everywhere* ***dense*** *in* $\mathcal{H}$. *This means that given any* $\epsilon > 0$ *and any* $\psi \in \mathcal{H}$, *there is an* f_j *such that* $\| f_j - \psi \| < \epsilon$, *that is, any vector in* $\mathcal{H}$ *is arbitrarily close to some member of the sequence* $\{ f_i \}$.

The properties of completeness and separability of a Hilbert space are intimately connected to the notion of the **completeness of a set of orthonormal vectors** in Hilbert space, which is crucially important in applications to quantum mechanics.

Definition 6.5. *A set of vectors* $\mathcal{D} \in \mathcal{H}$ *is said to be an orthonormal set if, for any* $\phi, \psi \in \mathcal{D}$,

$$(\phi, \psi) = \begin{cases} 1\,, & \textit{if } \psi = \phi \\ 0\,, & \textit{otherwise} \end{cases} \quad .$$

Definition 6.6. *An orthonormal set* $\mathcal{D} \subset \mathcal{H}$ *is said to be* ***complete*** *if the only vector in* $\mathcal{H}$ *orthogonal to every vector in* $\mathcal{D}$ *is the zero vector.*

The property of separability implies the following theorem.

Theorem 6.2. *Every orthonormal set in* $\mathcal{H}$ *is either finite or countably infinite; if it is complete, then it is certainly infinite.*

Completeness of $\mathcal{H}$ implies the following:

Theorem 6.3. *Let $\{\phi_i\}$ be an infinite orthonormal set in a Hilbert space $\mathcal{H}$. Then the series*

$$\sum_{\nu=1}^{\infty} x_\nu \phi_\nu \,, \quad x_\nu \in \mathbb{C}$$

converges (in the norm) if and only if the series $\sum_{\nu=1}^{\infty} |\, x_\nu \,|^2$ converges.

Finally we state the following useful theorem:

Theorem 6.4. *Let $\{\phi_i\}$ be an orthonormal set in a Hilbert space $\mathcal{H}$. Each of the following conditions is a necessary and sufficient condition for the set to be complete:*

i) The closed linear manifold $[\phi_1, \phi_2, \ldots]$ spanned by $\{\phi_i\}$ is $\mathcal{H}$ itself;

ii) For any $\psi \in \mathcal{H}$, $\psi = \sum_\nu (\phi_\nu, \psi)\, \phi_\nu$,

iii) For any $\psi_1, \psi_2 \in \mathcal{H}$, $(\psi_1, \psi_2) = \sum_\nu (\psi_1, \phi_\nu)(\phi_\nu, \psi_2)$.

As we will see in later chapters, this theorem furnishes the mathematical basis of many direct applications of Hilbert space theory to quantum mechanics. Conditions ii) and iii), in particular, will be rewritten in a somewhat more notationally transparent manner using the Dirac bracket notation [as (7.9) and (7.10), respectively]. We will begin to study this all-important notation in the next chapter.

Chapter 7

The Dirac Bra Ket Notation and the Riesz Theorem

We will introduce in this chapter the so-called Dirac Bra Ket (or bracket) notation, which finds almost universal use in the physics literature on quantum theory because of its great elegance and practicability. In essence, it is a very convenient notational device encoding the mathematical content of an important fact in Hilbert space theory known as the Riesz Theorem.

To explain this theorem we first need to introduce the very important notion of the **dual space** to a linear vector space.

Definition 7.1. *The dual space $\mathcal{H}^*$ to a linear vector space $\mathcal{H}$ is the vector space of all bounded, linear functions on $\mathcal{H}$.*

The term "bounded, linear function" simply means the following. Suppose f is a linear, bounded function on $\mathcal{H}$. Then, for any $\psi, \psi_1, \psi_2 \in \mathcal{H}$ and any $z_1, z_2 \in \mathbb{C}$, we have (assuming that $\mathcal{H}$ is a complex vector space)

(i) $|\, f(\psi)\, | < \infty \, ,$ (boundedness) ;

(ii) $f(z_1\psi_1 + z_2\psi_2) = z_1 f(\psi_1) + z_2 f(\psi_2)\, ,$ (linearity) .

It can be shown relatively easily (see the alternative proof of the Riesz Theorem presented below) that any finite-dimensional vector space is isomorphic to its dual. The generalization of this fact to a Hilbert space (by definition infinite-dimensional) constitutes the Riesz theorem.

Theorem 7.1 (Riesz Theorem). *Let $\mathcal{H}$ be a Hilbert space and $\mathcal{H}^*$ its dual space. Then a hermitian inner product (,) [satisfying conditions i) through iv) above Eq. (6.5)] induces a* **conjugate isomorphism** $G : \mathcal{H} \to \mathcal{H}^*$:

$$G(\psi) = \psi^* \in \mathcal{H}^*\, , \tag{7.1}$$

such that

$$\psi^*(\phi) = (\psi, \phi)\,, \quad \textit{for all } \phi \in \mathcal{H}\,, \tag{7.2}$$

and

$$G(\alpha\psi) = \alpha^*\psi^*\,, \quad \alpha \in \mathbb{C}\,. \tag{7.3}$$

Proof. We first note that ψ^* as defined by (7.2) is a continuous linear function on $\mathcal{H}$. In fact, for any $\phi_1, \phi_2 \in \mathcal{H}$ and $\alpha, \beta \in \mathbb{C}$,

$$\psi^*(\alpha\phi_1 + \beta\phi_2) = (\psi, \alpha\phi_1 + \beta\phi_2) = \alpha(\psi, \phi_1) + \beta(\psi, \phi_2) = \alpha\psi^*(\phi_1) + \beta\psi^*(\phi_2)\,,$$

where the second equality follows from condition (ii) following Definition 7.1. Continuity of ψ^* follows from the Schwarz inequality [Eq. (6.7)]:

$$|\psi^*(\phi)| = |(\psi, \phi)| \leq (\psi, \psi)(\phi, \phi)\,.$$

Now choose an arbitrary non-zero $\psi^* \in \mathcal{H}^*$. We wish to prove the existence of a unique $\psi \in \mathcal{H}$ such that $\psi^*(\phi) = (\psi, \phi)$ for all $\phi \in \mathcal{H}$.

Let $ker(\psi^*) \subset \mathcal{H}$ be the **kernel** of ψ^*:

$$ker(\psi^*) = \{\phi \in \mathcal{H} \,|\, \psi^*(\phi) = 0\}\,.$$

The set $ker(\psi^*)$ is a closed subspace of $\mathcal{H}$. In fact, consider a sequence $\{\phi_n\} \subset ker(\psi^*)$. Thus $\psi^*(\phi_n) = 0$ for all n. If $\lim_{n\to\infty} \phi_n = \phi$ [recall Eq. (6.9)], then the continuity of ψ^* implies that $\psi^*(\phi) = 0$.

Now there exists a non-zero $\phi_0 \in (ker(\psi^*))^\perp$ [ϕ_0 is in the orthogonal complement of $ker(\psi^*)$]. Otherwise $(ker(\psi^*))^\perp = \{0\}$, implying that $ker(\psi^*) = \mathcal{H}$. But this is impossible because ψ^* is not identically zero by assumption. Since $\psi^*(\phi_0) \neq 0$, we can, without loss of generality, set $\psi^*(\phi_0) = 1$ [by replacing ϕ_0 by $\phi_0/\psi^*(\phi_0)$]. Then

$$\psi^*(\phi - \psi^*(\phi)\phi_0) = 0\,, \quad \text{for all } \phi \in \mathcal{H}\,,$$

which implies that

$$\phi' \equiv \phi - \psi^*(\phi)\phi_0 \in ker(\psi^*)\,.$$

Hence we have the orthogonal decomposition

$$\phi = \phi' + \psi^*(\phi)\phi_0\,; \quad \phi' \in ker(\psi^*)\,,\ \phi_0 \in (ker(\psi^*))^\perp\,,$$

for all $\phi \in \mathcal{H}$. Taking the hermitian inner product (ϕ_0, ϕ), we have

$$(\phi_0, \phi) = (\phi_0, \phi') + \psi^*(\phi)(\phi_0, \phi_0) = \psi^*(\phi)(\phi_0, \phi_0)\,.$$

The sought-for $\psi \in \mathcal{H}$ is given by

$$\psi = \frac{\phi_0}{(\phi_0, \phi_0)}\,.$$

Indeed, (7.2) follows from

$$\psi^*(\phi) = \frac{(\phi_0, \phi)}{(\phi_0, \phi_0)} = (\psi, \phi)\,.$$

Uniqueness of ψ can be established as follows. Suppose there exists a $\psi' \neq \psi$ such that

$$(\psi, \phi) = (\psi', \phi) \quad \text{for all } \phi \in \mathcal{H} .$$

Then

$$(\psi - \psi', \phi) = 0 \quad \text{for all } \phi \in \mathcal{H} .$$

Choosing $\phi = \psi - \psi'$, we have

$$(\psi - \psi', \psi - \psi') = 0 ,$$

which implies $\psi = \psi'$.

Finally, (6.5) implies that

$$G(\alpha\psi)(\phi) = (\alpha\psi, \phi) = \alpha^*(\psi, \phi) = \alpha^*\psi^*(\phi) = \alpha^* G(\psi)(\phi) ,$$

for all $\phi \in \mathcal{H}$. Thus (7.3) follows. □

The above elegant proof of the Riesz theorem does not assume the property of separability of the Hilbert space $\mathcal{H}$. If this property is assumed, so that any $\psi \in \mathcal{H}$ can be expanded in terms of a complete, orthonormal, countably infinite set in $\mathcal{H}$, an alternative (and perhaps more computationally useful) proof can be given as follows.

Proof. (*of the Riesz Theorem assuming separability*). For an arbitrary $\psi \in \mathcal{H}$, define a function f_ψ on $\mathcal{H}$ such that

$$f_\psi(\phi) \equiv (\psi, \phi) , \quad \text{for any } \phi \in \mathcal{H} .$$

This is clearly a linear, bounded function.

Conversely, given an arbitrary $f \in \mathcal{H}^*$, there corresponds a unique $\psi \in \mathcal{H}$ such that $f(\phi) = (\psi, \phi)$, for all $\phi \in \mathcal{H}$. To see this, choose an orthonormal basis $\{\phi_n\}$ in $\mathcal{H}$. It follows from statement ii) of Theorem 6.4 that any given $\phi \in \mathcal{H}$ can be expressed as

$$\phi = \sum_n x_n \phi_n , \quad x_n \in \mathbb{C} .$$

Then, by linearity, $f(\phi) = \sum_n x_n f(\phi_n)$. Now construct the following vector $\psi \in \mathcal{H}$:

$$\psi \equiv \sum_n [f(\phi_n)]^* \phi_n .$$

We then have

$$\begin{aligned} (\psi, \phi) &= \sum_n ([f(\phi_n)]^* \phi_n, \phi) \\ &= \sum_n f(\phi_n)(\phi_n, \phi) \qquad \text{[by (6.5)]} \\ &= \sum_n \sum_m x_m f(\phi_n)(\phi_n, \phi_m) = \sum_{n,m} x_m f(\phi_n) \delta_{nm} \\ &= \sum_n x_n f(\phi_n) = f(\phi) \quad . \end{aligned} \tag{7.4}$$

If ψ' is another vector in $\mathcal{H}$ such that $(\psi', \phi) = (\psi, \phi)$ for all $\phi \in \mathcal{H}$, then $(\psi' - \psi, \phi) = 0$ for all $\phi \in \mathcal{H}$, which implies that $\psi' = \psi$. Hence ψ is unique. □

We are now ready to introduce the Dirac notation.

In this notation, vectors $\psi \in \mathcal{H}$ are labelled $|\psi\rangle$, called a **ket vector**, or simply a **ket**. According to Riesz's Theorem, for every $\psi \in \mathcal{H}$, there corresponds a unique $f_\psi \in \mathcal{H}^*$, which is labelled $\langle\psi|$, called a **bra vector**, or simply a **bra**. We will indicate this fact by writing

$$\langle\psi| \in \mathcal{H}^* \longleftrightarrow |\psi\rangle \in \mathcal{H} \quad . \tag{7.5}$$

The **pairing** between vectors in $\mathcal{H}^*$ and $\mathcal{H}$ is then written as $\langle\psi|\phi\rangle$. It is given by

$$\langle\psi|\phi\rangle = (\psi, \phi) \quad , \tag{7.6}$$

where the quantity ψ in the hermitian product on the right-hand side is the vector in $\mathcal{H}$ uniquely corresponding to $\langle\psi| \in \mathcal{H}^*$ by virtue of Riesz's Theorem. Property ii) of the Hermitian product [above (6.5)] then implies

$$\boxed{\langle\psi|\phi\rangle^* = \langle\phi|\psi\rangle} \quad . \tag{7.7}$$

The Dirac notation makes it clear whether it is a vector in $\mathcal{H}$ or a vector in its dual space $\mathcal{H}^*$ that one is referring to: $|\ \rangle \in \mathcal{H}$, $\langle\ | \in \mathcal{H}^*$. Equation (7.3) also implies the anti-linear correspondence

$$c_1^*\langle\psi_1| + c_2^*\langle\psi_2| \longleftrightarrow c_1|\psi_1\rangle + c_2|\psi_2\rangle \quad . \tag{7.8}$$

Conditions ii) and iii) in Theorem 6.4 can be written in the Dirac notation as, respectively,

$$|\psi\rangle = \sum_n \langle\phi_n|\psi\rangle|\phi_n\rangle = \sum_n |\phi_n\rangle\langle\phi_n|\psi\rangle \quad , \tag{7.9}$$

and

$$\langle\psi_1|\psi_2\rangle = \sum_n \langle\psi_1|\phi_n\rangle\langle\phi_n|\psi_2\rangle \quad . \tag{7.10}$$

Both of the above equations imply that

$$\boxed{\textstyle\sum_n |\phi_n\rangle\langle\phi_n| = 1 \qquad \text{(the identity operator)}} \quad , \tag{7.11}$$

which expresses the **completeness condition** for the orthonormal set $\{|\phi_n\rangle\}$. It is to be noted that the sum in the above equation is over all members of the complete set, so it is usually an infinite sum. The completeness condition as expressed by (7.10) demonstrates one aspect of the great elegance and convenience of the Dirac notation.

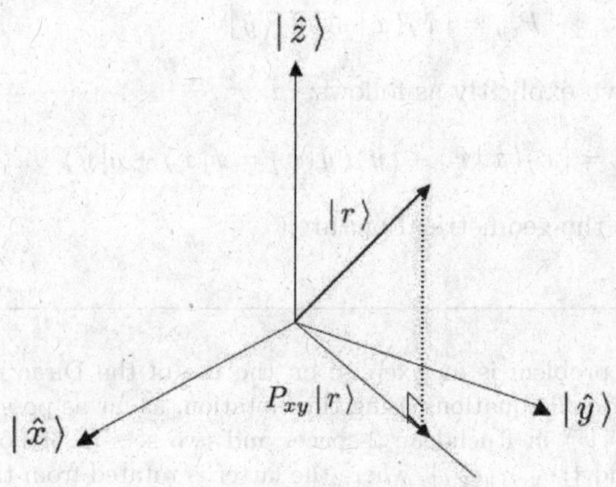

Fig. 7.1

A **projection operator** P in a vector space is one that satisfies the condition $P^2 = P$. It is easily seen, using the Dirac notation, that an operator P' of the form

$$P' \equiv \sum_{\{i\}} |\,\phi_i\,\rangle\langle\,\phi_i\,\rangle \quad ,$$

is a projection operator, where the sum is over a set $\{i\} \subset \{n\}$ of the full range of indices n labelling all members of the complete orthonormal set $\{\phi_n\}$. Indeed,

$$\begin{aligned}(P')^2 &= \left(\sum_{\{i\}} |\,\phi_i\,\rangle\langle\,\phi_i\,|\right)\left(\sum_{\{j\}} |\,\phi_j\,\rangle\langle\,\phi_j\,|\right) \\ &= \sum_{\{i\}}\sum_{\{j\}} |\,\phi_i\,\rangle\langle\,\phi_i\,|\,\phi_j\,\rangle\langle\,\phi_j\,| = \sum_{\{i\}}\sum_{\{j\}} \delta_{ij}|\,\phi_i\,\rangle\langle\,\phi_j\,| \\ &= \sum_{\{i\}} |\,\phi_i\,\rangle\langle\,\phi_i\,| = P' \quad .\end{aligned} \tag{7.12}$$

A projection operator is so-called because it projects an arbitrary vector in $\mathcal{H}$ onto a particular subspace. This can be intuitively understood by considering the familiar 3-dimensional Euclidean space $\mathbb{R}^3$ equipped with the usual scalar product (which is the special case of a hermitian product on a real vector space). Writing the unit vectors along the x, y and z axes using the Dirac notation, the completeness relation (7.10) becomes

$$|\,\hat{x}\,\rangle\langle\,\hat{x}\,| + |\,\hat{y}\,\rangle\langle\,\hat{y}\,| + |\,\hat{z}\,\rangle\langle\,\hat{z}\,| = 1 \quad . \tag{7.13}$$

The projection operator that projects an arbitrary vector $|\,\boldsymbol{r}\,\rangle \in \mathbb{R}^3$ onto the xy-plane, for example, is given by

$$P_{xy} = |\,\hat{x}\,\rangle\langle\,\hat{x}\,| + |\,\hat{y}\,\rangle\langle\,\hat{y}\,| \quad , \tag{7.14}$$

and its action is shown explicitly as follows:

$$P_{xy}|\,\boldsymbol{r}\,\rangle = |\,\hat{x}\,\rangle\langle\,\hat{x}\,|\,\boldsymbol{r}\,\rangle + |\,\hat{y}\,\rangle\langle\,\hat{y}\,|\,\boldsymbol{r}\,\rangle = x|\,\hat{x}\,\rangle + y|\,\hat{y}\,\rangle \quad . \tag{7.15}$$

Figure 7.1 illustrates the geometrical picture.

Problem 7.1 This problem is an exercise on the use of the Dirac notation in a familiar situation. Write all equations using this notation, as far as possible.

Consider a vector $|\,\psi\,\rangle$ in Euclidean 2-space, and two sets of orthonormal basis vectors $\{|\,e_1\,\rangle, |\,e_2\,\rangle\}$ and $\{|\,e'_1,\rangle, |\,e'_2\,\rangle\}$, where the latter is rotated from the first by an angle θ in the counterclockwise sense.

(a) Write the completeness conditions for $\{|\,e_i\,\rangle\}$ and $\{|\,e'_i\,\rangle\}$.

(b) Let the matrix

$$(a^i_j) = \begin{pmatrix} a^1_1 & a^1_2 \\ a^2_1 & a^2_2 \end{pmatrix}$$

be defined by $a^i_j = \langle\,e'_i\,|\,e_j\,\rangle$. Display the matrix explicitly (in terms of θ). Use one of the completeness conditions in (a) to show how the $|\,e_i\,\rangle$ can be expressed in terms of the $|\,e'_j\,\rangle$ with the help of the matrix (a^i_j). Write the basis vectors as row matrices to display this relationship.

(c) Use the other completeness condition to express the $|\,e'_i\,\rangle$ in terms of the $|\,e_j\,\rangle$. Let the relevant matrix in this expression be

$$(b^i_j) = \begin{pmatrix} b^1_1 & b^1_2 \\ b^2_1 & b^2_2 \end{pmatrix} .$$

Display the matrix (b^j_i) explicitly (in terms of θ). How are the matrices (a^j_i) and (b^j_i) related to each other? Is the matrix (a^j_i) **orthogonal**?

(d) Let $\psi^i = \langle\,e_i\,|\,\psi\,\rangle$ and $(\psi')^j = \langle\,e'_j\,|\,\psi\,\rangle$. Use the completeness conditions in (a) and the result in (b) to express the $(\psi')^i$ in terms of the ψ^j with the help of the matrix (a^j_i). Also display the matrix equation explicitly, this time writing the ψ^i and $(\psi')^i$ as column matrices.

Problem 7.2 Let $\{|\,e_1\,\rangle, |\,e_2\,\rangle, |\,e_3\,\rangle\}$ be an orthonormal complete set in a linear vector space, and an operator A and vector $|\,\psi\,\rangle$ be given by

$$A = |\,e_1\,\rangle\langle\,e_1\,| + |\,e_2\,\rangle\langle\,e_2\,| \quad \text{and} \quad |\,\psi\,\rangle = 2|\,e_1\,\rangle + 3|\,e_2\,\rangle + 4|\,e_3\,\rangle ,$$

respectively. Calculate $A\,|\,\psi\,\rangle$.

Problem 7.3 In Fourier analysis one uses a complete set of complex-valued functions on the unit circle given by

$$\psi_n(\phi) = \frac{1}{\sqrt{2\pi}}\, e^{in\phi} \,, \quad n = 0, \pm 1, \pm 2, \dots \,,$$

where $0 \leq \phi \leq \pi$. With the hermitian product between two such functions given by

$$\langle\, \psi_n \,|\, \psi_m \,\rangle = \int_0^{2\pi} d\phi \, \psi_n^* \, \psi_m \,,$$

show that this is an orthonormal set as well.

Chapter 8

Self-Adjoint Operators in Hilbert Space

In this chapter we go to the heart of the abstract formalism of quantum mechanics by studying the so-called self-adjoint operators in Hilbert space, which represent **observables** in a physical theory. In the process, we will learn to appreciate the power and elegance of the Dirac notation, which we will be using exclusively from this point on.

First we recall the notion of a **linear operator** in a Hilbert space $\mathcal{H}$. Such an operator A satisfies the following conditions:

i)
$$A\,(c_1|\,\psi_1\,\rangle + c_2|\,\psi_2\,\rangle) = c_1 A\,|\,\psi_1\,\rangle + c_2 A\,|\,\psi_2\,\rangle \quad , \tag{8.1}$$

ii)
$$(A+B)|\,\psi\,\rangle = A\,|\,\psi\,\rangle + B\,|\,\psi\,\rangle \quad , \tag{8.2}$$

iii)
$$(AB)\,|\,\psi\,\rangle = A\,(B\,|\,\psi\,\rangle) \quad , \tag{8.3}$$

where $|\,\psi\rangle, |\,\psi_1\,\rangle\,|\,\psi_2\,\rangle \in \mathcal{H}$, $c_1, c_2 \in \mathbb{C}$ and B is any linear operator in $\mathcal{H}$.

Choose a complete orthonormal basis set $\{|\,\phi_n\,\rangle\}$ in $\mathcal{H}$ with the corresponding completeness relation

$$\sum_n |\,\phi_n\,\rangle\langle\,\phi_n\,| = 1 \quad . \tag{8.4}$$

Suppose $A\,|\,\psi\,\rangle = |\,\phi\,\rangle$. We can expand the vectors $|\,\psi\rangle$ and $|\,\phi\,\rangle$ in terms of the $|\,\phi_n\,\rangle$, using the completeness relation (7.11), as follows:

$$|\,\psi\,\rangle = 1\,|\,\psi\,\rangle = \sum_n |\,\phi_n\,\rangle\langle\,\phi_n\,|\,\psi\,\rangle = \sum_n x_n|\,\phi_n\,\rangle \quad , \tag{8.5}$$

$$|\,\phi\,\rangle = 1\,|\,\phi\,\rangle = \sum_n |\,\phi_n\,\rangle\langle\,\phi_n\,|\,\phi\,\rangle = \sum_n y_n|\,\phi_n\,\rangle \quad , \tag{8.6}$$

where

$$x_n \equiv \langle \phi_n \,|\, \psi \rangle \,, \qquad y_n \equiv \langle \phi_n \,|\, \phi \rangle \quad . \tag{8.7}$$

Then, making use of (8.1),

$$\sum_j x_j A \,|\, \phi_j \rangle = \sum_j y_j |\, \phi_j \rangle \quad . \tag{8.8}$$

Multiplying on the left by $\langle \phi_i \,|$, we have

$$\sum_j \langle \phi_i \,|\, A \,|\, \phi_j \rangle \, x_j = \sum_j y_j \langle \phi_i \,|\, \phi_j \rangle = \sum_j \delta_{ij} y_j = y_i \quad .$$

We can write

$$y_i = \sum_j A_{ij} x_j \quad , \tag{8.9}$$

where the **matrix element**, A_{ij}, of the operator A is given, with respect to the basis $\{|\, \phi_i \rangle\}$, by

$$\boxed{A_{ij} \equiv \langle \phi_i \,|\, A \,|\, \phi_j \rangle} \quad . \tag{8.10}$$

An operator A in a vector space $\mathcal{H}$ is completely specified by its **matrix representation** with respect to a particular basis set $\{\phi_i\}$ of $\mathcal{H}$, with matrix elements given by the above equation.

In (8.10), the right-hand side means $\langle \phi_i \,|\, (A \,|\, \phi_j \rangle)$, with the operator A acting to the right, that is, acting on vectors in $\mathcal{H}$. But the Dirac notation $\langle \phi_i \,|\, A \,|\, \phi_j \rangle$ suggests that we can define the action of A to the left also, by requiring

$$(\langle \phi \,|\, A) \,|\, \psi \rangle = \langle \phi \,|\, (A \,|\, \psi \rangle) \quad . \tag{8.11}$$

In other words, on the left-hand side of the above equation, A is an operator on the dual space $\mathcal{H}^*$. According to Riesz's theorem, there is then a unique ket $|\, \chi \rangle \in \mathcal{H}$ with a conjugate $\langle \chi \,| \in \mathcal{H}^*$ [in the sense of (7.5)] such that

$$\langle \chi \,| = \langle \phi \,|\, A \quad . \tag{8.12}$$

Now $|\, \chi \rangle$ is uniquely determined by the operator A and the ket $|\, \phi \rangle$. So there must exist a unique linear operator $A^\dagger$ on $\mathcal{H}$, called the **adjoint** of A, such that

$$|\, \chi \rangle = A^\dagger |\, \phi \rangle \equiv |\, A^\dagger \phi \rangle \quad . \tag{8.13}$$

Hence (8.11) can be rewritten as

$$\boxed{\langle \phi \,|\, A \,|\, \psi \rangle = \langle A^\dagger \phi \,|\, \psi \rangle \,, \qquad \text{for any } |\, \phi \rangle, |\, \psi \rangle \in \mathcal{H}} \quad . \tag{8.14}$$

This equation is often used as the definition of the adjoint operator. Furthermore, it follows from (7.7) that

$$\boxed{\langle \phi \,|\, A \,|\, \psi \rangle^* = \langle \psi \,|\, A^\dagger \,|\, \phi \rangle \,, \qquad \text{for any } |\, \phi \rangle, |\, \psi \rangle \in \mathcal{H}} \quad . \tag{8.15}$$

Using (7.7) again, we can rewrite this equation as

$$\langle \psi | A^\dagger | \phi \rangle = \langle A\psi | \phi \rangle \quad . \tag{8.16}$$

Hence

$$\langle A\psi | = \langle \psi | A^\dagger \quad , \tag{8.17}$$

which implies that, under the conjugate isomorphism of the Riesz Theorem,

$$\boxed{\langle \psi | A^\dagger \longleftrightarrow A | \psi \rangle} \quad . \tag{8.18}$$

Using (8.14) on the right-hand side of (8.15) and (7.7) on the left-hand side, we obtain

$$\langle (A^\dagger)^\dagger \psi | \phi \rangle = \langle A\psi | \phi \rangle \quad . \tag{8.19}$$

Since this equation is true for arbitrary $|\psi\rangle$ and $|\phi\rangle \in \mathcal{H}$, we conclude that

$$\boxed{(A^\dagger)^\dagger = A} \quad . \tag{8.20}$$

From the above development, we see that the Dirac notation conveniently encodes the mathematical consequences of the Riesz Theorem.

We now introduce the important notion of a self-adjoint operator.

Definition 8.1. *If an operator A is equal to its adjoint, that is, if $A = A^\dagger$, then A is said to be* ***self-adjoint****, or* ***hermitian****.*

(Note: In the mathematical literature there is a technical distinction between self-adjointness and hermiticity, the details of which need not concern us here. In general self-adjointness is the stronger condition. In this book we will use the two terms interchangeably.)

A useful criterion for the self-adjointness of an operator is given by the following theorem.

Theorem 8.1. *If $\langle \psi | A | \psi \rangle = \langle \psi | A | \psi \rangle^*$ for all $|\psi\rangle \in \mathcal{H}$, then $\langle \phi_1 | A | \phi_2 \rangle = \langle \phi_2 | A | \phi_1 \rangle^*$ for all $|\phi_1\rangle, |\phi_2\rangle \in \mathcal{H}$, and hence $A = A^\dagger$.*

Proof. Choose any $|\phi_1\rangle, |\phi_2\rangle \in \mathcal{H}$. Let $|\psi\rangle = a|\phi_1\rangle + b|\phi_2\rangle$, where $a, b \in \mathbb{C}$. Then

$$\langle \psi | A | \psi \rangle = |a|^2 \langle \phi_1 | A | \phi_1 \rangle + |b|^2 \langle \phi_2 | A | \phi_2 \rangle \\ + a^* b \langle \phi_1 | A | \phi_2 \rangle + a b^* \langle \phi_2 | A | \phi_1 \rangle$$

is real by hypothesis. The first two terms on the right-hand side are also real by hypothesis. Hence the sum of the last two terms must be real also, that is,

$$a^* b \langle \phi_1 | A | \phi_2 \rangle + a b^* \langle \phi_2 | A | \phi_1 \rangle$$

is real. On setting $a = b = 1$ in the above expression and equating it to its complex conjugate, we obtain

$$\langle \phi_1 | A | \phi_2 \rangle + \langle \phi_2 | A | \phi_1 \rangle = \langle \phi_1 | A | \phi_2 \rangle^* + \langle \phi_2 | A | \phi_1 \rangle^* \quad .$$

On setting $a = 1$ and $b = i$ we can similarly obtain

$$i \langle \phi_1 | A | \phi_2 \rangle - i \langle \phi_2 | A | \phi_1 \rangle = -i \langle \phi_1 | A | \phi_2 \rangle^* + i \langle \phi_2 | A | \phi_1 \rangle^* \quad .$$

Multiplying the first of the above pair of equations by i and adding, we get

$$\langle \phi_1 | A | \phi_2 \rangle = \langle \phi_2 | A | \phi_1 \rangle^* \quad .$$

On the other hand, it follows from (8.15) that it is always true that

$$\langle \phi_1 | A | \phi_2 \rangle = \langle \phi_2 | A^\dagger | \phi_1 \rangle^* \quad .$$

It then follows from the above two equations that $A = A^\dagger$. □

Problem 8.1 Show that the operator

$$p_m = \frac{\hbar}{i} \frac{\partial}{\partial x^m}$$

acting on the Hilbert space of square-integrable functions is a **self-adjoint** operator. What is the **adjoint** of the operator $\dfrac{\partial}{\partial x^m}$?

Problem 8.2 For a charged particle of mass m and charge e moving in a magnetic field $\boldsymbol{B} = \nabla \times \boldsymbol{A}$, where $\boldsymbol{A}$ is the vector potential, the Hamiltonian is given by

$$H = \frac{\left(\boldsymbol{p} - \frac{e}{c}\boldsymbol{A}\right)^2}{2m} = \frac{1}{2m}\left\{\boldsymbol{p}^2 - \frac{e}{c}\left(\boldsymbol{p}\cdot\boldsymbol{A} + \boldsymbol{A}\cdot\boldsymbol{p}\right) + \left(\frac{e}{c}\right)^2 \boldsymbol{A}^2\right\} .$$

Show that H is self-adjoint by showing that $\boldsymbol{p}\cdot\boldsymbol{A} + \boldsymbol{A}\cdot\boldsymbol{p}$ is self-adjoint. Note that $\boldsymbol{p}$ and $\boldsymbol{A}$ *do not* commute as operators.
Hint: Use integration by parts (partial integration) and assume that

$$\phi(\boldsymbol{x}) \, , \, \psi(\boldsymbol{x}) \xrightarrow[|\boldsymbol{x}|\to\infty]{} 0 \, .$$

Also remember that $\boldsymbol{p} \longrightarrow -i\hbar\nabla$.

Next we recall two fundamental notions in the study of linear operators on vector spaces [cf. discussion following (4.16)].

Definition 8.2. *Suppose A is a linear operator on a vector space $\mathcal{H}$. If*

$$A\,|\,\phi\,\rangle = \lambda\,|\,\phi\,\rangle\,, \quad |\,\phi\,\rangle \in \mathcal{H}\,,\ \lambda \in \mathbb{C}\,, \tag{8.21}$$

then $|\,\phi\,\rangle$ is called an **eigenvector**, *and λ an* **eigenvalue** *of the operator A.*

We note that the adjoint of the above equation is

$$\langle\,\phi\,|\,A^\dagger = \lambda^*\,\langle\,\phi\,| \quad . \tag{8.22}$$

Self-adjoint operators play a fundamental role in quantum theory because they satisfy the following two remarkable properties (Theorems 8.2 and 8.3).

Theorem 8.2. *If $A = A^\dagger$ is a self-adjoint operator in a Hilbert space $\mathcal{H}$, then all eigenvalues of A are real.*

Proof. Let $A\,|\,\phi\,\rangle = \lambda\,|\,\phi\,\rangle\,, \lambda \in \mathbb{C}\,, |\,\phi\,\rangle \in \mathcal{H}\,, |\,\phi\,\rangle \neq 0$. Then

$$\begin{aligned} A = A^\dagger &\Longrightarrow \langle\,\phi\,|\,A\,|\,\phi\,\rangle = \langle\,\phi\,|\,A\,|\,\phi\,\rangle^* \qquad [\text{ by } (8.15)\ ] \\ &\Longrightarrow \lambda\,\langle\,\phi\,|\,\phi\,\rangle = \lambda^*\,\langle\,\phi\,|\,\phi\,\rangle \qquad [\text{ by } (7.7)\ ] \\ &\Longrightarrow \lambda = \lambda^* \quad . \end{aligned}$$

The last step follows from the assumption that $|\,\phi\,\rangle \neq 0$, which implies that $\langle\,\phi\,|\,\phi\,\rangle \neq 0$. □

The above theorem is the fundamental reason why in quantum theory observables (measurabe quantities) are represented by self-adjoint operators: measurements must always yield real values (in order to make physical sense), and within the theory measured results of an observable represented by an operator A are given by the eigenvalues of A.

Theorem 8.3. *Eigenvectors corresponding to distinct eigenvalues of a self-adjoint operator are orthogonal to each other.*

Proof. Let $A = A^\dagger$ be a self-adjoint operator on a Hilbert space $\mathcal{H}$. Suppose $A\,|\,\phi_1\,\rangle = \lambda_1\,|\,\phi_1\,\rangle$ and $A\,|\,\phi_2\,\rangle = \lambda_2\,|\,\phi_2\,\rangle$, $|\,\phi_1\,\rangle\,, |\,\phi_2\,\rangle \in \mathcal{H}$, $\lambda_1\,, \lambda_2 \in \mathbb{C}$. Then by Theorem 8.2, λ_1 and λ_2 are both real. Thus

$$\begin{aligned} 0 &= \langle\,\phi_1\,|\,A\,|\,\phi_2\,\rangle - \langle\,\phi_2\,|\,A\,|\,\phi_1\,\rangle^* = \lambda_2\,\langle\,\phi_1\,|\,\phi_2\,\rangle - \lambda_1\,\langle\,\phi_2\,|\,\phi_1\,\rangle^* \\ &= \lambda_2\,\langle\,\phi_1\,|\,\phi_2\,\rangle - \lambda_1\,\langle\,\phi_1\,|\,\phi_2\,\rangle = (\lambda_2 - \lambda_1)\,\langle\,\phi_1\,|\,\phi_2\,\rangle \quad . \end{aligned}$$

Hence, if $\lambda_1 \neq \lambda_2$, $\langle\,\phi_1\,|\,\phi_2\,\rangle = 0$. □

We conclude this chapter with some more useful facts on operators and their adjoints:

$$(cA)^\dagger = c^*\,A^\dagger\,,\ c \in \mathbb{C} \quad , \tag{8.23}$$

$$(A+B)^\dagger = A^\dagger + B^\dagger \quad , \tag{8.24}$$

$$(AB)^\dagger = B^\dagger\,A^\dagger \quad , \tag{8.25}$$

$$(\,|\,\phi\,\rangle\langle\,\psi\,|\,)^\dagger = |\,\psi\,\rangle\langle\,\phi\,| \quad . \tag{8.26}$$

Proof. [*of (8.23)*] We have, for any $|\,\phi\,\rangle\,,\,|\,\psi\,\rangle \in \mathcal{H}$,

$$\begin{aligned}\langle\,\phi\,|\,(cA)^\dagger\,|\,\psi\,\rangle &= \langle\,\psi\,|\,cA\,|\,\phi\,\rangle^* = c^*\,\langle\,\psi\,|\,A\,|\,\phi\,\rangle^* \\ &= c^*\,\langle\,\phi\,|\,A^\dagger\,|\,\psi\,\rangle = \langle\,\phi\,|\,c^*A^\dagger\,|\,\psi\,\rangle \quad .\end{aligned}$$

The result follows. □

Proof. [*of (8.24)*] We have, for any $|\,\phi\,\rangle\,,\,|\,\psi\,\rangle \in \mathcal{H}$,

$$\begin{aligned}&\langle\,\phi\,|\,(A+B)^\dagger\,|\,\psi\,\rangle = \langle\,\psi\,|\,(A+B)\,|\,\phi\,\rangle^* \\ &= \langle\,\psi\,|\,A\,|\,\phi\,\rangle^* + \langle\,\psi\,|\,B\,|\,\phi\,\rangle^* = \langle\,\phi\,|\,A^\dagger\,|\,\psi\,\rangle + \langle\,\phi\,|\,B^\dagger\,|\,\psi\,\rangle \\ &= \langle\,\phi\,|\,(A^\dagger + B^\dagger)\,|\,\psi\,\rangle \quad .\end{aligned}$$

The result follows. □

Proof. [*of (8.25)*] We have, for any $|\,\phi\,\rangle\,,\,|\,\psi\,\rangle \in \mathcal{H}$,

$$\begin{aligned}\langle\,\phi\,|\,(AB)^\dagger\,|\,\psi\,\rangle &= \langle\,\psi\,|\,AB\,|\,\phi\,\rangle^* = \langle\,A^\dagger\,\psi\,|\,B\,|\,\phi\,\rangle^* \\ &= (\,\langle\,\phi\,|\,B^\dagger A^\dagger\,|\,\psi\,\rangle^*\,)^* = \langle\,\phi\,|\,B^\dagger A^\dagger\,|\,\psi\,\rangle \quad ,\end{aligned}$$

where the second equality follows from (8.14). The result follows. □

Proof. [*of (8.26)*] We have, for any $|\,\alpha\,\rangle\,,\,|\,\beta\,\rangle \in \mathcal{H}$,

$$\langle\,\alpha\,|\,(\,|\,\phi\,\rangle\langle\,\psi\,|\,)^\dagger\,|\,\beta\,\rangle = (\,\langle\,\beta\,|\,\phi\,\rangle\langle\,\psi\,|\,\alpha\,\rangle\,)^* = \langle\,\alpha\,|\,\psi\,\rangle\langle\,\phi\,|\,\beta\,\rangle \quad .$$

The result follows. □

From (8.23) and (8.24), it immediately follows that

$$\boxed{(A+iB)^\dagger = A^\dagger - iB^\dagger} \quad . \tag{8.27}$$

This result shows that the operation of taking the adjoint of operators is analogous to the complex conjugation of complex numbers.

Problem 8.3 Recall that the hermitian adjoint of a matrix A (denoted by $A^\dagger$) is defined as the complex conjugate of its transpose:

$$A^\dagger = (A^T)^*\ , \quad \text{or} \quad (A^\dagger)_{ij} = (A_{ji})^*\ ;$$

that a matrix A is hermitian if $A^\dagger = A$; and that a matrix U is **unitary** if $U^\dagger = U^{-1}$. Show that if U is a unitary matrix and A is an arbitrary matrix of the same size, then

$$U^\dagger e^A U = e^{U^\dagger A U} \quad .$$

Problem 8.4 Show that if the matrix A is hermitian, then the matrix e^{iA} is unitary.

Problem 8.5 Show that a matrix of order N

(a) is necessarily diagonal if it commutes with all diagonal matrices of order N;

(b) is necessarily a constant times the identity matrix if it commutes with all matrices of order N.

Statement (b) is a special case of the so-called **Schur's lemma**, a theorem of fundamental importance in group representation theory.

Problem 8.6 Assume that any hermitian matrix can be diagonalized by a unitary matrix. Show that two hermitian matrices can be diagonalized by the *same* unitary matrix if and only if they commute.

Problem 8.7 This problem introduces formally the notions of **mixed states** and **density matrices** used in quantum statistics. Consider a system of particles and a Hilbert space $\mathcal{H}$ whose elements describe the states of an individual particle in the system. In practice it is impossible to prepare the system so that every particle is in the same state in $\mathcal{H}$. Instead, one always has a mixed state, which is characterized, at some initial time $t = 0$, by a certain probability p_i of finding the particles in a particular state $|\, i\,\rangle$, where $\{|\, i\,\rangle\}$ is a complete set of states spanning $\mathcal{H}$. The mixed state can then be described by a (time-dependent) density matrix $\rho_{ij}(t)$ such that $\rho_{ij}(0) = p_i\, \delta_{ij}$ $(\sum_i p_i = 1)$.

The density matrix $\rho_{ij}(t)$ can be considered as the matrix representation (with respect to the basis set $\{|\, i\,\rangle\}$) of the **density operator** $\rho(t)$:

$$\rho(t) = \sum_i |\, i,\, t\,\rangle\, p_i\, \langle\, i,\, t\,| = \sum_{ij} \rho_{ij}(t)\, |\, i\,\rangle\langle\, j\,|\,,$$

where $|\, i,\, t\,\rangle$ satisfies the Schrödinger equation

$$H|\, i,\, t\,\rangle = i\hbar \frac{d}{dt} |\, i,\, t\,\rangle\,,$$

with the initial condition $|\, i,\, 0\,\rangle = |\, i\,\rangle$.

(a) Show that the density operator is hermitian and positive-definite (eigenvalues all ≥ 0).

(b) Show that the density operator (as well as the density matrix) satisfies the equation of motion

$$i\hbar \frac{d\rho}{dt} = [\,H,\, \rho\,]\,.$$

Compare this with the Heisenberg equation of motion for an observable [cf. (2.13)]. (Hint: Use the above Schrödinger equation and its adjoint equation.)

(c) Show that $Tr\, \rho(t) = 1$. Interpret this result physically.

(d) Show that $Tr\, \rho^2 \leq 1$. Under what conditions would the equality sign apply? (When $Tr\, \rho^2 = 1$, the system is said to be in a **pure state**.)

(e) Consider an observable represented by the hermitian operator O. Its average (expectation) value at $t = 0$ is naturally given by $\langle\, O\,\rangle_{t=0} = \sum_i p_i \langle\, i\,|\, O\,|\, i\,\rangle$. As the system evolves in time, we have, at arbitrary time t,

$$\langle\, O\,\rangle_t = \sum_i p_i\, \langle\, i,\, t\,|\, O\,|\, i,\, t\,\rangle\,.$$

Show that $\langle\, O\,\rangle_t = Tr\, (\rho(t) O)$.

Problem 8.8 Consider a two-state system whose Hilbert space is spanned by the "spin up" and "spin down" states $|\uparrow\rangle$ and $|\downarrow\rangle$, respectively. Suppose the density matrix is initially given in the $\{|\uparrow\rangle, |\downarrow\rangle\}$ representation by

$$\rho(0) = \begin{pmatrix} p_\uparrow & 0 \\ 0 & p_\downarrow \end{pmatrix}, \quad p_\uparrow + p_\downarrow = 1\,, \;\; p_\uparrow \geq 0\,, p_\downarrow \geq 0\,.$$

Show that $\rho(0)$ can also be written as

$$\rho(0) = \begin{pmatrix} \frac{1}{2}(1+P) & 0 \\ 0 & \frac{1}{2}(1-P) \end{pmatrix},$$

where the **polarization** P is defined by $P \equiv p_\uparrow - p_\downarrow$. Under what conditions would $Tr\,(\rho(0))^2 = Tr\,\rho(0) = 1$?

Chapter 9

The Spectral Theorem, Discrete and Continuous Spectra

The set of all eigenvalues of an operator A on a vector space $\mathcal{H}$ is called the **spectrum** of A. More formally, we have the following definition.

Definition 9.1. *The spectrum of an operator A on a complex vector space is the set of all complex numbers λ for which $A-\lambda$ is not invertible.*

To see that Def. 9.1 makes sense, we need to show that if $A-\lambda$ is invertible, then λ cannot be an eigenvalue of A. Now suppose that $A-\lambda$ is invertible *and* λ is an eigenvalue. Then there exists a non-zero vector $|\,v\,\rangle$ such that $(A-\lambda)\,|\,v\,\rangle = 0$. In addition,

$$(A-\lambda)^{-1}(A-\lambda)\,|\,v\,\rangle = 1\cdot|\,v\,\rangle = |\,v\,\rangle \neq 0 \quad .$$

On the other hand

$$(A-\lambda)^{-1}(A-\lambda)\,|\,v\,\rangle = (A-\lambda)^{-1}\cdot 0 = 0 \quad \text{(the zero vector)} \quad .$$

Thus we arrive at a contradiction, which proves that λ cannot be an eigenvalue of A if $A-\lambda$ is invertible.

Suppose the eigenvectors $\{|\,\phi_i\,\rangle\}$ of A form a complete orthonormal set in $\mathcal{H}$, that is,

$$A\,|\,\phi_i\,\rangle = \lambda_i\,|\,\phi_i\,\rangle \;,\lambda_i \in \mathbb{C}, \qquad \text{and} \qquad \sum_i |\,\phi_i\,\rangle\langle\,\phi_i\,| = 1 \quad .$$

Then

$$\begin{aligned} A = A\cdot 1 = A\sum_i |\,\phi_i\,\rangle\langle\,\phi_i\,| &= \sum_i A\,|\,\phi_i\,\rangle\langle\,\phi_i\,| \\ &= \sum_i \lambda_i\,|\,\phi_i\,\rangle\langle\,\phi_i\,| \equiv \sum_i \lambda_i\,P_i \quad , \end{aligned} \tag{9.1}$$

where

$$P_i \equiv |\,\phi_i\,\rangle\langle\,\phi_i\,| \tag{9.2}$$

is the projection operator onto the subspace of $\mathcal{H}$ spanned by the eigenvetor $|\,\phi_i\,\rangle$. Equation (9.1) gives the so-called **spectral representation** of an operator A when all its eigenvalues are discrete. It should be noted that a self-adjoint operator on a Hilbert space $\mathcal{H}$ may not necessarily possess a complete set of eigenvectors in $\mathcal{H}$.

If the spectrum of A is **degenerate**. that is, when the subspace of eigenvectors corresponding to a particular eigenvalue is of dimension higher than one, but still discrete, the projection operator onto the entire subspace spanned by all the orthonormal (degenerate) eigenvectors corresponding to an eigenvalue λ can be written as

$$P(\lambda) = \sum_i |\,\phi_i\,\rangle\langle\,\phi_i\,|\,\delta_{\lambda\lambda_i} \;. \tag{9.3}$$

Then

$$A = \sum_\lambda \lambda P(\lambda) \qquad \text{(for discrete spectrum)} \quad , \tag{9.4}$$

where the sum is over the discrete values of the spectrum of A.

A well-known and basic fact in linear operator theory is the following, which we will state without proof.

Theorem 9.1. *If A is self-adjoint, then any of its matrix representations can be diagonalized. Furthermore, the diagonal elements of the diagonalized matrix, not all of which are distinct in case of degeneracies, are the eigenvalues of A.*

A significant complication arises if an operator has a continuous spectrum, or a spectrum that is partially continuous. In this case we have the so-called spectral theorem as a generalization of (9.4).

Theorem 9.2 (Spectral Theorem). *If A is a self-adjoint operator on a Hilbert space, then there exists a unique compact, complex* ***spectral measure*** *E, called the spectral measure of A, such that*

$$\boxed{A = \int \lambda\, dE(\lambda)} \quad . \tag{9.5}$$

A few explanations of the terminology of this important theorem are in order, but we shall not be concerned with all the technical details. $E(\lambda)$ is a projection operator onto the eigen-subspace of $\mathcal{H}$ with eigenvalue λ. Thus the spectral measure E is an operator-valued function whose domain is the set of real numbers. The integral in (9.5) is to be interpreted as a **Stieltjes integral**:

$$\int_a^b g(x)\, d\sigma(x) = \lim_{n\to\infty} \sum_{k=1}^{n} g(x_k)\,[\,\sigma(x_k) - \sigma(x_{k-1})\,] \quad , \tag{9.6}$$

where the function σ is called the **measure**. If $\sigma(x) = x$, then the Stieltjes integral reduces to the Riemann integral.

Equation (9.5) is a general statement valid for both discrete and continuous spectra. If the spectrum is discrete, we claim that the spectral measure is given by

$$E(\lambda) = \sum_i |\,\phi_i\,\rangle\langle\,\phi_i\,|\,\theta(\lambda - \lambda_i) \quad , \tag{9.7}$$

where the sum is over a complete orthonormal set, and $\theta(x)$ is the step function. Indeed,

$$\begin{aligned}\int \lambda\, dE(\lambda) &= \int \lambda \sum_i |\,\phi_i\,\rangle\langle\,\phi_i\,|\,\frac{d\theta(\lambda-\lambda_i)}{d\lambda}\, d\lambda \\ &= \int \lambda \sum_i |\,\phi_i\,\rangle\langle\,\phi_i\,|\,\delta(\lambda - \lambda_i)\, d\lambda \\ &= \sum_i \lambda_i\,|\,\phi_i\,\rangle\langle\,\phi_i\,| = \sum_i \lambda P(\lambda) \quad .\end{aligned} \tag{9.8}$$

To study a continuous spectrum we consider the **position operator** Q acting on the Hilbert space of square-integrable (L^2) functions $\psi(x)$ defined by

$$Q\,\psi(x) = x\psi(x) \quad . \tag{9.9}$$

It is obvious that $Q = Q^\dagger$. Indeed

$$\begin{aligned}\langle\,\phi\,|\,Q\,|\,\psi\,\rangle &= \int dx\,\phi^*(x)Q\psi(x) = \int dx\,\phi^* x\psi \\ &= \left(\int dx\,\psi^* x\phi\right)^* = \left(\int dx\,\psi^* Q\phi\right)^* = \langle\,\psi\,|\,Q\,|\,\phi\,\rangle^* \quad .\end{aligned}$$

What are the eigenvalues and eigenfunctions of the operator Q?

On writing

$$Q\phi(x) = x\phi(x) = \lambda\phi(x)\,, \quad \lambda \in \mathbb{R} \quad , \tag{9.10}$$

we see that the formal solution for the eigenfunction is the Dirac delta function $\delta(x - \lambda)$ where λ is a real number. We thus have a situation where the eigenfunction of Q does not exist, since $\delta(x - \lambda)$ is not a function in the proper mathematical sense of the word, let alone a square-integrable function. In other words, the eigenfunctions of Q, which acts on the Hilbert space $\mathcal{H}$ of square-integrable functions, do not belong to $\mathcal{H}$ at all! On the other hand, the spectral measure $E(\lambda)$ for Q is well-defined. We claim that it is given by

$$E(\lambda)\,\psi(x) = \theta(\lambda - x)\,\psi(x) \quad . \tag{9.11}$$

Indeed, on using the above equation in (9.5), we have

$$\begin{aligned}Q\psi(x) &= \int_{-\infty}^{\infty} \lambda\, dE(\lambda)\psi(x) = \int_{-\infty}^{\infty} \lambda\, d\,[\,\theta(\lambda - x)\psi(x)\,] \\ &= \int_{-\infty}^{\infty} \lambda\,\frac{d\theta(\lambda - x)}{d\lambda}\,\psi(x)\,d\lambda = \int_{-\infty}^{\infty} d\lambda\,\lambda\psi(x)\delta(\lambda - x) = x\psi(x) \quad .\end{aligned}$$

Dirac still considers the delta-functions as eigenfunctions of Q anyway, and formally associates kets with them. Thus we have the following correspondences:

$$\begin{aligned} \text{normalizable functions:} \quad & \psi(x) \longrightarrow |\,\psi\,\rangle \\ \text{non-normalizable functions:} \quad & \delta(x-\lambda) \longrightarrow |\,\lambda\,\rangle \quad . \end{aligned}$$

He also writes the eigenvalue equation for Q as

$$Q\,|\,\lambda\,\rangle = \lambda\,|\,\lambda\,\rangle \quad , \tag{9.12}$$

where $|\,\lambda\,\rangle$ is the "eigenvector" of Q corresponding to the eigenvalue λ. Now, according to the definition of Q [cf. (9.9)],

$$\langle\,\lambda'\,|\,Q\,|\,\lambda\,\rangle = \int_{-\infty}^{\infty} dx\,\delta(x-\lambda')\,x\,\delta(x-\lambda) = \lambda\,\delta(\lambda'-\lambda) \quad . \tag{9.13}$$

On the other hand, according to (9.12),

$$\langle\,\lambda'\,|\,Q\,|\,\lambda\,\rangle = \lambda\,\langle\,\lambda'|\,\lambda\,\rangle \quad . \tag{9.14}$$

Thus we obtain the following important "normalization" condition for eigenvectors corresponding to continuous eigenvalues:

$$\boxed{\langle\,\lambda'\,|\,\lambda\,\rangle = \delta(\lambda'-\lambda)} \quad . \tag{9.15}$$

The delta-function on the right-hand side implies that the continuous eigenket $|\,\lambda\,\rangle$ is not normalizable in the usual sense, since $\langle\,\lambda\,|\,\lambda\,\rangle = \delta(0) = \infty$, whereas all vectors in the Hilbert space of L^2 functions have finite length.

Instead of (9.5) for the spectral representation of Q, Dirac writes the more transparent formula

$$\boxed{Q = \textstyle\int_{-\infty}^{\infty} d\lambda\,\lambda\,|\,\lambda\,\rangle\langle\,\lambda\,|} \tag{9.16}$$

as a generalization of (9.4). The above equation makes sense since then

$$Q\,|\,\lambda'\,\rangle = \int_{-\infty}^{\infty} d\lambda\,\lambda\,|\,\lambda\,\rangle\langle\,\lambda\,|\,\lambda'\,\rangle = \int_{-\infty}^{\infty} d\lambda\,\lambda\,|\,\lambda\,\rangle\,\delta(\lambda-\lambda') = \lambda'\,|\,\lambda'\,\rangle \quad ,$$

which verifies (9.12). Using (9.16) the spectral measure $E(\lambda)$ in (9.5) can be identified as follows

$$dE(\lambda) = |\,\lambda\,\rangle\langle\,\lambda\,|\,d\lambda \quad , \tag{9.17}$$

which implies

$$E(\lambda) = \int_{-\infty}^{\lambda} d\lambda'\,|\,\lambda'\rangle\langle\,\lambda'\,| \quad . \tag{9.18}$$

Now consider the inner product $\langle\,\lambda\,|\,\psi\,\rangle$, where $|\,\psi\,\rangle$ corresponds to a normalizable function. We have

$$\langle \lambda \,|\, \psi \rangle = \int_{-\infty}^{\infty} dx\, \delta(x-\lambda)\psi(x) = \psi(\lambda) \quad . \tag{9.19}$$

Thus the inner product $\langle x \,|\, \psi \rangle$ is precisely the Schrödinger wave function:

$$\boxed{\psi(x) = \langle x \,|\, \psi \rangle} \quad . \tag{9.20}$$

For any $|\phi\rangle, |\psi\rangle \in \mathcal{H}$, we have

$$\langle \phi | \left(\int_{-\infty}^{\infty} dx \,|\, x \rangle\langle x \,| \right) | \psi \rangle$$
$$= \int_{-\infty}^{\infty} dx \,\langle \phi \,|\, x \rangle\langle x \,|\, \psi \rangle = \int_{-\infty}^{\infty} dx\, \phi^*(x)\psi(x) = \langle \phi \,|\, \psi \rangle \quad .$$

Hence we can establish the following completeness condition for the continuous eigenkets $|\, x \rangle$ [compare with (7.11)]:

$$\boxed{\int_{-\infty}^{\infty} dx \,|\, x \rangle\langle x \,| = 1} \quad . \tag{9.21}$$

Using (9.18) for the spectral measure $E(\lambda)$ and (9.15) (the normalization condition for continuous eigenkets) we can also corroborate (9.11) as follows:

$$\begin{aligned}\langle x \,|\, E(\lambda) \,|\, \psi \rangle &= \int_{-\infty}^{\lambda} d\lambda' \,\langle x \,|\, \lambda' \rangle\langle \lambda' \,|\, \psi \rangle \\ &= \int_{-\infty}^{\lambda} d\lambda'\, \delta(\lambda' - x)\, \psi(\lambda') = \begin{cases} 0 & , \quad x \geq \lambda \\ \psi(x) & , \quad x \leq \lambda \end{cases} \\ &= \theta(\lambda - x)\, \psi(x) \quad .\end{aligned}$$

For N-dimensional configuration space we can generalize the Dirac delta function as follows. Let $\boldsymbol{q} = (q^1, \ldots, q^N)$. Then

$$|\, \boldsymbol{q} \rangle = |\, q^1 \rangle \ldots |\, q^N \rangle \quad , \tag{9.22}$$
$$\langle \boldsymbol{q}' \,|\, \boldsymbol{q} \rangle = \delta^N(\boldsymbol{q}' - \boldsymbol{q}) \quad , \tag{9.23}$$

where

$$\delta^N(\boldsymbol{q}' - \boldsymbol{q}) = \delta((q')^1 - q^1) \ldots \delta((q')^N - q^N) \quad . \tag{9.24}$$

We also have the completeness condition

$$\int_{\Omega} d^N q \,|\, \boldsymbol{q} \rangle\langle \boldsymbol{q} \,| = 1 \quad , \tag{9.25}$$

where Ω is the domain of the N-dimensional configuration space. Finally

$$\psi(\boldsymbol{q}) = \langle \boldsymbol{q} \,|\, \psi \rangle \quad . \tag{9.26}$$

To summarize: If the spectrum of a self-adjoint operator has a continuous part, then there are no eigenvectors in a Hilbert space consisting of vectors of finite norm corresponding to continuous eigenvalues. One approach to remedy this situation is to use the spectral theorem and to formulate quantum mechanics in terms of projection operators $E(\lambda)$ when dealing with the continuous part of the spectrum. This approach was adopted by the mathematician von Neumann. Another approach, pioneered by Dirac, is to admit eigenvectors of infinite norm, such as the $|\lambda\rangle$, into the theory, and enlarge Hilbert space to what is called **rigged Hilbert space** (also known as **Gel'fand triplet**)(Gel'fand and Vilenkin 1964). The Dirac approach, because of the convenience, elegance, and power of the Dirac notation, is almost universally favored by physicists.

Chapter 10

Coordinate and Momentum Representations of Quantum States, Fourier Transforms

The position eigenket $|\,\lambda\,\rangle$, $\lambda \in \mathbb{R}$, can be interpreted as describing the state of a particle located sharply at the position $x = \lambda$, since $\langle\, x\,|\,\lambda\,\rangle = \delta(x-\lambda)$ is supposed to be the wave function corresponding to such a state. The Dirac delta function has the following useful integral representation [cf. (5.23)]:

$$\delta(x-\lambda) = \frac{1}{2\pi}\int_{-\infty}^{\infty} dk\, e^{ik(x-\lambda)} \quad . \tag{10.1}$$

What is the physical interpretation of this equation, and how can it be justified physically? We will attempt to provide answers to these questions in the present chapter, which will lead to a deeper understanding of the relationship between the position and momentum as observables in quantum mechanics. We will see that this relationship gives a physical manifestation of the important mathematical notion of the Fourier transform.

Let us introduce a complete set of continuous eigenkets $|\,k\,\rangle$ in the momentum variable k, $-\infty < k < \infty$, such that

$$\psi_k(x) = \langle\, x\,|\,k\,\rangle = \frac{1}{\sqrt{2\pi}}\, e^{ikx} \,, \tag{10.2}$$

which are wave functions describing **plane waves** of wave number k. The momentum eigenkets $|\,k\,\rangle$ satisfy the eigenvalue equation for the **momentum operator** P [analogous to (9.12) for the position eigenket $|\,\lambda\,\rangle$]:

$$P\,|\,k\,\rangle = \hbar k\,|\,k\,\rangle \quad ; \tag{10.3}$$

while the wave functions $\psi_k(x)$ satisfy

$$\frac{\hbar}{i}\frac{d}{dx}\,\psi_k(x) = \hbar k\,\psi_k(x) \quad . \tag{10.4}$$

Thus the momentum operator P acting on wave functions (considered as functions of the position coordinate x) is given by [recall (1.33)]

$$P = \frac{\hbar}{i}\,\frac{d}{dx} \quad . \tag{10.5}$$

The completeness condition for the $|\,k\,\rangle$ [analogous to (9.21)] is

$$\boxed{\int_{-\infty}^{\infty} dk\,|\,k\,\rangle\langle\,k\,| = 1} \quad . \tag{10.6}$$

Problem 10.1 Given $P\,|\,k\,\rangle = \hbar k\,|\,k\,\rangle$, derive the translation key (from classical to quantum mechanics):

$$P \longrightarrow -i\hbar\,\frac{d}{dx}$$

using the completeness condition for the momentum eigenstates

$$\int_{-\infty}^{\infty} dk\,|\,k\,\rangle\langle\,k\,| = 1\,.$$

The momentum eigenket $|\,k\,\rangle$ describes the state of a particle (wave) having a sharp momentum k. Its wave function, the coordinate representation of $|\,k\,\rangle$, is precisely $\psi_k(x)$ in (10.2). Using the completeness condition for $|\,k\,\rangle$ [(10.6)] we can now justify the integral representation of the Dirac delta function given by (10.1). Indeed

$$\begin{aligned}\delta(x-\lambda) = \langle\,x\,|\,\lambda\,\rangle &= \int_{-\infty}^{\infty} dk\,\langle\,x\,|\,k\,\rangle\langle\,k\,|\,\lambda\,\rangle \\ &= \frac{1}{2\pi}\int_{-\infty}^{\infty} dk\,e^{ikx}\,e^{-ik\lambda} = \frac{1}{2\pi}\int_{-\infty}^{\infty} dk\,e^{ik(x-\lambda)} \quad .\end{aligned} \tag{10.7}$$

For a general state vector $|\,\psi\,\rangle \in \mathcal{H}$, $\langle\,x\,|\,\psi\,\rangle = \psi(x)$ is called its **wave function in the coordinate representation**, or simply its (Schrödinger) wave function; and $\langle\,k\,|\,\psi\,\rangle = \psi(k)$ is called its **wave function in the momentum representation**. The completeness condition (10.6) also implies

$$|\,x\,\rangle = \int_{-\infty}^{\infty} dk\,|\,k\,\rangle\langle\,k\,|\,x\,\rangle \quad , \tag{10.8}$$

or, in view of (10.2),

$$| x \rangle = \frac{1}{\sqrt{2\pi}} \int_{-\infty}^{\infty} dk \, e^{-ikx} \, | k \rangle \quad . \tag{10.9}$$

Thus the sharp-position state $| x \rangle$ can be expanded in terms of the complete set of sharp-momentum states $| k \rangle$. Conversely, using the completeness condition for the $| x \rangle$ (9.21), we have

$$| k \rangle = \int_{-\infty}^{\infty} dx \, | x \rangle \langle x | k \rangle \quad , \tag{10.10}$$

or, in view of (10.2) again,

$$| k \rangle = \frac{1}{\sqrt{2\pi}} \int_{-\infty}^{\infty} dx \, e^{ikx} \, | x \rangle \quad . \tag{10.11}$$

More generally, taking the conjugates of (10.9) and (10.11) [under the conjugate isomorphism of the Riesz Theorem as represented by (7.5)] with the help of (7.8), and forming inner products with the ket $| \psi \rangle$ on the right, we obtain

$$\psi(x) = \frac{1}{\sqrt{2\pi}} \int_{-\infty}^{\infty} dk \, e^{ikx} \, \psi(k) \, , \quad \psi(k) = \frac{1}{\sqrt{2\pi}} \int_{-\infty}^{\infty} dx \, e^{-ikx} \, \psi(x) \quad . \tag{10.12}$$

Thus the coordinate representation $\psi(x) = \langle x | \psi \rangle$ and the momentum representation $\psi(k) = \langle k | \psi \rangle$ of a state vector $| \psi \rangle$ in Hilbert space are **Fourier transforms** of each other. Note the difference in sign in the exponents in the above **Fourier transform pair**. The wave function $\psi(x)$ as expressed by the right-hand side of the first equation of (10.12) may be normalizable, and thus may represent a spatially localized **wave packet**, provided that the **Fourier amplitude** $\psi(k)$ suitably approaches zero as $k \to \pm\infty$. The wave packet can be considered as being built from a superposition of the plane waves $\exp(ikx)$, each of which is non-normalizable.

Because of the symmetry of the formulas in (10.12), all equations remain valid when *all* of the following substitutions are made:

$$\psi(x) \leftrightarrow \psi(k) \, , \quad x \leftrightarrow k \, , \quad i \leftrightarrow -i \quad . \tag{10.13}$$

In particular we have the **Parseval Theorem**:

$$\int_{-\infty}^{\infty} dx \, \psi^*(x)\psi(x) = \int_{-\infty}^{\infty} dk \, \psi^*(k)\psi(k) \quad . \tag{10.14}$$

Problem 10.2 Show formally that

$$\delta(x - y) = \frac{1}{2\pi} \int_{-\infty}^{\infty} dk \, e^{ik(x-y)}$$

with the following steps.

i) Note that

$$\int_{-\infty}^{\infty} g(z)\delta(z-y)dz = g(y)\,.$$

ii) Consider the Fourier transform pair

$$v(k) = \int_{-\infty}^{\infty} dx\, e^{-ikx}\, u(x)\,, \quad \text{for all } k \in \mathbb{R}\,,$$

$$u(x) = \frac{1}{2\pi}\int_{-\infty}^{\infty} dk\, e^{ikx}\, v(k)\,, \text{ for all } x \in \mathbb{R}\,.$$

Then set $u(x) = \delta(x-y)$ and apply (1) to the above two equations.

Problem 10.3 In the coordinate representation, the Hamiltonian for an electron moving in the field of a nucleus of charge Ze fixed at the origin is

$$H = -\frac{\hbar^2}{2m}\nabla^2 - \frac{Ze^2}{|\boldsymbol{x}|}\,.$$

Derive the Schrödinger equation for the wave function $\psi(\boldsymbol{p}, t)$ in the momentum representation.

Problem 10.4 Using the Schrödinger equation in the momentum representation, calculate the bound state energy eigenvalue and the coresponding eigenfunction for the one-dimensional attractive delta-function potential

$$V(x) = -g\,\delta(x)\,, \quad g > 0\,.$$

Chapter 11

The Uncertainty Principle

Heisenberg's Uncertainty Principle is a most fundamental and far-reaching principle in modern physics and lies at the heart of quantum theory. In this chapter we will study in some detail how it can be understood and derived within the framework of the formalism of quantum mechanics that we have developed. In particular, we will see how it arises from the commutation relations between certain pairs of non-commuting observables.

For a wave packet described by a normalized wave function $\psi(x)$ we define the **expectation value** (mean value) of a function $f(x)$ by

$$\langle f(x) \rangle \equiv \langle \psi \,|\, f(Q) \,|\, \psi \rangle = \int dx\, \psi^*(x) f(x) \psi(x) \quad , \tag{11.1}$$

where Q is the position operator [cf. (9.9)]. Similarly we define

$$\langle f(k) \rangle \equiv \langle \psi \,|\, f(P/\hbar) \,|\, \psi \rangle = \int dx\, \psi^*(k) f(k) \psi(k) \quad , \tag{11.2}$$

where P is the momentum operator [cf. (10.3)].

Expectation values of $f(x)$ can also be expressed in terms of momentum representation wave functions; and conversely, those of $f(k)$ can be expressed in terms of coordinate representation wave functions. Let us first consider the case $f(x) = x$. We have

$$\begin{aligned}
\langle x \rangle = \langle \psi | Q | \psi \rangle &= \int_{-\infty}^{\infty} dk \int_{-\infty}^{\infty} dx \, \langle \psi | x \rangle \langle x | Q | k \rangle \langle k | \psi \rangle \\
&= \int_{-\infty}^{\infty} dx \int_{-\infty}^{\infty} dk \, \psi^*(x) \, x \, \langle x | k \rangle \, \psi(k) = \frac{1}{\sqrt{2\pi}} \int_{-\infty}^{\infty} dx \, \psi^*(x) \int_{-\infty}^{\infty} dk \, x e^{ikx} \psi(k) \\
&= \frac{1}{\sqrt{2\pi}} \int_{-\infty}^{\infty} dx \, \psi^*(x) \int_{-\infty}^{\infty} dk \, (-i)(ix) e^{ikx} \psi(k) \\
&= \frac{1}{\sqrt{2\pi}} \int_{-\infty}^{\infty} dx \, \psi^*(x) \int_{-\infty}^{\infty} dk \, (-i) \, \psi(k) \, \frac{\partial}{\partial k} \left(e^{ikx} \right) \\
&= \frac{1}{\sqrt{2\pi}} \int_{-\infty}^{\infty} dx \, \psi^*(x) \int_{-\infty}^{\infty} dk \, i \, \frac{\partial \psi(k)}{\partial k} \, e^{ikx} \\
&= \int_{-\infty}^{\infty} dk \, i \, \frac{\partial \psi(k)}{\partial k} \, \frac{1}{\sqrt{2\pi}} \int_{-\infty}^{\infty} dx \, \psi^*(x) \, e^{ikx} = \int_{-\infty}^{\infty} dk \, \psi^*(k) \, i \frac{\partial \psi(k)}{\partial k} \quad ,
\end{aligned} \tag{11.3}$$

that is,

$$\langle x \rangle = \int_{-\infty}^{\infty} dk \, \psi^*(k) \left(i \frac{\partial}{\partial k} \right) \psi(k) \quad . \tag{11.4}$$

On the right-hand side of the seventh equality sign of (11.3), we have performed an integration by parts (over k), and made use of the fact that the boundary terms vanish [$\psi(k) \to 0$ as $k \to \pm\infty$]. By n steps of integration by parts, we have

$$\langle x^n \rangle = \int_{-\infty}^{\infty} dk \, \psi^*(k) \left(i \frac{\partial}{\partial k} \right)^n \psi(k) \quad . \tag{11.5}$$

Thus, for any analytic function of x (one that is expressible as a power series in x), we have

$$\langle f(x) \rangle = \int_{-\infty}^{\infty} dk \, \psi^*(k) \, f \left(i \frac{\partial}{\partial k} \right) \psi(k) \quad . \tag{11.6}$$

By repeating the above procedure with $f(k)$ substituted for $f(x)$ we obtain similarly,

$$\langle f(k) \rangle = \int_{-\infty}^{\infty} dx \, \psi^*(x) \, f \left(-i \frac{\partial}{\partial x} \right) \psi(x) \quad . \tag{11.7}$$

It is convenient to summarize in a table (top of p. 65) our results so far for the coordinate and momentum representations of a state $| \psi \rangle$, the position operator Q, and the momentum operator $P/\hbar$.

Let us define the following square-deviations:

$$(\delta x)^2 \equiv (x - \langle x \rangle)^2 , \quad (\delta k)^2 \equiv (k - \langle k \rangle)^2 \quad . \tag{11.8}$$

For simplicity, let $\langle x \rangle = 0$. This can always be achieved by a coordinate transformation. It then follows that $\langle k \rangle$ also vanishes. Indeed, using the Heisenberg equation of motion (2.20) and the fundamental commutation relations (2.14),

	$\vert\psi\rangle$	Q	$P/\hbar$
coordinate representation	$\langle x\vert\psi\rangle = \psi(x)$	x	$-i\partial/\partial x$
momentum representation	$\langle k\vert\psi\rangle = \psi(k)$	$i\partial/\partial k$	k

and assuming that the Hamiltonian H can be written in the form (1.34), we have

$$\begin{aligned} i\hbar\frac{dx}{dt} &= [x, H] = \frac{1}{2m}[x, p^2] + [x, V(x)] \\ &= \frac{1}{2m}[x, p^2] = \frac{1}{2m}2i\hbar p = i\hbar\frac{p}{m} \quad . \end{aligned} \tag{11.9}$$

Hence

$$\frac{d\langle x\rangle}{dt} = \frac{\langle p\rangle}{m} , \tag{11.10}$$

from which it follows that, if $\langle x\rangle$ is independent of the time t, then $\langle p\rangle = 0$. In particular, $\langle x\rangle = 0$ implies $\langle p\rangle = 0$.

We wish to calculate the **mean-square-deviations**

$$\langle(\delta x)^2\rangle = \langle x^2\rangle \qquad \text{and} \qquad \langle(\delta k)^2\rangle = \langle k^2\rangle \tag{11.11}$$

(under the assumption that $\langle x\rangle = \langle k\rangle = 0$). Using (11.6) and (11.7), and integrating by parts, we have

$$\langle x^2\rangle = \int_{-\infty}^{\infty} dk\,\psi^*(k)\left(-\frac{\partial^2}{\partial k^2}\right)\psi(k) = \int_{-\infty}^{\infty} dk\,\frac{\partial\psi^*(k)}{\partial k}\frac{\partial\psi(k)}{\partial k} , \tag{11.12a}$$

$$\langle k^2\rangle = \int_{-\infty}^{\infty} dx\,\psi^*(x)\left(-\frac{\partial^2}{\partial x^2}\right)\psi(x) = \int_{-\infty}^{\infty} dx\,\frac{\partial\psi^*(x)}{\partial x}\frac{\partial\psi(x)}{\partial x} \quad . \tag{11.12b}$$

To derive the uncertainty relation, we have to compute the product $\langle x^2\rangle\langle k^2\rangle$. We start from the inequality

$$D(x) \equiv \left|\frac{x}{2\langle x^2\rangle}\psi(x) + \frac{\partial\psi}{\partial x}\right|^2 \geq 0 \quad . \tag{11.13}$$

The real quantity $D(x)$ can be expanded as follows:

$$\begin{aligned} D(x) &= \frac{x^2}{4\langle x^2 \rangle^2} \,|\, \psi(x) \,|^2 + \frac{x}{2\langle x^2 \rangle} \left(\psi^* \frac{\partial \psi}{\partial x} + \psi \frac{\partial \psi^*}{\partial x} \right) + \left| \frac{\partial \psi}{\partial x} \right|^2 \\ &= \frac{1}{4} \left(\frac{x}{\langle x^2 \rangle} \right)^2 \psi^*(x)\psi(x) + \frac{1}{2} \frac{\partial}{\partial x} \left(\frac{x}{\langle x^2 \rangle} \psi^*(x)\psi(x) \right) \\ &\quad - \frac{1}{2\langle x^2 \rangle} \psi^*(x)\psi(x) + \frac{\partial \psi^*(x)}{\partial x} \frac{\partial \psi(x)}{\partial x} \\ &= \frac{1}{4\langle x^2 \rangle^2} \left(x^2 - 2\langle x^2 \rangle \right) \psi^*(x)\psi(x) \\ &\quad + \frac{1}{2} \frac{\partial}{\partial x} \left(\frac{x}{\langle x^2 \rangle} \psi^*(x)\psi(x) \right) + \frac{\partial \psi^*(x)}{\partial x} \frac{\partial \psi(x)}{\partial x} \quad . \end{aligned} \tag{11.14}$$

By using (11.12b), the definition (11.1), and the fact that (as a normalizable wave function) $\psi(x) \xrightarrow[x \to \pm\infty]{} 0$, it follows that

$$\begin{aligned} 0 \leq \int_{-\infty}^{\infty} dx \, D(x) &= \frac{1}{4\langle x^2 \rangle^2} \left(\langle x^2 \rangle - 2\langle x^2 \rangle \right) + \langle k^2 \rangle \\ &= \langle k^2 \rangle - \frac{1}{4\langle x^2 \rangle} \quad . \end{aligned} \tag{11.15}$$

Hence

$$\langle k^2 \rangle \langle x^2 \rangle \geq \frac{1}{4} \quad . \tag{11.16}$$

More generally, for $\langle x \rangle$, $\langle k \rangle$ not necessarily zero,

$$\langle (\delta x)^2 \rangle \langle (\delta k)^2 \rangle \geq \frac{1}{4} \quad . \tag{11.17}$$

The **uncertainties** Δx and Δk associated with the measurements of the position and momentum, respectively, of a physical system in a certain quantum state are defined by the following **root-mean-square deviations**:

$$\Delta x \equiv \sqrt{\langle (\delta x)^2 \rangle} \,, \qquad \Delta k \equiv \sqrt{\langle (\delta k)^2 \rangle} \quad . \tag{11.18}$$

Thus, from (11.17), we have **Heisenberg's Uncertainty Principle**:

$$\boxed{(\Delta x)(\Delta p) \geq \frac{\hbar}{2}} \quad , \tag{11.19}$$

where $p = \hbar k$.

When the equality sign holds in the above equation, we have the minimum-uncertainty condition. Equation (11.15) implies that this condition holds only when $D(x) = 0$, that is [on using the definition of D in (11.13)], when

$$\frac{\partial \psi(x)}{\partial x} = -\frac{x}{2\langle x^2 \rangle} \psi(x) \quad . \tag{11.20}$$

The solution of this equation is a **Gaussian distribution**, which, when normalized, can be expressed as

$$\psi(x) = \left(\frac{2a}{\pi}\right)^{1/4} e^{-ax^2} \quad , \tag{11.21}$$

where

$$a \equiv \frac{1}{4\langle x^2 \rangle} = \left(\frac{1}{2(\Delta x)}\right)^2 \quad . \tag{11.22}$$

Figure 11.1(a) shows a schematic representation of a Gaussian distribution and Fig. 11.1(b) its Fourier transform. From the values of Δx and Δk shown in the figures we see that a Gaussian wave packet is indeed a **minimum uncertainty wave packet**.

We can verify this statement directly by computing the Fourier transform $\psi(k)$ of the Gaussian $\psi(x)$, and then calculating the product $(\Delta x)(\Delta k)$. We have

$$\begin{aligned}
\psi(k) &= \frac{1}{\sqrt{2\pi}} \int_{-\infty}^{\infty} dx\, e^{-ikx} \left(\frac{2a}{\pi}\right)^{1/4} e^{-ax^2} = \frac{1}{\sqrt{2\pi}} \left(\frac{2a}{\pi}\right)^{1/4} \int_{-\infty}^{\infty} dx\, e^{-ikx-ax^2} \\
&= \frac{1}{\sqrt{2\pi}} \left(\frac{2a}{\pi}\right)^{1/4} \int_{-\infty}^{\infty} dx \exp\left\{-a\left(x + \frac{ik}{2a}\right)^2\right\} \exp\left\{-\frac{k^2}{4a}\right\} \\
&= \frac{1}{\sqrt{2\pi}} \left(\frac{2a}{\pi}\right)^{1/4} \sqrt{\frac{\pi}{a}} \exp\left\{-\frac{k^2}{4a}\right\} \quad ,
\end{aligned} \tag{11.23}$$

that is,

$$\psi(k) \doteq \left(\frac{1}{2\pi a}\right)^{1/4} \exp\left\{-\frac{k^2}{4a}\right\} \quad , \tag{11.24}$$

which is also a Gaussian. From (11.22) we easily see that $\Delta x = 1/(2\sqrt{a})$ and $\Delta k = \sqrt{a}$. Thus $(\Delta x)(\Delta k) = 1/2$ for a Gaussian wave packet.

Problem 11.1 Using the uncertainty relation, but not the explicit solutions of the eigenvalue problem, show that the expectation value of the energy of a harmonic oscillator can never be less than the zero-point energy $\hbar\omega/2$. The Hamiltonian of the harmonic oscillator is given by (4.1).

Problem 11.2 Assuming a particle to be in one of the stationary states of an infinitely high one-dimensional box, calculate Δx and Δp, and show that Heisenberg's uncertainty relation is obeyed. Also show that in the limit of very large quantum numbers, Δx is equal to the rms deviation of the position of a particle moving classically in the enclosure.

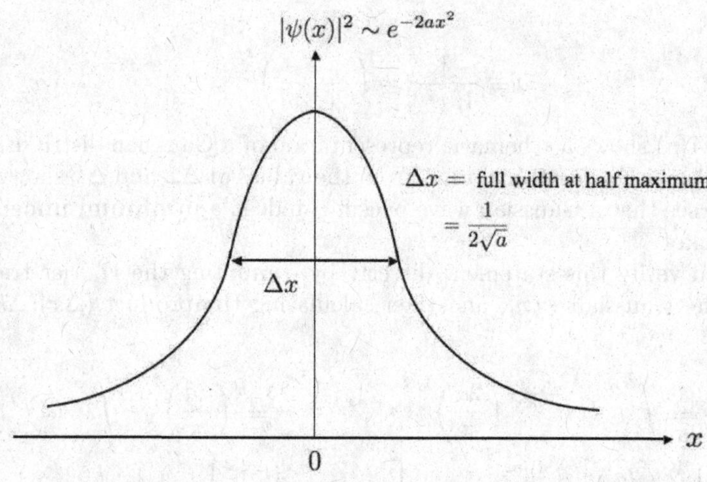

Fig. 11.1(a)

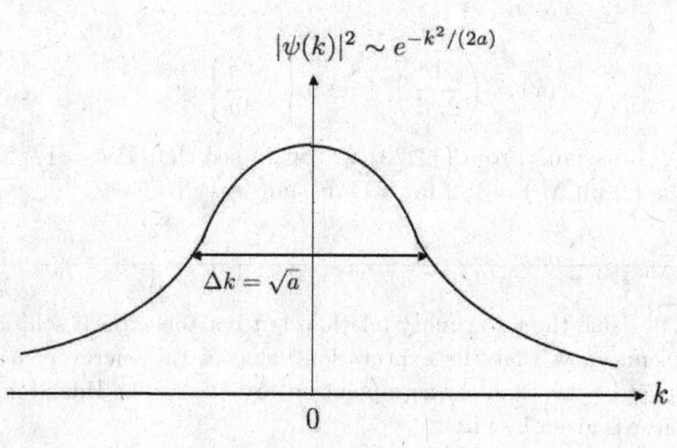

Fig. 11.1(b)

We will now derive Heisenberg's Uncertainty Principle again without explicit use of any wave functions, and show that this principle is a direct consequence of the commutation relation $[x, p] = i\hbar$, where x and p are regarded as the position operator and the momentum operator, respectively, in rigged Hilbert space (formerly written as Q and P).

Again, by definition,

$$\Delta x \equiv \sqrt{\langle (x - \langle x \rangle)^2 \rangle} = \sqrt{\langle x^2 \rangle - \langle x \rangle^2} \quad , \tag{11.25a}$$

$$\Delta p \equiv \sqrt{\langle (p - \langle p \rangle)^2 \rangle} = \sqrt{\langle p^2 \rangle - \langle p \rangle^2} \quad . \tag{11.25b}$$

Introduce the observables

$$\hat{x} \equiv x - \langle x \rangle \, , \qquad \hat{p} \equiv p - \langle p \rangle \quad . \tag{11.26}$$

Clearly

$$[\hat{x}, \hat{p}] = i\hbar \quad , \tag{11.27}$$

and

$$\Delta \hat{x} = \Delta x = \sqrt{\langle \hat{x}^2 \rangle} \, , \qquad \Delta \hat{p} = \Delta p = \sqrt{\langle \hat{p}^2 \rangle} \, . \tag{11.28}$$

Assume that the state vector $|\,\psi\,\rangle$ (with respect to which expectation values are taken) is normalized, that is, $\langle\,\psi\,|\,\psi\,\rangle = 1$. We then apply the Schwarz inequality [cf. (6.7)]

$$|\langle\,\psi_1\,|\,\psi_2\,\rangle| \le \sqrt{\langle\,\psi_1\,|\,\psi_1\,\rangle\langle\,\psi_2\,|\,\psi_2\,\rangle} \tag{11.29}$$

to the vectors $\hat{x}\,|\,\psi\,\rangle$ and $\hat{p}\,|\,\psi\,\rangle$. We thus obtain

$$\begin{aligned}(\Delta x)^2(\Delta p)^2 &= (\Delta \hat{x})^2(\Delta \hat{p})^2 = \langle\,\hat{x}^2\,\rangle\langle\,\hat{p}^2\,\rangle \\ &= \langle\,\psi\,|\,\hat{x}\hat{x}\,|\,\psi\,\rangle\langle\,\psi\,|\,\hat{p}\hat{p}\,|\,\psi\,\rangle \ge |\langle\,\psi\,|\,\hat{x}\,\hat{p}\,|\,\psi\,\rangle|^2 \quad .\end{aligned} \tag{11.30}$$

Next we separate $\hat{x}\,\hat{p}$ into the sum of a hermitian (self-adjoint) part and an anti-hermitian part:

$$\hat{x}\,\hat{p} = \frac{\hat{x}\,\hat{p} + \hat{p}\,\hat{x}}{2} + \frac{\hat{x}\,\hat{p} - \hat{p}\,\hat{x}}{2} = \frac{\hat{x}\,\hat{p} + \hat{p}\,\hat{x}}{2} + \frac{i\hbar}{2} \quad , \tag{11.31}$$

where the second term on the right-hand side of the last equality results from the commutation relation (11.27). Taking expectation values on both sides of the above equation we immediately obtain

$$\langle\,\psi\,|\,\hat{x}\,\hat{p}\,|\,\psi\,\rangle = \left\langle \frac{\hat{x}\,\hat{p} + \hat{p}\,\hat{x}}{2} \right\rangle + \frac{i\hbar}{2} \quad . \tag{11.32}$$

Now the expectation value of a hermitian operator is always real. Indeed, for any hermitian operator A and any $|\,\psi\,\rangle \in \mathcal{H}$,

$$\langle\,\psi\,|\,A\,|\,\psi\,\rangle = \langle\,\psi\,|\,A^\dagger\,|\,\psi\,\rangle^* = \langle\,\psi\,|\,A\,|\,\psi\,\rangle^* \quad .$$

Thus (11.30) implies

$$(\Delta x)^2(\Delta p)^2 \ge \left\langle \frac{\hat{x}\,\hat{p} + \hat{p}\,\hat{x}}{2} \right\rangle^2 + \frac{\hbar^2}{4} \ge \frac{\hbar^2}{4} \quad , \tag{11.33}$$

and we again arrive at Heisenberg's Uncertainty Principle (11.19).

For the minimum uncertainty condition $(\Delta x)(\Delta p) = \hbar/2$ to hold, we must have: (1) The equality sign in the Schwarz inequality (11.30) holds, which implies $\hat{x}\,|\,\psi\,\rangle = c\,\hat{p}\,|\,\psi\,\rangle$, $c \in \mathbb{C}$, and (2) $\langle\,\hat{x}\,\hat{p} + \hat{p}\,\hat{x}\,\rangle = 0$. These two conditions imply

$$\begin{aligned}\langle\,\psi\,|\,\hat{x}\,\hat{p} + \hat{p}\,\hat{x}\,|\,\psi\,\rangle &= \langle\,\psi\,|\,\hat{x}\,\hat{p}\,|\,\psi\,\rangle + \langle\,\psi\,|\,\hat{p}\,\hat{x}\,|\,\psi\,\rangle \\ &= (c^* + c)\,\langle\,\psi\,|\,\hat{p}^2\,|\,\psi\,\rangle = 2\,Re\,(c)\,\langle\,\psi\,|\,\hat{p}^2\,|\,\psi\,\rangle = 0 \quad ,\end{aligned} \tag{11.34}$$

or $Re\,(c) = 0$.

To sum up, the necessary and sufficient condition for $|\,\psi\,\rangle$ to be a minimum uncertainty state is that

$$(x - \langle\,x\,\rangle)\,|\,\psi\,\rangle = i\gamma\,(p - \langle\,p\,\rangle)\,|\,\psi\,\rangle\,, \quad \gamma \in \mathbb{R} \quad . \tag{11.35}$$

As before, we can always make a coordinate transformation so that $\langle\,x\,\rangle = \langle\,p\,\rangle = 0$. Then the above equation reduces to

$$x\,|\,\psi\,\rangle = i\gamma\,p\,|\,\psi\,\rangle \quad , \tag{11.36}$$

which becomes, when expressed in the coordinate representation,

$$\langle\,x\,|\,x\,|\,\psi\,\rangle = i\gamma\,\langle\,x\,|\,p\,|\,\psi\,\rangle \quad . \tag{11.37}$$

The left-hand side of the above equation can be expressed in terms of the wave function $\psi(x)$:

$$\begin{aligned}\langle\,x\,|\,x\,|\,\psi\,\rangle &= \int_{-\infty}^{\infty} dx'\,\langle\,x\,|\,x\,|\,x'\,\rangle\langle\,x'\,|\,\psi\,\rangle \\ &= \int_{-\infty}^{\infty} dx'\,x'\,\delta(x - x')\,\psi(x') = x\psi(x) \quad ,\end{aligned} \tag{11.38}$$

while the matrix element $\langle\,x\,|\,p\,|\,\psi\,\rangle$ on the right-hand side can also be so expressed:

$$\begin{aligned}\langle\,x\,|\,p\,|\,\psi\,\rangle &= \int_{-\infty}^{\infty} dk\,\langle\,x\,|\,k\,\rangle\langle\,k\,|\,p\,|\,\psi\,\rangle = \frac{\hbar}{\sqrt{2\pi}}\int_{-\infty}^{\infty} dk\,k\,e^{ikx}\,\psi(k) \\ &= -\frac{i\hbar}{\sqrt{2\pi}}\int_{-\infty}^{\infty} dk\,\frac{\partial}{\partial x}\left(e^{ikx}\right)\,\psi(k) = -i\hbar\,\frac{\partial}{\partial x}\left\{\frac{1}{\sqrt{2\pi}}\int_{-\infty}^{\infty} dk\,e^{ikx}\,\psi(k)\right\} \\ &= -i\hbar\,\frac{\partial\psi(x)}{\partial x} \quad .\end{aligned} \tag{11.39}$$

Thus (11.37) is equivalent to

$$x\psi(x) = \gamma\hbar\,\frac{\partial\psi(x)}{\partial x} \quad , \tag{11.40}$$

which is of the same form as (11.20).

Problem 11.3 The Heisenberg equation of motion implies the following equation for the expectation value of an arbitrary observable A:

$$i\hbar\,\frac{d\langle A\rangle}{dt} = \langle\,[\,A\,,\,H\,]\,\rangle + i\hbar\left\langle\frac{\partial A}{\partial t}\right\rangle .$$

Use this equation and the commutation relation $[\,x\,,\,p\,] = i\hbar$ to derive the following time development of $(\Delta x)^2$ for a free particle moving in one dimension:

$$(\Delta x)_t^2 = (\Delta x)_0^2 + \frac{2}{\mu}\left[\frac{\langle xp + p\,x\rangle_0}{2} - \langle x\rangle_0\langle p\,\rangle\right] t + \frac{(\Delta p)^2 t^2}{\mu^2} ,$$

where the subscripts 0 and t denote the times 0 and t, respectively, and μ is the mass of the particle. This equation describes the **spreading of a wave packet**.

Problem 11.4 The **virial theorem** states that, for a bound state of a system,

$$2\langle\,T\,\rangle = \langle\,\boldsymbol{r}\cdot\nabla V\,\rangle ,$$

where T and V are the kinetic and potential energy operators, respectively. Show that for a central force associated with the potential function $V(r) = \alpha r^n$ (α = constant),

$$\langle\,V\,\rangle = \frac{2}{n+2}\,E ,$$

where $E \equiv \langle\,H\,\rangle = \langle\,T\,\rangle + \langle\,V\,\rangle$.

In general, an uncertainty relationship is obtained for every pair of observables whose corresponding (self-adjoint) operators do not commute with each other. Suppose $A = A^\dagger$ and $B = B^\dagger$ are two arbitrary self-adjoint operators such that $[\,A\,,\,B\,] \neq 0$. Let

$$\hat{A} \equiv A - \langle\,A\,\rangle\,, \qquad \hat{B} \equiv B - \langle\,B\,\rangle \tag{11.41}$$

for a particular normalized $|\,\psi\,\rangle$. We have, as before,

$$\Delta\hat{A} = \Delta A = \sqrt{\langle\,\hat{A}^2\,\rangle}\,, \qquad \Delta\hat{B} = \Delta B = \sqrt{\langle\,\hat{B}^2\,\rangle}\quad . \tag{11.42}$$

Applying the Schwarz inequality to $\hat{A}\,|\,\psi\,\rangle$ and $\hat{B}\,|\,\psi\,\rangle$, we obtain

$$\begin{aligned}(\Delta A)^2(\Delta B)^2 = (\Delta\hat{A})^2(\Delta\hat{B})^2 &= \langle\,\hat{A}^2\,\rangle\langle\,\hat{B}^2\,\rangle = \langle\,\psi\,|\,\hat{A}\hat{A}\,|\,\psi\,\rangle\langle\,\psi\,|\,\hat{B}\hat{B}\,|\,\psi\,\rangle \\ &\geq |\,\langle\,\psi\,|\,\hat{A}\hat{B}\,|\,\psi\,\rangle\,|^2 = |\,\langle\,\hat{A}\hat{B}\,\rangle\,|^2 \quad .\end{aligned} \tag{11.43}$$

Again, analogous to (11.31), we break $\hat{A}\hat{B}$ up into a sum of hermitian and anti-hermitian parts:

$$\hat{A}\hat{B} = \frac{\hat{A}\hat{B} + \hat{B}\hat{A}}{2} + \frac{[\hat{A}, \hat{B}]}{2} \quad , \tag{11.44}$$

from which it follows that

$$\langle \hat{A}\hat{B} \rangle = \frac{1}{2} \langle \hat{A}\hat{B} + \hat{B}\hat{A} \rangle + \frac{1}{2} \langle [\hat{A}, \hat{B}] \rangle = \frac{1}{2} \langle \hat{A}\hat{B} + \hat{B}\hat{A} \rangle + \frac{1}{2} \langle [A, B] \rangle \quad . \tag{11.45}$$

Since $\hat{A}\hat{B} + \hat{B}\hat{A}$ is hermitian, we can suppose

$$\langle \hat{A}\hat{B} + \hat{B}\hat{A} \rangle = \alpha \in \mathbb{R} \quad . \tag{11.46}$$

On the other hand, since $[A, B]$ is anti-hermitian, its expectation value is purely imaginary:

$$\langle [A, B] \rangle = i\beta \, , \quad \beta \in \mathbb{R} \quad . \tag{11.47}$$

Indeed, if $A = -A^\dagger$, then

$$\langle A \rangle = \langle \psi | A | \psi \rangle = \langle \psi | A^\dagger | \psi \rangle^* = -\langle \psi | A | \psi \rangle^* = -\langle A \rangle^* \quad .$$

Thus

$$\langle \hat{A}\hat{B} \rangle = \frac{\alpha}{2} + i\,\frac{\beta}{2} \, , \qquad \alpha, \beta \in \mathbb{R} \quad , \tag{11.48}$$

and

$$| \langle \hat{A}\hat{B} \rangle |^2 = \frac{1}{4} (\alpha^2 + \beta^2) \geq \frac{\beta^2}{4} \quad . \tag{11.49}$$

Finally, the uncertainty relationship for the pair of non-commuting observables A and B can be written

$$\boxed{(\Delta A)^2 (\Delta B)^2 \geq \left(\frac{\langle [A, B] \rangle}{2i} \right)^2} \quad . \tag{11.50}$$

We note that (11.47) guarantees that the right-hand side of the above inequality is real. For $A = x$ and $B = p$, the above inequality yields Heisenberg's Uncertainty Principle (11.19) on using the commutation relation (1.37).

Problem 11.5 Given the commutation relation (see Chapters 17 and 20 for more details):

$$[L_x \, , \, L_y] = i\hbar L_z \, ,$$

where L_x, L_y and L_z are the three Cartesian components of the angular momentum operator, compute the uncertainty relationship for $(\Delta L_x)(\Delta L_y)$ in terms of expectation values.

Chapter 12

Commutator Algebra

In this chapter we will develop some basic calculational techniques on the commutators of operators on Hilbert space. These techniques will prove to be very useful in a variety of quantum mechanical applications.

Starting with the commutation relations [cf. (1.37)]

$$[q^i \,,\, q^j] = [p_i \,,\, p_j] = 0 \quad , \tag{12.1}$$

we obtain

$$[q^i \,,\, F(q^1 \,,\, \dots \,,\, q^N)] = 0 \quad , \tag{12.2}$$
$$[p_i \,,\, G(p_1 \,,\, \dots \,,\, p_N)] = 0 \quad , \tag{12.3}$$

where F and G are analytic functions of the canonical coordinates q^i and the canonical momenta p_j, respectively. We also have the following important and useful commutation relations:

$$[q^i \,,\, G(p_1 \,,\, \dots \,,\, p_N)] = i\hbar \frac{\partial G}{\partial p_i} \quad , \tag{12.4}$$

$$[p_i \,,\, F(q^1 \,,\, \dots \,,\, q^N)] = -i\hbar \frac{\partial F}{\partial q^i} \quad . \tag{12.5}$$

We will prove (12.4) for the case $N = 1$. First we note that, for arbitrary operators A, B and C,

$$[A \,,\, BC] = [A \,,\, B]\, C + B\, [A \,,\, C] \quad . \tag{12.6}$$

Indeed,

$$[A \,,\, BC] = ABC - BAC + BAC - BCA = [A \,,\, B]\, C + B\, [A \,,\, C] \quad .$$

Next we let, since $G(p)$ is assumed to be analytical,

$$G(p) = \sum_{n=0}^{\infty} \alpha_n \, p^n \quad . \tag{12.7}$$

Then

$$[q\,,\,G(p)\,] = \sum_{n=0}^{\infty} \alpha_n\,[\,q\,,\,p^n\,] \quad . \tag{12.8}$$

Repeatedly using (12.6) and the basic commutation relation $[\,q\,,\,p\,] = i\hbar$, we have

$$\begin{aligned}
[\,q\,,\,p^n\,] &= [\,q\,,\,p^{\,n-1}\,p\,] \\
&= [\,q\,,\,p^{\,n-1}\,]\,p + p^{\,n-1}\,[\,q\,,\,p\,] \\
&= [\,q\,,\,p^{\,n-2}\,p\,]\,p + i\hbar\,p^{\,n-1} \\
&= \{[\,q\,,\,p^{\,n-2}\,]\,p + p^{\,n-2}\,[\,q\,,\,p\,]\}\;p + i\hbar\,p^{\,n-1} \\
&= [\,q\,,\,p^{\,n-2}\,]\,p^{\,2} + 2i\hbar\,p^{\,n-1} \\
&\;\;\vdots \\
&= [\,q\,,\,p^{\,n-(n-1)}\,]\,p^{\,n-1} + (n-1)\,i\hbar\,p^{\,n-1} \\
&= i\hbar\,np^{\,n-1} \quad .
\end{aligned}$$

Hence

$$[\,q\,,\,G(p)\,] = i\hbar \sum_n \alpha_n\,np^{\,n-1} = i\hbar\,\frac{\partial G}{\partial p} \quad .$$

The proof of the multi-dimensional case is a straightforward generalization of the above procedure. Equation (12.5) can also be proved in a similar fashion, and in fact can be read off from (12.4) by the following interchanges:

$$q^i \longleftrightarrow p_i \;, \qquad i \longleftrightarrow -i \quad .$$

We collect together here a list of some general and useful commutator rules for arbitrary operators. Equation (12.11) is the same as (12.6) and has already been proved. The others are very readily demonstrated.

$$[\,A\,,\,B\,] = -[\,B\,,\,A\,] \quad , \tag{12.9}$$

$$[\,A\,,\,B + C\,] = [\,A\,,\,B\,] + [\,A\,,\,C\,] \quad , \tag{12.10}$$

$$[\,A\,,\,BC\,] = [\,A\,,\,B\,]\,C + B\,[\,A\,,\,C\,] \quad , \tag{12.11}$$

$$[\,AB\,,\,C\,] = A\,[\,B\,,\,C\,] + [\,A\,,\,C\,]\,B \quad , \tag{12.12}$$

$$[\,A\,,\,[\,B\,,\,C\,]] + [\,B\,,\,[\,C\,,\,A\,]] + [\,C\,,\,[\,A\,,\,B\,]] = 0 \quad . \tag{12.13}$$

The last equation [Eq. (12.13)] is an important identity called the **Jacobi identity**. It qualifies the commutator bracket as a special kind of non-commutative product in the algebra of linear operators in Hilbert space. Such algebras are called **Lie algebras**, which play a very important role in the applications of group theory to physics (see Chapter 30).

As a generalization of the procedure used above for the calculation of $[\,q\,,\,p^{\,n}\,]$, we will derive the following useful expression for $[\,A\,,\,B^n\,]$ by repeated use of

(12.11), where A and B are arbitrary operators:

$$\boxed{[A, B^n] = \sum_{s=0}^{n-1} B^s [A, B] B^{n-s-1}} \quad . \tag{12.14}$$

Note that $[q, p]$ is a special case where $[A, B]$ is equal to a number times the identity operator.

Proof. [of (12.14)] We will use mathematical induction. The statement is obviously true for $n = 1$. Assume that it is true for n. Then

$$\begin{aligned}[A, B^{n+1}] &= [A, B^n B] = [A, B^n] B + B^n [A, B] \\ &= \sum_{s=0}^{n-1} B^s [A, B] B^{n-s} + B^n [A, B] \\ &= \sum_{s=0}^{n} B^s [A, B] B^{n-s} \quad .\end{aligned}$$

Hence it is true for $n + 1$ also. It follows by induction that (12.14) is true for all positive integers n. □

The commutator $[q, p^n]$, for example, can be calculated directly from (12.14) as follows:

$$[q, p^n] = i\hbar \sum_{s=0}^{n-1} p^s p^{n-s-1} = i\hbar \sum_{s=0}^{n-1} p^{n-1} = i\hbar n p^{n-1} \quad .$$

$[p, q^n]$ can be similarly calculated. We reproduce these last two important results for ease of reference.

$$\boxed{[q, p^n] = i\hbar n p^{n-1} \quad , \quad [p, q^n] = -i\hbar n q^{n-1}} \quad . \tag{12.15}$$

Problem 12.1 Prove the **Baker-Cambell-Hausdorff Expansion** (given to 4-th order):

$$C = A + B + \frac{1}{2}[A, B] + \frac{1}{12}\{A, B^2\} + \frac{1}{12}\{B, A^2\} + \frac{1}{24}\{\{B, A^2\}, B\} + \dots ,$$

where the operators A, B and C are related by

$$e^A e^B = e^C ,$$

and the **string commutator** $\{A, B^n\}$ is defined by

$$\{A, B^n\} \equiv \underbrace{[[\dots[}_{n \text{ brackets}} A, B \underbrace{], B], \dots, B]}_{n \text{ brackets}} .$$

This is a hard problem. You may want to consult Magnus (1954) and Weiss and Maradudin (1962).

Chapter 13

Ehrenfest's Theorem

Ehrenfest's Theorem asserts a formal similarity between Hamilton's equations of motion and certain results derived from Heisenberg's equation of motion. It is thus, in addition to the developments presented in Chapters 1 and 2, another manifestation of the deep ties between classical and quantum mechanics.

Recall Heisenberg's equation of motion for an arbitrary operator u [cf. (2.13)]:

$$\frac{du}{dt} = \frac{1}{i\hbar}[\, u \, , \, H \,] + \frac{\partial u}{\partial t} \quad , \tag{13.1}$$

where H is the Hamiltonian. For operators u which do not depend on time explicitly, it follows from the above equation that

$$i\hbar \frac{du}{dt} = [\, u \, , \, H \,] \quad . \tag{13.2}$$

As special cases, we have

$$i\hbar \frac{dq^i}{dt} = [\, q^i \, , \, H \,] \quad , \tag{13.3}$$

$$i\hbar \frac{dp_i}{dt} = [\, p_i \, , \, H \,] \quad . \tag{13.4}$$

Taking the expectation value on both sides of (13.1) with respect to a fixed $|\,\psi\,\rangle$, we have

$$\frac{d\langle\, u \,\rangle}{dt} = \frac{1}{i\hbar}\langle\, [\, u \, , \, H \,] \,\rangle + \left\langle \frac{\partial u}{\partial t} \right\rangle \quad . \tag{13.5}$$

The above equation is also valid as it stands if the expectation value $\langle\, u \,\rangle$ is understood to mean $\langle\, \psi(t) \,|\, u \,|\, \psi(t) \,\rangle$, where $|\, \psi(t) \,\rangle$ satisfies the time-dependent Schrödinger equation [cf. (1.32)]

$$H \,|\, \psi(t) \,\rangle = i\hbar \frac{\partial \,|\, \psi(t) \,\rangle}{\partial t} \quad . \tag{13.6}$$

This can be verified directly as follows:

$$\begin{aligned} i\hbar\frac{d\langle u\rangle}{dt} &= i\hbar\frac{d}{dt}\langle\,\psi(t)\,|\,u\,|\,\psi(t)\,\rangle = i\hbar\frac{d}{dt}\int_{-\infty}^{\infty}dx\,\psi^*(x,t)\,u\,\psi(x,t) \\ &= i\hbar\int_{-\infty}^{\infty}dx\,\frac{\partial\psi^*}{\partial t}\,u\psi + i\hbar\int_{-\infty}^{\infty}dx\,\psi^* u\,\frac{\partial\psi}{\partial t} + i\hbar\int_{-\infty}^{\infty}dx\,\psi^*\,\frac{\partial u}{\partial t}\psi \\ &= -\int_{-\infty}^{\infty}dx\,H\psi^*\,u\psi + \int_{-\infty}^{\infty}dx\,\psi^* uH\psi + i\hbar\int_{-\infty}^{\infty}dx\,\psi^*\,\frac{\partial u}{\partial t}\psi \\ &= \int_{-\infty}^{\infty}dx\,\psi^*(uH - Hu)\psi + i\hbar\int_{-\infty}^{\infty}dx\,\psi^*\,\frac{\partial u}{\partial t}\psi \\ &= \langle\,[\,u\,,\,H\,]\,\rangle + i\hbar\left\langle\,\frac{\partial u}{\partial t}\,\right\rangle \quad , \end{aligned} \tag{13.7}$$

where in the fifth equality we have used (8.14) and the fact that H is hermitian.

From (13.3) and (13.4) we have

$$i\hbar\frac{d\langle\,q^i\,\rangle}{dt} = \langle\,[\,q^i\,,\,H\,]\,\rangle \quad , \tag{13.8}$$

$$i\hbar\frac{d\langle\,p_i\,\rangle}{dt} = \langle\,[\,p_i\,,\,H\,]\,\rangle \quad ; \tag{13.9}$$

while from (12.4) and (12.5),

$$[\,q^i\,,\,H\,] = i\hbar\frac{\partial H}{\partial p_i} \quad , \tag{13.10}$$

$$[\,p_i\,,\,H\,] = -i\hbar\frac{\partial H}{\partial q^i} \quad . \tag{13.11}$$

Thus

$$\boxed{\frac{d\langle\,q^i\,\rangle}{dt} = \left\langle\,\frac{\partial H}{\partial p_i}\,\right\rangle \quad , \quad \frac{d\langle\,p_i\,\rangle}{dt} = -\left\langle\,\frac{\partial H}{\partial q^i}\,\right\rangle} \quad . \tag{13.12}$$

The above equations, known as **Ehrenfest's equations**, constitute **Ehrenfest's Theorem**. Their forms resemble Hamilton's canonical equations of motion in classical mechanics.

As an example, consider a Hamiltonian given by

$$H = \frac{\boldsymbol{p}^2}{2m} + V(\boldsymbol{x}) \quad . \tag{13.13}$$

Then,

$$\frac{\partial H}{\partial p_i} = \frac{p_i}{m} \quad , \quad -\frac{\partial H}{\partial q^i} = -\frac{\partial V}{\partial q^i} \equiv F_i \quad . \tag{13.14}$$

Thus Ehrenfest's equations applied to this case become

$$m\frac{d\langle q^i \rangle}{dt} = \langle p_i \rangle \quad , \tag{13.15}$$

$$\frac{d\langle p_i \rangle}{dt} = -\left\langle \frac{\partial V}{\partial q^i} \right\rangle = \langle F_i \rangle \quad . \tag{13.16}$$

Equation (13.16) is the quantum mechanical analog of Newton's second law; but it is not exactly Newton's second law, which would have required $\langle F_i(\boldsymbol{x}) \rangle = F_i(\langle \boldsymbol{x} \rangle)$. For forces depending linearly on the coordinates $\boldsymbol{x}$, such as in the case of the simple harmonic oscillator (which we will study in the next chapter), this condition is true. For other cases, the wave function with respect to which the expectation values are calculated has to be sufficiently localized in space so that the force remains practically constant over the localized region.

Chapter 14

The Simple Harmonic Oscillator

The problem of the simple harmonic oscillator is one of the most fundamental in quantum mechanics. Its solution provides a starting point to a very important procedure called second quantization, which is at the basis of various quantum field theories finding applications in quantum statistical mechanics, condensed matter physics, and high energy physics. The common feature in all these applications is that the number of particles in the system under study is not conserved. We shall, however, understand here that we are treating a one-particle problem.

We have introduced this problem already in Chapter 4, where it was used as an example to provide matrix representations of the position and momentum operators x and p that satisfy the fundamental quantum conditions (commutation relation) (2.14). In this chapter we will start with these conditions and present a very elegant algebraic method [due to Dirac (see Dirac 1967)] to construct both the spectrum of energy eigenvalues [already given in (4.13)] and the corresponding eigenstates. The latter will also be given in their coordinate representations, that is, wave functions, as solutions to the Schrödinger equation. In this procedure we will also study more fully the properties of the very important creation and annihilation operators [cf. (4.9) and (4.8)].

Let us first recall the set-up of the problem already presented in Chapter 4 [given again below as Eqs. (14.1) to (14.6)]. The problem is specified completely by giving the Hamiltonian [cf. (4.1)]:

$$H = \frac{p^2}{2m} + \frac{1}{2}m\omega^2 q^2 \quad . \tag{14.1}$$

Define the dimensionless coordinate and momentum as follows [cf. (4.2)]:

$$Q \equiv \sqrt{\frac{m\omega}{\hbar}}\, q \ , \quad P \equiv \frac{1}{\sqrt{m\hbar\omega}}\, p \quad . \tag{14.2}$$

Then [cf. (4.3)]

$$H = \frac{\hbar\omega}{2}(P^2 + Q^2) \quad . \tag{14.3}$$

Further define [cf. (4.4) and (4.5)] the **annihilation operator** a and its adjoint, the **creation operator** $a^\dagger$, by (this terminology will be justified later):

$$a \equiv \frac{1}{\sqrt{2}}(Q + iP)\,, \qquad a^\dagger = \frac{1}{\sqrt{2}}(Q - iP) \quad . \tag{14.4}$$

Then [cf. (4.6)]

$$Q = \frac{1}{\sqrt{2}}(a + a^\dagger) \quad , \quad P = -\frac{i}{\sqrt{2}}(a - a^\dagger) \quad , \tag{14.5}$$

from which it follows that [cf. (4.7)]

$$q = \sqrt{\frac{\hbar}{m\omega}}\,\frac{1}{\sqrt{2}}(a + a^\dagger) \quad , \quad p = \sqrt{m\hbar\omega}\left(-\frac{i}{\sqrt{2}}\right)(a - a^\dagger) \quad . \tag{14.6}$$

Now we have

$$[Q, P] = \frac{1}{\hbar}[q, p] = \frac{i\hbar}{\hbar} = i \quad . \tag{14.7}$$

Thus

$$\begin{aligned}[a\,, a^\dagger\,] &= \frac{1}{2}[\,Q + iP\,, Q - iP\,] = \frac{1}{2}(i\,[\,P\,, Q\,] - i\,[\,Q\,, P\,]) \\ &= \frac{1}{2}(i(-i) - i(i)) \quad ,\end{aligned}$$

or

$$\boxed{[\,a\,, a^\dagger\,] = 1} \quad . \tag{14.8}$$

The above equation represents the basic quantum mechanical result satisfied by **bosonic** creation and annihilation operators.

Using (14.5) in (14.3) for the Hamiltonian H, we have, remembering that a and $a^\dagger$ do not commute,

$$\begin{aligned}H &= \frac{\hbar\omega}{2}(P^2 + Q^2) = \left(\frac{\hbar\omega}{2}\right)\left(\frac{1}{2}\right)(-(a - a^\dagger)(a - a^\dagger) + (a + a^\dagger)(a + a^\dagger)) \\ &= \left(\frac{\hbar\omega}{2}\right)\left(\frac{1}{2}\right)(-a^2 + a^\dagger a + aa^\dagger - a^\dagger a^\dagger + a^2 + a^\dagger a + aa^\dagger + a^\dagger a^\dagger) \quad ,\end{aligned}$$

or

$$H = \frac{\hbar\omega}{2}(aa^\dagger + a^\dagger a) \quad . \tag{14.9}$$

Equation (14.8) then implies

$$\boxed{H = \hbar\omega\left(a^\dagger a + \frac{1}{2}\right) \equiv \hbar\omega\left(N + \frac{1}{2}\right)} \quad , \tag{14.10}$$

where the self-adjoint (hermitian) operator

$$N \equiv a^\dagger a \tag{14.11}$$

is called the **number operator** (this name will be justified below). Since

$$Na = a^\dagger a a = (aa^\dagger - 1)a = aa^\dagger a - a\ , \tag{14.12}$$
$$Na^\dagger = a^\dagger a a^\dagger = a^\dagger(1 + a^\dagger a)\ , \tag{14.13}$$

the number operator satisfies the following relations:

$$Na = a(N-1)\ , \quad Na^\dagger = a^\dagger(N+1) \quad . \tag{14.14}$$

These will be crucial in determining the eigenvalues and eigenvectors of N, which, according to (14.10), will yield those of the Hamiltonian H. The procedure is based on the following theorem.

Theorem 14.1. *If ν is an eigenvalue of N, and $|\,\nu\,\rangle$ the corresponding non-zero eigenvector, then:*

i) $\nu \geq 0$,

ii) $\nu = 0$ *if and only if* $a\,|\,\nu\,\rangle = 0$; *if* $\nu \neq 0$, *then* $a\,|\,\nu\,\rangle$ *is a non-zero vector whose norm is* $\nu\,\langle\,\nu\,|\,\nu\,\rangle$, *and it is an eigenvector of* N *with eigenvalue* $\nu - 1$;

iii) $a^\dagger\,|\,\nu\,\rangle$ *is not the zero-vector; its norm is* $(\nu+1)\,\langle\,\nu\,|\,\nu\,\rangle$, *and it is an eigenvector of* N *with eigenvalue* $\nu + 1$.

Proof. By assumption

$$N\,|\,\nu\,\rangle = \nu\,|\,\nu\,\rangle\ , \qquad \langle\,\nu\,|\,\nu\,\rangle > 0\ . \tag{14.15}$$

The inequality follows from the fact that $\langle\,\psi\,|\,\psi\,\rangle = 0$ if and only if $|\,\psi\,\rangle$ is the zero vector. The norms of $a\,|\,\nu\,\rangle$ and $a^\dagger\,|\,\nu\,\rangle$ are then given, respectively, by

$$\langle\,\nu\,|\,a^\dagger a\,|\,\nu\,\rangle = \langle\,\nu\,|\,N\,|\,\nu\,\rangle = \nu\,\langle\,\nu\,|\,\nu\,\rangle\ , \tag{14.16}$$
$$\langle\,\nu\,|\,aa^\dagger\,|\,\nu\,\rangle = \langle\,\nu\,|\,1 + a^\dagger a\,|\,\nu\,\rangle = \langle\,\nu\,|\,N+1\,|\,\nu\,\rangle = (\nu+1)\,\langle\,\nu\,|\,\nu\,\rangle \quad . \tag{14.17}$$

The first of the above two equations implies that $\nu \geq 0$, and that if $\nu = 0$, $a\,|\,\nu\,\rangle$ must be the zero vector. Conversely, if $a\,|\,\nu\,\rangle = 0$ (the zero vector), then $\nu = 0$. Now, from (14.14), we have

$$Na\,|\,\nu\,\rangle = a(N-1)\,|\,\nu\,\rangle = (\nu - 1)\,a\,|\,\nu\,\rangle\ , \tag{14.18}$$
$$Na^\dagger\,|\,\nu\,\rangle = a^\dagger(N+1)\,|\,\nu\,\rangle = (\nu+1)\,a^\dagger\,|\,\nu\,\rangle\ . \tag{14.19}$$

Therefore, $a\,|\,\nu\,\rangle$ is an eigenvector of N with eigenvalue $\nu - 1$, and $a^\dagger\,|\,\nu\,\rangle$ is an eigenvector of N with eigenvalue $\nu + 1$. The vector $a^\dagger\,|\,\nu\,\rangle$ is definitely not the zero vector since, according to (14.17), its norm is $(\nu+1)\,\langle\,\nu\,|\,\nu\,\rangle > 0$. □

We are now ready to determine the spectrum (eigenvalues) of N. Suppose $\nu > 0$ is an eigenvalue of N. Then ii) of the above theorem implies that $a\,|\,\nu\,\rangle$ is not the zero vector. Then the theorem can be applied to it, that is, to $a\,|\,\nu\,\rangle$, which, also according to ii) of the theorem, is an eigenvector of N with eigenvalue $\nu - 1$. Then i) of the theorem implies that $\nu - 1 \geq 0$, that is, $\nu \geq 1$. So we conclude that if ν is an eigenvalue of N and if $\nu > 0$, then $\nu \geq 1$. Suppose further that $\nu > 1$. Then $a^2\,|\,\nu\,\rangle$ is not the zero vector and the theorem applies to it also. It is an eigenvector of N with eigenvalue $\nu - 2$. So from i) we conclude that $\nu \geq 2$. Repeating the above argument one can obtain successively the sequence of eigenvectors of N

$$a\,|\,\nu\,\rangle\,,\ a^2\,|\,\nu\,\rangle\,,\ a^3\,|\,\nu\,\rangle, \ldots ,\ a^p\,|\,\nu\,\rangle, \ldots$$

belonging to the eigenvalues

$$\nu - 1\,,\ \nu - 2\,,\ \nu - 3\,, \ldots ,\ \nu - p\,, \ldots$$

But by i) of the theorem, the spectrum of N has a lower bound of zero. Therefore the above sequence of eigenvectors will vanish from a certain value of p, say $p = n + 1$, onwards. Thus, by ii) of the theorem, the action of the annihilation operator a on the non-zero eigenvector $a^n\,|\,\nu\,\rangle$ belonging to the eigenvalue $\nu - n$ yields the zero vector. ii) then also requires that $\nu = n$.

In the same manner, one can apply the theorem to the infinite sequence of non-zero vectors

$$a^\dagger\,|\,\nu\,\rangle\,,\ (a^\dagger)^2\,|\,\nu\,\rangle\,, \ldots$$

which are eigenvectors of N belonging to the eigenvalues $\nu + 1, \nu + 2\,, \ldots$

In conclusion, the spectrum of N consists of the set of non-negative integers

$$n = 0,\ 1,\ 2,\ 3,\ \ldots$$

It is thus discrete and also non-degenerate (there is only one eigenvector belonging to a particular eigenvalue). The corresponding eigenvectors must be orthogonal to each other according to Theorem 8.3. Each of these can be normalized by multiplication by a suitable constant. From now on we will designate the set of orthonormal eigenvectors of N by

$$\{\,|\,n\,\rangle\,\} = \{\,|\,0\,\rangle\,,\ |\,1\,\rangle\,,\ |\,2\,\rangle, \ldots\} \tag{14.20}$$

Orthonormality of this set is expressed by the condition

$$\langle\,n\,|\,m\,\rangle = \delta_{nm}\,. \tag{14.21}$$

The set is also complete since, according to the Spectral Theorem (Theorem 9.2) applied to the case of a discrete spectrum,

$$N = \sum_0^\infty n\,|\,n\rangle\langle n\,| = \sum_0^\infty N\,|\,n\rangle\langle n\,|\,, \tag{14.22}$$

that is, $\sum_n |\, n \,\rangle\langle\, n \,| = 1$.

We summarize below the most important properties obtained thus far for the set $\{ |\, n \,\rangle \}$:

$$N \,|\, n \,\rangle = n \,|\, n \,\rangle \,, \tag{14.23}$$

$$a \,|\, 0 \,\rangle = 0 \,, \qquad \text{(Note: } |\, 0 \,\rangle \neq 0) \,, \tag{14.24}$$

$$a \,|\, n \,\rangle = \sqrt{n} \,|\, n-1 \,\rangle \,, \qquad (n > 0) \,, \tag{14.25}$$

$$a^\dagger \,|\, n \,\rangle = \sqrt{n+1} \,|\, n+1 \,\rangle \,, \tag{14.26}$$

$$|\, n \,\rangle = \frac{1}{\sqrt{n!}} (a^\dagger)^n \,|\, 0 \,\rangle \,, \tag{14.27}$$

$$\sum_0^\infty |\, n \,\rangle\langle\, n \,| = 1 \,, \qquad \text{(completeness)} \,. \tag{14.28}$$

Equation (14.27) follows directly from (14.26); while (14.25) and (14.26) follow from (14.16) and (14.17), respectively.

The operators a, $a^\dagger$ and $N = a^\dagger a$ can be given matrix representations with respect to the orthonormal set $\{ |\, n \,\rangle \}$, with matrix elements given by [cf. (8.1)]

$$a_{m1} = \langle\, m-1 \,|\, a \,|\, 0 \,\rangle = 0 \,, \tag{14.29}$$

$$a_{mn} = \langle\, m-1 \,|\, a \,|\, n-1 \,\rangle = \sqrt{n-1}\; \delta_{m-1,\,n-2} \,, \quad n = 2, 3, 4, \ldots \,, \tag{14.30}$$

$$(a^\dagger)_{mn} = \langle\, m-1 \,|\, a^\dagger \,|\, n-1 \,\rangle = \sqrt{n}\; \delta_{m-1,\,n} \,, \tag{14.31}$$

$$N_{mn} = \langle\, m-1 \,|\, N \,|\, n-1 \,\rangle = (n-1)\; \delta_{mn} \,. \tag{14.32}$$

We note that the n-th row (column) of the matrices corresponds to the eigenstate $|\, n-1 \,\rangle$. Equations (14.29) through (14.31) give rise to precisely the matrices given by (4.8) and (4.9) in Chapter 4. We reproduce them here for convenience.

$$a = \begin{pmatrix} 0 & \sqrt{1} & 0 & 0 & \ldots & \ldots \\ 0 & 0 & \sqrt{2} & 0 & \ldots & \ldots \\ 0 & 0 & 0 & \sqrt{3} & 0 & \ldots \\ & & \vdots & & \vdots & \\ & & \vdots & & \vdots & \end{pmatrix} , \quad a^\dagger = \begin{pmatrix} 0 & 0 & 0 & 0 & \ldots & \ldots \\ \sqrt{1} & 0 & 0 & 0 & \ldots & \ldots \\ 0 & \sqrt{2} & 0 & 0 & \ldots & \ldots \\ 0 & 0 & \sqrt{3} & 0 & \ldots & \ldots \\ & & \vdots & & \vdots & \\ & & \vdots & & \vdots & \end{pmatrix} . \tag{14.33}$$

Equation (14.32) yields

$$N = \begin{pmatrix} 0 & 0 & 0 & 0 & \dots \\ 0 & 1 & 0 & 0 & \dots \\ 0 & 0 & 2 & 0 & \dots \\ 0 & 0 & 0 & 3 & \dots \\ & & \vdots & & \end{pmatrix} . \tag{14.34}$$

From (14.10) the energy eigenvalues of the simple harmonic oscillator are then given by

$$\boxed{E = \hbar\omega\left(n + \frac{1}{2}\right) , \quad n = 0, 1, 2, \dots} . \tag{14.35}$$

This result has already been given in (4.13).

Instead of viewing n as labelling the quantum state of a single particle, as we have in this chapter, in the so-called **second quantization** procedure, one can use the vectors $|\,0\,\rangle$, $|\,1\,\rangle$, $|\,2\,\rangle, \dots$ to represent quantum states with $0, 1, 2, \dots$ particles, etc., each with energy $\hbar\omega$. The representation generated by the complete orthonormal set $\{\,|\,n\,\rangle\,\}$ is then called the **number representation** or the **Fock representation** for a system of **identical bosons**, such as photons. This representation is the origin of the terminology of "creation" and "annihilation operators".

Problem 14.1 Find the exact energy eigenvalues for the simple harmonic oscillator when a constant force is added, that is, when the potential energy function is given by

$$V(x) = \frac{1}{2} m\omega^2 x^2 + \alpha x ,$$

where α is a constant.

Problem 14.2 Instead of the commutation rule (14.8) for bosons, the creation and annihilation operators $a^\dagger$ and a for **fermions** obey the following **anticommutation** rules:

$$\{\, a\,, a^\dagger \,\} = 1\,, \quad \{\, a\,, a\,\} = \{\, a^\dagger\,, a^\dagger\,\} = 0\,,$$

where $\{\, A\,, B\,\} \equiv AB + BA$. Show that for fermions

$$a\,|\,0\,\rangle = 0\,, \quad a\,|\,1\,\rangle = |\,0\,\rangle\,, \quad a^\dagger\,|\,0\,\rangle = |\,1\,\rangle\,, \quad a^\dagger\,|\,1\,\rangle = 0\,.$$

Thus for a system of identical fermions (such as electrons), a particular quantum state cannot accommodate more than one particle. This is known as the **Pauli Exclusion Principle**.

For the remainder of the chapter we will look at the solution of the simple harmonic oscillator problem in the Schrödinger picture, that is, we wish to solve the time-independent Schrödinger equation [cf. (1.35)]

$$-\frac{\hbar^2}{2m}\frac{d^2\psi(q)}{dq^2}+\frac{1}{2}m\omega^2q^2\psi(q)=E\psi(q)\ . \tag{14.36}$$

We will make use of the results that we have already obtained above by the algebraic method. We thus expect the solutions to (14.36) to be the eigenfunctions $\psi_n(q)=\langle\, q\,|\, n\,\rangle$.

To simplify the Schrödinger equation (14.36) we use the dimensionless coordinate $Q=\sqrt{m\omega/\hbar}\,q$ [cf. (14.2)] to define

$$u_n(Q)\equiv\langle\, Q\,|\, n\,\rangle=\left(\frac{\hbar}{m\omega}\right)^{1/4}\psi_n(q)\ . \tag{14.37}$$

The constant coefficient $(\hbar/(m\omega))^{1/4}$ of $\psi_n(q)$ ensures the correct normalization

$$\int_{-\infty}^{\infty}dQ\,u_n^*(Q)u_n(Q)=\int_{-\infty}^{\infty}dq\,\psi_n^*(q)\psi_n(q)=1\ , \tag{14.38}$$

$$\text{or}\quad\int_{-\infty}^{\infty}dQ\,|\,Q\,\rangle\langle\,Q\,|=\int_{-\infty}^{\infty}dq\,|\,q\,\rangle\langle\,q\,|=1\ . \tag{14.39}$$

Expressed in the Q-representation the Schrödinger equation (14.36) becomes

$$\frac{1}{2}\left(-\frac{d^2}{dQ^2}+Q^2\right)u(Q)=\varepsilon\,u(Q)\ , \tag{14.40}$$

where the dimensionless energy ε is defined by

$$\varepsilon\equiv\frac{E}{\hbar\omega}\ . \tag{14.41}$$

Rather than solving (14.40) directly, we appeal to (14.24) and solve for the ground state $u_0(Q)$ first. From (14.24) we have

$$\langle\, Q\,|\, a\,|\, 0\,\rangle=0\ , \tag{14.42}$$

which, with the help of (14.5), implies

$$\frac{1}{\sqrt{2}}\langle\, Q\,|\,(Q+iP)\,|\,0\,\rangle=0\ , \tag{14.43}$$

or, equivalently,

$$Q\,u_0(Q)+i\,\langle\, Q\,|\, P\,|\, 0\,\rangle=0\ . \tag{14.44}$$

Recalling the action of the momentum operator P in the coordinate representation as given in the table on p. 65 (Chapter 11), the above equation is actually the differential equation

$$\frac{d\,u_0(Q)}{dQ}+Q\,u_0(Q)=0\ . \tag{14.45}$$

Note that in the present context P and Q are both dimensionless quantities, so that

$$P \longrightarrow -i \frac{d}{dQ} \quad . \tag{14.46}$$

The normalized solution of (14.45) is easily seen to be

$$u_0(Q) = \frac{1}{\pi^{1/4}} e^{-Q^2/2} \quad . \tag{14.47}$$

From this ground state we can construct the $u_n(Q)$ $(n \neq 0)$ by applying $a^\dagger$ successively according to (14.27). Thus

$$u_n(Q) = \langle Q \,|\, n \rangle = \frac{1}{\sqrt{n!}} \langle Q \,|\, (a^\dagger)^n \,|\, 0 \rangle \quad . \tag{14.48}$$

Recalling (14.4) for $a^\dagger$ and (14.46) for P the above equation becomes

$$\boxed{u_n(Q) = \frac{1}{(\sqrt{\pi}\, 2^n\, n!)^{1/2}} \left(Q - \frac{d}{dQ} \right)^n e^{-Q^2/2}} \quad . \tag{14.49}$$

According to (14.37) the normalized harmonic oscillator wave functions in the coordinate representation are then given by

$$\psi_n(q) = \left(\frac{m\omega}{\hbar} \right)^{1/4} u_n(Q) \quad , \tag{14.50}$$

where Q and q are related by (14.2).

We now proceed to write $u_n(Q)$ in terms of Hermite polynomials, a class of special functions. First we note that

$$Q - \frac{d}{dQ} = -\, e^{Q^2/2} \frac{d}{dQ} e^{-Q^2/2} . \tag{14.51}$$

Indeed, using the operator on the right-hand side to act on an arbitrary function $u(Q)$,

$$\begin{aligned} -\, e^{Q^2/2} \frac{d}{dQ} \left(e^{-Q^2/2} u \right) &= -\, e^{Q^2/2} \left(e^{-Q^2/2} \frac{du}{dQ} - Q e^{-Q^2/2} u \right) \\ &= \left(Q - \frac{d}{dQ} \right) u \quad . \end{aligned} \tag{14.52}$$

Thus, from (14.49),

$$u_n(Q) = \frac{1}{(\sqrt{\pi}\, 2^n\, n!\,)^{1/2}} \left(-\, e^{Q^2/2} \frac{d}{dQ} e^{-Q^2/2} \right)^n e^{-Q^2/2} . \tag{14.53}$$

Now the **Hermite polynomials** $H_n(Q)$ are defined by the following formula (known as Rodriques formula):

$$\boxed{H_n(Q) = (-1)^n e^{Q^2} \frac{d^n}{dQ^n} e^{-Q^2} \; ; \quad n = 0, 1, 2, 3, \ldots} \quad , \tag{14.54}$$

and we assert that

$$\left(- e^{Q^2/2} \frac{d}{dQ} e^{-Q^2/2} \right)^n e^{-Q^2/2} = e^{-Q^2/2} H_n(Q) \; ; \quad n = 0, 1, 2, 3, \ldots \; . \tag{14.55}$$

This can be proved by induction. The assertion is obviously true for $n = 0$, since

$$\left(- e^{Q^2/2} \frac{d}{dQ} e^{-Q^2/2} \right)^0 e^{-Q^2/2} = e^{-Q^2/2} \, ,$$

while, from (14.54),

$$e^{-Q^2/2} H_0(Q) = e^{-Q^2/2} (-1)^0 e^{Q^2} \left(\frac{d^0}{dQ^0} e^{-Q^2} \right) = e^{-Q^2/2} \, .$$

Assume that the assertion is true for n, then

$$\begin{aligned} &\left(- e^{Q^2/2} \frac{d}{dQ} e^{-Q^2/2} \right)^{n+1} e^{-Q^2/2} \\ &= \left(- e^{Q^2/2} \frac{d}{dQ} e^{-Q^2/2} \right) \left\{ e^{-Q^2/2} (-1)^n e^{Q^2} \left(\frac{d^n}{dQ^n} e^{-Q^2} \right) \right\} \\ &= e^{Q^2/2} (-1)^{n+1} \left(\frac{d^{n+1}}{dQ^{n+1}} e^{-Q^2} \right) = e^{-Q^2/2} (-1)^{n+1} e^{Q^2} \left(\frac{d^{n+1}}{dQ^{n+1}} e^{-Q^2} \right) \\ &= e^{-Q^2/2} H_{n+1}(Q) \, , \end{aligned} \tag{14.56}$$

that is, it is also true for $n+1$. Thus (14.55) is valid for all non-negative integers n. We can then write the following expression for the simple harmonic oscillator normalized wave function in the Q-representation:

$$\boxed{u_n(Q) = \frac{1}{(\sqrt{\pi}\, 2^n \, n!\,)^{1/2}} \, e^{-Q^2/2} H_n(Q)} \quad , \tag{14.57}$$

where $H_n(Q)$ is the n-th Hermite polynomial given by (14.54). The differential equation satisfied by the $u_n(Q)$ is (14.40), which, in view of the fact that $\varepsilon_n = E_n/(\hbar\omega) = n + 1/2$, now reads

$$\frac{1}{2} \left(- \frac{d^2}{dQ^2} + Q^2 \right) u_n(Q) = \left(n + \frac{1}{2} \right) u_n(Q) \, . \tag{14.58}$$

From (14.4) and (14.46), the pair of equations (14.25) and (14.26) are equivalent to the following recursion relations for the $u_n(Q)$:

$$\frac{1}{\sqrt{2}} \left(Q + \frac{d}{dQ} \right) u_n(Q) = \sqrt{n}\; u_{n-1}(Q) \, , \tag{14.59}$$

$$\frac{1}{\sqrt{2}} \left(Q - \frac{d}{dQ} \right) u_n(Q) = \sqrt{n+1}\; u_{n+1}(Q) \, . \tag{14.60}$$

Also, the first of the equations (14.5) implies the additional recursion relation

$$Q\,u_n(Q) = \sqrt{\frac{n+1}{2}}\;u_{n+1}(Q) + \sqrt{\frac{n}{2}}\;u_{n-1}(Q)\,. \tag{14.61}$$

From the definition (14.54) of the Hermite polynomials we see that

$$H_n(-Q) = (-1)^n\,H_n(Q)\,. \tag{14.62}$$

Equation (14.57) then implies the following **parity** condition for the $u_n(Q)$:

$$u_n(-Q) = (-1)^n\,u_n(Q)\,. \tag{14.63}$$

From the orthonormality condition $\langle\, n\,|\,m\,\rangle = \delta_{nm}$ [cf. (14.21)] and the completeness condition for the $|\,Q\,\rangle$ [cf. (14.39)] we have

$$\int_{-\infty}^{\infty} dQ\,\langle\, n\,|\,Q\,\rangle\langle\, Q\,|\,m\,\rangle = \delta_{nm}\,,$$

that is,

$$\int_{-\infty}^{\infty} dQ\,u_n^*(Q)u_m(Q) = \delta_{nm}\,, \tag{14.64}$$

as the orthonormality condition for the $u_n(Q)$. From the completeness condition $\sum_n |\,n\,\rangle\langle\, n\,| = 1$ it follows that

$$\sum_{n=0}^{\infty}\langle\, Q\,|\,n\,\rangle\langle\, n\,|\,Q'\,\rangle = \langle\, Q\,|\,Q'\,\rangle\,,$$

that is,

$$\sum_{n=0}^{\infty} u_n^*(Q')u_n(Q) = \delta(Q-Q')\,, \tag{14.65}$$

which is the completeness condition for the $u_n(Q)$.

We can use (14.54) to easily generate the first few Hermite polynomials:

$$H_0(Q) = 1\,, \tag{14.66a}$$
$$H_1(Q) = 2Q \tag{14.66b}$$
$$H_2(Q) = 4Q^2 - 2\,, \tag{14.66c}$$
$$H_3(Q) = 8Q^3 - 12Q\,, \tag{14.66d}$$
$$H_4(Q) = 16Q^4 - 48Q^2 + 12\,, \tag{14.66e}$$
$$H_5(Q) = 32Q^5 - 160Q^3 + 120Q\,. \tag{14.66f}$$

The differential equation satisfied by $H_n(Q)$ is

$$\left(\frac{d^2}{dQ^2} - 2Q\frac{d}{dQ} + 2n\right) H_n(Q) = 0\,. \tag{14.67}$$

The individual $H_n(Q)$ can also be obtained by a **generating function**:

$$e^{-s^2+2sQ} = \sum_{n=0}^{\infty} \frac{s^n}{n!} H_n(Q) . \tag{14.68}$$

Some recursion relations for the $H_n(Q)$ [analogous to (14.59), (14.60) and (14.61) for the $u_n(Q)$] are

$$\frac{d}{dQ} H_n(Q) = 2n\, H_{n-1}(Q) , \tag{14.69}$$

$$\left(2Q - \frac{d}{dQ}\right) H_n(Q) = H_{n-1}(Q) , \tag{14.70}$$

$$2Q\, H_n(Q) = H_{n+1}(Q) + 2n\, H_{n-1}(Q) . \tag{14.71}$$

Sketches of the first few $u_n(Q)$ are shown in Figs. 14.1[(a) to (d)] . Note the parity of the wave functions according to (14.63).

Problem 14.3 For a complex number $z \in \mathbb{C}$ define a **coherent state** by

$$|\, z \,\rangle \equiv \exp(za^\dagger - z^* a)\, |\, 0 \,\rangle ,$$

where a and $a^\dagger$ are annihilation and creation operators, respectively.

(a) Show that

$$|\, z \,\rangle = e^{-|z|^2/2} \sum_{n=0}^{\infty} \frac{z^n}{n!} \,|\, n \,\rangle .$$

(b) Show that a coherent state is always normalized:

$$\langle\, z \,|\, z \,\rangle = 1 .$$

(c) Show that for any two complex numbers z and w,

$$\langle\, z \,|\, w \,\rangle = e^{-|z-w|^2} ,$$

so that coherent states corresponding to different complex numbers are not orthogonal, but nearly so if they are far away from each other in the complex plane.

(d) Show that coherent states satisfy the following completeness relation:

$$\int |\, z \,\rangle\langle\, z \,|\, \frac{d^2 z}{\pi} = 1 , \qquad (d^2 z = r dr d\theta) .$$

(e) Show that coherent states are eigenstates of annihilation operators:

$$a\,|\, z \,\rangle = z\,|\, z \,\rangle .$$

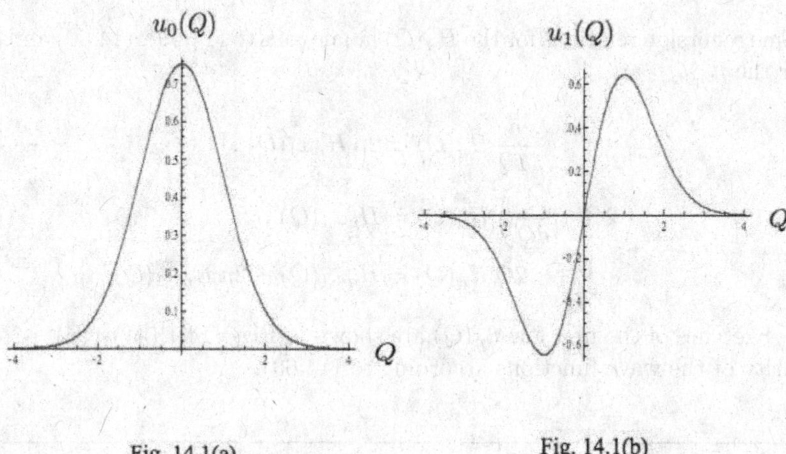

Fig. 14.1(a)

Fig. 14.1(b)

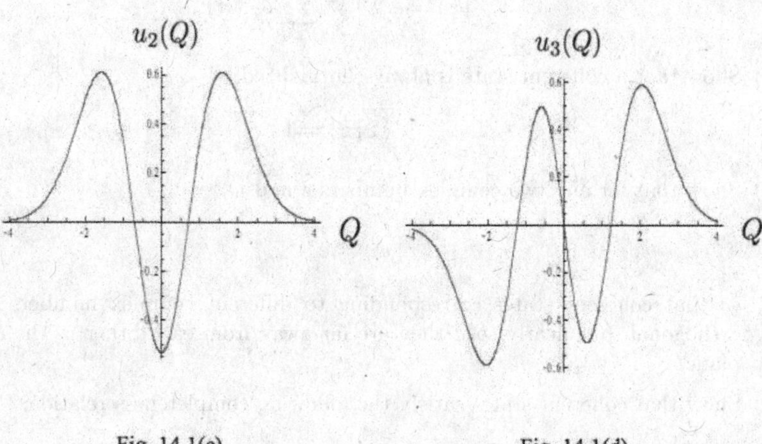

Fig. 14.1(c)

Fig. 14.1(d)

(f) Show that the mean number of bosons in the coherent state $|\,z\,\rangle$ is given by $|z|^2$:

$$\langle\, z\,|\, a^\dagger a\,|\, z\,\rangle = |z|^2 \,.$$

(g) Show that the probability of finding n bosons in the coherent state $|\,z\,\rangle$ is a **Poisson distribution** about the mean $|z|^2$:

$$|\langle\, n\,|\, z\,\rangle|^2 = e^{-|z|^2}\,\frac{|z|^{2n}}{n!} \,.$$

(h) Show that the coherent state represents a minimum uncertainty wave (satisfying the condition $\Delta p \Delta q = \hbar/2$), where p and q are given by (14.6).

Coherent states were originally developed in the study of quantum optics, but have subsequently found wide applications in many areas of physics and mathematical physics.

Chapter 15

Complete Set of Commuting Observables

In this chapter we will study a central theorem (Theorem 15.1) concerning a pair of commuting self-adjoint operators, that is, those representing observables in the formalism of quantum mechanics. This theorem is of fundamental importance in quantum theory, and, as we shall see, leads to the possibility of the identification of a quantum system by its **quantum numbers**.

Theorem 15.1. *Two observables commute if and only if they possess a complete orthonormal set of eigenvectors.*

Proof. First we will prove the "if" part. Consider two observables represented by two self-adjoint operators A and B. Suppose A and B possess a complete orthonormal set of eigenvectors $|\,\psi_i\,\rangle$. Let

$$A\,|\,\psi_i\,\rangle = a_i\,|\,\psi_i\,\rangle\,, \qquad B\,|\,\psi_i\,\rangle = b_i\,|\,\psi_i\,\rangle\,, \tag{15.1}$$

where the eigenvalues in either set $\{a_i\}$ and $\{b_i\}$ need not be distinct. In other words we include the possibility of degeneracies. Since the set $\{|\,\psi_i\,\rangle\}$ is complete we can expand an arbitrary vector $|\,\psi\,\rangle$ as

$$|\,\psi\,\rangle = \sum_i c_i\,|\,\psi_i\,\rangle\,. \tag{15.2}$$

Then

$$(AB-BA)\,|\,\psi\,\rangle = \sum_i c_i\,(AB-BA)\,|\,\psi_i\,\rangle = \sum_i c_i(a_ib_i-b_ia_i)\,|\,\psi_i\,\rangle = 0\,. \tag{15.3}$$

Since $|\,\psi\,\rangle$ is arbitrary, it follows that $[\,A\,,\,B\,] = 0$.

Now we will prove the "only if" part (which is much harder). We will first do so under the assumption that all eigenvalues are non-degenerate, and then generalize to include degeneracies. Suppose two observables A and B commute, that is, $[A, B] = 0$. The Spectral Theorem (Theorem 9.2) ensures that each of A and B possesses a complete set of eigenvectors in rigged Hilbert space, which may include those belonging to the continuous parts of the spectra, if there are any. Let $|a\rangle$ be an eigenvector of A with eigenvalue a, so that

$$A|a\rangle = a|a\rangle . \tag{15.4}$$

Note that we can label an eigenvector by its eigenvalue here because we have assumed non-degeneracy. The vector $|a\rangle$ can be expanded in terms of a complete set of orthonormal eigenvectors of B:

$$|a\rangle = \sum_m |b_m\rangle\langle b_m|a\rangle \equiv \sum_m |a; b_m\rangle . \tag{15.5}$$

(The above equation assumes that the spectrum is entirely discrete. For the remainder of the proof we will continue to make this assumption for ease of presentation. The generalization to include continuous parts of the spectra involves no essential difficulties.) Obviously $|a; b_m\rangle \equiv \langle b_m|a\rangle|b_m\rangle$ is an eigenvector of B with eigenvalue b_m. We will show that it is also an eigenvector of A with eigenvalue a, that is,

$$A|a; b_m\rangle = a|a; b_m\rangle . \tag{15.6}$$

First consider the vector $(A - a)|a; b_m\rangle$. We have, under the assumption that A and B commute,

$$\begin{aligned} B(A-a)|a; b_m\rangle &= (A-a)B|a; b_m\rangle \\ = (A-a)b_m|a; b_m\rangle &= b_m(A-a)|a; b_m\rangle . \end{aligned} \tag{15.7}$$

Thus $(A - a)|a; b_m\rangle$ is an eigenvector of B with eigenvalue b_m. Since the b_m are distinct (by the assumption of non-degeneracy), the $(A - a)|a; b_m\rangle$ are mutually orthogonal (by Theorem 8.3), and hence are linearly independent. On the other hand, (15.4) and (15.5) imply

$$(A-a)|a\rangle = \sum_m (A-a)|a; b_m\rangle = 0 . \tag{15.8}$$

Thus

$$(A-a)|a; b_m\rangle = 0 \tag{15.9}$$

for all m, that is, $|a; b_m\rangle$ is an eigenvector of A with eigenvalue a. We then arrive at the result that the vectors $|a; b_m\rangle$ are simultaneous eigenvectors of A and B.

The above argument can be repeated with degeneracies incorporated. Let $\{|a_n^{(r_n)}\rangle\}$ and $\{|b_m^{(s_m)}\rangle\}$ be complete sets of eigenvectors for A and B, respectively, where r_n and s_m are degeneracy indices. We may assume them to be orthonormalized sets. These eigenvectors satisfy

$$A|a_n^{(r_n)}\rangle = a_n|a_n^{(r_n)}\rangle , \qquad B|b_m^{(s_m)}\rangle = b_m|b_m^{(s_m)}\rangle . \tag{15.10}$$

The completeness conditions are

$$\sum_{n,r_n} |\, a_n^{(r_n)} \rangle\langle\, a_n^{(r_n)} \,| = 1\,, \qquad \sum_{m,s_m} |\, b_m^{(s_m)} \rangle\langle\, b_m^{(s_m)} \,| = 1\,. \tag{15.11}$$

As before we may expand

$$|\, a_n^{(r_n)} \rangle = \sum_m \sum_{s_m} |\, b_m^{(s_m)} \rangle\langle\, b_m^{(s_m)} \,|\, a_n^{(r_n)} \rangle \equiv \sum_m |\, a_n^{(r_n)}\,;\, b_m \rangle\,, \tag{15.12}$$

where

$$|\, a_n^{(r_n)}\,;\, b_m \rangle \equiv \sum_{s_m} |\, a_n^{r_n)}\,;\, b_m^{(s_m)} \rangle \equiv \sum_{s_m} \langle\, b_m^{(s_m)} \,|\, a_n^{(r_n)} \rangle \,|\, b_m^{(s_m)} \rangle\,. \tag{15.13}$$

Obviously, $|\, a_n^{(r_n)}\,;\, b_m \rangle$ is an eigenvector of B with eigenvalue b_m. It is also an eigenvector of A, with eigenvalue a_n. To show this we note that the vector $(A - a_n)\,|\, a_n^{(r_n)}\,;\, b_m \rangle$ is an eigenvector of B with eigenvalue b_m. Indeed,

$$B(A - a_n)\,|\, a_n^{(r_n)}\,;\, b_m \rangle = (A - a_n)B\,|\, a_n^{(r_n)}\,;\, b_m \rangle = b_m\,(A - a_n)\,|\, a_n^{(r_n)}\,;\, b_m \rangle\,. \tag{15.14}$$

Now the first of the equations (15.10) and Eq. (15.12) imply that

$$\sum_m (A - a_n)\,|\, a_n^{(r_n)}\,;\, b_m \rangle = 0\,. \tag{15.15}$$

On the other hand the vectors $(A - a_n)\,|\, a_n^{(r_n)}\,;\, b_m \rangle$ (for different m) are linearly independent since they belong to distinct eigenvalues b_m (by Theorem 8.3). Thus

$$(A - a_n)\,|\, a_n^{(r_n)}\,;\, b_m \rangle = 0\,. \tag{15.16}$$

So $|\, a_n^{(r_n)}\,;\, b_m \rangle$ is a simultaneous eigenvector of A and B with eigenvalues a_n and b_m, respectively.

Now consider the set

$$\{|\, a_n^{(r_n)}\,;\, b_m \rangle\}\,, \quad r_n = 1, 2, \ldots, d_n\,,$$

where d_n is the degree of degeneracy of the eigenvalue a_n. The vectors in this set are not a priori linearly independent. But they all belong to the subspace of simultaneous eigenvectors of A and B with eigenvalues a_n and b_m. It is always possible to construct an orthonormal set $|\, \chi_{nm}^{(s)} \rangle$, $s = 1, 2, \ldots$, which generate this subspace, such that

$$|\, a_n^{(r_n)}\,;\, b_m \rangle = \sum_s \gamma_{r_n s}\,|\, \chi_{nm}^{(s)} \rangle\,. \tag{15.17}$$

The set $\{|\, \chi_{nm}^{(s)} \rangle\}$ (for different n, m and s) is a complete orthonormal set, since any $|\, \psi \rangle$ can be expanded as

$$\begin{aligned} |\, \psi \rangle &= \sum_n \sum_{r_n} \beta_{nr_n}\,|\, a_n^{(r_n)} \rangle = \sum_n \sum_{r_n} \sum_m \beta_{nr_n}\,|\, a_n^{(r_n)}\,;\, b_m \rangle \\ &= \sum_{nms} \left(\sum_{r_n} \beta_{nr_n} \gamma_{r_n s} \right) |\, \chi_{nm}^{(s)} \rangle\,. \end{aligned} \tag{15.18}$$

Orthonormality within the set for fixed n and m follows from construction. For distinct pairs (n, m) the eigenvalues a_n and b_m are distinct. Hence the corresponding eigenvectors must be orthogonal. □

The above theorem can be applied to the case of N pairwise commuting observables. The following definition introduces the important notion of a complete set of commuting observables.

Definition 15.1. *N observables $A_1, \ldots, A_N$ are said to form a* ***complete set of commuting observables*** *if they possess a unique complete orthonormal set of eigenvectors.*

The eigenvalues of a complete set of commuting observables can be used to specify completely the quantum state of a physical system. These eigenvalues are called the **quantum numbers** of the system. As a very important example consider a particle in a spherically symmetric potential $V(r)$ (where r is the magnitude of the position vector $\boldsymbol{r}$ of the particle). (This is the case, for example, in the highly simplified hydrogen-atom problem in which the electron is subjected only to the Coulomb potential due to the proton.) In this case the Hamiltonian H [given by (1.34)] commutes with L^2, L_z and P, where $\boldsymbol{L} = -i\hbar\boldsymbol{r} \times \nabla$ is the angular momentum operator, $L^2 = \boldsymbol{L} \cdot \boldsymbol{L}$, L_z is the z-component of $\boldsymbol{L}$, and P is the **parity** operator [defined by $P\,\psi(\boldsymbol{r}) = \psi(-\boldsymbol{r})$]. Then the operators H, L^2, L_z and P form a complete set of commuting observables. It will be shown later that these possess a unique complete set of eigenvectors represented by wave functions of the form

$$\psi_{nlm}(\boldsymbol{r}) = R_{nl}(r)\, Y_{lm}(\theta, \phi)\,, \tag{15.19}$$

where $n = 1, 2, \ldots$, $l = 0, 1, \ldots, n-1$ (for a given n), and $m = -l, -l+1, \ldots, l$ (for a given l). The eigenvalues are given by

$$H\,\psi_{nlm} = E_{nl}\,\psi_{nlm}\,, \tag{15.20}$$

$$L^2\,\psi_{nlm} = \hbar^2\, l(l+1)\,\psi_{nlm}\,, \tag{15.21}$$

$$L_z\,\psi_{nlm} = \hbar\, m\,\psi_{nlm}\,, \tag{15.22}$$

$$P\,\psi_{nlm} = (-1)^l\,\psi_{nlm}\,. \tag{15.23}$$

Chapter 16

Solving Schrödinger's Equation

In this chapter we will provide a brief introduction to the solutions of Schrödinger's equation. Our development will first be set up in the context of a single particle moving in three-dimensional space, under the influence of a spherically symmetric potential. A detailed treatment of the angular part of the equation, which is intimately connected to the conservation of angular momentum as a consequence of rotational symmetry, will be deferred until later (see Chapters 20 and 21). Here we will just focus our attention on the radial part of the equation, which is equivalent formally to the one-dimensional Schrödinger equation.

Recall the Schrödinger equation (1.35). Since we are considering a spherically symmetric potential $V(r)$, it is most convenient to write the Laplacian ∇^2 in spherical coordinates:

$$\nabla^2 = \frac{1}{r^2}\frac{\partial}{\partial r}\left(r^2\frac{\partial}{\partial r}\right) + \frac{1}{r^2\sin\theta}\frac{\partial}{\partial\theta}\left(\sin\theta\frac{\partial}{\partial\theta}\right) + \frac{1}{r^2\sin\theta}\frac{\partial}{\partial\phi}\left(\frac{1}{\sin\theta}\frac{\partial}{\partial\phi}\right) . \tag{16.1}$$

The Hamiltonian H can then be written as

$$H = -\frac{\hbar^2}{2mr^2}\frac{\partial}{\partial r}\left(r^2\frac{\partial}{\partial r}\right) + \frac{L^2}{2I} + V(r) \quad , \tag{16.2}$$

where $I = mr^2$ is the moment of inertia of the particle of mass m, and L denotes the **angular momentum operator** given by

$$L^2 = -\hbar^2\frac{1}{\sin\theta}\left\{\frac{\partial}{\partial\theta}\left(\sin\theta\frac{\partial}{\partial\theta}\right) + \frac{\partial}{\partial\phi}\left(\frac{1}{\sin\theta}\frac{\partial}{\partial\phi}\right)\right\} \quad . \tag{16.3}$$

The first term in H, involving only the radial distance r, represents the radial kinetic energy of the particle, while the term $L^2/(2I)$ represents the rotational kinetic energy of the particle with respect to the origin. The above two equations show explicitly that, for a spherically symmetric potential [$V(r)$ only depends on the radial distance r], L^2 commutes with the Hamiltonian H.

As noted at the end of the last chapter, the simultaneous eigenfunctions $\psi(\boldsymbol{r})$ of H and L^2 can be separated into radial and angular parts as follows:

$$\psi(\boldsymbol{r}) = R_l(r)Y_{lm}(\theta, \phi) \quad , \tag{16.4}$$

where the spherical harmonics $Y_{l,m}$ satisfy

$$L^2 Y_{lm} = \hbar^2\, l(l+1)\, Y_{lm} \quad . \tag{16.5}$$

Note that in the above equations [compare with (15.19) and (15.21)] we have suppressed the principal quantum number n in order to simplify the notation.

Substitution of (16.4) into the Schrödinger equation $H\psi = E\psi$ yields, on using (16.5) and on cancellation of Y_{lm} from both sides of the equation, the following radial equation:

$$-\frac{\hbar^2}{2m}\frac{1}{r^2}\frac{d}{dr}\left(r^2\frac{dR_l}{dr}\right) + \left\{V(r) + \frac{\hbar^2\, l(l+1)}{2mr^2}\right\} R_l = ER_l \quad . \tag{16.6}$$

This equation my be simplified by the substitution

$$R_l(r) = \frac{\chi_l(r)}{r} \quad . \tag{16.7}$$

It then follows by a simple calculation that

$$-\frac{\hbar^2}{2m}\frac{d^2\chi_l}{dr^2} + \left\{V(r) + \frac{\hbar^2\, l(l+1)}{2mr^2}\right\}\chi_l = E\,\chi_l \quad . \tag{16.8}$$

This is a one-dimensional Schrödinger equation (in the single variable r), with an **effective potential** depending only on r given by the expression within the brackets. The repulsive term depending on r^{-2} is called the **centrifugal potential**. Defining

$$V'(r) \equiv \frac{2m}{\hbar^2}V(r) \quad , \tag{16.9}$$

$$U_l(r) \equiv \frac{l(l+1)}{r^2} + V'(r) \quad , \tag{16.10}$$

$$k^2 \equiv \frac{2mE}{\hbar^2} \quad , \tag{16.11}$$

Eq. (16.8) can be rewritten as

$$\frac{d^2\chi_{kl}}{dr^2} + \{k^2 - U_l(r)\}\,\chi_{kl} = 0 \quad , \tag{16.12}$$

where we have labelled the wave function corresponding to a particular energy E by the parameter k [according to (16.11)]. Figure 16.1 shows a schematic picture for $V'(r)$ (for the case of an attractive Coulomb potential, where $V'(r) \propto -1/r$) and $U_l(r)$ (for $l \neq 0$).

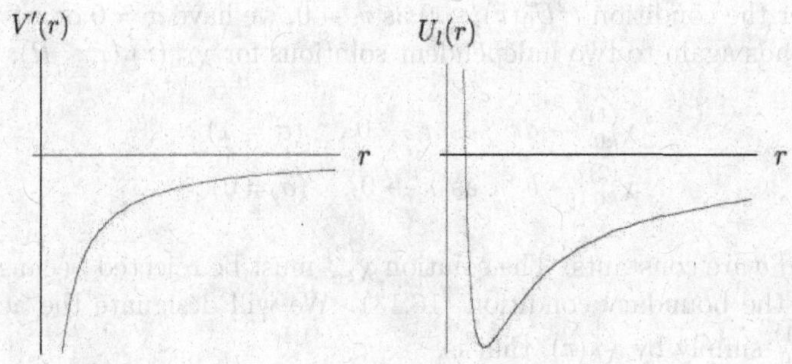

Fig. 16.1

If we assume that $U_l(r)$ is a smooth function of r, both χ_{kl} and its first derivative $d\chi_{kl}/dr$ should be continuous. For if $d\chi_{kl}/dr$ were not continuous, the second derivative $d^2\chi_{kl}/dr^2$ would have a δ-function component, which is not the case according to (16.12). Since the full wave function $\psi(\boldsymbol{r})$ is required to be bounded over all space, $R_{kl} = \chi_{kl}/r$ is bounded for all r, that is, for $0 \leq r < \infty$. As a consequence, χ_{kl} itself is required to satisfy

$$\chi_{kl} \longrightarrow 0 \quad \text{as} \quad r \longrightarrow 0 \quad . \tag{16.13}$$

The Coulomb potential is what is known as a **long range potential**. In practice one is more likely to encounter **short range potentials**, defined to be ones decreasing faster than $1/r$ as $r \to \infty$. In this case we may approximate the potential by imposing a cut-off range R such that $V(r) = 0$ for $r > R$. If we further assume $l = 0$, then (16.12) becomes

$$\begin{aligned} \frac{d^2\chi_{k0}}{dr^2} + (k^2 - U_0(r))\chi_{k0} = 0\,, \quad r < R\,, \\ \frac{d^2\chi_{k0}}{dr^2} + k^2\chi_{k0} = 0\,, \quad r \geq R\,. \end{aligned} \tag{16.14}$$

For $r > R$, we have the two independent solutions

$$\chi_k^{(\pm)} = e^{\pm ikr}\,. \tag{16.15}$$

For $r \leq R$, we seek a solution satisfying the boundary condition $\chi_{k0} \to 0$ as $r \to 0$ [cf. (16.13)], of the form

$$\chi_{k0} \sim r^\sigma\,, \quad \text{as} \quad r \to 0\,. \tag{16.16}$$

The first equation in (16.14) then implies

$$\sigma(\sigma-1) = -r^2(k^2 - U_0(r)) \,.$$

Thus, under the condition $r^2U_0(r) \to 0$ as $r \to 0$, we have $\sigma = 0$ or $\sigma = 1$. This situation leads again to two independent solutions for $\chi_{k0}(r)$ $(r \le R)$:

$$\begin{aligned} \chi_{k0}^{(1)} &\to ar \quad \text{as } r \to 0\,, \quad (\sigma = 1)\,, \\ \chi_{k0}^{(2)} &\to b \quad\ \text{as } r \to 0\,, \quad (\sigma = 0)\,, \end{aligned} \tag{16.17}$$

where a and b are constants. The solution $\chi_{k0}^{(2)}$ must be rejected because it does not satisfy the boundary condition (16.13). We will designate the acceptable solution $\chi_{k0}^{(1)}$ simply by $\chi_k(r)$, that is,

$$\chi_k(r) \longrightarrow ar\,, \quad \text{as } r \to 0\,. \tag{16.18}$$

Now consider the different cases of positive energies ($k^2 > 0$) and negative energies ($k^2 < 0$) [cf. (16.11)]. For $k^2 > 0$, both solutions $\chi_k^{(\pm)}$ are acceptable. Hence we can write the general solution for $r > R$ as a linear combination of them:

$$\chi_k(r) = A(k)(\chi_k^{(-)} - S(k)\chi_k^{(+)})\,, \quad r > R\,. \tag{16.19}$$

Since $\chi_k(r)$ and $\chi_k'(r)$ must both be continuous at $r = R$, the above equation implies that

$$A(k)(\chi_k^{(-)}(R) - S(k)\chi_k^{(+)}(R)) = \chi_k(R)\,, \tag{16.20}$$

$$A(k)\left((\chi_k^{(-)})'(R) - S(k)(\chi_k^{(+)})'(R)\right) = \chi_k'(R)\,, \tag{16.21}$$

where the primes denote first derivatives with respect to r. The above equations determine uniquely the quantities $A(k)$ and $S(k)$. In fact, simultaneous solution of them yields

$$A(k) = \frac{\chi_k(R)(\chi_k^{(+)})'(R) - \chi_k'(R)\chi_k^{(+)}(R)}{\chi_k^{(-)}(R)(\chi_k^{(+)})'(R) - \chi_k^{(+)}(R)(\chi_k^{(-)})'(R)}\,, \tag{16.22}$$

$$S(k) = \frac{\chi_k(R)(\chi_k^{(-)})'(R) - \chi_k'(R)\chi_k^{(-)}(R)}{\chi_k(R)(\chi_k^{(+)})'(R) - \chi_k'(R)\chi_k^{(+)}(R)}\,. \tag{16.23}$$

Using the explicit forms for $\chi_k^{(\pm)}$ given by (16.15), the above equations simplify to

$$A(k) = \frac{e^{ikR}}{2ik}\,(ik\chi_k(R) - \chi_k'(R))\,, \tag{16.24}$$

$$S(k) = -e^{-2ikR}\left(\frac{ik\chi_k(R) + \chi_k'(R)}{ik\chi_k(R) - \chi_k'(R)}\right)\,. \tag{16.25}$$

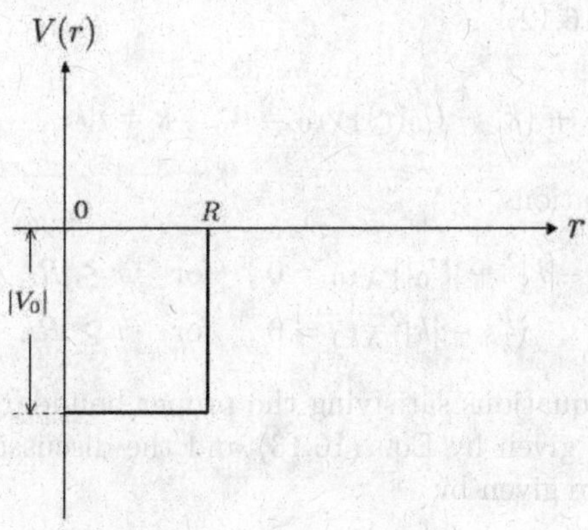

Fig. 16.2

We thus see that for each positive energy k^2, the Schrödinger equation (for $l = 0$) has a unique solution $\chi_k(r)$, given by (16.18), (16.19), (16.24), and (16.25). The physical meanings of $A(k)$ and $S(k)$ will be discussed below in more detail.

Now consider negative energies: $k^2 \equiv 2mE/\hbar^2 < 0$. Suppose $k = i|k|$. Then

$$\chi_k^{(+)} = e^{-|k|r} \quad \text{and} \quad \chi_k^{(-)} = e^{|k|r} \longrightarrow \infty \quad \text{as } r \to \infty\,. \tag{16.26}$$

(If k is in the lower half-plane, the roles of $\chi_k^{(+)}$ and $\chi_k^{(-)}$ are simply reversed.) Thus $\chi_k^{(-)}$ violates the boundary condition that χ_{kl}/r must be bounded as $r \to \infty$ [recall discussion above Eq. (16.13)], and must be rejected. The most general solution for $r > R$ is therefore

$$\chi_k(R) = A(k)\chi_k^{(+)}(r)\,, \quad r > R\,. \tag{16.27}$$

Matching the logarithmic derivative at $r = R$, one then obtains

$$\frac{\chi_k'(R)}{\chi_k(R)} = \frac{(\chi_k^{(+)})'(R)}{\chi_k^{(+)}(R)} = ik = -|k|\,. \tag{16.28}$$

This is in general a transcendental equation for k and is satisfied only by certain discrete imaginary values $k = k_n$, or correspondingly, negative energies E_n. Since $\chi_k(r)$ decreases exponentially as $r \to \infty$, it represents bound states.

To understand better how the above general development works we consider a specific example with an attractive potential. Suppose [recall Eq. (16.10)]

$$U_0(r) = -\frac{2m}{\hbar^2}\,|V_0|\,, \quad r \leq R\,, \qquad \text{and} \qquad U_0(r) = 0\,, \quad r \geq R\,, \tag{16.29}$$

where $|V_0|$ is a constant (see Fig. 16.2). Consider negative energies first. The Schrödinger equation [(16.12)]

$$\frac{d^2\chi_{k0}}{dr^2} + \{k^2 - U_0(r)\}\chi_{k0} = 0\,, \quad k = i|k|\,, \tag{16.30}$$

becomes the pair of equations

$$\chi''_{k0} + \{-|k|^2 + |U_0|\}\,\chi_{k0} = 0\,, \quad \text{for} \quad r \le R\,, \tag{16.31}$$

$$\chi''_{k0} - |k|^2\,\chi_{k0} = 0\,, \quad \text{for} \quad r \ge R\,. \tag{16.32}$$

The solutions of these equations satisfying the proper boundary conditions for $r \to 0$ and $r \to \infty$ [as given by Eq. (16.13) and the discussion immediately above that equation] are given by

$$\begin{aligned} &\chi_{k0}(r) = B\sin Kr\,, \quad \left(K \equiv \sqrt{|U_0| - |k|^2}\,, \quad |U_0| = \frac{2m}{\hbar^2}|V_0|\right)\,, \quad r \le R\,,\\ &\chi_{k0}(r) = Ae^{-|k|r}\,, \quad r \ge R\,. \end{aligned} \tag{16.33}$$

Matching boundary conditions for $\chi_{k0}(r)$ and $\chi'_{k0}(r)$ at $r = R$ gives

$$B\sin KR = Ae^{-|k|R}\,, \qquad BK\cos KR = -|k|Ae^{-|k|R}\,. \tag{16.34}$$

Dividing the second by the first of the above equations, we have the transcendental equation [cf. (16.28)] determining the discrete energy eigenvalues:

$$K\cot KR = -|k| = -\sqrt{|U_0| - K^2}\,. \tag{16.35}$$

It is interesting to determine the critical condition relating the strength of the potential $|U_0|$ and its range R in order for at least one discrete bound state to exist. Setting $|k| = 0$ in (16.34) (for the bound state to be just barely formed) we have

$$B\sin(\sqrt{|U_0|}R) = A\,, \qquad B\sqrt{|U_0|}\cos(\sqrt{|U_0|}R) = 0\,. \tag{16.36}$$

These imply $\tan(\sqrt{|U_0|}R) = \infty$ and hence $\sqrt{|U_0|}R = \pi/2$, from which $|U_0| = \pi^2/(4R^2)$, or, by the first of the equations in (16.29),

$$|V_0| = \frac{\pi^2\hbar^2}{8mR^2}\,. \tag{16.37}$$

This is at least how deep a potential well of range R has to be for a bound state to be formed.

Problem 16.1 Solve Eq. (16.35), that is,

$$K\cot KR = -\sqrt{|U_0| - K^2}\,,$$

by graphical means and numerically.

Consider now the continuous spectrum for the potential square well of Fig. 16.2. In this case $k^2 > 0$, and the particle is **scattered**. Outside the well, the Schrödinger equation is given by the second equation of (16.14), where k^2 is understood to be positive. This time we write the general solution in the following form [compare with (16.19)]:

$$\chi_{k0}(r) = A\sin(kr + \delta(k))\,, \quad r \geq R\,. \tag{16.38}$$

The quantity $\delta(k)$ is called the **scattering phase shift**. It is so-called for the following reason. In the absence of a potential ($V_0 = 0$), the above solution should be valid for all r. Hence, according to the boundary condition $\chi_{k0} \to 0$ as $r \to 0$, we should have $\delta(k) = 0$ for $V_0 = 0$. For a particle impinging on a potential well, we can match the logarithmic derivative of $\chi_{k0}(r)$ at $r = R$ to obtain

$$\left.\frac{d\,(\ln\chi_{k0})}{dr}\right|_{r=R} = k\cot(kR + \delta(k))\,, \tag{16.39}$$

where in the left-hand side, χ_{k0} is considered to be the wave function inside the well. For a narrow well ($R \to 0$) we have the approximate result

$$k\cot\delta(k) \sim \left.\frac{d\,(\ln\chi_{k0})}{dr}\right|_{r=0}\,. \tag{16.40}$$

To bring (16.38) into conformity with (16.19) we need to first understand the physical meanings of the free-particle solutions $\chi_k^{(\pm)} = e^{\pm ikr}$ [cf. (16.15)]. Using (16.4) for the full wave function and (16.7), together with the fact that $Y_{00}(\theta,\phi) = 1/\sqrt{4\pi}$, let us consider the free-particle wave functions

$$\psi^{(\pm)}(\boldsymbol{r}) = \frac{1}{\sqrt{4\pi}}\frac{e^{\pm ikr}}{r}\,, \quad (l = 0)\,. \tag{16.41}$$

Corresponding to these wave functions we would like to calculate the quantum mechanical current density [cf. Problem (1.5)]

$$\boldsymbol{j}(\boldsymbol{r}) = -\frac{i\hbar}{2m}(\psi^*\nabla\psi - \psi\nabla\psi^*)\,, \tag{16.42}$$

so-called because it satisfies the equation of continuity

$$\nabla\cdot\boldsymbol{j} + \frac{\partial\rho}{\partial t} = 0\,, \tag{16.43}$$

where $\rho = \psi^*\psi$ is the probability density, if the potential is real. Making use of $\nabla r = \boldsymbol{r}/r$ we have

$$\nabla\psi^{(\pm)}(\boldsymbol{r}) = \frac{1}{\sqrt{4\pi}}\,e^{\pm ikr}\,\frac{\boldsymbol{r}}{r}\left(\pm\frac{ik}{r} - \frac{1}{r^2}\right)\,, \quad \nabla(\psi^{(\pm)})^* = \frac{1}{4\pi}\,e^{\mp ikr}\,\frac{\boldsymbol{r}}{r}\left(\mp\frac{ik}{r} - \frac{1}{r^2}\right)\,. \tag{16.44}$$

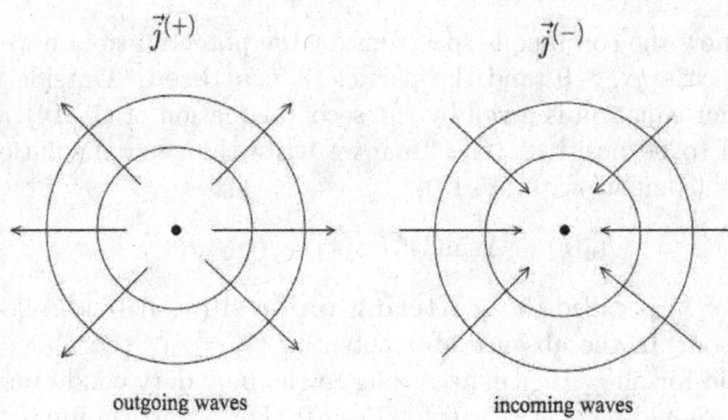

Fig. 16.3

Thus

$$(\psi^{(\pm)})^* \nabla \psi^{(\pm)} = \frac{1}{4\pi}\frac{\boldsymbol{r}}{r^2}\left(\pm\frac{ik}{r} - \frac{1}{r^2}\right),$$
$$\psi^{(\pm)} \nabla (\psi^{(\pm)})^* = \frac{1}{4\pi}\frac{\boldsymbol{r}}{r^2}\left(\mp\frac{ik}{r} - \frac{1}{r^2}\right). \tag{16.45}$$

Consequently,

$$\boldsymbol{j}^{(\pm)}(\boldsymbol{r}) = \left(-\frac{i\hbar}{2m}\right)\left(\frac{1}{4\pi}\right)\frac{\boldsymbol{r}}{r^2}\left(\pm\frac{2ik}{r}\right) = \pm\left(\frac{\hbar k}{m}\right)\left(\frac{1}{4\pi}\right)\frac{\boldsymbol{r}}{r^3}, \tag{16.46}$$

or

$$\boldsymbol{j}^{(\pm)}(\boldsymbol{r}) = \pm\frac{|\boldsymbol{v}|}{4\pi}\frac{\boldsymbol{r}}{r^3}, \tag{16.47}$$

if we set $|\boldsymbol{v}| = \hbar k/m$, which can be interpreted as the velocity of the particle. The above equation clearly shows that $\psi^{(+)}(\boldsymbol{r})$ and $\psi^{(-)}(\boldsymbol{r})$ are **outgoing** and **incoming** spherical waves, respectively, giving rise to the current densities $\boldsymbol{j}^{(+)}(\boldsymbol{r})$ and $\boldsymbol{j}^{(-)}(\boldsymbol{r})$ (number of particles crossing per unit area of spherical surfaces of radius r per unit time), with the $(+)$ sign for particles travelling outwards and the $(-)$ sign for particles travelling inwards (Fig. 16.3).

Let us consider (16.19) again, which gives the general solution for the wave function describing a particle coming in from infinity and scattered by a potential of range R. The coefficient $A(k)$ is the amplitude of the incoming wave and is clearly independent of the potential. It only depends on the flux of the particles "aimed" at $r = 0$ from infinity, that is, on the preparation of the initial state. $S(k)$, on the other hand, called the **scattering matrix**, is determined by the scattering power of the potential. In order to isolate the part of the wave function which appears as a consequence of the scattering, we note that if the potential vanishes, then the boundary condition $\chi_{kl} \to 0$ as $r \to 0$ dictates that the free-particle wave function is uniquely given by

$$\chi_k(r) = \chi_k^{(-)}(r) - \chi_k^{(+)}(r) = e^{-ikr} - e^{ikr} = -2i \sin kr \xrightarrow[r\to 0]{} -2ikr\,, \quad (16.48)$$

which conforms with (16.18). We then rewrite (16.19) as

$$\chi_k(r) = A(k)(\chi_k^{(-)} - \chi_k^{(+)} - (S(k)-1)\chi_k^{(+)})\,. \quad (16.49)$$

The first two terms inside the brackets then coincide with the wave function for free motion. The last term therefore describes the scattered wave. Assuming conservation of the number of particles due to scattering we require that $|S(k)| = 1$, so we can write in general

$$S(k) = e^{2i\delta(k)}\,, \quad (16.50)$$

where $\delta(k)$ is a real function. It is in fact the scattering phase shift already introduced in (16.38). This can be seen by using (16.50) to manipulate (16.19) as follows:

$$\begin{aligned} \chi_k(r) &= A(k)(\chi_k^{(-)} - e^{2i\delta}\chi_k^{(+)}) = A(k)e^{i\delta}(e^{-ikr}e^{-i\delta} - e^{ikr}e^{i\delta}) \\ &= -A(k)e^{i\delta}(e^{i(kr+\delta)} - e^{-i(kr+\delta)}) \quad (16.51) \\ &= -2iA(k)e^{i\delta(k)}\sin(kr+\delta(k))\,, \end{aligned}$$

which agrees with (16.38).

Problem 16.2 Consider the *hard-sphere scattering problem* (in three-dimensional space) where the spherically symmetric potential is given by

$$V(\boldsymbol{x}) = V(r) = \begin{cases} \infty\,, & 0 \le r \le a\,, \\ 0\,, & r > a\,. \end{cases}$$

Show that for $l = 0$, the scattering phase shift $\delta(k)$ is given by

$$\tan\delta(k) = \frac{j_0(ka)}{n_0(ka)}\,,$$

where $j_0(x)$ and $n_0(x)$ are the zeroth order **spherical Bessel function** and **spherical Neumann function**, respectively, with the asymptotic properties

$$j_0(x) \xrightarrow[x\to\infty]{} \frac{\sin x}{x}\,, \qquad n_0(x) \xrightarrow[x\to\infty]{} -\frac{\cos x}{x}\,.$$

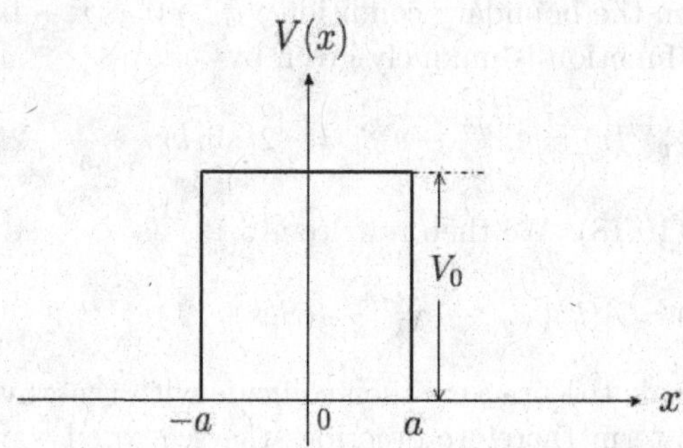

Fig. 16.4

To gain further understanding on the significance and physical meaning of the incoming and outgoing waves $\chi_k^{(-)}$ and $\chi_k^{(+)}$, respectively, we will finally consider the one-dimensional rectangular barrier potential (Fig. 16.4):

$$V(x) = \begin{cases} V_0 \,, & -a < x < a \,, \\ 0 \,, & x < -a, \; x > a \,. \end{cases} \tag{16.52}$$

(We have used x for the spatial coordinate instead of r here to distinguish the present context of a one-dimensional problem in contrast to the earlier situation where r is the radial coordinate in a three-dimensional situation, so that the boundary condition for $r \to 0$ used before no longer applies here.) We will consider the scenario where a wave is incident on the barrier from the left (Fig. 16.4) and then partially reflected and transmitted. The dynamics of this situation can be better described by using the solution to the time-dependent Schrödinger equation [cf. (1.32)]:

$$-\frac{\hbar^2}{2m}\frac{d^2\psi}{dx^2} + V(x)\psi = i\hbar\frac{d\psi}{dt} \,. \tag{16.53}$$

The solution in the regions where $V(x) = 0$ (outside the barrier) can be written as

$$\psi_E(x,t) = \begin{cases} \left(e^{ikx} - A(k)e^{-ikx}\right) e^{-iEt/\hbar} \,, & x < -a \,, \\ B(k)e^{ikx - iEt/\hbar} \,, & x > a \,, \end{cases} \tag{16.54}$$

where $E = \hbar^2 k^2/(2m)$. This solution describes an incident wave (of unit amplitude) and a reflected wave to the left of the barrier, and a transmitted wave

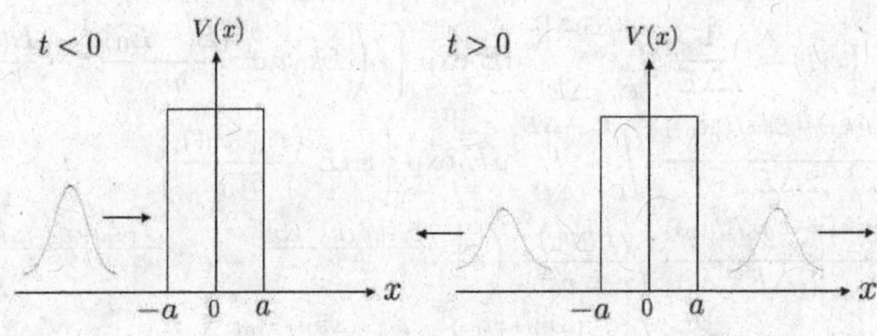

Fig. 16.5

to the right. $A(k)$ and $B(k)$ are called the **reflection coefficient** and **transmission coefficient**, respectively. To describe the more physical situation of a particle incident from the left and the subsequent state of affairs, we can build **localized wave packets** from the plane waves of (16.54). In what follows we will demonstrate that the wave packet

$$\psi(x,t) = \frac{1}{2\Delta E}\int_{E_0-\Delta E}^{E_0+\Delta E} \psi_E(x,t)dE\,, \quad (\Delta E \ll E_0) \tag{16.55}$$

will describe the time-evolution picture where, for $t < 0$, a packet is incident on the potential barrier from the left, and for $t > 0$, there will only a a reflected packet on the left and a transmitted packet on the right (Fig. 16.5).

Using the dispersion relation $E = \hbar^2k^2/(2m)$ we first make the expansion

$$k = k_0 + (E - E_0)\left.\frac{dk}{dE}\right|_{E_0} + \cdots \simeq k_0 + \frac{(E-E_0)}{\hbar v_0}\,, \tag{16.56}$$

where in the last (approximate) equality we have made use of the fact that $dk/dE = m/(\hbar^2 k) = 1/(\hbar v)$ and set $v_0 \equiv \hbar k_0/m$. Since ΔE is small, we assume that the k-dependences of $A(k)$ and $B(k)$ can be ignored in the range $k_0 - \Delta E < E < E_0 + \Delta E$, and put $A(k) \simeq A(k_0)$, $B(k) \simeq B(k_0)$. Then we have

$$\psi(x,t) = \begin{cases} \varphi^{(+)}(x,t) - A(k_0)\,\varphi^{(-)}(x,t)\,, & x < -a\,, \\ B(k_0)\,\varphi^{(+)}(x,t)\,, & x > a\,, \end{cases} \tag{16.57}$$

where

$$\varphi^{(\pm)}(x,t) \equiv \frac{1}{2\Delta E}\int_{E_0-\Delta E}^{E_0+\Delta E} dE\, e^{i(\pm kx - Et/\hbar)} \,. \tag{16.58}$$

Carrying out the integration after using (16.56) for k in the integrand, we have

$$\begin{aligned}
\varphi^{(\pm)}(x,t) &= \frac{1}{2\Delta E}\int_{E_0-\Delta E}^{E_0+\Delta E} dE\, \exp\left\{i\left(\pm k_0 x \pm \frac{(E-E_0)}{\hbar v_0}x - \frac{Et}{\hbar}\right)\right\} \\
&= \frac{e^{\pm i(k_0 x - E_0 x/(\hbar v_0))}}{2\Delta E}\int_{E_0-\Delta E}^{E_0+\Delta E} dE\, \exp\left\{\pm iE\,\frac{(x\mp v_0 t)}{\hbar v_0}\right\} \\
&= \frac{e^{\pm i(k_0 x - E_0 x/(\hbar v_0))}}{2i\Delta E}\,\frac{(\pm\hbar v_0)}{(x\mp v_0 t)}\left(e^{\pm i\frac{(x\mp v_0 t)(E_0+\Delta E)}{\hbar v_0}} - e^{\pm i\frac{(x\mp v_0 t)(E_0-\Delta E)}{\hbar v_0}}\right) \\
&= \frac{e^{i(\pm k_0 x - E_0 t/\hbar)}}{\pm\left(\frac{(\Delta E)(x\mp v_0 t)}{\hbar v_0}\right)}\left(\frac{e^{\pm i\frac{(\Delta E)(x\mp v_0 t)}{\hbar v_0}} - e^{\mp i\frac{(\Delta E)(x\mp v_0 t)}{\hbar v_0}}}{2i}\right).
\end{aligned} \tag{16.59}$$

Thus

$$\varphi^{(\pm)}(x,t) = e^{i(\pm k_0 x - E_0 t/\hbar)}\,\frac{\sin\left(\frac{(\Delta E)(x\mp v_0 t)}{\hbar v_0}\right)}{\left(\frac{(\Delta E)(x\mp v_0 t)}{\hbar v_0}\right)} \,. \tag{16.60}$$

The above functions are appreciably different from zero only when $x \mp v_0 t \simeq 0$. $\varphi^{(+)}(x,t)$, with the argument of the sine function proportional to $x - v_0 t$, describes a rightward-moving packet with velocity v_0, while $\varphi^{(-)}(x,t)$, with the corresponding argument proportional to $x + v_0 t$, describes a leftward-moving packet with the same velocity. Consider a wave packet initially incident from the left at $t < 0$. The right-moving packet $\varphi^{(+)}(x,t)$ is then appreciably different from zero only for $t < 0$ and $x < -a$ such that $-|x| + v_0|t| \simeq 0$ (incident packet), and for $t > 0$ and $x > a$ such that $x - v_0 t \simeq 0$ (transmitted packet). On the other hand, the left-moving packet $\varphi^{(-)}(x,t)$ is appreciably different from zero only for $t > 0$ and $x < -a$ such that $-|x| + v_0 t \simeq 0$ (reflected packet). This is exactly the situation depicted in Fig. 16.5.

In Chapters 35 and 36, we will develop more systematically the basic ideas in **scattering theory** introduced in this chapter. In Chapter 37, these ideas will be illustrated by means of a specific example.

Problem 16.3 A particle of mass μ is moving in the xy-plane in a circular orbit of radius ρ centered at the origin, but is otherwise free. Determine the energy eigenvalues and eigenfunctions.

Problem 16.4 A particle is confined in a two-dimensional square box whose sides are of length L and are oriented along the x and y axes with one corner at the origin.

Find the energy eigenvalues and eigenfunctions, and calculate the energy density of eigenstates (number of eigenstates per unit energy) for high energies.

Chapter 17

Symmetry, Invariance, and Conservation in Quantum Mechanics

It is a basic theme in physics that symmetries of certain attributes of a physical system lead to invariance of the physical laws describing the system. The invariance in turn gives rise to conserved quantities, important and familiar examples of which are linear and angular momentum, and energy. In modern physics, symmetry and invariance have been elevated to the status of a fundamental requirement in the search for the basic principles of physics. In Chapter 41 we will consider the all important principle of **gauge invariance**, which forms the basis of the laws governing the fundamental forces of nature. Starting with this chapter, we will introduce the more familiar spacetime symmetries of a physical system, known as Lorentz and Poincaré symmetries in a relativistic setting. More specifically, we will study time translation symmetries, and spatial symmetries such as spatial translations and rotations, which are all special cases of Lorentz symmetries.

We begin by considering the transformation of a scalar wave function $\psi(\boldsymbol{x})$ (considered as a **scalar field**, as opposed to general **tensor** and **spinor** fields) under rotations. We will use the so-called *active viewpoint*, where the physical system itself is rotated with the coordinate frame fixed. (The *passive viewpoint*, where the coordinate frame is rotated instead, but with the system fixed, can be used equally well.) For example, in Fig. 17.1, a system is rotated about the z-axis in the positive (counter-clockwise) sense by an angle θ. The effect of the rotation on a position vector $\boldsymbol{x}$ will be denoted by g:

$$\boldsymbol{x} \longrightarrow \boldsymbol{x}' = g\boldsymbol{x} \,. \tag{17.1}$$

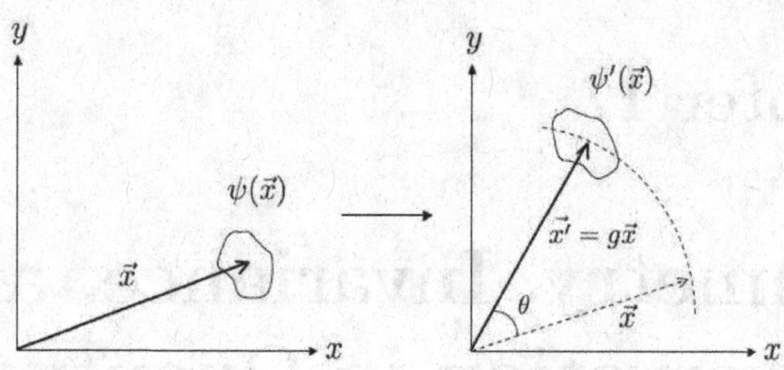

Fig. 17.1

In the example in Fig. 17.1, g is represented by a 3×3 matrix:

$$\boldsymbol{x}' = \begin{pmatrix} x' \\ y' \\ z' \end{pmatrix} = \begin{pmatrix} \cos\theta & -\sin\theta & 0 \\ \sin\theta & \cos\theta & 0 \\ 0 & 0 & 1 \end{pmatrix} \begin{pmatrix} x \\ y \\ z \end{pmatrix} , \tag{17.2}$$

where (x', y', z') are the new coordinates of the rotated point with respect to a fixed frame. g is in fact an element of the **group** $SO(3)$, the special orthogonal group of dimension 3, or the group of 3×3 orthogonal matrices with determinants equal to one. (The mathematical concept of a group will be introduced more formally in the following chapter.) Each matrix in this group uniquely describes a rotation in 3-dimensional Euclidean space. Suppose that, under the rotation g, the wave function transforms as

$$\psi(\boldsymbol{x}) \longrightarrow \psi'(\boldsymbol{x}) , \tag{17.3}$$

where $\psi'(\boldsymbol{x})$ is a *different* function of the *same* variables $\boldsymbol{x} = (x, y, z)$. Then it is obvious from Fig. 17.1, if we imagine the "glob" to represent the probability amplitude of some particle, that

$$\psi'(\boldsymbol{x}') = \psi(\boldsymbol{x}) , \tag{17.4}$$

that is, from (17.1), $\psi'(g\boldsymbol{x}) = \psi(\boldsymbol{x})$, or

$$\boxed{\psi'(\boldsymbol{x}) = \psi(g^{-1}\boldsymbol{x})} , \tag{17.5}$$

where g^{-1} is the inverse of g:

$$g^{-1}g = gg^{-1} = 1 \,. \tag{17.6}$$

Equation (17.5) gives the general transformation rule for a scalar field under a group acting on the domain space of the field. It is easily seen that for g given by the 3×3 matrix in (17.2), the inverse is given by

$$g^{-1} = \begin{pmatrix} \cos\theta & \sin\theta & 0 \\ -\sin\theta & \cos\theta & 0 \\ 0 & 0 & 1 \end{pmatrix} . \tag{17.7}$$

Problem 17.1 Under the one-dimensional translation $x \to x - 2$ a wave function transforms as $\psi(x) \to \psi'(x)$. Write an expression for $\psi'(x)$ in terms of ψ. If

$$\psi(x) = Ae^{-\alpha(x-3)^2} ,$$

what is $\psi'(x)$? Plot $\psi(x)$ and $\psi'(x)$.

The transformation of the wave function $\psi(\boldsymbol{x})$, when the latter is viewed as a vector in a Hilbert space $\mathcal{H}$, gives a **representation** of the group $SO(3)$ in terms of an operator $U(g)$ on $\mathcal{H}$:

$$\psi(\boldsymbol{x}) \longrightarrow \psi'(\boldsymbol{x}) = U(g)\psi(\boldsymbol{x}) \,. \tag{17.8}$$

We will give a more detailed discussion of the important mathematical concept of group representations in the next chapter. In (17.8), $U(g)$ is a linear operator on $\mathcal{H}$ corresponding to the group element g, and obviously has to satisfy the following normalization requirement for the wave functions $\psi(\boldsymbol{x})$ and $\psi'(\boldsymbol{x})$:

$$\int d^3x\, \psi^*(\boldsymbol{x})\psi(\boldsymbol{x}) = \int d^3x\, \psi'^*(\boldsymbol{x})\psi'(\boldsymbol{x}) \,, \tag{17.9}$$

or

$$\langle\, \psi \,|\, \psi \,\rangle = \langle\, U(g)\psi \,|\, U(g)\psi \,\rangle = \langle\, \psi \,|\, U^\dagger U \,|\, \psi \,\rangle \,. \tag{17.10}$$

The above equation implies $U^\dagger U = 1$, or

$$(U(g))^\dagger = (U(g))^{-1} \,, \quad \text{for all } g \in SO(3) \,, \tag{17.11}$$

that is, $U(g)$ is **unitary**. We say that $U(g)$ gives a **unitary representation** of the rotation group $SO(3)$.

Under the group action $\boldsymbol{x} \to \boldsymbol{x}' = g\boldsymbol{x}$, an observable A undergoes a corresponding transformation $A \to A'$, which is determined by the physical requirement of the invariance of the expectation value of A:

$$\langle \psi \,|\, A \,|\, \psi \rangle = \langle \psi' \,|\, A' \,|\, \psi' \rangle = \langle \psi \,|\, U^\dagger A' U \,|\, \psi \rangle \,. \tag{17.12}$$

Thus $A = U^\dagger A' U$, or

$$A' = U A U^\dagger \,. \tag{17.13}$$

A specially important case is when A is invariant under the above transformation, that is, when $UAU^\dagger = UAU^{-1} = A$. In this case, $AU = UA$, or $[A, U] = 0$, that is, A commutes with $U(g)$ for all group elements g in a particular group.

In non-relativistic quantum mechanics, the spatial transformations (those not involving time), and internal transformations (those not involving spacetime) play a crucial role. *The unitary transformations that leave the equations of motion (the Schrödinger equation or the Heisenberg equation) invariant are called* ***symmetry transformations***. The following is a basic fact concerning these transformations.

Theorem 17.1. *The necessary and sufficient condition for a unitary transformation to be a symmetry transformation is that it commutes with the Hamiltonian.*

Proof. We will prove the theorem for the Heisenberg equation of motion

$$\frac{dA}{dt} = \frac{[\,A\,,\,H\,]}{i\hbar} + \frac{\partial A}{\partial t} \tag{17.14}$$

for an observable A, and under the condition that the unitary operator $U(g)$ does not depend on time.

Suppose U commutes with H, that is $[\,U\,,\,H\,] = 0$. Then $H = U^{-}HU$, from which it follows that

$$[\,H\,,\,U^{-1}\,] = [\,H\,,\,U^\dagger\,] = 0\,. \tag{17.15}$$

We have

$$\begin{aligned} \frac{dA'}{dt} &= \frac{d(UAU^\dagger)}{dt} = U\,\frac{dA}{dt}\,U^\dagger = \frac{U\,[\,A\,,\,H\,]\,U^\dagger}{i\hbar} + U\,\frac{\partial A}{\partial t}\,U^\dagger \\ &= \frac{U\,[\,A\,,\,H\,]\,U^\dagger}{i\hbar} + \frac{\partial A'}{\partial t}\,. \end{aligned} \tag{17.16}$$

Using our hypothesis that U commutes with H and its consequence (17.15), we see that

$$U\,[\,A\,,\,H\,]\,U^\dagger = U\,(AH - HA)\,U^\dagger = (UAU^\dagger)\,H - H\,(UAU^\dagger) = A'H - HA'\,. \tag{17.17}$$

Thus the Heisenberg equation of motion (17.14) is transformed, under the action of U, into

$$\frac{dA'}{dt} = \frac{[\,A'\,,\,H\,]}{i\hbar} + \frac{\partial A'}{\partial t}\,, \tag{17.18}$$

which has exactly the same form as (17.14). Hence the Heisenberg equation of motion is invariant under the unitary transformation U.

Conversely, suppose that U is a symmetry transformation, that is, the Heisenberg equation of motion (17.14) and the Schrödinger equation are both invariant under U. With $A' = UAU^\dagger$, the transformed Heisenberg equation of motion (17.18) can be recast, on multiplying both sides on the left by $U^\dagger$ and on the right by U, as the following equation in A:

$$\frac{dA}{dt} = \frac{U^\dagger\,[\,UAU^\dagger\,,\,H\,]\,U}{i\hbar} + \frac{\partial A}{\partial t}\,. \tag{17.19}$$

But

$$\begin{aligned} U^\dagger\,[\,UAU^\dagger\,,\,H\,]\,U &= U^\dagger\,(UAU^\dagger H - HUAU^\dagger)\,U \\ &= A\,(U^\dagger HU) - (U^\dagger HU)\,A\,. \end{aligned} \tag{17.20}$$

Thus

$$\frac{dA}{dt} = \frac{[\,A\,,\,H'\,]}{i\hbar} + \frac{\partial A}{\partial t}\,, \tag{17.21}$$

where

$$H' \equiv U^\dagger HU \tag{17.22}$$

also appears as the Hamiltonian of the system. Note the difference between (17.22) and (17.13). Since A also satisfies (17.14), Eq. (17.21) implies $[\,A\,,\,H'\,] = [\,A\,,\,H\,]$, or

$$[\,A\,,\,H' - H\,] = 0 \tag{17.23}$$

for all observables, and indeed, for all operators A. By a theorem of linear algebra, the vanishing of the commutator (17.23) implies that $H' - H$ must be proportional to the identity operator:

$$H' - H = U^\dagger HU - H = \lambda I\,, \tag{17.24}$$

where λ is a constant. For an arbitrary normalized state vector $|\,\psi\,\rangle$, the expectation values of the Hamiltonians H' and H, which must describe the same system on account of the simultaneous validity of (17.14) and (17.21), are then related by

$$\langle\,\psi\,|\,H'\,|\,\psi\,\rangle = \langle\,\psi\,|\,H\,|\,\psi\,\rangle + \lambda\,. \tag{17.25}$$

Thus the spectra of H' and H just differ by a constant offset λ in the energy scale, which has no observable physical consequences. On requiring this offset to be zero ($\lambda = 0$), which is equivalent to maintaining the same zero of the potential energy under the transformation U, we have, from (17.24), $U^\dagger HU - H = 0$, which implies $[\,H\,,\,U\,] = 0$.

Note: A more direct argument to show that a symmetry transformation U necessarily commutes with the Hamiltonian H can be provided by using the invariance of the time-independent Schrödinger equation $H\,|\,\psi\,\rangle = E\,|\,\psi\,\rangle$ under U. Indeed, by the Spectral Theorem (Theorem 9.2), the Hamiltonian H, being a self-adjoint operator, possesses a complete set of eigenvectors. Denote these by $|\,\psi_i\,\rangle$, so that

$$H\,|\,\psi_i\,\rangle = E_i\,|\,\psi_i\,\rangle\,. \tag{17.26}$$

(In general, part of the spectrum of H may be continuous, but this fact does not affect the argument below, so, for ease of presentation, we will assume that the spectrum is discrete.) Expand an arbitrary vector $|\,\psi\,\rangle$ in the Hilbert space of the system in terms of the $|\,\psi_i\,\rangle$:

$$|\,\psi\,\rangle = \sum_i c_i\,|\,\psi_i\,\rangle\,. \tag{17.27}$$

Then

$$\begin{aligned}(HU - UH)\,|\,\psi\,\rangle &= \sum_i c_i(HU - UH)\,|\,\psi_i\,\rangle \\ = \sum_i c_i\,(HU\,|\,\psi_i\,\rangle - E_iU\,|\,\psi_i\,\rangle) &= \sum_i c_i\,(H\,|\,\psi_i'\,\rangle - E_i\,|\,\psi_i'\,\rangle) = 0\,,\end{aligned} \tag{17.28}$$

where the last equality is due to the fact that the Schrödinger equation $H\,|\,\psi_i\,\rangle = E_i\,|\psi_i\,\rangle$ is invariant under U. It follows that $[\,H\,,\,U\,] = 0$. □

As a general rule, symmetry transformations lead to conserved observables, or **constants of motion**. Let us first consider **discrete transformations** (those not depending on continuous parameters) that satisfy the condition

$$U^2 = 1\,. \tag{17.29}$$

Examples of this class of transformations are parity, charge conjugation, and exchange of two identical particles. The above equation and unitarity of U imply that $U = U^\dagger$, that is, U is self-adjoint. Hence U is an observable. Also, since it is a symmetry transformation, it commutes with the Hamiltonian H by Theorem 17.1. Thus, the Heisenberg equation of motion

$$\frac{dU}{dt} = \frac{[\,U\,,\,H\,]}{i\hbar} \tag{17.30}$$

implies that U is independent of time, that is, U is a constant of motion. We have an analogous situation in classical mechanics, where the Poisson bracket plays the role of the commutator [cf. (2.11)].

Let us now consider continuous unitary transformations, in geneal depending on n continuous real parameters $\boldsymbol{\alpha} = (\alpha_1, \ldots, \alpha_n)$ (for example, the three Euler angles in three-dimensional rotations), such that

$$U(\boldsymbol{\alpha} = 0) = 1\,, \tag{17.31}$$

where the "1" on the right-hand side stands for the identity operator. Near the neighborhood of $\boldsymbol{\alpha} = 0$ in parameter-space, we may expand $U(\boldsymbol{\alpha})$ in terms of small parameters $\delta\boldsymbol{\alpha} = (\delta\alpha_1, \ldots, \delta\alpha_n)$:

$$U(\delta\boldsymbol{\alpha}) = 1 + i\,\delta\boldsymbol{\alpha}\cdot\boldsymbol{G} + O(\delta\alpha_i^2)\,, \tag{17.32}$$

where $\delta\boldsymbol{\alpha}\cdot\boldsymbol{G} = (\delta\alpha_1)\,G_1 + \cdots + (\delta\alpha_n)\,G_n$ and $\boldsymbol{G} = (G_1, \ldots, G_n)$ represents n linear operators. (The reason for the introduction of the imaginary unit i will become clear shortly.) Considering only terms up to first order in $\delta\alpha_i$, we have

$$U(\delta\boldsymbol{\alpha}) \simeq 1 + i\,\delta\boldsymbol{\alpha}\cdot\boldsymbol{G}\,, \qquad U^\dagger(\delta\boldsymbol{\alpha}) \simeq 1 - i\,\delta\boldsymbol{\alpha}\cdot\boldsymbol{G}^\dagger\,. \tag{17.33}$$

Unitarity of U requires that

$$UU^\dagger = (1 + i\,\delta\boldsymbol{\alpha}\cdot\boldsymbol{G})(1 - i\,\delta\boldsymbol{\alpha}\cdot\boldsymbol{G}^\dagger) \simeq 1 + i\,\delta\boldsymbol{\alpha}\cdot(\boldsymbol{G} - \boldsymbol{G}^\dagger) = 1\,, \tag{17.34}$$

which implies that

$$G_i = G_i^\dagger\,, \tag{17.35}$$

that is, the G_i's are self-adjoint. Now, since U is a symmetry transformation, $[\,U\,,\,H\,] = 0$ by Theorem 17.1. It follows that

$$[\,G_i\,,\,H\,] = 0\,, \quad i = 1, \ldots, n\,, \tag{17.36}$$

also. Thus the G_i's are all symmetry transformations, again by Theorem 17.1. Further, because they are self-adjoint, they represent observables which are constants of motion, by virtue of Heisenberg's equation of motion. Because of (17.32) we say that the G_i's are the **infinitesimal generators** of the unitary transformation U. Mathematically, they form a basis of the **Lie algebra** (which is also a linear vector space) of the symmetry group represented by U, with the commutator [,] being the multiplication rule of the algebra [recall the discussion immediately after Eq. (12.13)].

In the following table, we list the most important space-time symmetry transformations in non-relativistic quantum mechanics, with the corresponding conserved observables.

symmetry transformation	conserved observable
time translation	Hamiltonian (total energy)
spatial translation	linear momentum
spatial rotation	angular momentum
spatial reflection	parity

Of the above, spatial reflection is a discrete symmetry transformation satisfying (17.29), while the other three are continuous symmetry transformations satisfying (17.34). We will now discuss each of the three continuous transformations in more detail.

Consider the time translation [cf. (17.1)]

$$t \longrightarrow t' = t + \tau = gt\,. \tag{17.37}$$

Equations (17.5) and (17.8) for this case then read

$$\psi(t) \longrightarrow \psi'(t) = \psi(t-\tau) = U(\tau)\psi(t)\,. \tag{17.38}$$

Now consider the infinitesimal transformation [cf. (17.32)]

$$U(\delta\tau) \simeq 1 + i\,(\delta\tau)G\,. \tag{17.39}$$

From (17.38) we then have by Taylor expansion

$$(1 + i\,(\delta\tau)\,G)\,\psi(t) = \psi(t-\delta\tau) \simeq \psi(t) - (\delta\tau)\,\frac{\partial\psi}{\partial t}\,. \tag{17.40}$$

Hence

$$i\,(\delta\tau)\,G\psi(t) = -(\delta\tau)\,\frac{\partial\psi}{\partial t}\,, \tag{17.41}$$

or $G\psi(t) = i\,\partial\psi/\partial t$. If we let $\hbar G \equiv H$ and identify H as the Hamiltonian, we immediately have the time-dependent Schrödinger equation $H\psi = i\hbar\,\partial\psi/\partial t$. Equation (17.39) can also be written as

$$U(\delta\tau) \simeq 1 + \frac{i}{\hbar}(\delta\tau)\,H\,. \tag{17.42}$$

The Hamiltonian is thus the infinitesimal generator of and the conserved dynamical variable corresponding to time translations.

Since time translation is an additive group [cf. (17.37)], $U(t+\delta t) = U(\delta t)U(t)$. (The justification of this equation actually requires an understanding of the mathematical concepts of group homomorphisms and group representations, which we will present in the next chapter.) Equation (17.42) then gives

$$U(t+\delta t) \simeq \left(1 + \frac{i}{\hbar}\,(\delta t)\,H\right) U(t)\,. \tag{17.43}$$

Thus $U(t)$ satisfies the differential equation

$$\frac{dU(t)}{dt} = \frac{i}{\hbar}\,HU\,, \tag{17.44}$$

which has the solution

$$\boxed{U(t) = \exp\left(\frac{i}{\hbar}\,Ht\right)}\,. \tag{17.45}$$

The spatial translations can be dealt with in largely the same manner as time translations. Consider a spatial translation

$$\boldsymbol{x} \longrightarrow \boldsymbol{x}' = \boldsymbol{x} + \boldsymbol{a} = g\,\boldsymbol{x}\,. \tag{17.46}$$

Then, according to (17.5) and (17.8), the (scalar) wave function transforms as

$$\psi(\boldsymbol{x}) \longrightarrow \psi'(\boldsymbol{x}) = \psi(\boldsymbol{x} - \boldsymbol{a}) = U(\boldsymbol{a})\psi(\boldsymbol{x})\,. \tag{17.47}$$

For the infinitesimal translation $\boldsymbol{x} \to \boldsymbol{x}' = \boldsymbol{x}+\delta\boldsymbol{a}$, we obtain by Taylor expansion

$$\begin{aligned} U(\delta\boldsymbol{a})\psi(\boldsymbol{x}) &= \psi(\boldsymbol{x}-\delta\boldsymbol{a}) = \psi(\boldsymbol{x}) - (\delta\boldsymbol{a})\cdot\nabla\psi + O(\delta a_i^2) \\ &= \left(1 - \frac{i}{\hbar}(\delta\boldsymbol{a})\cdot(-i\hbar\nabla) + \dots\right)\psi(\boldsymbol{x}) , \end{aligned} \tag{17.48}$$

that is,

$$U(\delta\boldsymbol{a}) \simeq 1 - \frac{i}{\hbar}(\delta\boldsymbol{a})\cdot\boldsymbol{p} + \dots \ , \tag{17.49}$$

where we have identified

$$\boldsymbol{p} = -i\hbar\nabla \tag{17.50}$$

as the linear momentum operator. Since the spatial translations form an additive group, we again have

$$\begin{aligned} U(\boldsymbol{a}+\delta\boldsymbol{a})\,\psi(\boldsymbol{x}) &= \psi(\boldsymbol{x}-\boldsymbol{a}-\delta\boldsymbol{a}) = \psi(\boldsymbol{x}-\delta\boldsymbol{a}-\boldsymbol{a}) = U(\boldsymbol{a})\,\psi(\boldsymbol{x}-\delta\boldsymbol{a}) \\ &= U(\boldsymbol{a})U(\delta\boldsymbol{a})\,\psi(\boldsymbol{x}) = U(\boldsymbol{a})\left(1 - \frac{i}{\hbar}(\delta\boldsymbol{a})\cdot\boldsymbol{p}\right)\psi(\boldsymbol{x}) . \end{aligned} \tag{17.51}$$

Hence

$$U(\boldsymbol{a}+\delta\boldsymbol{a}) - U(\boldsymbol{a}) = -\frac{i}{\hbar}U(\boldsymbol{a})\,(\delta\boldsymbol{a})\cdot\boldsymbol{p} . \tag{17.52}$$

On letting $\delta\boldsymbol{a} \to 0$ we arrive at a differential equation for $U(\boldsymbol{a})$ whose solution is

$$\boxed{U(\boldsymbol{a}) = \exp\left(-\frac{i}{\hbar}\,\boldsymbol{a}\cdot\boldsymbol{p}\right)} \ . \tag{17.53}$$

The components of the linear momentum operator are thus infinitesimal generators of and the conserved conserved dynamical variables corresponding to spatial translations. Since spatial translations are commutative, the unitary operators $U(\boldsymbol{a})$ commute, and so do the linear momentum operators:

$$\boxed{[p_i\,, p_j\,] = 0\,, \quad (i,j=1,2,3)} \ . \tag{17.54}$$

Finally we will study the more involved case of spatial rotations. Consider the rotation about the z-axis given by (17.2) (cf. Fig. 17.1). For an infinitesimal angle $\delta\theta$, Eq. (17.7) gives

$$g^{-1}\,\boldsymbol{x} = \begin{pmatrix} \cos(\delta\theta) & \sin(\delta\theta) & 0 \\ -\sin(\delta\theta) & \cos(\delta\theta) & 0 \\ 0 & 0 & 1 \end{pmatrix}\begin{pmatrix} x \\ y \\ z \end{pmatrix} \simeq \begin{pmatrix} 1 & \delta\theta & 0 \\ -\delta\theta & 1 & 0 \\ 0 & 0 & 1 \end{pmatrix}\begin{pmatrix} x \\ y \\ z \end{pmatrix} = \begin{pmatrix} x+(\delta\theta)\,y \\ y-(\delta\theta)\,x \\ z \end{pmatrix} . \tag{17.55}$$

Hence it follows from (17.5) and (17.8) that, under this infinitesimal rotation,

$$\psi'(\boldsymbol{x}) = U_z(\delta\theta)\,\psi(\boldsymbol{x}) = \psi(x + (\delta\theta)\,y\,,\, y - (\delta\theta)\,x\,,\, z)$$
$$\simeq \psi(x,y,z) + (\delta\theta)\,y\,\frac{\partial\psi}{\partial x} - (\delta\theta)\,x\,\frac{\partial\psi}{\partial y} = \psi(\boldsymbol{x}) + (\delta\theta)\left(y\frac{\partial}{\partial x} - x\frac{\partial}{\partial y}\right)\psi(\boldsymbol{x})\,. \tag{17.56}$$

We can then write

$$U_z(\delta\theta) \simeq 1 - i\frac{(\delta\theta)}{\hbar}\,L_z\,, \tag{17.57}$$

where

$$L_z = -i\hbar\left(x\frac{\partial}{\partial y} - y\frac{\partial}{\partial x}\right) = xp_y - yp_x \tag{17.58}$$

can be identified as the z-component of the angular momentum operator. Analogous to (17.53) we have

$$U_z(\theta) = \exp\left(-\frac{i}{\hbar}\,\theta\,\hat{\boldsymbol{z}}\cdot\boldsymbol{L}\right)\,, \tag{17.59}$$

where $\hat{\boldsymbol{z}}$ is the unit vector along the z-axis. For a general rotation about the axis $\hat{\boldsymbol{n}}$ by an angle θ, we have

$$\boxed{U(\hat{\boldsymbol{n}},\theta) = \exp\left(-\frac{i}{\hbar}\,\theta\,\hat{\boldsymbol{n}}\cdot\boldsymbol{L}\right)}\,. \tag{17.60}$$

The angular momentum operator is the infinitesimal generator of finite rotations. If the Hamiltonian H is rotationally invariant, then it commutes with U and hence with $\boldsymbol{L}$ also. By Heisenberg's equation of motion, $\boldsymbol{L}$ is a constant of motion. Thus *the angular momentum is the conserved dynamical variable corresponding to spatial rotations.*

Since rotations, as elements of the group $SO(3)$, do not commute, the different components of the angular momentum do not commute with each other. In fact, using the explicit expressions [cf. (17.58)]

$$\boxed{L_x = -i\hbar\left(y\frac{\partial}{\partial z} - z\frac{\partial}{\partial y}\right)\,,\; L_y = -i\hbar\left(z\frac{\partial}{\partial x} - x\frac{\partial}{\partial z}\right)\,,\; L_z = -i\hbar\left(x\frac{\partial}{\partial y} - y\frac{\partial}{\partial x}\right)} \tag{17.61}$$

we can show directly that

$$\boxed{[L_x\,,\,L_y] = i\hbar\,L_z\,,\quad [L_y\,,\,L_z] = i\hbar\,L_x\,,\quad [L_z\,,\,L_x] = i\hbar\,L_y}\,, \tag{17.62}$$

by using both sides to operate on an arbitrary wave function $\psi(\boldsymbol{x})$. Using the general commutation equalities (12.10) and (12.11), one can also show that

$$\boxed{[L^2\,,\,L_x] = [L^2\,,\,L_y] = [L^2\,,\,L_z] = 0}\,, \tag{17.63}$$

where

$$L^2 = L_x^2 + L_y^2 + L_z^2 \,. \tag{17.64}$$

For a rotationally invariant system, that is, one for which $[\, H \,,\, \boldsymbol{L} \,] = 0$, one usually chooses the eigenvalues of L^2 and L_z to be quantum numbers of the system [cf. (15.21) and (15.22)]. The algebra generated by these angular momentum operators and their spectrum will be studied in Chapter 20. The general mathematical theorem which (loosely) states that corresponding to each symmetry there is a conserved quantity is called **Noether's Theorem**.

Problem 17.2 Consider a particle of mass m moving in a one-dimensional lattice with spacing d. Show that the Hamiltonian

$$H = \frac{p^2}{2m} + V(x) \,,$$

where $V(x)$ is a periodic potential energy function with period d, is invariant under the group of linear translations

$$G = \{\, g(n) \mid g(n)x = x + nd \,,\; n = 0, \pm 1, \pm 2, \ldots \,\} \,.$$

Chapter 18

Why is Group Theory Useful in Quantum Mechanics?

In the previous chapter we discussed the importance of symmetries in quantum physics. Since the natural mathematical tool for the study of symmetries is group theory, we will in this chapter provide a brief introduction to the mathematics of groups, and in particular, group representations. In the process, the question posed in the title of this chapter will be answered in part. We begin with a few mathematical definitions.

Definition 18.1. *A set G with a multiplication rule that satisfies the following properties is called a* ***group:***

(1) if $g_1 \in G$ and $g_2 \in G$, then $g_1 g_2 \in G$ (closure under multiplication);

(2) $g_1(g_2 g_3) = (g_1 g_2) g_3$, for all $g_1, g_2, g_3 \in G$ (associativity);

(3) there exists an element $e \in G$, called the ***identity****, such that*

$$eg = ge = g \quad \text{for all} \quad g \in G \quad ;$$

(4) for every $g \in G$, there exists an element $g^{-1} \in G$, called the ***inverse*** *of g, such that*

$$gg^{-1} = g^{-1}g = e \quad .$$

Definition 18.2. *A* ***group homomorphism*** *from a group G to another group G' is a mapping (not necessarily one-to-one) $f : G \to G'$ which preserves group multiplication, that is, if $g_1, g_2 \in G$ and $g_1 g_2 = g_3$, then $f(g_1)f(g_2) = f(g_3)$.*

It follows immediately from this definition that if e and e' are the identity elements in G and G', respectively, and g is any element in G, then $f(e) = e'$ and $f(g^{-1}) = (f(g))^{-1}$.

Definition 18.3. *A* ***group representation*** *of a group G is a group homomorphism $f : G \to \mathcal{O}(V)$, where $\mathcal{O}(V)$ is a group of linear operators on a vector space V. A representation is said to be* ***faithful*** *if the homomorphism is also an* ***isomorphism*** *[one-to-one (injective) and onto (surjective)].*

Definition 18.4. *A subspace $\mathbb{V}_1 \subset \mathbb{V}$ is said to be an* ***invariant subspace*** *of $\mathbb{V}$ with respect to a group representation $D : G \to \mathcal{O}(\mathbb{V})$ if $v \in \mathbb{V}_1$ implies $D(g)v \in \mathbb{V}_1$ also, for all $g \in G$. An invariant subspace is said to be* ***proper*** *(or* ***minimal****) if it does not contain any invariant subspace with respect to D other than itself and $\{0\}$, that is, if it contains no non-trivial invariant subspace.*

Now let $U : G \to \mathcal{U}(\mathcal{H})$ be a unitary representation of a symmetry group G [such as $SO(3)$] on a Hilbert space $\mathcal{H}$, where $\mathcal{U}(\mathcal{H})$ is a group of unitary operators on $\mathcal{H}$. Recall that G is a **symmetry group** of the Hamiltonian H means that it leaves H invariant [cf. Theorem 17.1]:

$$U^\dagger(g) H U(g) = H \,, \quad \text{or} \quad U(g)H = HU(g) \,, \tag{18.1}$$

for all $g \in G$. Suppose $|\,\psi\,\rangle$ is an eigenvector of H with eigenvalue E, that is, $H\,|\,\psi\,\rangle = E\,|\,\psi\,\rangle$. Then, as a result of the symmetry property of G, any $U(g)|\,\psi\rangle$ is also an eigenvector of H with the same eigenvalue. Indeed

$$HU(g)|\,\psi\rangle = U(g)U^\dagger(g)HU(g)|\,\psi\rangle = U(g)H|\,\psi\rangle = U(g)E|\,\psi\rangle = E(U(g)|\,\psi\rangle) \,. \tag{18.2}$$

In general, an energy eigenvalue E may be degenerate. It is said to be r-fold degenerate if a set of r vectors $\{|\,\psi_1\rangle, \dots, |\,\psi_r\rangle\}$ in $\mathcal{H}$ span the eigenspace V_E for a particular E. The above discussion implies that if $|\,\psi\rangle \in V_E$, then $U(g)|\,\psi\rangle \in V_E$ also, for all $g \in G$. In other words, any eigensubspace V_E of $\mathcal{H}$ is an invariant subspace of $\mathcal{H}$ under $U(g)$, for all $g \in G$. We also say that *the group action of a symmetry group G leaves any energy eigensubspace invariant.*

We can easily verify that the action of $U(g)$ $(g \in G)$ on a wave function $\psi(\boldsymbol{x}) \in \mathcal{H}$ as given by (17.5) and (17.8), that is,

$$U(g)\,\psi(\boldsymbol{x}) = \psi(g^{-1}\boldsymbol{x}) \,, \tag{18.3}$$

indeed gives a representation of G. Let $g_1,\, g_2 \in G$, and, for any $\psi(\boldsymbol{x}) \in \mathcal{H}$ define a function $\psi'(\boldsymbol{x})$ by $\psi'(\boldsymbol{x}) \equiv \psi(g_2^{-1}\boldsymbol{x})$. Then, for any $\psi(\boldsymbol{x})$,

$$\begin{aligned} &U(g_1)U(g_2)\,\psi(\boldsymbol{x}) = U(g_1)\,\psi(g_2^{-1}\,\boldsymbol{x}) = U(g_1)\,\psi'(\boldsymbol{x}) = \psi'(g_1^{-1}\boldsymbol{x}) \\ &= \psi(g_2^{-1}g_1^{-1}\,\boldsymbol{x}) = \psi((g_1g_2)^{-1}\,\boldsymbol{x}) = U(g_1g_2)\,\psi(\boldsymbol{x}) \,. \end{aligned} \tag{18.4}$$

Hence

$$U(g_1)U(g_2) = U(g_1g_2) \,, \tag{18.5}$$

and so, according to Def. 18.2, $U(g)$ is a group homomorphism. The fact that $U(g)$ is a linear operator on a Hilbert space makes it a group representation also.

The usefulness of group representation theory in quantum mechanics is then based on the following consideration. If the Hamiltonian H commutes with

$U(g)$ for all $g \in G$, (in which case H is said to be invariant under the group G), then $U(g)$ gives a unitary representation of G on any eigensubspace of H. Thus, if we succeed in finding all the *inequivalent, irreducible representations* (see definitions immediately below) of a group G under which H is invariant– a completely mathematical problem– then we will have found a complete classification of all the energy eigenvalues of H, without solving the Schrödinger equation explicitly. This kind of classification using group representation theory finds important applications in much of atomic, molecular, condensed matter, and particle physics.

It remains to define the basic concepts of equivalence and irreducibility in group representation theory.

Definition 18.5. *Two representations $D_1 : G \to \mathcal{O}(\mathbb{V}_1)$ and $D_2 : G \to \mathcal{O}(\mathbb{V}_2)$ of a group G are said to be* ***equivalent group representations*** *if they are related by a similarity transformation*

$$D_2(g) = SD_1(g)S^{-1} \tag{18.6}$$

for all $g \in G$, where S is an invertible linear transformation from $\mathbb{V}_1$ to $\mathbb{V}_2$.

Definition 18.6. *A representation $D : G \to \mathcal{O}(\mathbb{V})$ of a group G on a vector space $\mathbb{V}$ is said to be* ***irreducible*** *if $\mathbb{V}$ does not contain a non-trivial invariant subspace with respect to D; otherwise D is said to be* ***reducible***. *In the latter case, if the orthogonal complement of the non-trivial invariant subspace is also invariant with respect to D, then D is said to be* ***fully reducible*** *(or* ***decomposable****).*

Expressed as a matrix (with respect to some choice of basis vectors in $\mathbb{V}$), an operator $D(g)$ of a reducible representation (up to similarity transformations) has the following block form:

$$D(g) = \begin{pmatrix} D_1(g) & D'(g) \\ 0 & D_2(g) \end{pmatrix} . \tag{18.7}$$

If the representation space $\mathbb{V}$ is n-dimensional, then $D(g)$ is an $n \times n$ matrix. If the non-trivial invariant subspace $\mathbb{V}_1$ is n_1-dimensional, $n_1 < n$, then the block $D_1(g)$ is $n_1 \times n_1$. To see the form of (18.7), we choose a basis set of $\mathbb{V}$, $\{e_1, \ldots, e_{n_1}, e_{n_1+1}, \ldots, e_n\}$ such that the subset $\{e_1, \ldots, e_{n_1}\}$ is a basis of $\mathbb{V}_1$. The fact that $\mathbb{V}_1$ is invariant implies that

$$D(g)e_i = (D(g))^j_i \, e_j \in \mathbb{V}_1 \,, \quad i = 1, \ldots, n_1 \,. \tag{18.8}$$

Thus

$$(D(g))^j_i = 0 \,, \quad i = 1, \ldots, n_1 \,; j = n_1 + 1, \ldots, n \,. \tag{18.9}$$

These are precisely elements in the lower left zero-block of the RHS of (18.7). (Our convention here for matrix elements is that in a^j_i, i is the column index and

j is the row index. The **Einstein summation convention**, in which repeated indices – one upper and one lower – are summed over, is also followed.)

A fully reducible representation matrix, up to similarity transformations, has the following block-form:

$$D(g) = \begin{pmatrix} D_1(g) & 0 \\ 0 & D_2(g) \end{pmatrix} . \tag{18.10}$$

In this case, the space $\mathbb{V}_2 = \mathbb{V}_1^{\perp}$ spanned by $\{e_{n_1+1}, \dots, e_n\}$ is also invariant under $D(g)$. Thus

$$D(g)e_i = (D(g))^j_i \, e_j \in \mathbb{V}_2 \,, \quad i = n_1 + 1, \dots, n \,. \tag{18.11}$$

This implies that

$$(D(g))^j_i = 0 \,, \quad i = n_1 + 1, \dots, n \,; \; j = 1, \dots, n_1 \,. \tag{18.12}$$

These are precisely elements in the upper right zero-block of the RHS of (18.10).

To conclude this chapter we will go back to the example (rotations on a plane) introduced at the beginning of Chapter 17. Consider the group $SO(2)$–*the special orthogonal group of dimension* 2–which is the group of 2×2 orthogonal matrices with determinants equal to one. The so-called defining representation of this one-parameter Lie group is given by

$$(D(\theta))^j_i = \begin{pmatrix} \cos\theta & -\sin\theta \\ \sin\theta & \cos\theta \end{pmatrix} , \quad 0 \le \theta < 2\pi \,, \tag{18.13}$$

where the angle θ is the single parameter of the group and the representation space is the real vector space $\mathbb{R}^2$ (the Euclidean plane). This representation is irreducible on $\mathbb{R}^2$: we cannot find an invertible linear transformation S on this space such that $\mathcal{D}(\theta) = SD(\theta)S^{-1}$ becomes a 2×2 matrix of the form (18.7) for all θ in the given range. However, there exists a fully reducible representation of $SO(2)$ on complexified $\mathbb{R}^2$, namely $\mathbb{C}^2$ – complex 2-dimensional space – whose elements consist of pairs of ordered complex numbers.

Suppose we pick an orthonormal basis in the xy-plane ($\mathbb{R}^2$):

$$|\, e_1 \rangle \equiv |\, \hat{\boldsymbol{x}} \rangle \,, \quad |\, e_2 \rangle \equiv |\, \hat{\boldsymbol{y}} \rangle \,, \tag{18.14}$$

where vectors are written using the Dirac notation, and $\hat{\boldsymbol{x}}$ and $\hat{\boldsymbol{y}}$ are the unit vectors along the x and y axes, respectively. Then the matrix elements in (18.13) can be written as [cf. (8.10)]

$$(D(\theta))^j_i = \langle\, e_j \,|\, D(\theta) \,|\, e_i \,\rangle \,, \tag{18.15}$$

where j is the row index and i the column index. We recognize that $\{ |\, e_1 \rangle, |\, e_2 \rangle \}$ is an orthonormal basis in $\mathbb{C}^2$ also. Consider the following linear transformation from $\mathbb{R}^2$ to $\mathbb{C}^2$:

$$|\,e_1'\,\rangle = \frac{1}{\sqrt{2}}\left(|\,e_1\,\rangle + i\,|\,e_2\,\rangle\right), \qquad |\,e_2'\,\rangle = \frac{1}{\sqrt{2}}\left(|\,e_1\,\rangle - i\,|\,e_2\,\rangle\right). \tag{18.16}$$

The above two equations can be written in matrix form:

$$\begin{pmatrix} |\,e_1'\,\rangle & |\,e_2'\,\rangle \end{pmatrix} = \begin{pmatrix} |\,e_1\,\rangle & |\,e_2\,\rangle \end{pmatrix} \begin{pmatrix} \frac{1}{\sqrt{2}} & \frac{1}{\sqrt{2}} \\ \frac{i}{\sqrt{2}} & -\frac{i}{\sqrt{2}} \end{pmatrix}, \tag{18.17}$$

or equivalently,

$$|\,e_j'\,\rangle = S_j^l\,|\,e_l\,\rangle\,, \tag{18.18}$$

where the Einstein summation convention has been used (the repeated index l is summed over), and the matrix (S_i^j), with the upper index being the row index and the lower index being the column index, is given explicitly by the 2×2 matrix on the right-hand side of (18.17). Recalling (7.8), (18.8) implies

$$\langle\, e_i'\,| = (S_i^k)^*\,\langle\, e_k\,|\,. \tag{18.19}$$

Thus, with respect to the new basis $\{\,|\,e_1'\,\rangle, |\,e_2'\,\rangle\}$ in $\mathbb{C}^2$, the matrix representation of $D(\theta)$ is given by

$$(D'(\theta))^i_j = \langle\, e_i'\,|\,D(\theta)\,|\,e_j'\,\rangle = (S_i^k)^*\,\langle\, e_k\,|\,D(\theta)\,|\,e_l\,\rangle\,S_j^l = (S^\dagger)^i_k\,(D(\theta))^k_l\,S^l_j\,. \tag{18.20}$$

We recognize this to be the matrix multiplication equation

$$D'(\theta) = S^\dagger\,D(\theta)\,S\,, \tag{18.21}$$

which is a similarity transformation of the matrix D induced by a change of basis of the representation space. With D given by (18.13) and S by (18.17), the matrix multiplication can be displayed explicitly as follows

$$D'(\theta) = \frac{1}{2}\begin{pmatrix} 1 & -i \\ 1 & i \end{pmatrix}\begin{pmatrix} \cos\theta & -\sin\theta \\ \sin\theta & \cos\theta \end{pmatrix}\begin{pmatrix} 1 & 1 \\ i & -i \end{pmatrix} = \begin{pmatrix} e^{-i\theta} & 0 \\ 0 & e^{i\theta} \end{pmatrix}. \tag{18.22}$$

Thus the 2-dimensional irreducible representation $D(\theta)$ of the group $SO(2)$ on the real vector space $\mathbb{R}^2$ is completely reduced to the 2 inequivalent 1-dimensional representations contained in $D'(\theta)$ on the complex vector space $\mathbb{C}^2$. These 1-dimensional representations on complex space have very useful applications in quantum physics, for example, in the description of the effects of rotations on polarization states of photons, or spinor-states of spin-1/2 particles.

Chapter 19

$SO(3)$ and $SU(2)$

The 2×2 matrix [cf. (18.22)]

$$g(\theta) = \begin{pmatrix} e^{-i\theta} & 0 \\ 0 & e^{i\theta} \end{pmatrix} , \quad \theta \in \mathbb{R} , \tag{19.1}$$

is an element of the group $SU(2)$, or the **special unitary group** of dimension 2. The term "special" (or **unimodular**) refers to the fact that all matrices in the group have determinants equal to one. We saw in the last chapter that $g(\theta)$ also constitutes a (complex) two-dimensional representation of the group $SO(2)$. The two groups $SU(2)$ and $SO(3)$ are in fact intimately related. In this chapter, we will explore this relationship, which is of fundamental importance in quantum physics.

$SO(3)$, the special orthogonal group of dimension 3, is the group of rotations in 3-dimensional Euclidean space $\mathbb{R}^3$. We recall that an invertible matrix A is orthogonal if $A^{-1} = A^T$, where A^T is the transpose of A. The condition of orthogonality is required by the fact that if $\boldsymbol{y} = R\boldsymbol{x}$, where R is a matrix that rotates a vector $\boldsymbol{x}$ to another vector $\boldsymbol{y}$, then R preserves the length of $\boldsymbol{x}$, that is, $|\,\boldsymbol{y}\,| = |\,\boldsymbol{x}\,|$, where $|\ \ |$ denotes the length of a vector. Indeed, if we write

$$\begin{pmatrix} y^1 \\ y^2 \\ y^3 \end{pmatrix} = \begin{pmatrix} R^1_1 & R^1_2 & R^1_3 \\ R^2_1 & R^2_2 & R^2_3 \\ R^3_1 & R^3_2 & R^3_3 \end{pmatrix} \begin{pmatrix} x^1 \\ x^2 \\ x^3 \end{pmatrix} , \tag{19.2}$$

then length-invariance implies

$$\sum_j x^j x^j = \sum_i y^i y^i = \sum_i R^i_j x^j \, R^i_k x^k = \sum_k R^i_j (R^T)^k_i \, x^j x^k = \sum_k (R^T R)^k_j \, x^j x^k \; , \tag{19.3}$$

where the Einstein summation convention has been used. This can only be true if $(R^T R)_j^k = \delta_j^k$, or $R^T R = 1$, or $R^T = R^{-1}$. Thus a rotation matrix must be orthogonal. Now, since the determinant of a square matrix is equal to the determinant of its transpose, and $\det(AB) = \det(A)\det(B)$, where A and B are square matrices of the same size, the condition $A^T A = 1$ (orthogonality) only implies $(\det(A))^2 = 1$, or $\det(A) = \pm 1$. The choice of $+1$ for the determinant ensures that the orthogonal transformation is also **orientation preserving**, that is, a **proper rotation**. When $\det(A) = -1$, the orthogonal transformation A corresponds to either an inversion, or an inversion composed with rotations. The group of orthogonal 3×3 matrices, of which $SO(3)$ is a subgroup, is designated $O(3)$.

To see the relationship between $SO(3)$ and $SU(2)$ the first step is to realize that a vector $\boldsymbol{x} = (x^1, x^2, x^3) \in \mathbb{R}^3$ can be represented uniquely by a traceless 2×2 hermitian matrix. We express this fact mathematically by saying that there is an isomorphism $f : \mathbb{R}^3 \to \mathbb{H}_2$, where $\mathbb{H}_2$ is the set of traceless 2×2 hermitian matrices, given by

$$f(\boldsymbol{x}) = \begin{pmatrix} x^3 & x^1 - ix^2 \\ x^1 + ix^2 & -x^3 \end{pmatrix} . \tag{19.4}$$

The mapping f has the property that

$$\det(f(\boldsymbol{x})) = -(\,(x^1)^2 + (x^2)^2 + (x^3)^2\,) = -|\,\boldsymbol{x}\,|^2 . \tag{19.5}$$

Significantly, f maps the unit vectors along the x, y and z axes to the **Pauli spin matrices** σ_i. Thus, if $e_1 = \hat{\boldsymbol{x}} = (1,0,0)$, $e_2 = \hat{\boldsymbol{y}} = (0,1,0)$, and $e_3 = \hat{\boldsymbol{z}} = (0,0,1)$, then, by (19.4),

$$f(e_1) = \sigma_1 = \begin{pmatrix} 0 & 1 \\ 1 & 0 \end{pmatrix}, \quad f(e_2) = \sigma_2 = \begin{pmatrix} 0 & -i \\ i & 0 \end{pmatrix}, \quad f(e_3) = \sigma_3 = \begin{pmatrix} 1 & 0 \\ 0 & -1 \end{pmatrix} . \tag{19.6}$$

As we shall see, the Pauli matrices are essential for the study of the quantum mechanics of spin 1/2 particles (such as the electron). We list here for reference some of their most important properties (which can all be demonstrated directly from the definitions given above).

$$\sigma_i = \sigma_i^\dagger , \tag{19.7a}$$

$$tr\,\sigma_i = 0 , \tag{19.7b}$$

$$\sigma_i\sigma_j + \sigma_j\sigma_i = 2\delta_{ij} , \tag{19.7c}$$

$$[\,\sigma_i\,,\,\sigma_j\,] = 2i\varepsilon_{ij}{}^l\,\sigma_l\,, \quad \text{or} \quad \sigma_j\sigma_k = \delta_{jk} + i\varepsilon_{jk}{}^l\,\sigma_l\,, \tag{19.7d}$$

$$(\boldsymbol{a}\cdot\boldsymbol{\sigma})\,(\boldsymbol{b}\cdot\boldsymbol{\sigma}) = \boldsymbol{a}\cdot\boldsymbol{b} + i\,(\boldsymbol{a}\times\boldsymbol{b})\cdot\boldsymbol{\sigma}\,, \tag{19.7e}$$

$$e^{i\,\theta\,\sigma_j} = \cos\theta + i\,\sin\theta\,\sigma_j\,, \tag{19.7f}$$

where in (19.7d) $\varepsilon_{jk}{}^{l}$ is the Levi-Civita (completely antisymmetric) tensor, and the Einstein summation convention has been used; and in (19.7e), $\boldsymbol{a}$ and $\boldsymbol{b}$ are ordinary vectors in $\mathbb{R}^3$, and $\boldsymbol{\sigma}$ is understood to be a quantity with the three components $\sigma_1, \sigma_2, \sigma_3$ [as defined in (19.6)] so that the "scalar product" $\boldsymbol{a} \cdot \boldsymbol{\sigma}$, for example, is understood to mean $a_1\sigma_1 + a_2\sigma_2 + a_3\sigma_3$.

Problem 19.1 Prove the properties of the Pauli spin matrices specified by Eqs. (19.7a) through (19.7f).

Problem 19.2 Show that the 2×2 identity matrix σ_0 together with the 3 Pauli spin matrices σ_1, σ_2 and σ_3 form a complete set in the 4-dimensional real vector space of 2×2 hermitian matrices, that is, any 2×2 hermitian matrix can be written as a linear combination of $\sigma_0, \sigma_1, \sigma_2$ and σ_3 with real coefficients.

Problem 19.3 Recall the (2×2) density matrix $\rho(t)$ introduced in Problem 8.8.

(a) Use the general properties of density matrices (cf. Problem 8.7) and the Pauli spin matrices to show that

$$\rho = \frac{1}{2}\left(1 + \boldsymbol{\sigma} \cdot \boldsymbol{P}\right),$$

where the **polarization** $\boldsymbol{P}$ is given by the expectation values of the Pauli spin matrices:

$$\boldsymbol{P} = \langle\, \boldsymbol{\sigma} \,\rangle = Tr\,(\rho\boldsymbol{\sigma})\,.$$

(b) With respect to the representation that diagonalizes ρ, namely $\{|\uparrow\rangle, |\downarrow\rangle\}$, $\boldsymbol{\sigma} \cdot \boldsymbol{P}$ must also be diagonal. Show that when $\boldsymbol{P}$ is along the z-axis,

$$\boldsymbol{\sigma} \cdot \boldsymbol{P}\,|\uparrow\rangle = P\,|\uparrow\rangle\,, \quad \boldsymbol{\sigma} \cdot \boldsymbol{P}\,|\downarrow\rangle = -P\,|\downarrow\rangle\,,$$

where P is the magnitude of $\boldsymbol{P}$. Show that in this case we recover the result of Problem 8.8.

(c) Show that

$$\rho^2 = \frac{1}{4}\left(1 + P^2 + 2\boldsymbol{\sigma} \cdot \boldsymbol{P}\right).$$

Under what conditions would $\rho^2 = \rho$? Interpret these physically. [Hint: Use (19.7e).]

Problem 19.4 A beam of neutrons moves through a homogeneous and static magnetic field $\boldsymbol{B} = B\boldsymbol{u}$, $\boldsymbol{u}$ being a unit vector. At time $t = 0$ the polarization of the beam is $\boldsymbol{P}$. Show that the time evolution of the polarization is given by

$$\boldsymbol{P}(t) = \boldsymbol{P}\,\cos^2(\omega t) + (\boldsymbol{P} \times \boldsymbol{u})\,\sin(2\omega t) + [\boldsymbol{u}(\boldsymbol{u} \cdot \boldsymbol{P}) - (\boldsymbol{u} \times \boldsymbol{P}) \times \boldsymbol{u}]\,\sin^2(\omega t)\,,$$

where $\hbar\omega = \mu_0 B$, with μ_0 being the neutron's magnetic moment. Compare this result with the motion of a classical magnetic moment in the same field. Follow the steps below.

(a) Since the neutron is a spin 1/2 particle, the Hamiltonian can be written as the 2×2 matrix

$$H = H_0 - \hbar\omega\boldsymbol{\sigma} \cdot \boldsymbol{u} \,,$$

where H_0 does not involve $\boldsymbol{\sigma}$ and hence commutes with the second term. Use the equation of motion for the density matrix ρ [cf. Problem 8.7(b)] to show that

$$\frac{d\rho}{dt} = -\omega\boldsymbol{\sigma} \cdot (\boldsymbol{u} \times \boldsymbol{P}) \,.$$

(b) Use the result in the previous problem for ρ and differentiate directly to obtain

$$\frac{d\rho}{dt} = \frac{1}{2}\boldsymbol{\sigma} \cdot \frac{d\boldsymbol{P}}{dt} \,.$$

Compare the above two equations to get

$$\frac{d\boldsymbol{P}}{dt} = 2\omega(\boldsymbol{P} \times \boldsymbol{u}) \,,$$

which is the equation of motion for a precessing classical vector $\boldsymbol{P}$.

(c) The above equation is hard to solve directly in the quantum mechanical case. But we can make use of the result in the previous problem that $\boldsymbol{P}(t) = \langle \boldsymbol{\sigma}(t) \rangle$, where, from Heisenberg's equation of motion for $\boldsymbol{\sigma}$,

$$\boldsymbol{\sigma}(t) = \exp(iHt/\hbar)\, \boldsymbol{\sigma}\, \exp(-iHt/\hbar) = e^{-i\omega(\boldsymbol{\sigma}\cdot\boldsymbol{u})t}\, \boldsymbol{\sigma}\, e^{i\omega\boldsymbol{\sigma}\cdot\boldsymbol{u}} \,.$$

Use the properties of the Pauli spin matrices to show that

$$e^{\pm i\omega(\boldsymbol{\sigma}\cdot\boldsymbol{u})t} = \cos(\omega t) \pm i(\boldsymbol{\sigma} \cdot \boldsymbol{u}) \sin(\omega t) \,.$$

(d) Use the above result in the expansion for $\boldsymbol{\sigma}(t)$. Terms proportional to $\boldsymbol{\sigma}$, $(\boldsymbol{\sigma}\cdot\boldsymbol{u})\boldsymbol{\sigma}$, $\boldsymbol{\sigma}(\boldsymbol{\sigma} \cdot \boldsymbol{u})$ and $(\boldsymbol{\sigma} \cdot \boldsymbol{u})\boldsymbol{\sigma}(\boldsymbol{\sigma} \cdot \boldsymbol{u})$ will be obtained. Show that

$$\begin{aligned}
(\boldsymbol{\sigma} \cdot \boldsymbol{u})\, \boldsymbol{\sigma} &= \boldsymbol{u} - i(\boldsymbol{u} \times \boldsymbol{\sigma}) \,, \\
\boldsymbol{\sigma}\, (\boldsymbol{\sigma} \cdot \boldsymbol{u}) &= \boldsymbol{u} + i(\boldsymbol{u} \times \boldsymbol{\sigma}) \,, \\
(\boldsymbol{\sigma} \cdot \boldsymbol{u})\, \boldsymbol{\sigma}\, (\boldsymbol{\sigma} \cdot \boldsymbol{u}) &= \boldsymbol{u}(\boldsymbol{\sigma} \cdot \boldsymbol{u}) - (\boldsymbol{u} \times \boldsymbol{\sigma}) \times \boldsymbol{u} \,.
\end{aligned}$$

(e) Use the last three results for $\boldsymbol{\sigma}(t)$. Then take expectation values to obtain the stated result for $\boldsymbol{P}(t)$.

We can now define a group homomorphism (cf. Def. 18.2) $\phi : SU(2) \to SO(3)$ as follows. For $g \in SU(2)$ and $\boldsymbol{x} \in \mathbb{R}^3$, ϕ is given by

$$\boxed{\phi(g)(\boldsymbol{x}) = f^{-1}(gf(\boldsymbol{x})g^\dagger)} \quad . \tag{19.8}$$

To see that the definition makes sense, we have to check that (1) $gf(\boldsymbol{x})g^\dagger \in \mathbb{H}_2$, (2) $\phi(g_1 g_2) = \phi(g_1)\phi(g_2)$ for any $g_1, g_2 \in SU(2)$ (group homomorphism requirement), and (3) $\phi(g) \in SO(3)$. Indeed, since $f(\boldsymbol{x})$ is hermitian by definition, $(gf(\boldsymbol{x})g^\dagger)^\dagger = (g^\dagger)^\dagger f(\boldsymbol{x})^\dagger g^\dagger = gf(\boldsymbol{x})g^\dagger$. Thus $gf(\boldsymbol{x})g^\dagger$ is hermitian. It is also traceless since $tr(AB) = tr(BA)$ implies that $tr(gf(\boldsymbol{x})g^\dagger) = tr(g^\dagger gf(\boldsymbol{x})) = tr(g^{-1}gf(\boldsymbol{x})) = tr(f(\boldsymbol{x}))$, and $f(\boldsymbol{x})$ is traceless by definition. So (1) is established. Fact (2) is checked as follows. By (19.8), we have, for all $\boldsymbol{x} \in \mathbb{R}^3$,

$$\begin{aligned} \phi(g_1 g_2)(\boldsymbol{x}) &= f^{-1}(g_1 g_2 f(\boldsymbol{x}) g_2^\dagger g_1^\dagger) = f^{-1}(g_1 f(\phi(g_2)(\boldsymbol{x})) g_1^\dagger) \\ &= f^{-1} f(\phi(g_1)\phi(g_2)(\boldsymbol{x})) = \phi(g_1)\phi(g_2)(\boldsymbol{x}) \,. \end{aligned} \tag{19.9}$$

To check (3) we have to show that $|\,\phi(g)(\boldsymbol{x})\,| = |\,\boldsymbol{x}\,|$ and also that $\det(\phi(g)) = 1$. By (19.5), we have

$$\begin{aligned} |\,\phi(g)(\boldsymbol{x})\,|^2 &= -\det[f(\phi(g)(\boldsymbol{x}))] = -\det(gf(\boldsymbol{x})g^\dagger) \\ &= -\det(g)\det(f(\boldsymbol{x}))\det(g^\dagger) = -\det(f(\boldsymbol{x})) = |\,\boldsymbol{x}\,|^2 \,. \end{aligned} \tag{19.10}$$

Thus $\phi(g)$ leaves the length of $\boldsymbol{x}$ invariant, and so $\phi(g) \in O(3)$. The argument that $\phi(g)$ is a proper rotation (with determinant equal to $+1$), that is, $\phi(g) \in SO(3)$, is a bit more involved, and requires an understanding of some basic topological properties of $SU(2)$, $SO(3)$ and $O(3)$ (when these groups are regarded as topological manifolds). Let us now digress briefly to investigate these properties.

A general element of $SU(2)$ can be written as

$$g = \begin{pmatrix} a & b \\ -b^* & a^* \end{pmatrix}, \quad a, b \in \mathbb{C}\,, \tag{19.11}$$

with the complex numbers a and b, called the **Cayley-Klein parameters**, satisfying the condition $|\,a\,|^2 + |\,b\,|^2 = 1$, which is required by the defining condition $\det(g) = 1$. If we write $a = x^1 + ix^2$, $b = x^3 + ix^4$, where $x^1, x^2, x^3, x^4 \in \mathbb{R}$, this condition is equivalent to

$$(x^1)^2 + (x^2)^2 + (x^3)^2 + (x^4)^2 = 1\,. \tag{19.12}$$

This is the equation of the unit 3-sphere embedded in $\mathbb{R}^4$ (with the Euclidean metric), usually denoted S^3. Hence, as a topological manifold, $SU(2)$ is *homeomorphic* (topologically equivalent) to S^3. We write $SU(2) \sim S^3$. In fact, under the map

$$g = \begin{pmatrix} a & b \\ -b^* & a^* \end{pmatrix} \mapsto (x^1, x^2, x^3, x^4)\,, \tag{19.13}$$

where x^1, x^2, x^3 and x^4 are understood as Cartesian coordinates of a point in $\mathbb{R}^4$, we see that g and $-g$ are mapped to antipodal points on S^3. Now it is a well-known topological fact that the unit n-sphere S^n for $n \geq 2$ is **simply connected**, which roughly means that any closed loop on S^n can be shrunk to a point. Hence $SU(2)$ *as a topological manifold is simply connected.* $SO(3)$, however, does not share this property. We will show later that $SO(3)$ *is doubly* *connected.* The group $O(3)$, on the other hand, is not even connected. It contains two disconnected pieces: $SO(3)$ (a subgroup of $O(3)$), and $sSO(3)$, where s is the space inversion operator matrix (notice that it has determinant -1):

$$s = \begin{pmatrix} -1 & 0 & 0 \\ 0 & -1 & 0 \\ 0 & 0 & -1 \end{pmatrix}. \tag{19.14}$$

The notation $sSO(3)$ stands for the set of matrices obtained by multiplying s by all matrices in $SO(3)$. It is obvious that every matrix in $sSO(3)$ has determinant equal to -1. Unlike $SO(3)$ this set is not a subgroup of $O(3)$ because it does not contain the identity matrix.

The argument that the image of the group homomorphism ϕ must lie in $SO(3)$ runs roughly as follows. We already know from the above discussion that $\phi(g) \in O(3)$, since $\phi(g)$ preserves the lengths of vectors. Suppose, for the moment, that there exists a $g_0 \in SU(2)$ such that $\phi(g_0) \in sSO(3)$. Since $SU(2)$ is connected (in fact simply connected), we can choose a continuous path S in $SU(2)$ running from g_0 to the identity element e. Since ϕ is a continuous map, the image of S under ϕ ought to be a continuous path running from $\phi(g_0)$ to the identity $e' \in SO(3)$. But since $\phi(g_0) \in sSO(3)$ by assumption, and $sSO(3)$ is a disconnected component of $O(3)$, no such path in $O(3)$ exists (see Fig. 19.1). Hence the assumption that there exists an element in $SU(2)$ which is mapped by ϕ to an element in $sSO(3)$ must be invalid, and the image of the group homomorphism ϕ must lie entirely in $SO(3)$.

We will now show that the group homomorphism $\phi : SU(2) \to SO(3)$ is *two-to-one*, that is, for every element R in $SO(3)$, there are exactly two elements g and g' in $SU(2)$ such that $\phi(g) = \phi(g') = R$. First we demonstrate the following fact.

Lemma 19.1. *The* ***kernel*** *of the group homomorphism* $\phi : SU(2) \to SO(3)$ *consists of exactly two elements in* $SU(2)$: e *(the identity) and* $-e$. *Equivalently, if* $\phi(g) = e' \in SO(3)$, *(the identity in* $SO(3)$*), then* $g = \pm e \in SU(2)$. *We express this fact as:* $ker(\phi) = \{e, -e\}$.

Proof. Suppose $g \in ker(\phi)$. Then $\phi(g) = e'$ (the identity in $SO(3)$), which implies $\phi(g)(\boldsymbol{x}) = \boldsymbol{x}$ for all $\boldsymbol{x} \in \mathbb{R}^3$. By (19.8), then, $f(\boldsymbol{x}) = gf(\boldsymbol{x})g^{\dagger}$. Using (19.4) for $f(\boldsymbol{x})$ and (19.11) for g, this equation can be written out explicitly in matrix form as

$$\begin{pmatrix} x^3 & x^1 - ix^2 \\ x^1 + ix^2 & -x^3 \end{pmatrix} = \begin{pmatrix} a & b \\ -b^* & a^* \end{pmatrix} \begin{pmatrix} x^3 & x^1 - ix^2 \\ x^1 + ix^2 & -x^3 \end{pmatrix} \begin{pmatrix} a^* & -b \\ b^* & a \end{pmatrix}. \tag{19.15}$$

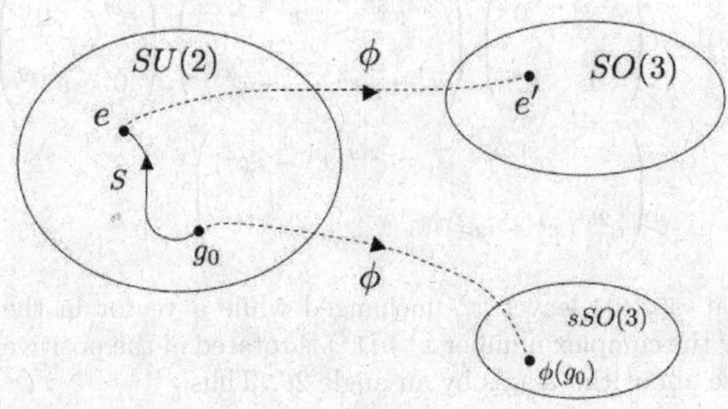

Fig. 19.1

Doing the matrix multiplication on the right-hand side and equating corresponding matrix elements on both sides, we see that

$$x^3 = (|\,a\,|^2 - |\,b\,|^2)\,x^3 + ab^*(x^1 - ix^2) + a^*b(x^1 + ix^2)\,, \tag{19.16}$$

$$x^1 - ix^2 = -2ab(x^3) + a^2(x^1 - ix^2) - b^2(x^1 + ix^2)\,. \tag{19.17}$$

Since x^1, x^2 and x^3 are arbitrary real numbers, (19.16) implies $|\,a\,|^2 - |\,b\,|^2 = 1$, on comparing coefficients for x^3. Subtracting this equation from the unimodularity condition $|\,a\,|^2 + |\,b\,|^2 = 1$, we get $|\,b\,|^2 = 0$, which implies $b = 0$. On comparing coefficients for x^1 in (19.17), we get $a^2 - b^2 = 1$. Thus $a^2 = 1$, and $a = \pm 1$. It follows from (19.11) that g is either e or $-e$. □

Now suppose two elements $g,\, g' \in SU(2)$ are such that $\phi(g) = \phi(g')$. We then have $\phi(g)(\boldsymbol{x}) = \phi(g')(\boldsymbol{x})$ for all $\boldsymbol{x} \in \mathbb{R}^3$. Thus

$$\begin{aligned} &\phi^{-1}(g)\phi(g')(\boldsymbol{x}) = \boldsymbol{x} \Longrightarrow \phi(g^{-1})\phi(g')(\boldsymbol{x}) = \boldsymbol{x} \\ &\Longrightarrow \phi(g^{-1}g')(\boldsymbol{x}) = \boldsymbol{x} \Longrightarrow g^{-1}g' = \pm\, e \Longrightarrow g' = \pm\, g\,, \end{aligned} \tag{19.18}$$

where the first and second implications follow from the general properties of group homomorphisms (cf. Def. 18.2 and the statement immediately following), and the third follows from an application of Lemma 19.1. This establishes the two-to-one nature of the group homomorphism ϕ, and that the pre-images of a particular $R \in SO(3)$ under ϕ are negatives of each other in $SU(2)$.

Let us now consider specific one-parameter subgroups of $SU(2)$. First look at $g \in SU(2)$ of the form given by (19.1), with $0 \le \theta < 2\pi$, and compute explicitly the effect of $\phi(g(\theta))$ on an arbitrary $\boldsymbol{x} \in \mathbb{R}^3$. We have

$$\begin{aligned} g(\theta)f(\boldsymbol{x})(g(\theta))^\dagger &= \begin{pmatrix} e^{-i\theta} & 0 \\ 0 & e^{i\theta} \end{pmatrix} \begin{pmatrix} x^3 & x^1 - ix^2 \\ x^1 + ix^2 & -x^3 \end{pmatrix} \begin{pmatrix} e^{i\theta} & 0 \\ 0 & e^{-i\theta} \end{pmatrix} \\ &= \begin{pmatrix} x^3 & e^{-2i\theta}(x^1 - ix^2) \\ e^{2i\theta}(x^1 + ix^2) & -x^3 \end{pmatrix} . \end{aligned} \tag{19.19}$$

This shows that $\phi(g(\theta))$ leaves x^3 unchanged while a vector in the xy-plane (represented by the complex number x^1+ix^2) is rotated in the positive (counterclockwise) sense about the z-axis by an angle 2θ. Thus

$$\phi(g(\theta)) = R_3(2\theta) , \quad \text{if} \quad g(\theta) = \begin{pmatrix} e^{-i\theta} & 0 \\ 0 & e^{i\theta} \end{pmatrix} , \tag{19.20}$$

where $R_3(\theta)$ denotes a rotation about the x^3-axis (or z-axis) by an angle θ. Note that

$$-g(\theta) = \begin{pmatrix} -e^{-i\theta} & 0 \\ 0 & -e^{i\theta} \end{pmatrix} = \begin{pmatrix} e^{-i(\theta+\pi)} & 0 \\ 0 & e^{i(\theta+\pi)} \end{pmatrix} = g(\theta + \pi) . \tag{19.21}$$

Thus

$$\phi(-g(\theta)) = R_3(2\theta + 2\pi) = R_3(2\theta) = \phi(g(\theta)) . \tag{19.22}$$

This verifies the two-to-one nature of ϕ restricted to the subgroup of $SU(2)$ to which $g(\theta)$ belongs.

Next consider another one-parameter subgroup of $SU(2)$, whose elements have the form [cf. (18.13)]

$$g'(\alpha) = \begin{pmatrix} \cos\alpha & -\sin\alpha \\ \sin\alpha & \cos\alpha \end{pmatrix}, \qquad 0 \le \alpha < 2\pi . \tag{19.23}$$

Note that

$$(g'(\alpha))^\dagger = g'(-\alpha) . \tag{19.24}$$

The effect of $\phi(g'(\alpha))$ on an arbitrary $\boldsymbol{x} \in \mathbb{R}^3$ is given by

$$g'(\alpha)f(\boldsymbol{x})(g'(\alpha))^\dagger = \begin{pmatrix} \cos\alpha & -\sin\alpha \\ \sin\alpha & \cos\alpha \end{pmatrix} \begin{pmatrix} x^3 & x^1 - ix^2 \\ x^1 + ix^2 & -x^3 \end{pmatrix} \begin{pmatrix} \cos\alpha & \sin\alpha \\ -\sin\alpha & \cos\alpha \end{pmatrix} \tag{19.25}$$

Applying this to $\boldsymbol{x} = e_2 = (0,1,0)$ (the unit y-axis) and recalling the second equation of (19.6), we see that

$$g'(\alpha)f(e_2)(g'(\alpha))^\dagger = g'(\alpha)\sigma_2(g'(\alpha))^\dagger = \sigma_2\ , \tag{19.26}$$

where σ_2 is a Pauli spin matrix. So the rotation $\phi(g'(\alpha))$ leaves e_2 invariant, and hence must correspond to a rotation about the e_2 axis (or y-axis). To see what the angle of rotation is we apply (19.25) to $\boldsymbol{x} = e_3 = (0,0,1)$ (the unit z-axis). Thus

$$\begin{aligned} &g'(\alpha)f(e_3)(g'(\alpha))^\dagger = g'(\alpha)\sigma_3(g'(\alpha))^\dagger \\ &= \begin{pmatrix} \cos\alpha & -\sin\alpha \\ \sin\alpha & \cos\alpha \end{pmatrix}\begin{pmatrix} 1 & 0 \\ 0 & -1 \end{pmatrix}\begin{pmatrix} \cos\alpha & \sin\alpha \\ -\sin\alpha & \cos\alpha \end{pmatrix} = \begin{pmatrix} \cos 2\alpha & \sin 2\alpha \\ \sin 2\alpha & -\cos 2\alpha \end{pmatrix} \\ &= f(\sin 2\alpha, 0, \cos 2\alpha)\ , \end{aligned} \tag{19.27}$$

which shows that $\phi(g'(\alpha))$ is a rotation about the y-axis by an angle 2α:

$$\phi(g'(\alpha)) = R_2(2\alpha)\ . \tag{19.28}$$

We single out the subgroups of $SU(2)$ giving rise under the homomorphism ϕ to the rotations R_3 and R_2 in $SO(3)$ because of the following important fact concerning general rotations.

Theorem 19.1 (Euler's Theorem). *Every rotation $R \in SO(3)$ can be written as a product of rotations*

$$R = R_3(\phi)\, R_2(\theta)\, R_3(\psi)\ , \tag{19.29}$$

where R_2 and R_3 are rotations about the y- and z- axes, respectively, and the angles ϕ, θ, ψ (called the **Euler angles***) fall in the ranges $0 \le \phi, \psi < 2\pi$, $0 \le \theta \le \pi$.*

Problem 19.5 Prove Euler's Theorem as stated above.

It follows from this theorem and the equations (19.20), (19.22) and (19.28) that ϕ is a *surjective* (onto) map, that is, for every $R \in SO(3)$, there is a $g \in SU(2)$ such that $\phi(g) = R$. Since ϕ is also two-to-one, with $\phi(g) = \phi(-g)$ [as concluded from (19.18)], the following two equations give completely the relationship between $SU(2)$ and $SO(3)$ under the two-to-one and surjective group homomorphism $\phi : SU(2) \to SO(3)$:

$$\phi^{-1}\left(R_3(\psi)\right) = \pm \begin{pmatrix} \exp\left(-i\frac{\psi}{2}\right) & 0 \\ 0 & \exp\left(i\frac{\psi}{2}\right) \end{pmatrix} = \pm \exp\left(-i\frac{\psi}{2}\sigma_3\right), \tag{19.30}$$

$$\phi^{-1}\left(R_2(\theta)\right) = \pm \begin{pmatrix} \cos\frac{\theta}{2} & -\sin\frac{\theta}{2} \\ \sin\frac{\theta}{2} & \cos\frac{\theta}{2} \end{pmatrix} = \pm \exp\left(-i\frac{\theta}{2}\sigma_2\right), \tag{19.31}$$

where σ_2 and σ_3 are Pauli spin matrices [cf. (19.6)]. These two equations in fact generalize to

$$\boxed{\phi^{-1}\left(R_{\boldsymbol{n}}(\theta)\right) = \pm \exp\left(-i\frac{\theta}{2}\boldsymbol{n}\cdot\boldsymbol{\sigma}\right)}, \tag{19.32}$$

where $\boldsymbol{n}$ is a unit vector along an arbitrary direction [compare with (17.60)].

To conclude this chapter we will look briefly at the topological properties of the group manifold of $SO(3)$. Since every rotation in $\mathbb{R}^3$ can be specified by an axis of rotation (given by an arbitrary unit vector $\boldsymbol{n}$) and an angle of rotation α about that axis, we can represent the manifold pictorially as a solid sphere of radius π in $\mathbb{R}^3$ with antipodal (diametrically opposite) points on the spherical surface identified (see Fig. 19.2). This is so because each $g \in SO(3)$ can be specified by three parameters: $0 \le \theta \le \pi$, $0 \le \varphi < 2\pi$ (which are the polar and azimuth angles specifying the orientation of the rotation axis $\boldsymbol{n}$), and $0 \le \alpha \le \pi$ (the angle of rotation). Thus each point in the interior of this sphere represents a distinct rotation $g \in SO(3)$. The identity (no rotation at all) corresponds to the center of the sphere. A rotation by α about $\boldsymbol{n}$ is the same as a rotation by $2\pi - \alpha$ about $-\boldsymbol{n}$. This limits the range of α as specified above and implies that a rotation by π about $\boldsymbol{n}$ is the same as a rotation by π about $-\boldsymbol{n}$. Thus antipodal points on the spherical surface (of radius π) have to be identified. Because of this identification, the shape of the group manifold of $SO(3)$ is a bit hard to grasp intuitively, but one can readily see that it is a *compact, three-dimensional differentiable manifold.*

Now consider the closed curve $R_3(\theta), 0 \le \theta \le 2\pi$, in $SO(3)$. The preimage of this curve under ϕ, such that $\phi^{-1}(e') = e$, is given by [cf. (19.30)]

$$g(\theta) = + \begin{pmatrix} \exp\left(-i\frac{\theta}{2}\right) & 0 \\ 0 & \exp\left(i\frac{\theta}{2}\right) \end{pmatrix}, \tag{19.33}$$

where the overall + sign is required by the above condition on ϕ^{-1}. Under the map given by (19.13),

$$g(\theta) \overset{h}{\mapsto} (\cos\frac{\theta}{2}, \sin\frac{\theta}{2}, 0, 0). \tag{19.34}$$

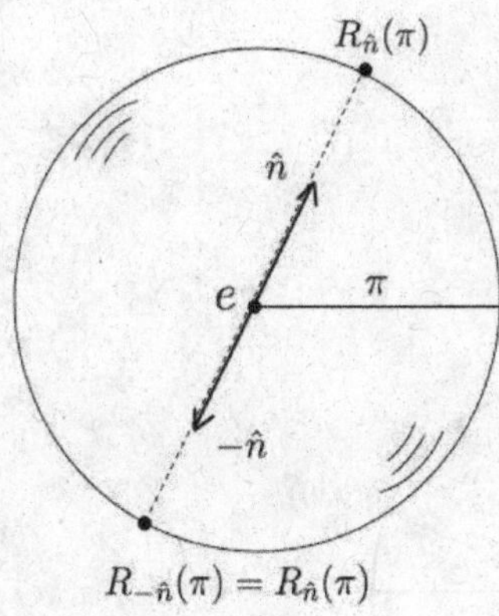

Fig.19.2

As θ varies from 0 to 2π, $h(g(\theta))$ traces out the semicircle $x^3 = x^4 = 0, x^1 = \cos(\theta/2), x^2 = \sin(\theta/2)$ on the projection of the unit 3-sphere S^3 on the x^1x^2-plane of $\mathbb{R}^4$, with the end-point corresponding to $g(2\pi) = -e$ (see Fig. 19.3, which also shows the image under ϕ of $g(\theta)$ in $SO(3)$). This obviously does not correspond to a closed curve in $SU(2)$. Hence, under ϕ, the corresponding closed curve $R_3(\theta)$ in $SO(3), 0 \leq \theta \leq 2\pi$, (Fig. 19.4) cannot be shrunk to a point in $SO(3)$.

The preimage of the double loop in $SO(3)$ under ϕ (shown in Fig. 19.5), however, is a closed path in $SU(2)$ corresponding to a full circle on the x^1x^2-plane (Fig. 19.3) in $\mathbb{R}^4$. Since S^3 is simply connected, this closed path in $SU(2)$ can be shrunk to a point (in $SU(2)$). Thus the double loop of Fig. 19.5 in $SO(3)$ can be shrunk to a point also. This can be seen pictorially in the sequence of deformations of the double loop in $SO(3)$ shown in Fig. 19.6.

$SO(3)$ as a manifold is said to be **doubly connected**. $SU(2)$ is called the universal covering group of $SO(3)$. The general definition of a universal covering group is given as follows.

Definition 19.1. *The **universal covering group** of a connected topological group G is a **simply connected** topological group G_{UC} such that there is a continuous homomorphism $\rho : G_{UC} \to G$ which is locally one-to-one. (See Fig. 19.7.)*

A universal covering group exists for every connected topological group (and in particular, for every connected Lie group), and is unique up to isomorphism. Our discussion in this chapter in fact shows that $SU(2)$ *is the universal covering group of the proper rotation group* $SO(3)$.

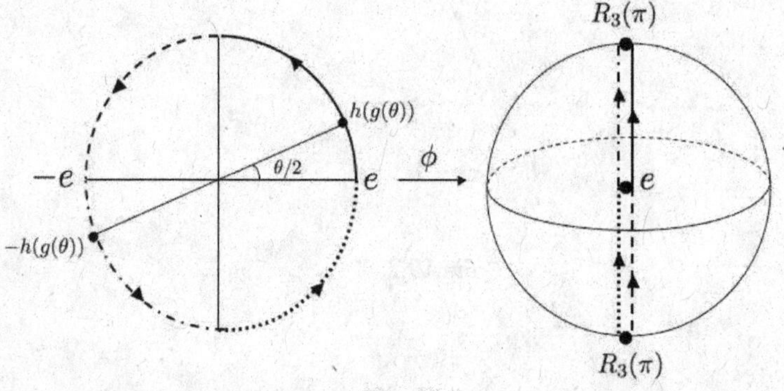

Fig. 19.3

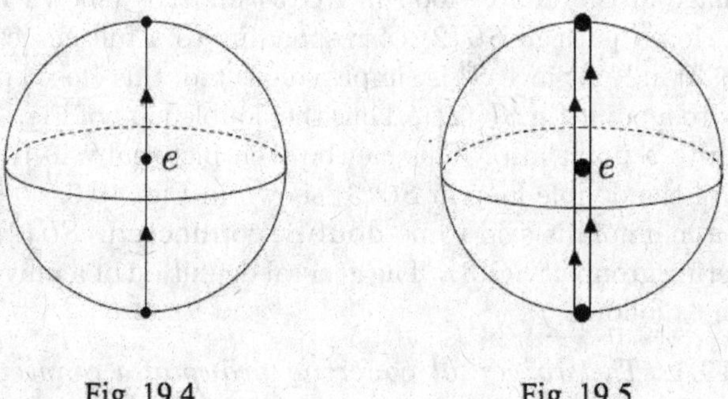

Fig. 19.4 Fig. 19.5

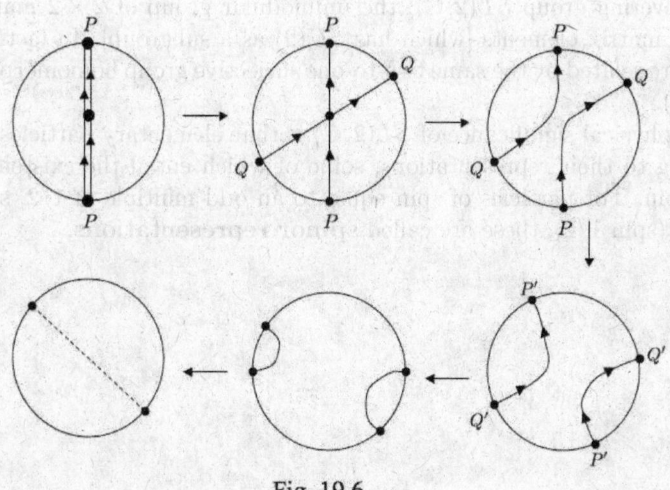

Fig. 19.6

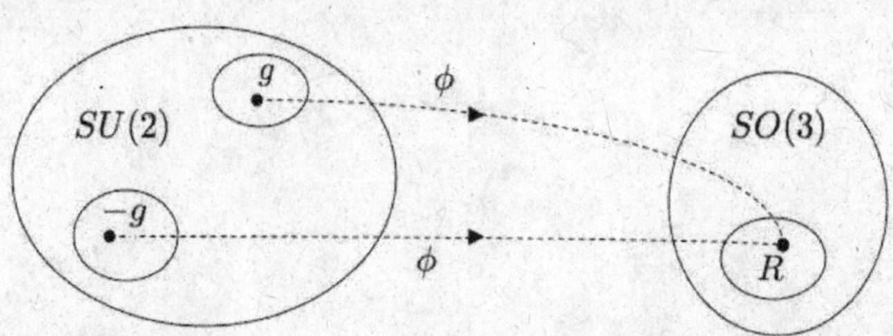

Fig. 19.7

A remarkable development of the discussion presented in this chapter is that, with relatively minor changes, it can be generalized to the **proper orthochronous Lorentz group** [which has $SO(3)$ as a subgroup] and its universal covering group $SL(2,\mathbb{C})$, the unimodular group of 2×2 matrices with complex matrix elements [which has $SU(2)$ as a subgroup]. In fact, these two groups are related by the same two-to-one surjective group homomorphism given by (19.8).

The physical significance of $SL(2,\mathbb{C})$ is that elementary particles transform according to their representations, some of which entail the existence of fractional spin. For particles of spin equal to an odd multiple of $1/2$, such as the electron (spin $1/2$), these are called **spinor representations**.

Chapter 20

The Spectrum of the Angular Momentum Operators

In Chapter 17 we saw that the angular momentum operators L_x, L_y, L_z are the infinitesimal generators of finite rotations in $\mathbb{R}^3$ [cf. (17.60)]. Mathematically, they form a basis of the **Lie algebra** $\mathcal{SO}(3)$ corresponding to the Lie group $SO(3)$. Let us explain this terminology in a bit more detail. *A* ***Lie group*** *is by definition both a group and a differentiable manifold.* The tangent space at any point in a differentiable manifold is a linear vector space whose dimension is the same as the dimensionality of the manifold. The tangent space at the identity element of a Lie group, endowed with a certain internal multiplication [,] satisfying (12.9), (12.10), and (12.13) (the Jacobi identity), is called the Lie algebra corresponding to the Lie group. As we saw in the last chapter, the Lie group $SO(3)$, in addition to being a group, also happens to be a three-dimensional compact differentiable manifold. So its Lie algebra is a three-dimensional linear vector space (as well as being an algebra). $\{L_x, L_y, L_z\}$ is one choice for a basis of this vector space, and the structure of the Lie algebra is determined by the commutation relations between the operators in this basis. As seen in the last chapter, a neighborhood of the identity in $SU(2)$ is locally isomorphic (topologically and in group structure) to a neighborhood of the identity in $SO(3)$. Thus there is a local isomorphism between their Lie algebras as well. A study of the spectrum of the operators in the Lie algebra of $SO(3)$ then reveals not only the local irreducible representations (near the identity) of $SO(3)$ but those of $SU(2)$ also. The global representations of these groups are then obtained by exponentiation, via (17.60) or (19.32).

We will first define the dimensionless angular momentum operators (to avoid carrying extra factors of $\hbar$):

$$\boldsymbol{L} = \hbar \boldsymbol{J} \,. \tag{20.1}$$

Then (17.62) and (17.63), equations that specify the structure of the Lie algebra

$\mathcal{SO}(3)$, can be rewritten as

$$[J_i\,,\,J_j] = i\varepsilon_{ij}{}^k\,J_k\,, \qquad \text{and} \tag{20.2}$$

$$[J^2\,,\,\boldsymbol{J}] = 0\,, \tag{20.3}$$

where $\varepsilon_{ij}{}^k$ is the Levi-Civita tensor, the Einstein summation convention has been used (the index k is summed over), and $J^2 = J_x^2 + J_y^2 + J_z^2$. The operator J^2 is called a **Casimir operator** of the Lie algebra $\mathcal{SO}(3)$, namely, one which commutes with every element of the algebra. There is only one such for $\mathcal{SO}(3)$. Next we introduce two operators that are hermitian adjoints of each other:

$$J_+ \equiv J_x + iJ_y\,, \qquad J_- \equiv J_x - iJ_y\,. \tag{20.4}$$

These satisfy the following commutation relations:

$$[J_z\,,\,J_+] = J_+\,, \tag{20.5}$$

$$[J_z\,,\,J_-] = -J_-\,, \tag{20.6}$$

$$[J_+\,,\,J_-] = 2J_z\,. \tag{20.7}$$

Also,

$$[J^2\,,\,J_+] = [J^2\,,\,J_-] = 0\,. \tag{20.8}$$

The Casimir operator J^2 can alternatively be written as

$$J^2 = \frac{1}{2}(J_+J_- + J_-J_+) + J_z^2\,, \tag{20.9}$$

from which the following two relations can be deduced with the help of (20.7):

$$J_-J_+ = J^2 - J_z(J_z+1)\,, \qquad J_+J_- = J^2 - J_z(J_z-1)\,. \tag{20.10}$$

Equation (20.3) in particular implies that $[J^2\,,\,J_z] = 0$. It follows from Theorem 15.1 that J^2 and J_z possess a complete orthonormal set of eigenvectors. Now J^2 is a **positive definite** hermitian operator (since $J^2 = \sum J_i^2$ and for any $|\,u\,\rangle \in \mathcal{H}$, $\langle\, u\,|\,J_i^2\,|\,u\,\rangle = \langle J_i u\,|\,J_i u\,\rangle \geq 0$). Thus the eigenvalues of J^2 must be larger than or equal to zero. We will label them by $j(j+1)$, where $j \geq 0$. Let the complete orthonormal set of eigenvectors of J^2 and J_z be labelled by $|\,j\,m\,\rangle$, such that

$$J^2\,|\,j\,m\,\rangle = j(j+1)\,|\,j\,m\,\rangle \qquad \text{and} \tag{20.11}$$

$$J_z\,|\,j\,m\,\rangle = m\,|\,j\,m\,\rangle\,. \tag{20.12}$$

Consider the actions of $J_\pm$ on $|\,j\,m\,\rangle$. From (20.10) we have

$$J_-J_+\,|\,j\,m\,\rangle = \{j(j+1) - m(m+1)\}\,|\,j\,m\,\rangle = (j-m)(j+m+1)\,|\,j\,m\,\rangle\,, \tag{20.13}$$

$$J_+J_-\,|\,j\,m\,\rangle = \{j(j+1) - m(m-1)\}\,|\,j\,m\,\rangle = (j+m)(j-m+1)\,|\,j\,m\,\rangle\,. \tag{20.14}$$

Since $J_+^\dagger = J_-$ and $J_-^\dagger = J_+$, the square of the norms of the vectors $J_+ \,|\, j\, m\,\rangle$ and $J_- \,|\, j\, m\,\rangle$ are, respectively,

$$\langle\, j\, m \,|\, J_- J_+ \,|\, j\, m\,\rangle = (j-m)(j+m+1)\,\langle\, j\, m \,|\, j\, m\,\rangle \geq 0\,, \tag{20.15}$$

$$\langle\, j\, m \,|\, J_+ J_- \,|\, j\, m\,\rangle = (j+m)(j-m+1)\,\langle\, j\, m \,|\, j\, m\,\rangle \geq 0\,. \tag{20.16}$$

Hence, since $\langle\, j\, m \,|\, j\, m\,\rangle \geq 0$,

$$(j-m)(j+m+1) \geq 0\,, \qquad (j+m)(j-m+1) \geq 0\,. \tag{20.17}$$

These two inequalities imply that

$$-j \leq m \leq j\,. \tag{20.18}$$

Moreover, since the norm of a vector vanishes if and only if the vector is the zero-vector,

$$J_+ \,|\, j\, m\,\rangle = 0 \quad \text{iff} \quad (j-m)(j+m+1) = 0\,, \tag{20.19}$$

$$J_- \,|\, j\, m\,\rangle = 0 \quad \text{iff} \quad (j+m)(j-m+1) = 0\,. \tag{20.20}$$

Due to (20.18), the above conditions imply

$$J_+ \,|\, j\, m\,\rangle = 0 \quad \text{iff} \quad m = j\,, \tag{20.21}$$

$$J_- \,|\, j\, m\,\rangle = 0 \quad \text{iff} \quad m = -j\,. \tag{20.22}$$

Consider the case $m \neq j$. Then (20.21) implies $J_+ \,|\, j\, m\,\rangle \neq 0$. We have, from (20.8) and (20.11),

$$J^2 J_+ \,|\, j\, m\,\rangle = J_+ J^2 \,|\, j\, m\,\rangle = j(j+1) \,|\, j\, m\,\rangle\,. \tag{20.23}$$

From (20.5) we also have

$$J_z J_+ = J_+ (J_z + 1)\,, \tag{20.24}$$

from which it follows that

$$J_z J_+ \,|\, j\, m\,\rangle = J_+ (J_z + 1) \,|\, j\, m\,\rangle = (m+1)\, J_+ \,|\, j\, m\,\rangle\,. \tag{20.25}$$

Equations (20.23) and (20.25) together imply that if $m \neq j$, then $J_+ \,|\, j\, m\,\rangle$ is an eigenvector of J^2 and J_z with eigenvalues $j(j+1)$ and $m+1$, respectively. Similarly, for $m \neq -j$, (20.22) implies that $J_- \,|\, j\, m\,\rangle \neq 0$; and from (20.8),

$$J^2 J_- \,|\, j\, m\,\rangle = J_- J^2 \,|\, j\, m\,\rangle = j(j+1)\, J_- \,|\, j\, m\,\rangle\,. \tag{20.26}$$

Using

$$J_z J_- = J_- (J_z - 1)\,, \tag{20.27}$$

which follows from (20.6), we have

$$J_z J_- \,|\, j\, m\,\rangle = J_- (J_z - 1) \,|\, j\, m\,\rangle = (m-1)\, J_- \,|\, j\, m\,\rangle\,. \tag{20.28}$$

Equations (20.26) and (20.28) imply that if $m \neq -j$, then $J_- \,|\, j\, m\,\rangle$ is an eigenvector of J^2 and J_z with eigenvalues $J(j+1)$ and $m-1$, respectively.

We would now like to prove the following results, which specify the spectrum of J^2 and J_z:

(A) j is non-negative integral or half-integral:

$$j = 0\,,\ \frac{1}{2}\,,\ 1\,,\ \frac{3}{2}\,,\ 2\,,\ \ldots\,,\ \infty\,.$$

(B) $m = 0\,,\ \pm\frac{1}{2}\,,\ \pm 1\,,\ \pm\frac{3}{2}\,,\ \pm 2\,,\ \ldots\,,\ \pm\infty\,.$

(C) If $|\,j\,m\,\rangle$ is an eigenvector of J^2 and J_z, then there are $2j+1$ possible values for m: $j\,,\ j-1\,,\ \ldots\,,\ -j+1\,,\ -j\,.$

Consider the following sequence of vectors:

$$J_+\,|\,j\,m\,\rangle\,,\ J_+^2\,|\,j\,m\,\rangle\,,\ \ldots\,,\ J_+^p\,|\,j\,m\,\rangle\,,\ \ldots\,.$$

We already know from (20.18) that $j \geq m \geq -j$. If $m < j$, then $J_+\,|\,j\,m\,\rangle \neq 0$. It is an eigenvector of J^2 with eigenvalue $j(j+1)$ and an eigenvector of J_z with eigenvalue $m+1$. The above sequence must terminate after a finite number of terms, otherwise it will contain vectors which are eigenvectors of J_z with eigenvalues larger than j, which is impossible. Let p be the non-negative integer such that $J_+^p\,|\,j\,m\,\rangle \neq 0$, but $J_+^{p+1}\,|\,j\,m\,\rangle = 0$. Then we have

$$m + p = j\,. \tag{20.29}$$

Hence

$$j - m = p \geq 0\,, \qquad (p = \text{a non-negative integer})\,. \tag{20.30}$$

Considering the sequence

$$J_-\,|\,j\,m\,\rangle\,,\ J_-^2\,|\,j\,m\,\rangle\,,\ \ldots\,,\ J_-^q\,|\,j\,m\,\rangle\,,\ \ldots\,,$$

a similar argument shows that there exists a non-negative integer q such that

$$m - q = -j\,, \tag{20.31}$$

or

$$j + m = q \geq 0\,, \qquad (q = \text{a non-negative integer})\,. \tag{20.32}$$

Adding (20.30) and (20.32), we have

$$2j = p + q \geq 0\,, \tag{20.33}$$

where $p+q$ is a non-negative integer. This proves (A). Subtracting (20.30) from (20.32), we obtain

$$2m = q - p = \text{an integer}\,. \tag{20.34}$$

Statement (B) follows. Finally (20.29) and (20.18) together imply statement (C). We have thus determined the spectrum of J^2 and J_z entirely from the algebraic structure of the Lie algebra $\mathcal{SO}(3)$, as embodied in the commutation relations (20.2) and (20.3).

We will now construct the orthonormal eigenstates $|\, j\, m\,\rangle$. Suppose $m < j$, and we write

$$J_+ \,|\, j\, m\,\rangle = c_m \,|\, j\,,\, m+1\,\rangle\,, \tag{20.35}$$

where c_m is a constant to be determined. Normalization of the eigenkets requires that

$$\langle\, j\, m\,|\, j\, m\,\rangle = \langle\, j\,,\, m+1\,|\, j\,,\, m+1\,\rangle = 1\,.$$

Equation (20.15) then implies that

$$|\, c_m\,|^2 = (j-m)(j+m+1) = j(j+1) - m(m+1)\,. \tag{20.36}$$

We will fix the phase of $|\, j\,,\, m+1\,\rangle$ so that c_m is both real and positive. This yields

$$\boxed{J_+ \,|\, j\, m\,\rangle = \sqrt{j(j+1) - m(m+1)}\;|\, j\,,\, m+1\,\rangle}\quad. \tag{20.37}$$

We can apply J_+ again to $|\, j\,,,m+1\,\rangle$. If $m+1 < j$ we obtain $|\, j\,,\, m+2\,\rangle$ and so on, until we arrive at $|\, j\, j\,\rangle$.

Similarly, starting with $|\, j\, m\,\rangle$, we can apply J_- to it, if $m > -j$, to get

$$J_- \,|\, j\, m\,\rangle = c'_m \,|\, j\,,\, m-1\,\rangle\,. \tag{20.38}$$

With the normalization condition $\langle\, j\,,\, m-1\,|\, j\,,\, m-1\,\rangle = 1$, (20.16) implies

$$|\, c'_m\,|^2 = (j+m)(j-m+1) = j(j+1) - m(m-1)\,. \tag{20.39}$$

Again, we fix the phase of $|\, j\,,\, m-1\,\rangle$ so that c'_m is real and positive. Thus the analogous equation to (20.37) is

$$\boxed{J_- \,|\, j\, m\,\rangle = \sqrt{j(j+1) - m(m-1)}\;|\, j\,,\, m-1\,\rangle}\quad. \tag{20.40}$$

If $m-1 > -j$, we can repeatedly apply J_- until we arrive at $|\, j\,,\, -j\,\rangle$.

The $2j+1$ vectors $|\, j\, j\,\rangle, |\, j\,,\, j-1\,\rangle, \dots, |\, j\,,\, -j\,\rangle$ span a certain vector space $\mathbb{V}^{(j)}$ which is invariant under the actions of the operators J^2, J_z, J_+, J_- (equivalently J^2, J_z, J_x, J_y), and hence under the action of any element in the Lie algebra $\mathcal{SO}(3)$. Indeed $\mathbb{V}^{(j)}$ is an irreducible representation space of the group $SO(3)$ or $SU(2)$, which we will label as $D^{(j)}$. The explicit forms for these important finite-dimensional representations will be presented in Chapter 22.

Using (20.37) and (20.40), we have the following general formulas for the construction of particular angular momentum eigenstates $|\, j\, m\,\rangle$ from other such eigenstates:

$$|\, j\,,\, \pm m\,\rangle = \sqrt{\frac{(j+m)!}{(2j)!(j-m)!}}\; J_{\mp}^{j-m}\,|\, j\,,\, \pm j\,\rangle\,, \tag{20.41}$$

$$|\, j\,,\, \pm j\,\rangle = \sqrt{\frac{(j+m)!}{(2j)!(j-m)!}}\; J_{\pm}^{j-m}\,|\, j\,,\, \pm m\,\rangle\,. \tag{20.42}$$

The actions of $J_\pm$ may be compared with those of a and $a^\dagger$, the annihilation and creation operators, respectively, of the simple harmonic oscillator problem [cf. (14.25) and (14.26)]. The linear vector space spanned by the set $\{a, a^\dagger, \mathbb{I}\}$, where $\mathbb{I}$ is the identity operator, together with the internal multiplication rule given by the commutation relation (14.18), forms a complex algebra called the **Heisenberg algebra**. It has the infinite-dimensional representation given by (14.33).

Problem 20.1 Consider a spin 1/2 particle (such as an electron) in a central field (the potential is spherically symmetric). Using the fact that

$$\boldsymbol{J} = \boldsymbol{L} + \frac{\boldsymbol{\sigma}}{2},$$

where the components of $\boldsymbol{\sigma}$ are the Pauli spin matrices, show that the operators

$$\Lambda_l^+ = \frac{l+1+\boldsymbol{\sigma}\cdot\boldsymbol{L}}{2l+1}, \quad \Lambda_l^- = \frac{l-\boldsymbol{\sigma}\cdot\boldsymbol{L}}{2l+1}$$

are *projection operators* onto the states $|\, j = l \pm 1/2\,,\, m\,\rangle$, respectively, in the subspace of orbital angular momentum l.

Chapter 21

Whence the Spherical Harmonics?

The spherical harmonics are functions defined on the unit sphere S^2 (embedded in $\mathbb{R}^3$), and arise as solutions to **Laplace's equation**:

$$\nabla^2 \psi(\boldsymbol{r}) \equiv \left(\frac{\partial^2}{\partial x^2} + \frac{\partial^2}{\partial y^2} + \frac{\partial^2}{\partial z^2} \right) \psi(\boldsymbol{r}) = 0 \,. \tag{21.1}$$

We will see in the next chapter that they are also basis vectors in the finite-dimensional irreducible representation spaces of $SO(3)$, and taken together, constitute a basis for square-integrable functions defined on the unit sphere. In this chapter we will show that they are eigenfunctions of **orbital** angular momentum operators and discuss some of their more important properties.

The symbol $\boldsymbol{J}$ in the previous chapter can stand for either orbital or spin angular momentum, or the sum of these. From now on we will use $\boldsymbol{L}$ to denote specifically orbital angular momentum, and $\boldsymbol{S}$ to denote specifically spin angular momentum, while $\boldsymbol{J}$ will be reserved for the total angular momentum (sum of orbital and spin). These are the traditional symbols used in the physics literature.

For $\boldsymbol{L}$ we write [from (20.11) and (20.12)]

$$L^2 \,|\, l\, m \,\rangle = l(l+1) \,|\, l\, m \,\rangle \,, \tag{21.2}$$

$$L_z \,|\, l\, m \,\rangle = m \,|\, l\, m \,\rangle \,, \tag{21.3}$$

where l is a non-negative integer and the operators are still considered to be dimensionless [cf. (20.1)]. These equations can be written in coordinate representation. Equation (21.3), for exampe, can be recast as

$$\langle\, \boldsymbol{x} \,|\, L_z \,|\, l\, m \,\rangle = m \,\langle\, \boldsymbol{x} \,|\, l\, m \,\rangle \,. \tag{21.4}$$

Replacing L_z by the operator $xk_y - yk_x$ [(17.58)] and using the completeness condition (10.6), the left-hand side can be written

$$\int d^3k \, \langle\, \boldsymbol{x} \,|\, xk_y - yk_x \,|\, \boldsymbol{k} \,\rangle\langle\, \boldsymbol{k} \,|\, l\, m \,\rangle = \int d^3k \, (xk_y - yk_x) \, \langle\, \boldsymbol{x} \,|\, \boldsymbol{k} \,\rangle\langle\, \boldsymbol{k} \,|\, l\, m \,\rangle \,, \tag{21.5}$$

where on the left-hand side the quantity $xk_y - yk_x$ is treated as an operator while on the right-hand side it represents a c-number. On recalling the generalization of (10.2) to a three-dimensional context, namely,

$$\langle\, \boldsymbol{x} \,|\, \boldsymbol{k} \,\rangle = \frac{1}{(2\pi)^{3/2}} \, e^{i\boldsymbol{k}\cdot\boldsymbol{x}} \,, \tag{21.6}$$

we have

$$\begin{aligned} \langle\, \boldsymbol{x} \,|\, L_z \,|\, l\, m \,\rangle &= -i \left(x\frac{\partial}{\partial y} - y\frac{\partial}{\partial x} \right) \int d^3k \, \frac{1}{(2\pi)^{3/2}} \, e^{i\boldsymbol{k}\cdot\boldsymbol{x}} \, \langle\, \boldsymbol{k} \,|\, l\, m \,\rangle \\ = -i \left(x\frac{\partial}{\partial y} - y\frac{\partial}{\partial x} \right) \int d^3k \, \langle\, \boldsymbol{x} \,|\, \boldsymbol{k} \,\rangle\langle\, \boldsymbol{k} \,| l\, m \,\rangle &= -i \left(x\frac{\partial}{\partial y} - y\frac{\partial}{\partial x} \right) \langle\, \boldsymbol{x} \,|\, l\, m \,\rangle \,. \end{aligned} \tag{21.7}$$

Defining the wave functions

$$F_l^m(\boldsymbol{x}) \equiv \langle\, \boldsymbol{x} \,|\, l\, m \,\rangle \,, \tag{21.8}$$

Eq. (21.4) can then be rewritten as the following differential equation:

$$-i \left(x\frac{\partial}{\partial y} - y\frac{\partial}{\partial x} \right) F_l^m(\boldsymbol{x}) = mF_l^m(\boldsymbol{x}) \,. \tag{21.9}$$

In terms of the spherical coordinates (r, θ, φ),

$$L_z = -i\partial/\partial\varphi \,. \tag{21.10}$$

Thus (21.9) can be written as

$$-i\frac{\partial}{\partial\varphi} \, F_l^m(\boldsymbol{x}) = mF_l^m(\boldsymbol{x}) \,, \tag{21.11}$$

the solution of which is easily seen to be

$$F_l^m(\boldsymbol{x}) = f_l^m(r, \theta) \, e^{im\varphi} \,, \tag{21.12}$$

where $f_l^m(\theta, \varphi)$ is as yet an undetermined function of the stated variables. Since the wave function $F_l^m(\boldsymbol{x})$ must be single-valued in $\boldsymbol{x}$, we require that

$$F_l^m(r, \theta, \varphi) = F_l^m(r, \theta, \varphi + 2\pi) \,. \tag{21.13}$$

This implies

$$e^{2im\pi} = 1 \,, \tag{21.14}$$

or

$$m = 0, \pm 1, \pm 2, \pm 3, \ldots . \tag{21.15}$$

Then it follows from the results in the last chapter on the spectrum of J^2 and J_z [properties (A), (B) and (C) on p. 128] that

$$l = 0, 1, 2, 3, \ldots . \tag{21.16}$$

We thus have the important conclusion that all *orbital* angular momentum quantum numbers are integral.

To solve for $f_l^m(r,\theta)$ we have to use the following differential expressions for the components of the orbital angular momentum:

$$L_\pm = L_x \pm iL_y = e^{\pm i\varphi}\left(\pm\frac{\partial}{\partial\theta} + i\cot\theta\,\frac{\partial}{\partial\varphi}\right) . \tag{21.17}$$

Consider the wave function

$$F_l^l(\boldsymbol{x}) = f_l^l(r,\theta)\, e^{il\varphi} . \tag{21.18}$$

We have, from (20.21),

$$L_+ F_l^l = 0 , \tag{21.19}$$

or

$$e^{i\varphi}\left(\frac{\partial}{\partial\theta} + i\cot\theta\,\frac{\partial}{\partial\varphi}\right) f_l^l(r,\theta)\, e^{il\varphi} = e^{i\varphi}\, e^{il\varphi}\left(\frac{\partial}{\partial\theta} - l\cot\theta\right) f_l^l(r,\theta) = 0 , \tag{21.20}$$

that is,

$$\left(\frac{\partial}{\partial\theta} - l\cot\theta\right) f_l^l(r,\theta) = 0 . \tag{21.21}$$

The solution of this equation is

$$f_l^l(r,\theta) = R_l(r)\sin^l\theta , \tag{21.22}$$

where $R_l(r)$ is a function of r only. Thus

$$F_l^l(\boldsymbol{x}) = R_l(r)\,\sin^l\theta\, e^{il\varphi} . \tag{21.23}$$

Since $L_\pm$ only involve angular variables, the radial parts of the wave functions $F_l^m(r,\theta,\varphi)$ must be the same for the $2l+1$ possible values of m for a given l, and are thus all equal to $R_l(r)$. We will henceforth consider only the angular parts of $F_l^m(\boldsymbol{x})$, assume that they are normalized, and require that they conform to a special phase convention to be specified below. The resulting functions are called the **spherical harmonics**, denoted $Y_l^m(\theta,\varphi)$. The normalization condition is

$$\int d\Omega\,(Y_l^m(\Omega))^* Y_l^m(\Omega) = 1 , \tag{21.24}$$

where $d\Omega = \sin\theta d\theta d\varphi$ is the solid angle element and the integration is over the entire unit sphere (over all solid angles). The phases of the Y_l^m's will be specified by requiring that $Y_l^0(0,0)$ is real and positive, and that they satisfy equations analogous to (20.37) and (20.40):

$$L_+ Y_l^m = \sqrt{l(l+1)-m(m+1)}\, Y_l^{m+1}\,, \quad (m \neq l)\,, \tag{21.25}$$

$$L_- Y_l^m = \sqrt{l(l+1)-m(m-1)}\, Y_l^{m-1}\,, \quad (m \neq -l)\,. \tag{21.26}$$

The spherical harmonics thus defined then form a set of complete, orthonormal set of square-integrable functions on the unit sphere:

$$\int d\Omega\, (Y_l^m)^* Y_{l'}^{m'} = \delta_{ll'}\delta_{mm'}\,, \quad \text{(orthonormality)}\,, \tag{21.27}$$

$$\sum_{l=0}^{\infty}\sum_{m=-l}^{l} (Y_l^m(\theta,\varphi))^* Y_l^m(\theta',\varphi') = \delta(\Omega-\Omega') = \frac{\delta(\theta-\theta')\delta(\varphi-\varphi')}{\sin\theta}\,, \quad \text{(complete}$$
(21.28)

In fact, the above equations are simply consequences of

$$\langle\, l\, m \,|\, l'\, m' \,\rangle = \delta_{mm'}\delta_{ll'} \quad \text{(orthonormality)} \tag{21.29}$$

and

$$\sum_{l=0}^{\infty}\sum_{m=-l}^{l} |\, l\, m\,\rangle\langle\, l\, m\,| = 1 \quad \text{(completeness)}\,, \tag{21.30}$$

if we write

$$Y_l^m(\theta,\varphi) = \langle\, \hat{\boldsymbol{n}} \,|\, l\, m\,\rangle\,, \tag{21.31}$$

where $\hat{\boldsymbol{n}}$ is the unit vector along the (θ,φ) direction.

We can now construct the Y_l^m's explicitly. From (20.41) we have

$$Y_l^m(\theta,\varphi) = \sqrt{\frac{(l+m)!}{(2l)!(l-m)!}}\,(L_-)^{l-m}\, Y_l^l(\theta,\varphi) \tag{21.32}$$

$$= \sqrt{\frac{(l-m)!}{(2l)!(l+m)!}}\,(L_+)^{l+m}\, Y_l^{-l}(\theta,\varphi)\,, \tag{21.33}$$

where the differential operators $L_\pm$ are given by (21.17). The following equality, which can be checked by straightforward differentiation, will be useful as a calculational tool: For any value of μ and any function $f(\theta)$,

$$L_\pm\, e^{i\mu\varphi} f(\theta) = \mp\, e^{i(\mu\pm 1)\varphi}\left(\sin^{1\pm\mu}\theta\,\frac{d}{(\cos\theta)}\,\sin^{\mp\mu}\theta\right) f(\theta)\,. \tag{21.34}$$

Repeated application of $L_\pm$ to $e^{i\mu\varphi} f(\theta)$ then gives, for positive integer values of s,

$$(L_\pm)^s\, e^{i\mu\varphi} f(\theta) = (\mp 1)^s\, e^{i(\mu\pm s)\varphi}\left(\sin^{s\pm\mu}\theta\,\frac{d^s}{d(\cos\theta)^s}\,\sin^{\mp\mu}\theta\right) f(\theta)\,. \tag{21.35}$$

Now (21.23) implies

$$Y_l^l(\theta,\varphi) = c_l \sin^l\theta\, e^{il\varphi} \,, \tag{21.36}$$

where c_l is a normalization constant. Its absolute value can be determined from the normalization condition (21.24). Thus

$$(2\pi)|\, c_l\,|^2 \int_0^\pi d\theta\, \sin\theta \sin^{2l}\theta = (2\pi)|\, c_l\,|^2 \int_{-1}^1 dz (1-z^2)^l = 1 \,, \tag{21.37}$$

where in the first equality we have used the substitution $z = \cos\theta$. This gives

$$|\, c_l\,| = \frac{1}{\sqrt{4\pi}}\, \frac{\sqrt{(2l+1)!}}{2^l\, l!} \,. \tag{21.38}$$

The phase of c_l remains to be determined in accordance with the conditions (21.25) and (21.26), and the convention that $Y_l^0(0,0)$ is real and positive.

Using (21.36) in (21.32), and with the help of (21.35) (for L_-, with $\mu = l$ and $s = l-m$), we have

$$Y_l^m(\theta,\varphi) = c_l \sqrt{\frac{(l+m)!}{(2l)!(l-m)!}}\, e^{im\varphi} \sin^{-m}\theta\, \frac{d^{l-m}}{d(\cos\theta)^{l-m}} \sin^{2l}\theta \,. \tag{21.39}$$

When $m = -l$, this formula gives

$$Y_l^{-l}(\theta,\varphi) = (-1)^l\, c_l\, e^{-il\varphi} \sin^l\theta \,. \tag{21.40}$$

Substituting this in (21.33) and using (21.35) again (for L_+, with $\mu = -l$ and $s = l+m$), we obtain another expression for Y_l^m:

$$Y_l^m(\theta,\varphi) = (-1)^m c_l \sqrt{\frac{(l-m)!}{(2l)!(l+m)!}}\, e^{im\varphi} \sin^m\theta\, \frac{d^{l+m}}{d(\cos\theta)^{l+m}} \sin^{2l}\theta \,. \tag{21.41}$$

Setting $m = 0$ in either of (21.39) or (21.41) we find

$$Y_l^0(\theta,\varphi) = c_l \sqrt{\frac{1}{(2l)!}}\, \frac{d^l}{d(\cos\theta)^l} (1-\cos^2\theta)^l \,, \tag{21.42}$$

which is just a **Legendre polynomial** to within a constant. These polynomials, with proper normalization, are given by

$$\boxed{P_l(\cos\theta) = \frac{1}{2^l\, l!}\, \frac{d^l}{d(\cos\theta)^l} (\cos^2\theta - 1)^l} \,. \tag{21.43}$$

Thus we can write

$$Y_l^0(\theta,\varphi) = (-1)^l c_l\, \frac{2^l\, l!}{\sqrt{(2l)!}}\, P_l(\cos\theta) \,. \tag{21.44}$$

It then follows from (21.38) that

$$Y_l^0(\theta,\varphi) = (-1)^l \,\frac{c_l}{|\,c_l\,|}\sqrt{\frac{2l+1}{4\pi}}\; P_l(\cos\theta)\;. \tag{21.45}$$

The convention that Y_l^0 is real and positive then leads to the following choice for the phase of c_l:

$$\frac{c_l}{|\,c_l\,|} = (-1)^l\,. \tag{21.46}$$

Summing up our results so far we have the following expressions for the spherical harmonics with the proper normalization constants:

$$Y_l^0(\theta,\varphi) = \sqrt{\frac{2l+1}{4\pi}}\; P_l(\cos\theta)\,, \tag{21.47}$$

$$Y_l^l(\theta,\varphi) = (-1)^l\sqrt{\frac{(2l+1)}{4\pi}\,\frac{(2l)!}{2^{2l}(l!)^2}}\;\sin^l\theta\, e^{il\varphi}\;. \tag{21.48}$$

Eq. (21.39) [or (21.41)] immediately implies that

$$\boxed{Y_l^{-m}(\theta,\varphi) = (-1)^m\,(Y_l^m(\theta,\varphi))^*}\quad . \tag{21.49}$$

It also shows that the *parity* of Y_l^m is $(-1)^l$, that is,

$$\boxed{Y_l^m(\pi-\theta,\varphi+\pi) = (-1)^l\,Y_l^m(\theta,\varphi)}\quad , \tag{21.50}$$

where $(\pi-\theta,\varphi+\pi)$ is the spatial reflection of the point (θ,φ) on the unit sphere.

With the phase convention for the normalization constants c_l given by (21.38) and (21.46), we can finally use either (21.39) or (21.41) to express the spherical harmonics in terms of the so-called **associated Legendre polynomials**, $P_l^m(\cos\theta)$, as follows:

$$\boxed{Y_l^m(\theta,\varphi) = (-1)^m\sqrt{\frac{(2l+1)}{4\pi}\,\frac{(l-|\,m\,|)!}{(l+|\,m\,|)!}}\;P_l^m(\cos\theta)\,e^{im\varphi}}\quad , \tag{21.51}$$

where

$$\boxed{P_l^m(\cos\theta) = (1-\cos^2\theta)^{|\,m\,|/2}\,\frac{d^{|\,m\,|}}{d(\cos\theta)^{|\,m\,|}}\;P_l(\cos\theta)}\quad . \tag{21.52}$$

For reference, a few low-order (and most commonly encountered) spherical harmonics are given below [the expressions for $m < 0$ can be obtained from (21.49)]:

$$Y_0^0 = \frac{1}{\sqrt{4\pi}} , \tag{21.53a}$$

$$Y_1^0 = \sqrt{\frac{3}{4\pi}} \cos\theta , \quad Y_1^1 = -\sqrt{\frac{3}{8\pi}} \sin\theta \, e^{i\varphi} , \tag{21.53b}$$

$$Y_2^0 = \sqrt{\frac{5}{16\pi}} (3\cos^2\theta - 1) , \quad Y_2^1 = -\sqrt{\frac{15}{8\pi}} \sin\theta\cos\theta \, e^{i\varphi} ,$$

$$Y_2^2 = \sqrt{\frac{15}{32\pi}} \sin^2\theta \, e^{2i\varphi} , \tag{21.53c}$$

$$Y_3^0 = \sqrt{\frac{7}{16\pi}} (5\cos^3\theta - 3\cos\theta) , \quad Y_3^1 = -\sqrt{\frac{21}{64\pi}} \sin\theta \, (5\cos^2\theta - 1) \, e^{i\varphi} ,$$

$$Y_3^2 = \sqrt{\frac{105}{32\pi}} \, sin^2\theta \cos\theta \, e^{2i\varphi} , \quad Y_3^3 = -\sqrt{\frac{35}{64\pi}} \sin^3\theta \, e^{3i\varphi} . \tag{21.53d}$$

Problem 21.1 A charged spin zero particle is bound to a Coulomb potential due to a point charge at the origin. This atom is perturbed by charges uniformly distributed on a very large circle of radius R in the xy-plane and centered at the origin. From the symmetry of the problem give some arguments to show that the perturbation Hamiltonian can be approximated by the electrostatic potential

$$\Phi(\boldsymbol{r}) = cr^2 \, Y_2^0(\theta) , \quad (r < R) ,$$

where c is a constant. From this form of the perturbation potential justify that i) parity is conserved, ii) J_z is conserved, and iii) the states $|\, l \, m \rangle$ and $|\, l \, , -m \rangle$ are degenerate.

Problem 21.2 A particle of mass M is constrained to move on a circle of radius R on the xy-plane with center at the origin, but is otherwise free. The Hamiltonian is given by

$$H = \frac{L_z^2}{2MR^2} ,$$

where

$$L_z = -i\hbar \, \frac{\partial}{\partial\varphi} .$$

Construct a complete set of simultaneous, orthonormal eigenfunctions of H and L_z: $f_m(\varphi)$, $-\infty < m < \infty$. Write the completeness and orthonormality conditions of this set of eigenfunctions using *both* the Dirac notation and in terms of the functional forms of $f_m(\varphi)$. Explain how this problem is relevant to Fourier analysis. Also interpret your results in terms of group representation theory.

Chapter 22

Irreducible Representations of $SU(2)$ and $SO(3)$, Rotation Matrices

As discussed in Chapter 19, the groups $SU(2)$ and $SO(3)$ are compact Lie groups. It is a mathematical fact that *all irreducible representations of compact Lie groups are finite dimensional and are equivalent to unitary representations.* Due to the 2-to-1 homomorphism ϕ between $SU(2)$ and $SO(3)$ discussed in Chapter 19 [cf. (19.8)], the irreducible representations of $SO(3)$ are given exhaustively by those of $SU(2)$. In fact, if $D(g)$, $g \in SU(2)$, is a matrix representation of $SU(2)$, then $D(g)$ and $-D(g)$ represent the same element $R \in SO(3)$, where $\phi(g) = \phi(-g) = R$.

The irreducible representations of $SU(2)$ are each characterized by the total angular momentum eigenvalue j. From the results of Chapter 20, we know that j can assume the values $0, 1/2, 1, 3/2, 2, \dots$ For a particular value of j, the corresponding irreducible representation will be denoted by $D^{(j)}$, and the representation space, $\mathbb{V}^j$, is $(2j+1)$-dimensional. The orthonormal set of simultaneous eigenstates of J^2 and J_z, that is, the set $\{|\, j\, m\,\rangle\}$, with $m = j, j-1, \dots, -j$, can be chosen as a basis of $\mathbb{V}^j$. The matrix representation of each $R \in SO(3)$ will be given with respect to this basis. We will assume that the phases of the normalization constants of the eigenstates are such that (20.37) and (20.40) are satisfied. This phase convention is known as the **Condon-Shortley convention**.

From Theorem 19.1 (Euler's Theorem) and (17.60), an arbitrary rotation

$$R(\varphi, \theta, \psi) = R_3(\varphi) R_2(\theta) R_3(\psi) \in SO(3) \;,$$

where (ϕ, θ, ψ) are the Euler angles, can be represented by the unitary operator

$$U(\psi, \theta, \psi) = e^{-i\varphi J_3}\, e^{-i\theta J_2}\, e^{-i\psi J_3} \;, \tag{22.1}$$

acting on $\mathbb{V}^j$. The matrix representation of this operator with respect to the basis states $|\,j\,m\,\rangle$ is given by

$$U(\varphi,\theta,\psi)\,|\,j\,m\,\rangle = \sum_{m'} D^{(j)}(\varphi,\theta,\psi)^{m'}_{m}\,|\,j\,m'\,\rangle\,, \tag{22.2}$$

where $D^{(j)}(\varphi,\theta,\psi)^{m'}_{m}$ is the $m'm$ element of the so-called **rotation matrix** $D^{(j)}$. From (22.1) these matrix elements are given explicitly in the Dirac notation by

$$\begin{aligned}(D^{(j)})^{m'}_{m} &= \langle\,j\,m'\,|\,U(\varphi,\theta,\psi)\,|\,j\,m\,\rangle = \langle\,j\,m'\,|\,e^{-i\varphi J_3}\,e^{-i\theta J_2}\,e^{-i\psi J_3}\,|\,j\,m\,\rangle\\ &= e^{-im'\varphi}\,d^j(\theta)^{m'}_{m}\,e^{-im\psi}\,,\end{aligned} \tag{22.3}$$

where the last equality follows from the fact that $|\,j\,m\,\rangle$ is an eigenstate of J_z with eigenvalue m [cf. (20.12)], and the matrix element $d^j(\theta)^{m'}_{m}$ is defined to be

$$\boxed{d^{(j)}(\theta)^{m'}_{m} = \langle\,j\,m'\,|\,e^{-i\theta J_2}\,|\,j\,m\,\rangle}\;. \tag{22.4}$$

The Condon-Shortley phase convention [(20.37) and (20.40)] and the fact that [recall (20.4)]

$$J_2 = \frac{1}{2i}\,(J_+ - J_-) \tag{22.5}$$

imply that J_2 (with respect to the basis states $|\,j\,m\,\rangle$) is always represented by an imaginary anti-symmetric matrix. For example

$$\begin{aligned}&\langle\,j\,m\,|\,J_+ - J_-\,|\,j\,,\,m-1\,\rangle = \langle\,j\,m\,|\,J_+\,|\,j\,,\,m-1\,\rangle = \sqrt{j(j+1)-m(m-1)}\,,\\ &\langle\,j\,,\,m-1\,|\,J_+ - J_-\,|\,j\,m\,\rangle = -\langle\,j\,,\,m-1\,|\,J_-\,|\,j\,m\,\rangle = -\sqrt{j(j+1)-m(m-1)}\,.\end{aligned}$$

Thus $-i\theta J_2$ must be real. Since $\exp(-i\theta J_2)$ is unitary, it must also be real orthogonal [recall discussion immediately preceding (19.2)]. Hence *the $d^{(j)}$ matrices are all real orthogonal matrices* under the Condon-Shortley convention. We will now construct the rotation matrices for some specific values of j.

First consider the case $j = 1/2$. The representation space $\mathbb{V}^{1/2}$ is two-dimensional since m can only assume the values $1/2$ and $-1/2$. From (20.12) and (20.37) it follows immediately that

$$J_3 = \begin{pmatrix} 1/2 & 0 \\ 0 & -1/2 \end{pmatrix} = \frac{1}{2}\,\sigma_3\,, \tag{22.6}$$

where σ_3 is a Pauli spin matrix [cf. (19.6)], and

$$J_+ = \begin{pmatrix} 0 & 1 \\ 0 & 0 \end{pmatrix}\,. \tag{22.7}$$

Then

$$J_+^\dagger = J_- = \begin{pmatrix} 0 & 0 \\ 1 & 0 \end{pmatrix} . \tag{22.8}$$

Thus

$$J_1 = \frac{1}{2}(J_+ + J_-) = \frac{1}{2}\begin{pmatrix} 0 & 1 \\ 1 & 0 \end{pmatrix} = \frac{1}{2}\sigma_1 , \tag{22.9}$$

$$J_2 = -\frac{i}{2}(J_+ - J_-) = -\frac{i}{2}\begin{pmatrix} 0 & 1 \\ -1 & 0 \end{pmatrix} = \frac{1}{2}\begin{pmatrix} 0 & -i \\ i & 0 \end{pmatrix} = \frac{1}{2}\sigma_2 , \tag{22.10}$$

where σ_1 and σ_2 are Pauli spin matrices given by (19.6). The above results for matrix representation of the angular momentum operators for the spin 1/2 case are summarized by:

$$\boxed{j = 1/2 : \quad J_i = \frac{1}{2}\sigma_i , \quad i = 1, 2, 3} \quad . \tag{22.11}$$

The above equation gives the spin 1/2 representation of the Lie algebra $\mathcal{SO}(3)$ of the Lie group $SO(3)$. Thus, from (22.4) and (19.7f),

$$d^{(1/2)}(\theta) = e^{-i(\theta/2)\,\sigma_2} = \cos\frac{\theta}{2} - i\sigma_2 \sin\frac{\theta}{2} , \tag{22.12}$$

or, in explicit matrix form,

$$d^{(1/2)}(\theta) = \begin{pmatrix} \cos\frac{\theta}{2} & -\sin\frac{\theta}{2} \\ \sin\frac{\theta}{2} & \cos\frac{\theta}{2} \end{pmatrix} . \tag{22.13}$$

It follows from (22.3) that

$$D^{(1/2)}(\varphi, \theta, \psi) = \begin{pmatrix} e^{-i\varphi/2}\cos\frac{\theta}{2}\, e^{-i\psi/2} & -e^{-i\varphi/2}\sin\frac{\theta}{2}\, e^{i\psi/2} \\ e^{i\varphi/2}\sin\frac{\theta}{2}\, e^{-i\psi/2} & e^{i\varphi/2}\cos\frac{\theta}{2}\, e^{i\psi/2} \end{pmatrix} . \tag{22.14}$$

We can verify directly that this matrix is unitary and that its determinant is equal to one. Thus $D^{(1/2)}(\varphi, \theta, \psi) \in SU(2)$. Note that, as expected, $D^{(1/2)}(0,0,0) = d^{(1/2)}(0) = e$, the 2×2 identity matrix. If we represent an arbitrary rotation in $SO(3)$ by $R_{\hat{\boldsymbol{n}}}(\theta)$, where $\hat{\boldsymbol{n}}$ is the axis of rotation and θ is the angle of rotation, we can see more generally that

$$D^{(1/2)}(R_{\hat{\boldsymbol{n}}}(2m\pi)) = (-1)^m\, e , \quad m \in \mathbb{Z} . \tag{22.15}$$

Before we prove this fact let us discuss its significance. Equation (22.15) indicates that in the representation $D^{(1/2)}$ an even number of rotations by 2π about any axis $\hat{\boldsymbol{n}}$ is represented by the identity matrix e, while an odd number of such rotations is represented by $-e$. This is just an instance of the two-to-one correspondence between $SU(2)$ and $SO(3)$ under the group homomorphism ϕ of (19.8). Physicists call this representation, with some abuse of language, a **double-valued representation** of $SO(3)$. This double-valuedness is intimately tied to the double-connectedness of the group manifold of $SO(3)$. It is also directly responsible for the transformation properties of spinors under rotations. We will discuss this in more detail in Chapter 24.

To prove (22.15) we will first introduce a definition and a lemma.

Definition 22.1. *Given a group action of a group G on a manifold M (such as the action of the rotation group $SO(3)$ on $\mathbb{R}^3$) the* ***isotropy subgroup*** *G_x, for a given $x \in M$, is*

$$G_x = \{ g \in G \,|\, gx = x \} . \tag{22.16}$$

To check that G_x as defined is indeed a subgroup, we note the following facts.

(i) $e \in G_x$, that is, the identity of G is an element of G_x.

(ii) If $g_1 \in G_x$ and $g_2 \in G_x$, then

$$(g_1 g_2)x = g_1(g_2 x) = g_1 x = x ,$$

that is, $g_1 g_2 \in G_x$.

(iii) If $g \in G_x$, then $gx = x$. Acting on both sides by g^{-1} we have $g^{-1}x = x$. Thus $g^{-1} \in G_x$.

Lemma 22.1. *Given a group action of a group G on a manifold M, the following relationship holds between isotropy subgroups:*

$$G_{gx} = g\, G_x \, g^{-1} , \tag{22.17}$$

where $g \in G$ and $x \in M$.

Proof. We will prove this lemma by showing that $G_{gx} \subset g\, G_x\, g^{-1}$ and $g\, G_x\, g^{-1} \subset G_{gx}$. Suppose $g' \in G_{gx}$. Then $g'gx = gx$, implying $(g^{-1}g'g)x = g^{-1}gx = x$. Hence $g'' \equiv g^{-1}g'g \in G_x$, from which $g' = gg''g^{-1} \in g\, G_x\, g^{-1}$. It follows that $G_{gx} \subset g\, G_x\, g^{-1}$. Conversely, suppose $g'' \in G_x$. Then $g''x = x$, implying $(gg''g^{-1})gx = gg''x = gx$. Thus $gg''g^{-1} \in G_{gx}$, from which $g\, G_x\, g^{-1} \subset G_{gx}$. The lemma follows. □

Now suppose $R \in SO(3)$ is a rotation that rotates the unit vector $\hat{e}_2$ to the unit vector $\hat{\boldsymbol{n}}$, that is, $R\,\hat{e}_2 = \hat{\boldsymbol{n}}$. By the above lemma and (22.12), for $m \in \mathbb{Z}$,

$$\begin{aligned} D^{(1/2)}(R_{\hat{\boldsymbol{n}}}(2m\pi)) &= D^{(1/2)}(R)\, e^{-im\pi\sigma_2}\, (D^{(1/2)}(R))^{-1} \\ &= D^{(1/2)}(R)\, (\cos m\pi - i\sigma_2 \sin m\pi)\, (D^{(1/2)}(R))^{-1} \\ &= D^{(1/2)}(R)\, (-1)^m e\, (D^{(1/2)}(R))^{-1} = (-1)^m e , \end{aligned} \tag{22.18}$$

where e is the 2×2 identity matrix. Thus (22.15) is proved.

Problem 22.1 A spinor ψ is said to point in the z direction if

$$\psi = e^{i\delta}\, u_+ \,, \quad \text{where } u_+ = \begin{pmatrix} 1 \\ 0 \end{pmatrix} ,$$

and δ is any real number, so that J_z is certainly 1/2.

(a) Using the representation $D^{(1/2)}$ of the rotation group, determine in what direction the spinor

$$\psi' = a_+ u_+ + a_- u_- = \begin{pmatrix} a_+ \\ a_- \end{pmatrix}$$

is pointing, where a_+ and a_- are any two complex numbers satisfying $|\, a_+ \,|^2 + |\, a_- \,|^2 = 1$. Specify the direction by giving the polar and azimuthal angles θ and φ.

(b) What are a_+ and a_- for spinors pointing in the x and y directions? Remember that a_+ and a_- are determined only up to a common phase factor.

We will now consider the case of $j = 1$. Applying (20.12) for J_z and (20.37) and (20.40) for J_+ and J_-, respectively, we have in explicit matrix form with respect to the basis $|\, 1\, m \,\rangle$,

$$J_3 = \begin{pmatrix} 1 & 0 & 0 \\ 0 & 0 & 0 \\ 0 & 0 & -1 \end{pmatrix} , \tag{22.19}$$

$$J_+ = \begin{pmatrix} 0 & \sqrt{2} & 0 \\ 0 & 0 & \sqrt{2} \\ 0 & 0 & 0 \end{pmatrix} , \quad J_- = \begin{pmatrix} 0 & 0 & 0 \\ \sqrt{2} & 0 & 0 \\ 0 & \sqrt{2} & 0 \end{pmatrix} = J_+^\dagger \,. \tag{22.20}$$

The above equations give completely the spin-1 representation of the Lie algebra $\mathcal{SO}(3)$. Proceeding further we have

$$J_2 = \frac{1}{2i}\,(J_+ - J_-) = \frac{1}{2i} \begin{pmatrix} 0 & \sqrt{2} & 0 \\ -\sqrt{2} & 0 & \sqrt{2} \\ 0 & -\sqrt{2} & 0 \end{pmatrix} . \tag{22.21}$$

Hence

$$d^{(1)}(\theta) = e^{-i\theta J_2} = \exp\left(-\frac{\theta}{\sqrt{2}}\,A\right), \tag{22.22}$$

where

$$A = \begin{pmatrix} 0 & 1 & 0 \\ -1 & 0 & 1 \\ 0 & -1 & 0 \end{pmatrix}. \tag{22.23}$$

A simple calculation yields

$$A^2 = \begin{pmatrix} -1 & 0 & 1 \\ 0 & -2 & 0 \\ 1 & 0 & -1 \end{pmatrix} \equiv B\,, \tag{22.24}$$

$$A^3 = (-2)A\,, \quad A^4 = (-2)B\,, \quad A^5 = (-2)^2 A\,, \quad A^6 = (-2)^2 B\,, \ldots. \tag{22.25}$$

In general,

$$A^{2n+1} = (-2)^n A \quad (n = 0, 1, 2, \ldots)\,; \qquad A^{2n} = (-2)^{n-1} B \quad (n = 1, 2, 3, \ldots)\,. \tag{22.26}$$

Thus we can write

$$\begin{aligned} e^{-i\theta J_2} &= \exp\left(-\frac{\theta}{\sqrt{2}}\,A\right) \\ &= \left(1 + \frac{\theta^2}{2!\,2}\,A^2 + \frac{\theta^4}{4!\,2^2}\,A^4 + \frac{\theta^6}{6!\,2^3}\,A^6 + \ldots\right) \\ &\quad - \left(\frac{\theta}{2^{1/2}}\,A + \frac{\theta^3}{3!\,2^{3/2}}\,A^3 + \frac{\theta^5}{5!\,2^{5/2}}\,A^5 + \ldots\right) \\ &= \left(1 + \frac{\theta^2}{2!\,2}\,B + \frac{\theta^4}{4!\,2^2}\,(-2)B + \frac{\theta^6}{6!\,2^3}\,(-2)^2 B + \ldots\right) \\ &\quad - \left(\frac{\theta}{2^{1/2}}\,A + \frac{\theta^3}{3!\,2^{3/2}}\,(-2)A + \frac{\theta^5}{5!\,2^{5/2}}\,(-2)^2 A + \ldots\right) \\ &= 1 + \frac{B}{2}\left(\frac{\theta^2}{2!} - \frac{\theta^4}{4!} + \frac{\theta^6}{6!} - \ldots\right) - \frac{A}{\sqrt{2}}\left(\theta - \frac{\theta^3}{3!} + \frac{\theta^5}{5!} - \ldots\right) \\ &= 1 + \frac{B}{2}\,(1 - \cos\theta) - \frac{A}{\sqrt{2}}\,\sin\theta\,. \end{aligned} \tag{22.27}$$

Using (22.23) (for A) and (22.24) (for B) in the above, we obtain

$$d^{(1)}(\theta) = \begin{pmatrix} \frac{1}{2}(1+\cos\theta) & -\frac{\sin\theta}{\sqrt{2}} & \frac{1}{2}(1-\cos\theta) \\ \frac{\sin\theta}{\sqrt{2}} & \cos\theta & -\frac{\sin\theta}{\sqrt{2}} \\ \frac{1}{2}(1-\cos\theta) & \frac{\sin\theta}{\sqrt{2}} & \frac{1}{2}(1+\cos\theta) \end{pmatrix} . \tag{22.28}$$

The general expression for the 3×3 matrix $D^{(1)}$ can easily be obtained by using the above result and (22.3). The representation $D^{(1)}$ that we have just obtained is in fact the defining representation for the group $SO(3)$: the 3×3 matrices in $D^{(1)}$ are precisely the 3×3 matrices in $SO(3)$. Unlike $D^{(1/2)}$ it is thus a **single-valued representation.**

Problem 22.2 Let $\boldsymbol{S}$ be the angular momentum operator for a spin 1 particle. Show that for a rotation $R \in SO(3)$ through θ about the unit vector $\boldsymbol{n}$, the rotation operator is given by

$$D^{(1)}(R) = 1 - i(\boldsymbol{n}\cdot\boldsymbol{S})\sin\theta - (1-\cos\theta)(\boldsymbol{n}\cdot\boldsymbol{S})^2 .$$

Evaluate $D^{(1)}(R)$ in matrix form explicitly when $\boldsymbol{n}$ is along the y-axis. Show that this is the matrix that rotates a Euclidean 3-vector.

It is easily seen that *the irreducible representations $D^{(j)}$ of $SO(3)$ fall into two classes: (i) single-valued when j is a positive integer; and (ii) double-valued when j is half of a positive odd integer.* Indeed, by Lemma 22.1,

$$R_{\hat{\boldsymbol{n}}}(2p\pi) = R\, R_3(2p\pi)\, R^{-1}\,, \quad p = 0, \pm 1, \pm 2, \dots\,, \tag{22.29}$$

where R is the rotation which sends e_3 to $\hat{\boldsymbol{n}}$. Thus

$$D^{(j)}(R_{\hat{\boldsymbol{n}}}(2p\pi)) = D^{(j)}(R)\, D^{(j)}(R_3(2p\pi))\, (D^{(j)}(R))^{-1} . \tag{22.30}$$

But

$$(D^{(j)}(R_3(2p\pi)))^{m'}_{m} = (D^{(j)}(e^{-2ip\pi J_3}))^{m'}_{m} = \delta_{m'm}\, e^{-2imp\pi}\,, \tag{22.31}$$

where p is an integer and $m = j\,, j-1\,, \dots\,, -j$. If j is an integer, so is mp, and $\exp(-2imp\pi) = 1$. It follows that $D^{(j)}(R_{\hat{\boldsymbol{n}}}(2p\pi)) = e$, the $(2j+1)\times(2j+1)$ identity matrix, and the representation is single-valued. If, on the other hand, $j = 1/2 \times$ odd integer, we can write $m = \pm(2n+1)/2\,, n = 0,\, 1,\, 2,\, \dots$. Thus

$$e^{-2imp\pi} = e^{\pm i\pi p(2n+1)} = (-1)^p\,, \quad p = 0\,, \pm 1\,, \pm 2\,, \dots\,. \tag{22.32}$$

In this case, $D^{(j)}(R_{\hat{\boldsymbol{n}}}(2p\pi)) = (-1)^p\, e$, and so the representation is double-valued.

We summarize a few important general properties of rotation matrices as follows:

(1) *Unitarity.* For all allowed values of j,

$$(D^{(j)})^{\dagger}(\varphi,\theta,\psi) = (D^{(j)})^{-1}(\varphi,\theta,\psi) = D^{(j)}(-\psi,-\theta,-\varphi)\,. \tag{22.33}$$

(2) *Unit Determinant.*

$$\begin{aligned}\det\{D^{(j)}(R_{\hat{n}}(\psi))\} &= \det\{D^{(j)}(R\,R_3(\psi)\,R^{-1})\}\\ &= \det\{D^{(j)}(R_3(\psi))\} = \prod_{m=-j}^{j} e^{-im\,\psi} = 1\,.\end{aligned} \tag{22.34}$$

(3) $d^{(j)}(\beta) = D^{(j)}(R_2(\beta))$ *is real and orthogonal.* This fact has been noted in the discussion following (22.25). Orthogonality of the matrix implies

$$(d^{(j)}(\beta))^{-1} = d^{(j)}(-\beta) = (d^{(j)}(\beta))^{T}. \tag{22.35}$$

(4) *Transformation under complex conjugation.* We have the following transformation rule:

$$(D^{(j)}(\varphi,\theta,\psi))^{*} = d^{(j)}(\pi)\,D^{(j)}(\varphi,\theta,\psi)\,(d^{(j)}(\pi))^{-1}. \tag{22.36}$$

Proof. For ease of writing we will suppress the (j) in $D^{(j)}$ in this proof. We have, by Euler's theorem (Theorem 19.1),

$$\begin{aligned}D^{*}(\varphi,\theta,\psi) &= D^{*}(R_3(\varphi)\,R_2(\theta)\,R_3(\psi))\\ &= D^{*}(R_3(\varphi))\,D^{*}(R_2(\theta))\,D^{*}(R_3(\psi))\,.\end{aligned} \tag{22.37}$$

But

$$\begin{aligned}D^{*}(R_3(\psi)) &= D^{*}(e^{-i\psi J_3}) = (D^{\dagger})^{T}(e^{-i\psi J_3}) = (D^{-1}(e^{-i\psi J_3}))^{T}\\ &= D^{T}(e^{i\psi J_3}) = D(e^{i\psi J_3}) = D(R_3(-\psi)) = D(e^{-i(-\psi)\,J\cdot\hat{\mathbf{e}}_3})\\ &= D(e^{-i\psi\,J\cdot(-\hat{\mathbf{e}}_3)}) = D(R_2(\pi)\,R_3(\psi)\,(R_2(\pi))^{-1})\\ &= D(R_2(\pi)\,R_3(\psi)\,R_2(-\pi)) = d(\pi)\,D(R_3(\psi))\,d(-\pi)\,,\end{aligned} \tag{22.38}$$

where, in the second equality, we have used the fact that $D^{\dagger} = (D^{*})^{T}$, in the third, that D is unitary, in the fifth, that J_3 is diagonal, and in the nineth, Lemma 22.1 together with the fact that $R_2(\pi)$ is the rotation that takes $\hat{\mathbf{e}}_3$ to $-\hat{\mathbf{e}}_3$. Also, because of property (3) above,

$$d^{*}(\theta) = d(\theta) = d(\pi)\,d(\theta)\,d(-\pi)\,. \tag{22.39}$$

The last equality obtains since the d matrices commute as they all represent rotations about the $\hat{\mathbf{e}}_2$ axis. Finally, using (22.38) and (22.39) in (22.37), we have

$$\begin{aligned}D^{*}(\varphi,\theta,\psi) &= d(\pi)\,D(R_3(\varphi))\,d^{-1}(\pi)\;\; d(\pi)d(\theta)d^{-1}(\pi)\\ &\quad\cdot d(\pi)\,D(R_3(\psi))\,d^{-1}(\pi)\\ &= d(\pi)\,D(R_3(\varphi))\,d(\theta)\,D(R_3(\psi))\,d^{-1}(\pi)\,.\end{aligned} \tag{22.40}$$

The result (22.36) follows, on using (22.3). □

The matrices $d^{(j)}(\pi)$ satisfy the following property:

$$(d^{(j)}(\pi))_m^{m'} = \delta_{-m}^{m'}(-1)^{j-m} . \tag{22.41}$$

This fact will not be proved here but it can be seen to be a direct consequence of (22.48) below. The reader should check that it is satisfied by both $d^{(1/2)}(\pi)$ and $d^{(1)}(\pi)$ as given explicitly by (22.13) and (22.28), respectively.

(5) *Symmetry under sign reversal of the magnetic quantum numbers m and m′ of the matrix element* $(d^{(j)}(\beta))_m^{m'}$. We have the following symmetry property:

$$(d^{(j)}(\beta))_m^{m'} = (d^{(j)}(-\beta))_{m'}^{m} = (-1)^{m'-m}\,(d^{(j)}(\beta))_{-m}^{-m'} . \tag{22.42}$$

Proof. Since all $d^{(j)}$ matrices commute, we have

$$d^{(j)}(\beta) = d^{(j)}(\pi)\, d^{(j)}(\beta)\, d^{(j)}(-\pi) . \tag{22.43}$$

Thus

$$\begin{aligned}
(d^{(j)}(\beta))_m^{m'} &= \sum_{l,\,n}(d^{(j)}(\pi))_l^{m'}\,(d^{(j)}(\beta))_n^l\,(d^{(j)}(-\pi))_m^n \\
&= \sum_{l,\,n}\delta_{-l}^{m'}\,(-1)^{j-l}\,(d^{(j)}(\beta))_n^l\,(d^{(j)}(\pi))_n^m \\
&= \sum_{n}(-1)^{j+m'}\,(d^{(j)}(\beta))_n^{-m'}\,\delta_{-n}^m\,(-1)^{j-n} \\
&= (-1)^{j+m'}\,(d^{(j)}(\beta))_{-m}^{-m'}\,(-1)^{j+m} = (-1)^{2j+m+m'}\,(d^{(j)}(\beta))_{-m}^{-m'} ,
\end{aligned} \tag{22.44}$$

where in the second and third equalities we have used (22.41). Since $(-1)^{2(j+m)} = 1$, we have $(-1)^{2j+m} = (-1)^{-m}$, and thus $(-1)^{2j+m+m'} = (-1)^{m'-m}$. The result (22.42) follows. □

(6) *Relation to the spherical harmonics.* For $j = l$, l being an integer, we have the following relationships:

$$Y_l^m(\theta,\varphi) = (-1)^m\left(\frac{2l+1}{4\pi}\right)^{1/2}\{(D^{(l)}(\varphi,\theta,0))_0^m\}^* \tag{22.45}$$

$$P_l^m(\cos\theta) = \left(\frac{(l+|\,m\,|)!}{(l-|\,m\,|)!}\right)^{1/2}(d^{(l)}(\theta))_0^m , \tag{22.46}$$

$$P_l(\cos\theta) = P_l^0(\cos\theta) = (d^{(l)}(\theta))_0^0 . \tag{22.47}$$

Finally, we give (without proof) a general formula for $d^{(j)}(\theta)$. For all allowed values of j,

$$\boxed{\begin{aligned}\left(d^{(j)}(\theta)\right)^{m}_{m'} &= \sum_k \frac{(-1)^k \sqrt{(j+m)(j-m)(j+m')(j-m')}}{k!\,(j+m-k)!(j-m'-k)!(k+m'-m)!} \\ &\quad \times (\cos(\theta/2))^{2j+m-m'-2k}\,(\sin(\theta/2))^{2k+m'-m}\end{aligned}} \quad , \qquad (22.48)$$

where the sum over k runs through the $2j+1$ values $0, 1, 2, \ldots, 2j$; but includes only terms for which the factorials have meaning. This equation, together with (22.3), give the complete expression for all rotation matrices.

Problem 22.3 Work out the spin 3/2 representations of J_1, J_2 and J_3 by computing the matrix elements of $J_\pm = J_1 \pm iJ_2$. Express your answers as 4 by 4 matrices.

Chapter 23

Direct Product Representations, Clebsch-Gordon Coefficients

Direct product representations of a symmetry group G come into play in quantum theory whenever two subsystems, each with Hamiltonian invariant under G, interact with each other. A specially important case is when G is the rotation group $SO(3)$, where the direct product of irreducible representations of two angular momenta gives rise to irreducible representations of the sum of the angular momenta. We will first in this chapter consider the general theory of direct product representations, and return in later ones to study the case for $SO(3)$.

Let $D^\mu(g)$ and $D^\nu(g)$, $g \in G$, be finite dimensional irreducible representations of a group G, whose representation (vector) spaces are V and W, respectively. Suppose $dim(V) = n$ and $dim(W) = m$. Then the **direct product representation**, $D^\mu \otimes D^\nu$, is defined on the **tensor product space** of V and W, $V \otimes W$, which is nm-dimensional. If

$$V = span\{\,|\,v_1\,\rangle, \ldots, |\,v_n\,\rangle\,\}\,, \quad W = span\{\,|\,w_1\,\rangle, \ldots, |\,w_m\,\rangle\,\}\,, \tag{23.1}$$

where $\{ | v_i \rangle \}$, $\{ | w_j \rangle \}$ are orthonormal bases, then

$$V \otimes W = span \left\{ \begin{matrix} | v_1 \rangle \otimes | w_1 \rangle, & | v_1 \rangle \otimes | w_2 \rangle, & \ldots, & | v_1 \rangle \otimes | w_m \rangle, \\ | v_2 \rangle \otimes | w_1 \rangle, & \ldots & \ldots & | v_2 \rangle \otimes | w_m \rangle, \\ \ldots & \ldots & \ldots & \ldots \\ \ldots & \ldots & \ldots & \ldots \\ | v_n \rangle \otimes | w_1 \rangle, & \ldots & \ldots, & | v_n \rangle \otimes | w_m \rangle \end{matrix} \right\} . \quad (23.2)$$

From now on we will just use the standard Dirac notation

$$| v_i \rangle | w_j \rangle \equiv | v_i \rangle \otimes | w_j \rangle . \quad (23.3)$$

The basis set $\{ | z_k \rangle ; k = 1, \ldots, nm \} \equiv \{ | v_i \rangle | w_j \rangle \}$ is also orthonormal:

$$\langle z_{k'} | z_k \rangle = \langle w_{j'} | \langle v_{i'} | v_i \rangle | w_j \rangle = \delta_{i'i} \delta_{j'j} . \quad (23.4)$$

The action of the product representation $D^{(\mu)} \otimes D^{(\nu)}$ on the representation space $V \otimes W$ with respect to the basis $\{ | v_i \rangle | w_j \rangle \}$ is then given by the following matrix representation:

$$(D^{(\mu)} \otimes D^{(\nu)})^{k'}_{k} = (D^{(\mu)})^{i'}_{i} \, (D^{(\nu)})^{j'}_{j} , \quad (23.5)$$

where $k' = (i', j')$, $k = (i, j)$. Equivalently

$$\langle w_{j'} | \langle v_{i'} | D^{(\mu)} \otimes D^{(\nu)} | v_i \rangle | w_j \rangle = \langle v_{i'} | D^{(\mu)} | v_i \rangle \langle w_{j'} | D^{(\nu)} | w_j \rangle . \quad (23.6)$$

The important point is that even though $D^{(\mu)}$ and $D^{(\nu)}$ are separately irreducible, $D^{(\mu)} \otimes D^{(\nu)}$ is in general reducible. The reduction takes the general form

$$D^{(\mu)} \otimes D^{(\nu)} = \sum_{\oplus \lambda} n_{(\lambda)} D^{(\lambda)} . \quad (23.7)$$

This equation means that the product representation space is given by

$$V \otimes W = \sum_{\oplus \lambda} n_{(\lambda)} \mathcal{V}^{(\lambda)} , \quad (23.8)$$

that is, it is the direct sum, over all allowed values of the representation index λ, of $n_{(\lambda)}$ copies of each of the vector spaces $\mathcal{V}^{(\lambda)}$, with $\mathcal{V}^{(\lambda)}$ being the representation space of the irreducible representation $D^{(\lambda)}$. After the reduction, the matrix representation of $D^{(\mu)} \otimes D^{(\nu)}$ will appear in the block diagonal form

$$D^{(\mu)} \otimes D^{(\nu)} = diag \{ n_{\lambda_1}(\lambda_1), n_{\lambda_2}(\lambda_2), \ldots, n_{\lambda_p}(\lambda_p) \} , \quad (23.9)$$

where there are n_{λ_1} blocks of $D^{(\lambda_1)}$, n_{λ_2} blocks of $D^{(\lambda_2)}$, etc., with all the blocks assuming diagonal positions in the matrix.

Let us further use the notation

$$|\,ij\,\rangle \equiv |\,v_i\,\rangle |\,w_j\,\rangle\,, \quad 1 \le i \le n\,, 1 \le j \le m\,. \tag{23.10}$$

The completeness condition for the basis set $\{|\,ij\,\rangle\}$ can then be written

$$\sum_{ij} |\,ij\,\rangle\langle\,ij\,| = 1\,. \tag{23.11}$$

Label a basis vector of the α-th copy of $\mathcal{V}^{(\lambda)}$ in $D^{(\mu)} \otimes D^{(\nu)}$ by $|\,\alpha\lambda l\,\rangle$, where $1 \le l \le dim\,\mathcal{V}^{(\lambda)}$. These vectors also form a complete orthonormal set in $V \otimes W$, and hence satisfy the completeness condition

$$\sum_{\alpha\lambda l} |\,\alpha\lambda l\,\rangle\langle\,\alpha\lambda l\,| = 1\,. \tag{23.12}$$

Equation (23.11) implies

$$|\,\alpha\lambda l\,\rangle = \sum_{ij} |\,ij\,\rangle\langle\,ij\,|\,\alpha\lambda l\,\rangle\,. \tag{23.13}$$

The matrix elements $\langle\,ij\,|\,\alpha\lambda l\,\rangle$ specifying the transformation between the basis states $|\,\alpha\lambda l\,\rangle$ and $|\,ij\,\rangle$ of $V \otimes W$ are called **Clebsch-Gordon coefficients**. They obviously satisfy the following pair of conditions as a consequence of the fact that both $\{\,|\,ij\,\rangle\,\}$ and $\{\,|\,\alpha\lambda l\,\rangle\,\}$ are complete orthonormal sets:

$$\sum_{ij} \langle\,\alpha'\lambda' l'\,|\,ij\,\rangle\langle\,ij\,|\,\alpha\,\lambda l\,\rangle = \delta_{\alpha'\alpha}\delta_{\lambda'\lambda}\delta_{l'l}\,, \tag{23.14}$$

$$\sum_{\alpha\lambda l} \langle\,i'j'\,|\,\alpha\lambda l\,\rangle\langle\,\alpha\lambda l\,|\,ij\,\rangle = \delta_{i'i}\delta_{j'j}\,. \tag{23.15}$$

The following reciprocal statements relating a direct product representation to that of its irreducible components are very useful in calculations.

$$D^{(\mu)}(g)_i^{i'}\,D^{(\nu)}(g)_j^{j'} = \sum_{\alpha\lambda l l'} \langle\,i'j'\,|\,\alpha\lambda l'\,\rangle\,D^{(\lambda)}(g)_l^{l'}\,\langle\,\alpha\lambda l\,|\,ij\,\rangle\,, \tag{23.16}$$

$$\delta_\alpha^{\alpha'}\,\delta_\lambda^{\lambda'}\,D^{(\lambda)}(g)_l^{l'} = \sum_{i'j'ij} \langle\,\alpha'\lambda' l'\,|\,i'j'\,\rangle\,D^{(\mu)}(g)_i^{i'}\,D^{(\nu)}(g)_j^{j'}\,\langle\,ij\,|\,\alpha\lambda l\,\rangle\,. \tag{23.17}$$

These are usually referred to as **Clebsch-Gordon series.**

To prove (23.16) we first observe that

$$|\,ij\,\rangle = \sum_{\alpha\lambda l} |\,\alpha\lambda l\,\rangle\langle\,\alpha\lambda l\,|\,ij\,\rangle\,. \tag{23.18}$$

Let $U(g)$ $(g \in G)$ be a representation of G acting on the states $|\,ij\,\rangle$. Then

$$U(g)\,|\,ij\,\rangle = \sum_{i'j'} |\,i'j'\,\rangle\langle\,i'j'\,|\,U(g)\,|\,ij\,\rangle = \sum_{i'j'} D^{(\mu)}(g)_i^{i'}\,D^{(\nu)}(g)_j^{j'}\,|\,i'j'\,\rangle\,. \tag{23.19}$$

On the other hand,

$$U(g)\,|\,\alpha\lambda l\,\rangle = \sum_{l'} |\,\alpha\lambda l'\,\rangle\langle\,\alpha\lambda l'\,|\,U(g)\,|\,\alpha\lambda l\,\rangle = \sum_{l'} D^{(\lambda)}(g)_l^{l'}\,|\,\alpha\lambda l'\,\rangle\,. \qquad (23.20)$$

Using (23.18) on the left-hand side of (23.19) we obtain

$$\begin{aligned}\sum_{\alpha\lambda l} U(g)\,|\,\alpha\lambda l\,\rangle\langle\,\alpha\lambda l\,|\,ij\,\rangle &= \sum_{\alpha\lambda l l'} |\,\alpha\lambda l'\,\rangle\langle\,\alpha\lambda l'\,|\,U(g)\,|\,\alpha\lambda l\,\rangle\langle\,\alpha\lambda l\,|\,ij\,\rangle \\ = \sum_{\alpha\lambda l l'} D^{(\lambda)}(g)_l^{l'}\,\langle\,\alpha\lambda l\,|\,ij\,\rangle\,|\,\alpha\lambda l'\,\rangle &= \sum_{i'j'}\sum_{\alpha\lambda l l'} |\,i'j'\,\rangle\langle\,i'j'\,|\,\alpha\lambda l'\,\rangle\,D^{(\lambda)}(g)_l^{l'}\,\langle\,\alpha\lambda l\,|\,ij\,\rangle\end{aligned} \qquad (23.21)$$

On comparison with the right-hand side of (23.19), (23.16) follows.

To prove (23.17) we use (23.13) on the left-hand side of (23.20). Thus

$$U(g)\,|\,\alpha\lambda l\,\rangle = \sum_{ij} U(g)\,|\,ij\,\rangle\langle\,ij\,|\,\alpha\lambda l\,\rangle\,. \qquad (23.22)$$

Taking the inner product (on the left) with $\langle\,\alpha'\lambda' l'\,| = \sum_{i'j'}\langle\,\alpha'\lambda' l'\,|\,i'j'\,\rangle\langle\,i'j'\,|$, we have

$$\delta_\alpha^{\alpha'}\,\delta_\lambda^{\lambda'}\,D^{(\lambda)}(g)_l^{l'} = \sum_{i'j'\,ij}\langle\,\alpha'\lambda' l'\,|\,i'j'\,\rangle\langle\,i'\,j'\,|\,U(g)\,|\,ij\,\rangle\langle\,ij\,|\,\alpha\lambda l\,\rangle\,, \qquad (23.23)$$

which is precisely (23.17).

Chapter 24

Transformations of Wave Functions and Vector Operators under $SO(3)$

Let us begin by reestablishing the formula (17.5) using the Dirac notation. Suppose $g \in SO(3)$ such that $g\boldsymbol{x} = \boldsymbol{x}'$, where $\boldsymbol{x}, \boldsymbol{x}' \in \mathbb{R}^3$. Equivalently, $x'^i = g^i_j\, x^j$. Consider the group representation $g \mapsto U(g)$, where $U(g)$ is a linear operator on a rigged Hilbert space $\mathcal{H}$, such that

$$U(g)\,|\,\boldsymbol{x}\,\rangle = |\,\boldsymbol{x}'\,\rangle\,. \tag{24.1}$$

Then, for arbitrary $|\,\psi\,\rangle \in \mathcal{H}$, since

$$|\,\psi\,\rangle = \int d^3x\,|\,\boldsymbol{x}\,\rangle\langle\,\boldsymbol{x}\,|\,\psi\,\rangle = \int d^3x\,|\,\boldsymbol{x}\,\rangle\,\psi(\boldsymbol{x})\,, \tag{24.2}$$

we have

$$|\,\psi'\rangle \equiv U(g)\,|\,\psi\,\rangle = \int d^3x\,U(g)|\,\boldsymbol{x}\,\rangle\,\psi(\boldsymbol{x}) = \int d^3x\,|\,\boldsymbol{x}'\,\rangle\,\psi(\boldsymbol{x})$$

$$= \int d^3x\,|\,\boldsymbol{x}'\,\rangle\,\psi(g^{-1}\boldsymbol{x}') = \int \left|\,\frac{\partial x^i}{\partial x'^j}\,\right| d^3x'\,|\,\boldsymbol{x}'\,\rangle\,\psi(g^{-1}\boldsymbol{x}') = \int d^3x'\,|\,\boldsymbol{x}'\,\rangle\,\psi(g^{-1}\boldsymbol{x}')\,, \tag{24.3}$$

where in the last equality we have used the fact that, by definition, the Jacobian of the $SO(3)$ transformation $\boldsymbol{x} \to \boldsymbol{x}'$ given by $\left|\,\dfrac{\partial x^i}{\partial x'^j}\,\right|$ is equal to one. It follows that, on taking the inner product on the left with $\langle\,\boldsymbol{x}\,|$,

$$\langle\,\boldsymbol{x}\,|\,\psi'\,\rangle = \int d^3x'\,\langle\,\boldsymbol{x}\,|\,\boldsymbol{x}'\,\rangle\,\psi(g^{-1}\boldsymbol{x}')\,. \tag{24.4}$$

On noting that $\langle \boldsymbol{x} \,|\, \psi' \rangle = \psi'(\boldsymbol{x})$ and $\langle \boldsymbol{x} \,|\, \boldsymbol{x}' \rangle = \delta^3(\boldsymbol{x} - \boldsymbol{x}')$ we recover the fact that, under a transformation $\boldsymbol{x} \to g\boldsymbol{x} = \boldsymbol{x}'$, $g \in SO(3)$, a scalar wave function $\psi(\boldsymbol{x})$ transforms as follows

$$\psi'(\boldsymbol{x}) = \psi(g^{-1}\boldsymbol{x}) \,, \tag{24.5}$$

which is precisely (17.5).

As a first example, let us consider the plane-wave state $|\,\psi\,\rangle = |\,\boldsymbol{k}\,\rangle$ characterized by the wave vector $\boldsymbol{k}$, with the coordinate-representation wave function given by [cf. (21.6)]

$$\psi_{\boldsymbol{k}}(\boldsymbol{x}) = \langle \boldsymbol{x} \,|\, \psi \rangle = \langle \boldsymbol{x} \,|\, \boldsymbol{k} \rangle = \frac{1}{(2\pi)^{3/2}}\, e^{i\boldsymbol{k}\cdot\boldsymbol{x}} \,. \tag{24.6}$$

Then, since

$$\psi_{\boldsymbol{k}}(g^{-1}\boldsymbol{x}) = \frac{1}{(2\pi)^{3/2}}\, e^{i\boldsymbol{k}\cdot g^{-1}\boldsymbol{x}} = \frac{1}{(2\pi)^{3/2}}\, e^{i(g\boldsymbol{k})\cdot\boldsymbol{x}} \,, \tag{24.7}$$

the transformation rule (24.5) implies

$$\psi'_{\boldsymbol{k}}(\boldsymbol{x}) = \psi_{g\boldsymbol{k}}(\boldsymbol{x}) \,. \tag{24.8}$$

The same result can also be obtained as follows:

$$\psi'_{\boldsymbol{k}}(\boldsymbol{x}) = \langle \boldsymbol{x} \,|\, \psi'_{\boldsymbol{k}} \rangle = \langle \boldsymbol{x} \,|\, U(g) \,|\, \boldsymbol{k} \rangle = \langle \boldsymbol{x} \,|\, \boldsymbol{k}' \rangle = \langle \boldsymbol{x} \,|\, g\boldsymbol{k} \rangle = \psi_{g\boldsymbol{k}}(\boldsymbol{x}) \,. \tag{24.9}$$

For the next example we consider a rotationally invariant Hamiltonian H, so that

$$[\, H \,,\, U(g) \,] = 0 \,, \quad g \in SO(3) \,. \tag{24.10}$$

It follows from (17.36) and (17.60) that

$$[\, H \,,\, J_i \,] = 0 \,, \quad i = 1, 2, 3 \,, \tag{24.11}$$

where J_i is the i-th component of the dimensionless angular momentum operator $\boldsymbol{J}$ [cf. (20.1)]. If we label the simultaneous eigenstates of the complete, commuting set of observables $(H, J^2; J_3)$ by $|\, Elm \rangle$, we have [cf. (15.20) to (15.22)]

$$H \,|\, Elm \,\rangle = E \,|\, Elm \,\rangle \,, \tag{24.12}$$

$$J^2 \,|\, Elm \,\rangle = l(l+1) \,|\, Elm \,\rangle \,, \tag{24.13}$$

$$J_3 \,|\, Elm \,\rangle = m \,|\, Elm \,\rangle \,, \tag{24.14}$$

where the allowed values for l and m are $l = 0, 1, 2, 3, \ldots$, $m = l, l-1, \ldots, -l$. The Schrödinger wave function for $|\, Elm \rangle$ is given by

$$\psi_{Elm}(\boldsymbol{x}) = \langle \boldsymbol{x} \,|\, Elm \,\rangle = \langle r, \theta, \phi \,|\, Elm \,\rangle \,, \tag{24.15}$$

where in the last equation we have used spherical coordinates for $\boldsymbol{x}$. Since the vector $\boldsymbol{x}$ with spherical coordinates (r, θ, ϕ) can be obtained by rotating the vector $r\hat{\boldsymbol{z}}$ first about the y-axis by an angle θ and then about the z-axis by an angle ϕ, we can write

$$|\,\boldsymbol{x}\,\rangle = |\,r,\theta,\phi\,\rangle = R_3(\phi)R_2(\theta)\,|\,r\hat{\boldsymbol{z}}\,\rangle = e^{-i\phi J_3}e^{-i\theta J_2}\,|\,r,0,0\,\rangle = U(\phi,\theta,0)\,|\,r,0,0\,\rangle\,, \tag{24.16}$$

where in the third equality we have used (17.60) and in the fourth, (22.1) (with $\psi = 0$). Then, by (8.18),

$$\langle\,r,\theta,\phi\,|\,Elm\,\rangle = \langle\,r,0,0\,|\,U^\dagger(\phi,\theta,0)\,|\,Elm\,\rangle\,. \tag{24.17}$$

Since $\{|\,Elm\,\rangle\}$, for particular values of E and l, spans an irreducible representation subspace under $U(g), g \in SO(3)$, we can write [cf. (22.2)]

$$\psi_{Elm}(\boldsymbol{x}) = \sum_{m'}\langle\,r,0,0\,|\,Elm'\,\rangle\,(D^{(l)})^\dagger(\phi,\theta,0)^{m'}_{m}\,. \tag{24.18}$$

Now rotation about the z-axis must leave $|\,r\hat{\boldsymbol{z}}\,\rangle$ invariant, which implies

$$e^{-i\alpha J_3}\,|\,r,0,0\,\rangle = |\,r,0,0\,\rangle\,, \tag{24.19}$$

which in turn implies

$$J_3\,|\,r,0,0\,\rangle = 0\,. \tag{24.20}$$

On account of (20.12), we conclude that

$$|\,r,0,0\,\rangle = \sum_{E'l'} f_{E'l'}(r)\,|\,E'l'0\,\rangle\,, \tag{24.21}$$

where the $f_{El}(r)$ are undetermined functions of r only. Thus

$$\langle\,r,0,0\,|\,Elm'\,\rangle = f_{El}(r)\,\delta_{m'0}\,. \tag{24.22}$$

Putting this result in (24.18) we have

$$\psi_{Elm}(r,\theta,\phi) = f_{El}(r)(D^{(l)}(\phi,\theta,0)^m_0)^*\,. \tag{24.23}$$

Recalling (22.45) we can express the Schrödinger wave function $\psi_{Elm}(r,\theta,\phi)$ describing a system with rotational symmetry as follows

$$\boxed{\psi_{Elm}(r,\theta,\phi) = \psi_{El}(r)\,Y_{lm}(\theta,\phi)}\,, \tag{24.24}$$

where Y_{lm} is a spherical harmonic and

$$\psi_{El}(r) \equiv \left(\frac{4\pi}{2l+1}\right)^{1/2} f_{El}(r)\,. \tag{24.25}$$

Equation (24.24) is a direct consequence of the rotational symmetry of the Hamiltonian, and is independent of the specific form of the central potential function $V(r)$, which, in fact, determines $f_{El}(r)$.

Under a rotation $R \in SO(3)$, $|\,Elm\,\rangle$ transforms according to [cf. (22.2)]

$$|\,\psi'\,\rangle = U(R)\,|\,Elm\,\rangle = \sum_{m'} D^{(l)}(R)^{m'}_{m}\,|\,Elm'\,\rangle\,. \tag{24.26}$$

Hence,

$$\begin{aligned}\psi'(\boldsymbol{x}) &= \langle\,\boldsymbol{x}\,|\,\psi'\,\rangle = \sum_{m'} D^{(l)}(R)^{m'}_{m}\,\langle\,\boldsymbol{x}\,|\,Elm'\,\rangle \\ &= \sum_{m'} D^{(l)}(R)^{m'}_{m}\,\psi_{Elm'}(\boldsymbol{x}) = \sum_{m'} D^{(l)}(R)^{m'}_{m}\,\psi_{El}(r)Y_{lm'}(\theta,\phi)\,.\end{aligned} \tag{24.27}$$

On the other hand, according to the transformation rule (24.5),

$$\psi'(\boldsymbol{x}) = \psi(R^{-1}\boldsymbol{x}) = \psi_{El}(r)Y_{lm}(R^{-1}\boldsymbol{x})\,. \tag{24.28}$$

We thus obtain the following well-known property of spherical harmonics:

$$\boxed{Y_{lm}(R^{-1}\boldsymbol{x}) = \textstyle\sum_{m'} Y_{lm'}(\boldsymbol{x})D^{(l)}(R)^{m'}_{m}}\,. \tag{24.29}$$

Next we enlarge the Hilbert space $\mathcal{H}$ so that it describes spin 1/2 particles. We have the completeness condition

$$\sum_{\sigma}\int d^3x\,|\,\boldsymbol{x},\sigma\,\rangle\langle\,\boldsymbol{x},\sigma\,| = 1\,, \qquad \sigma = \pm 1/2\,. \tag{24.30}$$

Consider a representation of $SO(3)$ on $\mathcal{H}$, $R \in SO(3) \mapsto U(R)$, such that

$$U(R)\,|\,\boldsymbol{x},\sigma\,\rangle = \sum_{\lambda} |\,R\boldsymbol{x},\lambda\,\rangle\,D^{(1/2)}(R)^{\lambda}_{\sigma}\,. \tag{24.31}$$

On using (24.30) an arbitrary $|\,\psi\,\rangle \in \mathcal{H}$ is given by

$$|\,\psi\,\rangle = \sum_{\sigma}\int d^3x\,|\,\boldsymbol{x},\sigma\,\rangle\,\psi^{\sigma}(\boldsymbol{x})\,, \tag{24.32}$$

where

$$\psi^{\sigma}(\boldsymbol{x}) \equiv \langle\,\boldsymbol{x},\sigma\,|\,\psi\,\rangle\,, \quad \sigma = \pm 1/2\,. \tag{24.33}$$

The quantities $\psi^{\sigma}(\boldsymbol{x})$ constitute a two-component **Pauli spinor** characterizing the state of a spin 1/2 particle. We then have, by (24.31),

$$\begin{aligned}|\,\psi'\,\rangle &= U(R)\,|\,\psi\,\rangle = \sum_{\sigma}\int d^3x\,U(R)|\,\boldsymbol{x},\sigma\,\rangle\,\psi^{\sigma}(\boldsymbol{x}) \\ &= \sum_{\lambda\sigma}\int d^3x\,|\,R\boldsymbol{x},\lambda\,\rangle\,D^{(1/2)}(R)^{\lambda}_{\sigma}\,\psi^{\sigma}(\boldsymbol{x}) = \sum_{\lambda\sigma}\int d^3x\,|\,\boldsymbol{x},\lambda\,\rangle\,D^{(1/2)}(R)^{\lambda}_{\sigma}\,\psi^{\sigma}(R^{-1}\boldsymbol{x})\end{aligned} \tag{24.34}$$

Comparing with (24.32) we see that the transformation rule for the components of the Pauli spinor is given by

$$\boxed{\psi'^{\lambda}(\boldsymbol{x}) = \sum_{\sigma} D^{(1/2)}(R)^{\lambda}_{\sigma}\, \psi^{\sigma}(R^{-1}\boldsymbol{x})} \,. \tag{24.35}$$

Problem 24.1 A spin-up free electron travelling along the $+y$ direction is described by a two-component Pauli spinor given by

$$\begin{pmatrix} \psi^{\uparrow}(\boldsymbol{x}) \\ \psi^{\downarrow}(\boldsymbol{x}) \end{pmatrix} = \frac{1}{(2\pi)^{3/2}}\, e^{iky} \begin{pmatrix} 1 \\ 0 \end{pmatrix}, \quad \boldsymbol{k} = k\hat{\boldsymbol{y}}\,.$$

What is the wave function

$$\begin{pmatrix} \psi'^{\uparrow}(\boldsymbol{x}) \\ \psi'^{\downarrow}(\boldsymbol{x}) \end{pmatrix}$$

in the frame rotated from the original one about the y-axis by an angle θ in the positive sense?

We can immediately generalize to the so-called irreducible wave functions which transform according to different irreducible representations of $SO(3)$.

Definition 24.1. *A set of multi-component functions* $\{\psi^m(\boldsymbol{x})\}, m = j, j-1, \ldots, -j$, *is called an* ***irreducible wave function*** *or* ***irreducible field*** *of spin* j *if they transform under a rotation* $R \in SO(3)$ *according to the rule*

$$\boxed{\psi'^{m}(\boldsymbol{x}) = \sum_{n} D^{(j)}(R)^{m}_{n}\, \psi^{n}(R^{-1}\boldsymbol{x})} \,. \tag{24.36}$$

Physical examples are electric fields $E^i(\boldsymbol{x})$ and magnetic fields $B^i(\boldsymbol{x})$ (spin 1), the velocity field $v^i(\boldsymbol{x})$ (spin 1), the Pauli spinor $\psi^{\sigma}(\boldsymbol{x})$ (spin 1/2), and the Dirac spinor, which under rotations are reducible fields consisting of the direct sum of two spin 1/2 irreducible fields.

Aside from the transformation properties of wave functions, we also need to investigate those of operators (representing observables) in quantum mechanical calculations. Recall that [cf. (17.13)] under a coordinate transformation $\boldsymbol{x} \to g\boldsymbol{x}, g \in SO(3)$, an observable A transforms as $A \to U(g)AU^{\dagger}(g)$. This motivates the following definition, in analogy to Definition 24.1.

Definition 24.2. *If a set of operators* $A^i, i = 1, 2, 3$, *transforms under* $g \in SO(3)$ *according to*

$$\boxed{A^i \to A'^i = U(g)A^iU^{-1}(g) = (g^{-1})^i_j\, A^j = \sum_j g^j_i\, A^j} \,, \tag{24.37}$$

then the A^i *are called components of the* ***vector operator A***.

This terminology follows from the fact that the components of the vector (position) operator $\boldsymbol{X}$, defined by $X^i \,|\, \boldsymbol{x} \,\rangle = x^i \,|\, \boldsymbol{x} \,\rangle$, transform according to the above equation. Indeed, it follows from $U(g) \,|\, \boldsymbol{x} \,\rangle = |\, g\boldsymbol{x} \,\rangle = |\, \boldsymbol{x}' \,\rangle$ that

$$\begin{aligned} & U(g)X^iU^{-1}(g) \,|\, \boldsymbol{x}' \,\rangle = U(g)X^iU^{-1}(g)U(g) \,|\, \boldsymbol{x} \,\rangle = U(g)X^i \,|\, \boldsymbol{x} \,\rangle \\ & = x^iU(g) \,|\, \boldsymbol{x} \,\rangle = x^i \,|\, \boldsymbol{x}' \,\rangle \,. \end{aligned} \tag{24.38}$$

But $\boldsymbol{x}$ and $\boldsymbol{x}'$ are related by

$$x'^i = g^i_j \, x^j \,, \quad \text{or} \quad x^i = (g^{-1})^i_j \, x'^j \,, \tag{24.39}$$

under a rotation $g \in SO(3)$. Hence

$$U(g)X^iU^{-1}(g) \,|\, \boldsymbol{x}' \,\rangle = (g^{-1})^i_j x'^j \,|\, \boldsymbol{x}' \,\rangle = (g^{-1})^i_j X^j \,|\, \boldsymbol{x}' \,\rangle \,. \tag{24.40}$$

Replacing $\boldsymbol{x}'$ by $\boldsymbol{x}$ in the above equation, we have

$$U(g)X^iU^{-1}(g) \,|\, \boldsymbol{x} \,\rangle = (g^{-1})^i_j X^j \,|\, \boldsymbol{x} \,\rangle \,, \tag{24.41}$$

which implies

$$U(g)X^iU^{-1}(g) = (g^{-1})^i_j X^j \,, \tag{24.42}$$

in conformity with (24.37).

Both the linear momentum operator $\boldsymbol{P}$ and the angular momentum operator $\boldsymbol{J}$ are vector operators, transforming according to (24.37) under $SO(3)$. Vector operators are special cases of the so-called **irreducible tensor operators**, which we will study in the next chapter.

Chapter 25

Irreducible Tensor Operators and the Wigner-Eckart Theorem

As a generalization of the vector operators defined in the last chapter, we introduce the so-called irreducible tensor operators as follows.

Definition 25.1. *A set of operators* $\{O_i^{(\mu)}\}$, $i = 1, \ldots, 2\mu+1$, *on a vector space* $\mathbb{V}$ *transforming under* $g \in SO(3)$ *according to*

$$\boxed{U(g)O_i^{(\mu)}U^\dagger(g) = D^{(\mu)}(g)_i^j\, O_j^{(\mu)}} \tag{25.1}$$

is said to be a set of ***irreducible tensor operators*** *corresponding to the* μ*-representation of* $SO(3)$.

Note that, since $D^{(\mu)}(g)$ is unitary [cf. (22.33)], $D^{(\mu)}(g)_i^j = (D^{(\mu)}(g^{-1})_j^i)^*$. Thus, on taking the hermitian adjoint of (25.1), and supposing that $O_i^{(\mu)}$ is self-adjoint, we have

$$U(g)O_i^{(\mu)}U^{-1}(g) = \sum_j D^{(\mu)}(g^{-1})_j^i\, O_j^{(\mu)}\ , \tag{25.2}$$

which is seen to be a direct generalization of (24.37).

Let us denote the basis vectors in the irreducible μ-representation of $SO(3)$ by $|\,\mu, i\,\rangle$, $i = 1, \ldots, 2\mu + 1$. Given an irreducible tensor operator $O_i^{(\mu)}$ we wish to calculate matrix elements of the form $\langle\, \lambda l \,|\, O_i^{(\mu)} \,|\, \nu j\,\rangle$. Matrix elements of this type occur in a wide variety of problems in quantum mechanics. For example, they appear frequently in the calculation of transition amplitudes between atomic or nuclear angular momentum eigenstates (where $|\,\mu i\,\rangle = |\,jm\,\rangle$) induced by photon absorption or emission. In these applications, the relevant tensor operators $O_i^{(\mu)}$ represent **multipole operators** in electromagnetic interactions.

The Wigner-Eckart theorem gives a far-reaching and powerful method for the calculation of this type of matrix elements. As a preparation for its discussion we will need the following important result in group representation theory (for a proof, see, for example, Lam 2003).

Theorem 25.1 (Orthonormality of Irreducible Representation Matrices). *Let $D^{(\mu)}(g)$ and $D^{(\nu)}(g)$ be irreducible representation matrices of a group G and $g \in G$. Then*

$$\frac{n_\mu}{n_G}\sum_g (D^{(\mu)})^\dagger(g)_i^k\, D^{(\nu)}(g)_l^j = \delta_{\mu\nu}\delta_{ij}\delta_{kl}\,, \tag{25.3}$$

where the sum is over all $g \in G$, n_G is the order of G (the number of elements in G), and n_μ is the dimension of the irreducible μ-representation.

Note: Even though the above theorem has been stated in terms of finite groups, a generalized version of it is true for **compact groups** [such as $SO(3)$] as well. In this version, $\frac{1}{n_G}\sum_g$ is replaced by $\frac{1}{V(G)}\int d\mu(g)$, where $V(G)$ is the "volume" of G considered as a compact manifold, and $d\mu(g)$ is an integration measure over the group manifold known as the **Haar measure**. We will not be concerned with these details here.

Corollary 25.1. *The invariant subspaces corresponding to inequivalent irreducible representations of a compact group are orthogonal to each other.*

Proof. Let the orthonormal bases $\{|\,\mu i\,\rangle\}$, $i = 1,\ldots,n_\mu$, and $\{|\,\nu j\,\rangle\}$, $j = 1,\ldots,n_\nu$ span the invariant subspaces of the μ- and ν-representations, respectively. Then

$$\begin{aligned}
\langle\,\nu j\,|\,\mu i\,\rangle &= \langle\,\nu j\,|\,U^\dagger(g)U(g)\,|\,\mu i\,\rangle = \sum_{kl}\langle\,\nu j\,|\,U^\dagger(g)\,|\,\nu k\,\rangle\langle\,\nu k\,|\,\mu l\,\rangle\langle\,\mu l\,|\,U(g)\,|\,\mu i\,\rangle \\
&= \sum_{kl} D^{(\nu)\dagger}(g)_k^j\,\langle\,\nu k\,|\,\mu l\,\rangle\, D^{(\mu)}(g)_i^l = \frac{1}{n_G}\sum_{kl}\sum_g D^{(\nu)\dagger}(g)_k^j D^{(\mu)}(g)_i^l\,\langle\,\nu k\,|\,\mu l\,\rangle \\
&= \frac{1}{n_\mu}\sum_{kl}\delta_{\mu\nu}\delta_{kl}\delta_{ji}\,\langle\,\nu k\,|\,\mu l\,\rangle = \frac{1}{n_\mu}\,\delta_{\mu\nu}\delta_{ji}\,\sum_k\langle\,\nu k\,|\,\mu k\,\rangle\,,
\end{aligned} \tag{25.4}$$

where in the first equality we used the fact that $U(g)$ is unitary, in the fourth equality we have averaged over all group elements g since the LHS is independent of g, and in the fifth equality we have used Theorem 25.1. □

Now let us consider specifically the group $SO(3)$ and an irreducible tensor operator $T_q^{(k)}$. We are interested in the calculation of a matrix element of the form

$$\langle\,\tau',\, j'm'\,|\,T_q^{(k)}\,|\,\tau,\, jm\,\rangle\,,$$

where τ' and τ are state labels not related to the symmetry group in question [in our case $SO(3)$]. We will need the following result:

$$\int d\mu(g)(D^{(j')}(g)^{\mu'}_{m'})^* D^{(k)}(g)^{\sigma}_{q} D^{(j)}(g)^{\mu}_{m} = \frac{\langle\, kj\,;\, \sigma\mu \,|\, j'\mu' \,\rangle\langle\, kj\,;\, qm \,|\, j'm' \,\rangle}{(2j'+1)} \int d\mu(g)\,, \tag{25.5}$$

where the integral on the RHS can be identified as the "volume" of the group manifold, which can be denoted by $V(G)$. The matrix elements on the RHS are the Clebsch-Gordon coefficients introduced in Chapter 23 (see Chapters 26 and 27 also). The above equation can be proved as follows.

From the Clebsch-Gordon series (23.16) we have

$$D^{(k)}(g)^{\sigma}_{q} D^{(j)}(g)^{\mu}_{m} = \sum_{JMM'} \langle\, k,j\,;\, \sigma,\mu \,|\, JM' \,\rangle\, D^{(J)}(g)^{M'}_{M}\, \langle\, JM \,|\, k,j\,;\, q,m \,\rangle\,. \tag{25.6}$$

Thus

$$\begin{aligned}
&\int d\mu(g)\,(D^{(j')}(g)^{\mu'}_{m'})^* D^{(k)}(g)^{\sigma}_{q} D^{(j)}(g)^{\mu}_{m} \\
&= \sum_{JMM'} \langle\, k,j\,;\, \sigma,\mu \,|\, JM' \,\rangle\langle\, JM \,|\, k,j\,;\, q,m \,\rangle \int d\mu(g)\,(D^{(j')}(g)^{\mu'}_{m'})^* D^{(J)}(g)^{M'}_{M} \\
&= \sum_{JMM'} \langle\, k,j\,;\, \sigma,\mu \,|\, JM' \,\rangle\langle\, JM \,|\, k,j\,;\, q,m \,\rangle\, \frac{\int d\mu(g)}{2j'+1}\, \delta_{j'J}\delta_{\mu'M'}\delta_{m'M} \\
&= \frac{\langle\, k,j\,;\, \sigma,\mu \,|\, j'\mu' \,\rangle\langle\, k,j\,;\, q,m \,|\, j'm' \,\rangle}{2j'+1} \int d\mu(g)\,,
\end{aligned} \tag{25.7}$$

where we have used Theorem 25.1 in the second equality.

The matrix element of interest can now be calculated as follows. We have

$$\begin{aligned}
&\langle\, \tau', j', m' \,|\, T^{(k)}_{q} \,|\, \tau, j, m \,\rangle \\
&= \langle\, \tau', j', m' \,|\, U^{\dagger}(g)\, U(g) T^{(k)}_{q} U^{\dagger}(g)\, U(g) \,|\, \tau, j, m \,\rangle \\
&= \sum_{\mu'} \langle\, \tau', j', \mu' \,|\, (D^{j'})(g)^{\mu'}_{m'})^* \sum_{\sigma} D^{(k)}(g)^{\sigma}_{q} T^{(k)}_{\sigma} \sum_{\mu} D^{(j)}(g)^{\mu}_{m} \,|\, \tau, j, \mu \,\rangle \\
&= \sum_{\mu'\sigma\mu} (D^{(j')}(g)^{\mu'}_{m'})^* D^{(k)}(g)^{\sigma}_{q} D^{(j)}(g)^{\mu}_{m}\, \langle\, \tau', j', \mu' \,|\, T^{(k)}_{\sigma} \,|\, \tau, j, \mu \,\rangle\,.
\end{aligned} \tag{25.8}$$

Integrating both sides of the above equation over the volume of the group manifold of $SO(3)$, we have

$$
\begin{aligned}
&\int d\mu(g) \, \langle\, \tau', j', m' \,|\, T_q^{(k)} \,|\, \tau, j, m \,\rangle \\
&= \sum_{\mu'\sigma\mu} \langle\, \tau', j', \mu' \,|\, T_\sigma^{(k)} \,|\, \tau, j, \mu \,\rangle \int d\mu(g) \, (D^{(j')}(g)^{\mu'}_{m'})^* D^{(k)}(g)^\sigma_q D^{(j)}(g)^\mu_m \\
&= \left(\sum_{\mu'\sigma\mu} \frac{\langle\, \tau', j', \mu' \,|\, T_\sigma^{(k)} \,|\, \tau, j, \mu \,\rangle \langle\, k, j \,;\, \sigma, \mu \,|\, j' \, \mu' \,\rangle}{2j'+1} \right) \langle\, k, j \,;\, q, m \,|\, j', m' \,\rangle \int d\mu(g)
\end{aligned}
\tag{25.9}
$$

The quantity within the brackets is independent of the "magnetic" quantum numbers and will be denoted by $\langle\, \tau', j' \,\|\, T^{(k)} \,\|\, \tau, j \,\rangle$:

$$
\langle\, \tau', j' \,\|\, T^{(k)} \,\|\, \tau, j \,\rangle \equiv \frac{1}{2j'+1} \sum_{\mu'\sigma\mu} \langle\, \tau', j', \mu' \,|\, T_\sigma^{(k)} \,|\, \tau, j, \mu \,\rangle \langle\, k, j \,;\, \sigma, \mu \,|\, j' \, \mu' \,\rangle \,.
\tag{25.10}
$$

It is called the **reduced matrix element**, and depends only on the total angular momentum quantum numbers. The **Wigner-Eckart Theorem** is then given by the following equation.

$$
\boxed{\langle\, \tau', j', m' \,|\, T_q^{(k)} \,|\, \tau, j, m \,\rangle = \langle\, \tau', j' \,\|\, T^{(k)} \,\|\, \tau, j \,\rangle \langle\, k, j \,;\, q, m \,|\, j', m' \,\rangle} \,.
\tag{25.11}
$$

This is one of the most important theorems in quantum mechanics. The usefulness of this celebrated theorem lies in the fact that the multitude of matrix elements represented in the LHS of the above equation are all determined by a few reduced matrix elements. The specific properties of the states and the operator $T_q^{(k)}$ only enter in the reduced matrix elements, a particular one of which can be computed from the LHS of (25.11) for one choice of the values (m', q, m), and then dividing by the appropriate Clebsch-Gordon coefficient. We will see examples on the application of the Wigner-Eckart Theorem in Chapter 28.

Chapter 26

Reduction of Direct Product Representations of $SO(3)$: The Addition of Angular Momenta

Consider the direct product representation $D^{(j_1)} \otimes D^{(j_2)}$ of $SO(3)$. The invariant subspace of this representation is *span* $\{ \,|\, j_1, j_2\,;\, m_1, m_2 \,\rangle \}$, where the basis vectors $|\, j_1, j_2\,;\, m_1, m_2 \,\rangle$ are eigenstates of $(J^{(1)})^2, (J^{(2)})^2, J_z^{(1)}$ and $J_z^{(2)}$:

$$(J^{(1)})^2 \,|\, j_1, j_2\,;\, m_1, m_2 \,\rangle = j_1(j_1+1) \,|\, j_1, j_2\,;\, m_1, m_2 \,\rangle\,, \tag{26.1}$$

$$(J^{(2)})^2 \,|\, j_1, j_2\,;\, m_1, m_2 \,\rangle = j_2(j_2+1) \,|\, j_1, j_2\,;\, m_1, m_2 \,\rangle\,, \tag{26.2}$$

$$J_z^{(1)} \,|\, j_1, j_2\,;\, m_1, m_2 \,\rangle = m_1 \,|\, j_1, j_2\,;\, m_1, m_2 \,\rangle\,, \tag{26.3}$$

$$J_z^{(2)} \,|\, j_1, j_2\,;\, m_1, m_2 \,\rangle = m_2 \,|\, j_1, j_2\,;\, m_1, m_2 \,\rangle\,, \tag{26.4}$$

and they transform under $SO(3)$ according to [cf. (22.2)]

$$U(g) \,|\, j_1, j_2\,;\, m_1, m_2 \,\rangle = D^{(j_1)}(g)^l_{\;m_1}\, D^{(j_2)}(g)^k_{\;m_2} \,|\, j_1, j_2\,;\, l, k \,\rangle\,, \tag{26.5}$$

where the indices l and k on the right-hand side are summed over (by the Einstein summation convention). Note that the representation is single-valued if $j_1 + j_2$ is an integer, and double-valued if $j_1 + j_2$ is a half-integer. Unless either j_1 or j_2 is equal to zero, the product representation is reducible.

We will first consider the case $j_1 = j_2 = 1/2$. The four basis states of the invariant subspace in the $|\, j_1, j_2\,;\, m_1, m_2 \,\rangle$ representation are

$$\left|\, \frac{1}{2}, \frac{1}{2}; \frac{1}{2}, \frac{1}{2} \right\rangle, \left|\, \frac{1}{2}, \frac{1}{2}; \frac{1}{2}, -\frac{1}{2} \right\rangle, \left|\, \frac{1}{2}, \frac{1}{2}; -\frac{1}{2}, \frac{1}{2} \right\rangle, \left|\, \frac{1}{2}, \frac{1}{2}; -\frac{1}{2}, -\frac{1}{2} \right\rangle.$$

We will denote these by

$$|\uparrow\uparrow\rangle\,,\; |\uparrow\downarrow\rangle\,,\; |\downarrow\uparrow\rangle\,,\; |\downarrow\downarrow\rangle\,,$$

where only the m_1 and m_2 labels are retained, $\uparrow$ means "spin up" ($m = 1/2$) and $\downarrow$ means "spin down" ($m = -1/2$). It is easy to show that the antisymmetric state

$$|\,a\,\rangle \equiv |\uparrow\downarrow\rangle - |\downarrow\uparrow\rangle \tag{26.6}$$

is invariant under $U(g)$, $g \in SO(3)$, and hence generates a one-dimensional invariant subspace of the 4-dimensional space spanned by the above four basis states. Indeed, from (26.5),

$$\begin{aligned}
&U(g)\,|\,a\,\rangle = U(g)\,|\uparrow\downarrow\rangle - U(g)\,|\downarrow\uparrow\rangle \\
&= (\,D^{(\frac{1}{2})}(g)^k_{\uparrow}\, D^{(\frac{1}{2})}(g)^l_{\downarrow} - D^{(\frac{1}{2})}(g)^k_{\downarrow}\, D^{(\frac{1}{2})}(g)^l_{\uparrow}\,)\,|\,kl\,\rangle \\
&= (\,D^{\uparrow}_{\uparrow} D^{\uparrow}_{\downarrow} - D^{\uparrow}_{\downarrow} D^{\uparrow}_{\uparrow}\,)\,|\uparrow\uparrow\rangle + (\,D^{\downarrow}_{\uparrow} D^{\downarrow}_{\downarrow} - D^{\downarrow}_{\downarrow} D^{\downarrow}_{\uparrow}\,)\,|\downarrow\downarrow\rangle \\
&+ (\,D^{\uparrow}_{\uparrow} D^{\downarrow}_{\downarrow} - D^{\uparrow}_{\downarrow} D^{\downarrow}_{\uparrow}\,)\,|\uparrow\downarrow\rangle + (\,D^{\downarrow}_{\uparrow} D^{\uparrow}_{\downarrow} - D^{\downarrow}_{\downarrow} D^{\uparrow}_{\uparrow}\,)\,|\downarrow\uparrow\rangle \\
&= (\,D^{\uparrow}_{\uparrow} D^{\downarrow}_{\downarrow} - D^{\uparrow}_{\downarrow} D^{\downarrow}_{\uparrow}\,)\,(\,|\uparrow\downarrow\rangle - |\downarrow\uparrow\rangle\,) = det\,(D^{(\frac{1}{2})}(g))\,|\,a\,\rangle = |\,a\,\rangle\,.
\end{aligned} \tag{26.7}$$

The last equality holds since $det\,(D^{(\frac{1}{2})}(g)) = 1$ by virtue of the fact that $D^{(\frac{1}{2})}(g) \in SU(2)$. Thus the representation whose 1-dimensional invariant subspace is $span\,\{|\,a\,\rangle\}$ must be $D^{(0)}$ ($j = 0$). Later in this chapter we will see that

$$D^{(\frac{1}{2})} \otimes D^{(\frac{1}{2})} = D^{(0)} \oplus D^{(1)}\,, \tag{26.8}$$

where the invariant subspace of $D^{(1)}$ is 3-dimensional and spanned by the symmetric states

$$|\uparrow\uparrow\rangle\,, \quad |\downarrow\downarrow\rangle \quad \text{and} \quad |\uparrow\downarrow\rangle + |\downarrow\uparrow\rangle\,.$$

We will now find the infinitesimal generators corresponding to the product representation. We have, for an infinitesimal rotation by an angle $d\theta$ about an axis $\hat{\boldsymbol{n}}$,

$$\begin{aligned}
&D^{(j_1)} \otimes D^{(j_2)}(d\theta\hat{\boldsymbol{n}}) = (I^{(j_1)} - id\theta J^{(j_1)}_n) \otimes (I^{(j_2)} - id\theta J^{(j_2)}_n) \\
&= I^{(j_1)} \otimes I^{(j_2)} - id\theta\,(J^{(j_1)}_n \otimes I^{(j_2)} + I^{(j_1)} \otimes J^{(j_2)}_n)\,,
\end{aligned} \tag{26.9}$$

where $I^{(j_i)}$ is the identity operator in the (j_i) invariant subspace, and $J_n = \boldsymbol{J}\cdot\hat{\boldsymbol{n}}$. Hence

$$J^{(j_1 \otimes j_2)}_n = J^{(j_1)}_n \otimes I^{(j_2)} + I^{(j_1)} \otimes J^{(j_2)}_n\,. \tag{26.10}$$

Let us recall the general reduction formula [cf. (23.7)]:

$$D^{(j_1)} \otimes D^{(j_2)} = \sum_{\oplus j} a_j D^{(j)}\,, \tag{26.11}$$

where a_j is the number of times that the (j)-irreducible representation occurs in the direct product representation. The invariant subspace of the product representation is

$$V^{(j_1)} \otimes V^{(j_2)} = span\,\{|\,j_1, j_2\,;\, m_1, m_2\,\rangle\}\,. \tag{26.12}$$

Due to (26.10) we see that the basis states $|\, j_1, j_2\,; m_1, m_2\,\rangle$ are eigenstates of $J_z^{(j_1 \otimes j_2)}$, since

$$\begin{aligned} &J_z^{(j_1\otimes j_2)}\,|\, j_1, j_2\,; m_1, m_2\,\rangle \\ &= J_z^{(j_1)}\,|\, j_1 m_1\,\rangle \otimes |\, j_2 m_2\,\rangle + |\, j_1 m_1\,\rangle \otimes J_z^{(j_2)}\,|\, j_2 m_2\,\rangle \\ &= (m_1 + m_2)\,|\, j_1, j_2\,; m_1, m_2\,\rangle\,. \end{aligned} \tag{26.13}$$

On the other hand, the eigenstates of $J_z^{(j_1\otimes j_2)}$ must also be of the form $|\, jm\,\rangle$:

$$J_z^{(j_1\otimes j_2)}\,|\, jm\,\rangle = m\,|\, jm\,\rangle\,. \tag{26.14}$$

Hence each $|\, jm\,\rangle$ must be a linear combination of the $|\, j_1, j_2\,; m_1, m_2\,\rangle$ such that $m = m_1 + m_2$.

Let $n(m)$ be the degree of degeneracy of the eigenvalue m of J_z in $V^{(j_1)} \otimes V^{(j_2)}$. Then $n(m)$ is simply equal to the number of ways that m_1 and m_2 can sum up to m. The relation between a_j [as defined in (26.11)] and $n(m)$ is then

$$n(m) = \sum_{j \geq |m|} a_j\,. \tag{26.15}$$

Thus

$$n(j) = a_j + a_{j+1} + a_{j+2} + \dots\,, \quad n(j+1) = a_{j+1} + a_{j+2} + \dots\,, \tag{26.16}$$

from which it follows that

$$a_j = n(j) - n(j+1)\,. \tag{26.17}$$

Consider Figs. 26.1 and 26.2, which show in lattice form all possible values that m_1 and m_2 can assume for given pairs of values of j_1 and j_2 [$j_1 = 3$, $j_2 = 7/2$ for Fig. 26.1 and $j_1 = 3/2$, $j_2 = 4$ for Fig. 26.2]. Each dot in a lattice corresponds to a pair of values for m_1 and m_2, and all dots on the same diagonal have a common value for $m = m_1 + m_2$. Thus $n(m)$ can be read off from the number of dots on the diagonal corresponding to m. We notice that for both lattices, as we move through the diagonal lines from the top right corner to the bottom left corner, m decreases in value by steps of 1, starting with $m = j_1 + j_2$. At the same time $n(m)$ increases from 1 (when $m = j_1 + j_2$) by steps of 1 to a maximum value of $2j_{min} + 1$ (where j_{min} is the smaller of j_1 and j_2), when the diagonal line first meets the top left corner of the lattice, and when $m = |\, j_1 - j_2\,|$. For this range of m values, that is, for $j_1 + j_2 \geq |m| \geq |\, j_1 - j_2\,|$, $|m| = j_1 + j_2 - (n(m) - 1)$; hence $n(m) = j_1 + j_2 + 1 - |m|$. As we further move through the diagonal lines we encounter a range of m values for which $|\, j_1 - j_2\,| \geq |m| \geq 0$. In this range, which terminates at a diagonal that meets the bottom right corner of the lattice, $n(m)$ remains constant at the value $2j_{min} + 1$. Beyond this range, the diagonal lines correspond to m values that again lie in the interval $j_1 + j_2 \geq |m| \geq |\, j_1 - j_2\,|$, with $n(m)$ taking values according to the same expression given earlier for this interval. Finally we reach

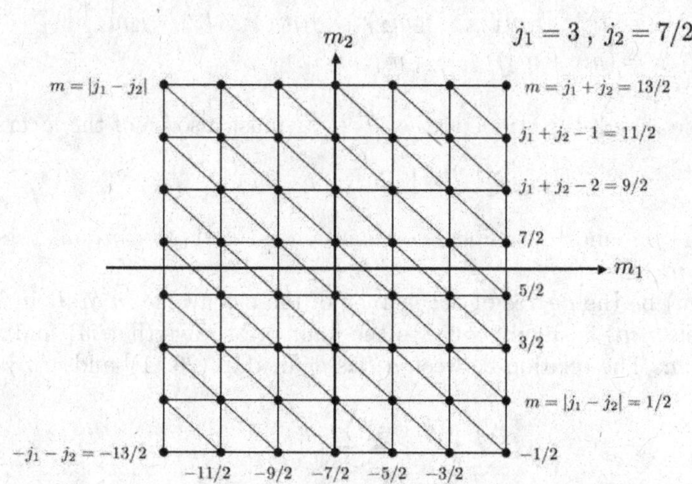

Fig. 26.1

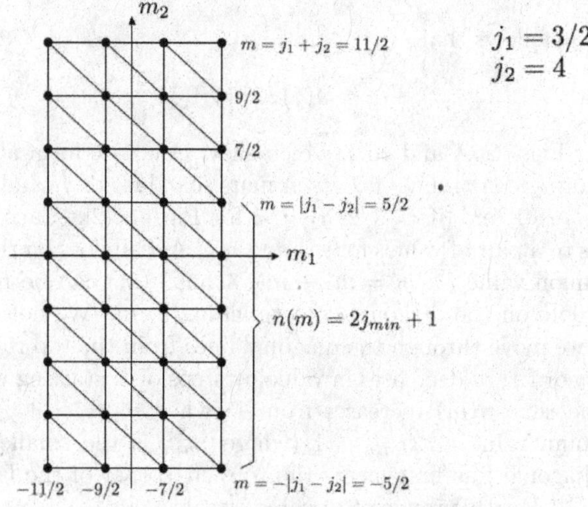

Fig. 26.2

the bottom left corner of the lattice, where $m = -j_1 - j_2$ and $n(m) = 1$ again. The above results can be summarized as follows.

$$\begin{aligned} n(m) &= 0 \quad &&\text{if} \quad |m| > j_1 + j_2 \,, \\ n(m) &= j_1 + j_2 + 1 - |m| \quad &&\text{if} \quad j_1 + j_2 \geq |m| \geq |\, j_1 - j_2 \,| \,, \\ n(m) &= 2j_{min} + 1 \quad &&\text{if} \quad |\, j_1 - j_2 \,| \geq |m| \geq 0 \,, \end{aligned} \tag{26.18}$$

where j_{min} = the smaller of j_1 and j_2. It then follows from (26.17) that

$$a_j = \begin{cases} 1 \,, & \text{for} \quad |\, j_1 - j_2 \,| \leq j \leq j_1 + j_2 \,, \\ 0 \,, & \text{otherwise} \,. \end{cases} \tag{26.19}$$

Finally we have the following important result: The direct product representations for $SO(3)$ can be reduced as

$$\boxed{D^{(j_1)} \otimes D^{(j_2)} = \sum\nolimits_{\oplus j} D^{(j)}} \quad , \tag{26.20}$$

where $j_1 + j_2 \geq j \geq |\, j_1 - j_2 \,|$ in steps of one. Equation (26.8) is a special case of this general result.

Problem 26.1 In the study of spectroscopy one must often go from one angular momentum coupling scheme to another. The simplest example is that of three commuting angular momenta:

$$\boldsymbol{J} = \boldsymbol{J}_1 + \boldsymbol{J}_2 + \boldsymbol{J}_3 \,.$$

This addition can be done in several ways. For example, one can add $\boldsymbol{J}_1$ to $\boldsymbol{J}_2$ first to give $\boldsymbol{J}_{12}$, and then add $\boldsymbol{J}_3$ to give $\boldsymbol{J}$. Let the state formed by this coupling scheme be denoted by $|\, j_1 j_2 (j_{12}) j_3;\, jm \,\rangle$, where $j_i(j_i+1)$ is the eigenvalue of $\boldsymbol{J}_i^2$, $j(j+1)$ is the eigenvalue of $\boldsymbol{J}^2$, and m is the eigenvalue of J_z. Using the angular momentum commutation rules show that the matrix elements

$$\langle\, j_1 j_2 (j_{12}) j_3;\, jm \,|\, j_1 j_3 (j_{13}) j_2;\, jm \,\rangle$$

do not depend on m. (Hint: Use the commutation rule $[\, J_+ ,\, J_- \,] = 2J_z$.)

Chapter 27

The Calculation of Clebsch-Gordon Coefficients: The 3-j Symbols

Recall that the Clebsch-Gordon (CG) coefficients are the hermitian products $\langle\, j_1 j_2\,;\, m_1 m_2 \,|\, jm \,\rangle$. Adopting the Condon-Shortley convention [(20.37) and (20.40)] and further imposing the requirement that the particular coefficient $\langle\, j_1 j_2;\, j_1, j - j_1 \,|\, jj \,\rangle$ be real and positive [henceforth referred to as (R)], we will verify that all CG coefficients become real quantities:

$$\langle\, j_1 j_2\,;\, m_1 m_2 \,|\, jm \,\rangle = \langle\, jm \,|\, j_1 j_2\,;\, m_1 m_2 \,\rangle\,. \tag{27.1}$$

Consider the case $j_1 = j_2 = 1/2$. We have

$$|\, jm \,\rangle = \sum_{m_1, m_2} |\, 1/2, m_1 \,\rangle\, |\, 1/2, m_2 \,\rangle\, \langle\, 1/2, 1/2\,;\, m_1\, m_2 \,|\, jm \,\rangle\,. \tag{27.2}$$

The possible values of j are 1 and 0; and it is always true that $m = m_1 + m_2$. In particular

$$|\, 11 \,\rangle = |\, 1/2\,,\, 1/2 \,\rangle\, |\, 1/2\,,\, 1/2 \,\rangle \langle\, 1/2\,,\, 1/2\,;\, 1/2\,,\, 1/2 \,|\, 11 \,\rangle\,. \tag{27.3}$$

Assuming that $|\, jm \,\rangle$, $|\, 1/2\,,\, m_1 \,\rangle$ and $|\, 1/2\,,\, m_2 \,\rangle$ are all normalized, the requirement (R) that

$$\langle\, j_1 j_2\,;\, j_1, j - j_1 \,|\, jj \,\rangle = \langle\, 1/2\,,\, 1/2\,;\, 1/2\,,\, 1/2 \,|\, 11 \,\rangle$$

be real and positive implies that

$$\langle\, 1/2\,,\, 1/2\,;\, 1/2\,,\, 1/2 \,|\, 11 \,\rangle = 1\,. \tag{27.4}$$

Next we apply

$$J_- = J_-^{(1)} + J_-^{(2)} \tag{27.5}$$

to (27.3). Recalling (20.40), we have

$$\sqrt{2}\,|\,1,0\,\rangle = |\,1/2\,,\,-1/2\,\rangle|\,1/2\,,\,1/2\,\rangle + |\,1/2\,,\,1/2\,\rangle|\,1/2\,,\,-1/2\,\rangle\,. \tag{27.6}$$

Applying J_- again to the above equation we obtain

$$|\,1\,,\,-1\,\rangle = |\,1/2\,,\,-1/2\,\rangle|\,1/2\,,\,-1/2\,\rangle\,. \tag{27.7}$$

Now recall that the singlet state $|\,0,0\,\rangle$ generates a one-dimensional invariant subspace under $U(g)\,,\; g \in SO(3)$ [cf. (26.7)] , and hence must be orthogonal to the triplet states $|\,1,1\,\rangle\,,\,|\,1,0\,\rangle$ and $|\,1\,,\,-1\,\rangle$. Thus

$$\langle\,0,0\,|\,1/2,\,1/2\,;\,1/2,\,1/2\,\rangle = 0\,, \tag{27.8}$$

$$\langle\,0,0\,|\,1/2,\,1/2\,;\,-1/2,\,-1/2\,\rangle = 0\,, \tag{27.9}$$

and

$$\begin{aligned} |\,0,0\,\rangle &= \frac{1}{\sqrt{2}}\left(|\,1/2,\,1/2\,\rangle|\,1/2,\,-1/2\,\rangle - |\,1/2,\,-1/2\,\rangle|\,1/2,\,1/2\,\rangle\right) \\ &\equiv \frac{1}{\sqrt{2}}\left(|\uparrow\downarrow\rangle - |\downarrow\uparrow\rangle\right), \end{aligned} \tag{27.10}$$

where, on the RHS of the second equality in the last equation, we have used the notation of (26.6). Note that in the above equation, the phase of $|\,0,0\,\rangle$ is determined by the requirement (R) that $\langle\, j_1, j_2\,;\, j_1, j - j_1\,|\, jj\,\rangle$ is real and positive.

Gathering the results of (27.3) to (27.10), we have the following matrix of values for the CG coefficients $\langle\,1/2,\,1/2\,;\,m_1,\,m_2\,|\,jm\,\rangle$:

	$\|\,1,1\,\rangle$	$\|\,1,0\,\rangle$	$\|\,1,-1\,\rangle$	$\|\,0,0\,\rangle$
$\langle\,1/2,\,1/2\,;\,1/2,\,1/2\,\|$	1	0	0	0
$\langle\,1/2,\,1/2\,;\,1/2,\,-1/2\,\|$	0	$1/\sqrt{2}$	0	$1/\sqrt{2}$
$\langle\,1/2,\,1/2\,;\,-1/2,\,1/2\,\|$	0	$1/\sqrt{2}$	0	$-1/\sqrt{2}$
$\langle\,1/2,\,1/2\,;\,-1/2,\,-1/2\,\|$	0	0	1	0

(27.11)

Before proceeding to a more general case we note that the CG coefficients vanish unless the following three conditions are all satisfied:

(i) $m = m_1 + m_2$,

(ii) $j_1 + j_2 \geq j \geq |\,j_1 - j_2\,|$ (the "triangle" inequality) ,

(iii) $j_1 + j_2 + j = \textit{an integer}.$

Conditions (i) and (ii) have already been proved in the last chapter. Condition (iii) can be proved as follows. Recall from (22.31) and (22.32) and the discussion surrounding these two equations that

$$U(R_3(2\pi))\,|\,jm\,\rangle = (-1)^{2j}\,|\,jm\,\rangle\,. \tag{27.12}$$

Applying $U(R_3(2\pi))$ to both sides of the following equation:

$$|\,jm\,\rangle = \sum_{m_1,m_2} |\,j_1,\,j_2\,;\,m_1,\,m_2\,\rangle\langle\,j_1,\,j_2\,;\,m_1,\,m_2\,|\,jm\,\rangle \tag{27.13}$$

we have

$$(-1)^{2j}\,|\,jm\,\rangle = (-1)^{2(j_1+j_2)} \sum_{m_1,m_2} |\,j_1,\,j_2\,;\,m_1,\,m_2\,\rangle\langle\,j_1,\,j_2\,;\,m_1,\,m_2\,|\,jm\,\rangle\,. \tag{27.14}$$

On comparing this equation with (27.13), we see that j_1+j_2-j must be an integer, which implies that j_1+j_2+j must also be an integer, since $2j$ is always an integer.

Now apply $J_- = J_-^{(1)} + J_-^{(2)}$ to (27.13). We obtain [using (20.40)]

$$\begin{aligned}
&\sqrt{(j+m)(j-m+1)}\,|\,j,\,m-1\,\rangle \\
&= \sum_{m_1,m_2} \Big[\sqrt{(j_1+m_1)(j_1-m_1+1)}\,|\,j_1,\,j_2\,;\,m_1-1,\,m_2\,\rangle \\
&\qquad \sqrt{(j_2+m_2)(j_2-m_2+1)}\,|\,j_1,\,j_2\,;\,m_1,\,m_2-1\,\rangle\Big]\;\langle\,j_1,\,j_2\,;\,m_1,\,m_2\,|\,jm\,\rangle\,.
\end{aligned} \tag{27.15}$$

On the other hand, replacing m by $m-1$ in (27.13), we have

$$|\,j,\,m-1\,\rangle = \sum_{m_1,m_2} |\,j_1,\,j_2\,;\,m_1,\,m_2\,\rangle\langle\,j_1,\,j_2\;\;m_1,\,m_2\,|\,j,\,m-1\,\rangle\,. \tag{27.16}$$

Substituting this in the LHS of (27.15) we obtain

$$\begin{aligned}
&\sum_{m_1,m_2} \sqrt{(j+m)(j-m+1)}\,\langle\,j_1,\,j_2\,;\,m_1,\,m_2\,|\,j,\,m-1\rangle\,|\,j_1,\,j_2\,;\,m_1,\,m_2\,\rangle \\
&= \sum_{m_1,m_2} \Big[\sqrt{(j_1+m_1)(j_1-m_1+1)}\,|\,j_1,\,j_2\,;\,m_1-1,\,m_2\,\rangle \\
&\qquad + \sqrt{(j_2+m_2)(j_2-m_2+1)}\,|\,j_1,\,j_2\,;\,m_1,\,m_2-1\,\rangle\Big]\;\langle\,j_1,\,j_2\,;\,m_1\,,m_2\,|\,jm\,\rangle\,.
\end{aligned} \tag{27.17}$$

On replacing m_1 by m_1+1 in the first term, and m_2 by m_2+1 in the second term of the RHS of the above equation, we obtain the following recursion relation for Clebsch-Gordon coefficients:

$$\boxed{\begin{aligned}&\sqrt{(j+m)(j-m+1)}\,\langle\, j_1,\, j_2\,;\, m_1,\, m_2\,|\, j,\, m-1\,\rangle\\&=\sqrt{(j_1+m_1+1)(j_1-m_1)}\,\langle\, j_1,\, j_2\,;\, m_1+1,\, m_2\,|\, j,\, m\,\rangle\\&\quad+\sqrt{(j_2+m_2+1)(j_2-m_2)}\,\langle\, j_1,\, j_2\,;\, m_1,\, m_2+1\,|\, j,\, m\,\rangle\,.\end{aligned}} \tag{27.18}$$

A similar recursion relation is obtained by applying $J_+ = J_+^{(1)} + J_+^{(2)}$ to (27.13):

$$\boxed{\begin{aligned}&\sqrt{(j+m+1)(j-m)}\,\langle\, j_1,\, j_2\,;\, m_1,\, m_2\,|\, j,\, m+1\,\rangle\\&=\sqrt{(j_1+m_1)(j_1-m_1+1)}\,\langle\, j_1,\, j_2\,;\, m_1-1,\, m_2\,|\, j,\, m\,\rangle\\&\quad+\sqrt{(j_2+m_2)(j_2-m_2+1)}\,\langle\, j_1,\, j_2\,;\, m_1,\, m_2-1\,|\, j,\, m\,\rangle\,.\end{aligned}} \tag{27.19}$$

The above recursion relations, together with the phase convention specified at the beginning of this chapter, can be used to calculate all CG coefficients, which, as noted before, can be taken to be all real [cf. (27.1)]. We will illustrate this procedure next for the important case of $j_1 = l$, $j_2 = 1/2$, which arises in the study of spin-orbit coupling.

Let $m_2 = 1/2$ (the largest value) in the right-hand side of (27.18). Then the second term vanishes, leaving

$$\begin{aligned}&\sqrt{(j-m+1)(j+m)}\,\langle\, l,\, 1/2\,;\, m-3/2,\, 1/2\,|\, j,\, m-1\,\rangle\\&=\sqrt{(l-m+3/2)(l+m-1/2)}\,\langle\, l,\, 1/2\,;\, m-1/2,\, 1/2\,|\, j,\, m\,\rangle\,.\end{aligned} \tag{27.20}$$

Replacing m by $m+1$, we have

$$\begin{aligned}&\langle\, l,\, 1/2\,;\, m-1/2,\, 1/2\,|\, j,\, m\,\rangle\\&=\sqrt{\frac{(l-m+1/2)(l+m+1/2)}{(j-m)(j+m+1)}}\;\langle\, l,\, 1/2\,;\, m+1/2,\, 1/2\,|\, j,\, m+1\,\rangle\,.\end{aligned} \tag{27.21}$$

Now let $j = l + 1/2$. Starting with an arbitrary value of m we can use this equation recursively until $m + 1 = l + 1/2$ (the maximum value) in the CG coefficient in the RHS. Thus

$$\begin{aligned}
&\langle\, l,\, 1/2\,;\, m-1/2,\, 1/2\,|\, l+1/2,\, m\,\rangle \\
&= \sqrt{\frac{(l-m+1/2)(l+m+1/2)}{(l-m+1/2)(l+m+3/2)}}\;\langle\, l,\, 1/2\,;\, m+1/2,\, 1/2\,|\, l+1/2,\, m+1\,\rangle \\
&= \sqrt{\frac{(l+m+1/2)}{(l+m+3/2)}}\sqrt{\frac{(l-m-1/2)(l+m+3/2)}{(l-m-1/2)(l+m+5/2)}} \\
&\quad \times \langle\, l,\, 1/2\,;\, m+3/2,\, 1/2\,|\, l+1/2,\, m+2\,\rangle \\
&\quad \vdots \\
&= \sqrt{\frac{l+m+1/2}{2l+1}}\;\langle\, l,\, 1/2\,;\, l,\, 1/2\,|\, l+1/2,\, l+1/2\,\rangle\,.
\end{aligned} \tag{27.22}$$

Since $m = m_1 + m_2$,

$$|\, l+1/2,\, l+1/2\,\rangle = |\, l,\, 1/2\,;\, l,\, 1/2\,\rangle\langle\, l,\, 1/2\,;\, l,\, 1/2\,|\, l+1/2,\, l+1/2\,\rangle\,. \tag{27.23}$$

Assuming that both kets in the above equation are normalized, the requirement (R) (recall discussion in the first paragraph of this chapter) then implies

$$\langle\, l,\, 1/2\,;\, l,\, 1/2\,|\, l+1/2,\, l+1/2\,\rangle = 1\,. \tag{27.24}$$

Hence (27.22) yields

$$\langle\, l,\, 1/2\,;\, m-1/2,\, 1/2\,|\, l+1/2,\, m\,\rangle = \sqrt{\frac{l+m+1/2}{2l+1}}\,. \tag{27.25}$$

Next consider the CG coefficient $\langle\, l,\, 1/2\,;\, m+1/2,\, -1/2\,|\, l+1/2,\, m\,\rangle$. This can be similarly obtained by setting $m = -1/2$ in (27.19). However, its magnitude can be more easily obtained by assuming all ket vectors appearing in the following equation to be normalized:

$$\begin{aligned}
|\, l+1/2,\, m\,\rangle = {} & |\, l,\, 1/2\,;\, m-1/2,\, 1/2\,\rangle\langle\, l,\, 1/2\,;\, m-1/2,\, 1/2\,|\, l+1/2,\, m\,\rangle \\
& + |\, l,\, 1/2\,;\, m+1/2,\, -1/2\,\rangle\langle\, l,\, 1/2\,;\, m+1/2,\, -1/2\,|\, l+1/2,\, m\,\rangle\,.
\end{aligned} \tag{27.26}$$

Thus

$$|\langle\, l,\, 1/2\,;\, m-1/2,\, 1/2\,|\, l+1/2,\, m\,\rangle|^2 + |\langle\, l,\, 1/2\,;\, m+1/2,\, -1/2\,|\, l+1/2,\, m\,\rangle|^2 = 1\,. \tag{27.27}$$

Hence

$$\langle\, l,\, 1/2\,;\, m+1/2,\, -1/2\,|\, l+1/2,\, m\,\rangle = \sqrt{\frac{l-m+1/2}{2l+1}}\,, \tag{27.28}$$

where the phase is again determined by the requirement (R).

The other two non-zero CG coefficients for the case $j_1 = l$, $j_2 = 1/2$,

$$a \equiv \langle\, l,\, 1/2\,;\, m - 1/2,\, 1/2 \,|\, l - 1/2,\, m \,\rangle \tag{27.29}$$

$$\text{and} \qquad b \equiv \langle\, l,\, 1/2\,;\, m + 1/2,\, -1/2 \,|\, l - 1/2,\, m \,\rangle\,, \tag{27.30}$$

are obtained by setting $j = l - 1/2$. These can be calculated again by resorting to the recursion relations (27.18) and (27.19). More simply, as in the calculation leading to (27.28), their relative magnitudes can be obtained by assuming all kets in the following equation to be normalized:

$$|\, l - 1/2,\, m \,\rangle = a \,|\, l,\, 1/2\,;\, m - 1/2,\, 1/2 \,\rangle + b \,|\, l,\, 1/2\,;\, m + 1/2,\, -1/2 \,\rangle\,, \tag{27.31}$$

and realizing that $|\, l - 1/2,\, m \,\rangle$ is orthogonal to $|\, l + 1/2,\, m \,\rangle$. Thus

$$|a|^2 + |b|^2 = 1\,; \tag{27.32}$$

$$a\sqrt{\frac{l + m + 1/2}{2l + 1}} + b\sqrt{\frac{l - m + 1/2}{2l + 1}} = 0\,. \tag{27.33}$$

These equations imply

$$\frac{a}{b} = -\sqrt{\frac{l - m + 1/2}{l + m + 1/2}}\,, \tag{27.34}$$

$$|a| = \sqrt{\frac{l - m + 1/2}{2l + 1}}\,, \qquad |b| = \sqrt{\frac{l + m + 1/2}{2l + 1}}\,. \tag{27.35}$$

The phases are determined by (R), which requires the CG coefficient

$$\langle\, l,\, 1/2\,;\, l,\, -1/2 \,|\, l - 1/2,\, l - 1/2 \,\rangle$$

to be real and positive. Thus

$$\langle\, l,\, 1/2\,;\, m - 1/2,\, 1/2 \,|\, l - 1/2,\, m \,\rangle = -\sqrt{\frac{l - m + 1/2}{2l + 1}}\,, \tag{27.36}$$

$$\langle\, l,\, 1/2\,;\, m + 1/2,\, -1/2 \,|\, l - 1/2,\, m \,\rangle = \sqrt{\frac{l + m + 1/2}{2l + 1}}\,. \tag{27.37}$$

We summarize below in matrix form the non-zero CG coefficients

$$\langle\, l,\, 1/2\,;\, m - m',\, m' \,|\, j,\, m \,\rangle$$

for the case $j_1 = l$, $j_2 = 1/2$:

$$\begin{array}{ccc} & m' = 1/2 & m' = -1/2 \\[1em] j = l + 1/2 & \sqrt{\dfrac{l + m + 1/2}{2l + 1}} & \sqrt{\dfrac{l - m + 1/2}{2l + 1}} \\[2em] j = l - 1/2 & -\sqrt{\dfrac{l - m + 1/2}{2l + 1}} & \sqrt{\dfrac{l + m + 1/2}{2l + 1}} \end{array} \,. \tag{27.38}$$

In many applications involving spin-orbit coupling, we use the **Pauli two-component spin-orbit wave function** $Y_l^{j,m}(\hat{\boldsymbol{n}})$ [$\hat{\boldsymbol{n}} \equiv (\theta, \phi)$ denoting a point on the unit sphere] defined by

$$(Y_l^{j,m})^\sigma(\theta,\phi) \equiv \langle\, \hat{\boldsymbol{n}},\, \sigma \,|\, j,\, m \,\rangle \equiv \langle\, \hat{\boldsymbol{n}} \,|\, \langle\, 1/2,\, \sigma \,|\, j,\, m \,\rangle$$

$$= \sum_{m_1, m_2} \langle\, \hat{\boldsymbol{n}} \,|\, \langle\, 1/2,\, \sigma \,|\, l,\, 1/2\, ;\, m_1,\, m_2 \,\rangle \langle\, l,\, 1/2\, ;\, m_1,\, m_2 \,|\, j,\, m \,\rangle$$

$$= \langle\, \hat{\boldsymbol{n}} \,|\, l,\, m-\sigma \,\rangle \langle\, l,\, 1/2\, ;\, m-\sigma,\, \sigma \,|\, j,\, m \,\rangle = Y_l^{m-\sigma}(\hat{\boldsymbol{n}}) \langle\, l,\, 1/2\, ;\, m-\sigma,\, \sigma \,|\, j,\, m \,\rangle\,, \tag{27.39}$$

where $|\, 1/2,\, \sigma = \pm 1/2 \,\rangle$ is a pure spin state and $Y_l^{m-\sigma}(\hat{\boldsymbol{n}})$ is a spherical harmonic. Hence we have

$$Y_l^{l+\frac{1}{2},m}(\hat{\boldsymbol{n}}) = \frac{1}{\sqrt{2l+1}} \begin{pmatrix} \sqrt{l+m+1/2}\; Y_l^{m-\frac{1}{2}}(\hat{\boldsymbol{n}}) \\ \sqrt{l-m+1/2}\; Y_l^{m+\frac{1}{2}}(\hat{\boldsymbol{n}}) \end{pmatrix}, \tag{27.40}$$

$$Y_l^{l-\frac{1}{2},m}(\hat{\boldsymbol{n}}) = \frac{1}{\sqrt{2l+1}} \begin{pmatrix} -\sqrt{l-m+1/2}\; Y_l^{m-\frac{1}{2}}(\hat{\boldsymbol{n}}) \\ \sqrt{l+m+1/2}\; Y_l^{m+\frac{1}{2}}(\hat{\boldsymbol{n}}) \end{pmatrix}. \tag{27.41}$$

The states $|\, j = l \pm 1/2,\, m \,\rangle$ as given by (27.26) and (27.31) are eigenstates of L^2, S^2, J^2 and J_z, where $\boldsymbol{S}$ is the spin angular momentum operator. Hence they are also eigenstates of the spin-orbit coupling operator $\boldsymbol{L} \cdot \boldsymbol{S}$, since

$$\boldsymbol{L} \cdot \boldsymbol{S} = \frac{1}{2}\,(J^2 - L^2 - S^2)\,. \tag{27.42}$$

We have

$$\boldsymbol{L} \cdot \boldsymbol{S}\,|\, j,\, m \,\rangle = \frac{1}{2}\,[j(j+1) - l(l+1) - 3/4]\,|\, j,\, m \,\rangle\,. \tag{27.43}$$

Thus,

$$\boldsymbol{L} \cdot \boldsymbol{S}\,|\, l+1/2,\, m \,\rangle = \frac{l}{2}\,|\, l+1/2,\, m \,\rangle\,, \tag{27.44}$$

$$\boldsymbol{L} \cdot \boldsymbol{S}\,|\, l-1/2,\, m \,\rangle = \left(\frac{l+1}{2}\right)\,|\, l-1/2,\, m \,\rangle\,. \tag{27.45}$$

A coefficient related to the CG coefficient but displaying more symmetry properties, called the **Wigner 3-j symbol**, is often used in quantum mechanical mechanical calculations. This symbol is defined as follows.

$$\begin{pmatrix} j_1 & j_2 & j_3 \\ m_1 & m_2 & m_3 \end{pmatrix} \equiv \frac{(-1)^{j_1-j_2-m_3}}{\sqrt{2j_3+1}}\; \langle\, j_1,\, j_2\, ;\, m_1,\, m_2 \,|\, j_3,\, -m_3 \,\rangle\,. \tag{27.46}$$

From the properties of the CG coefficients as described above it is straightforward to deduce the following symmetry properties for the 3-j symbols:

(1) A 3-j symbol vanishes unless $m_1 + m_2 + m_3 = 0$.

(2) A 3-j symbol vanishes unless $j_1 + j_2 + j_3 =$ an integer, and j_1, j_2, j_3 satisfy the triangle inequality: $j_1 + j_2 \geq j_3 \geq |j_1 - j_2|$.

(3)

$$\begin{pmatrix} j_{\sigma(1)} & j_{\sigma(2)} & j_{\sigma(3)} \\ m_{\sigma(1)} & m_{\sigma(2)} & m_{\sigma(3)} \end{pmatrix} = f(\sigma) \begin{pmatrix} j_1 & j_2 & j_3 \\ m_1 & m_2 & m_3 \end{pmatrix}, \tag{27.47}$$

where

$$f(\sigma) = \begin{cases} 1 & \text{if } \sigma \text{ is an even permutation of (123),} \\ (-1)^{j_1+j_2+j_3} & \text{if } \sigma \text{ is an odd permutation of (123).} \end{cases} \tag{27.48}$$

(4)

$$\begin{pmatrix} j_1 & j_2 & j_3 \\ -m_1 & -m_2 & -m_3 \end{pmatrix} = (-1)^{j_1+j_2+j_3} \begin{pmatrix} j_1 & j_2 & j_3 \\ m_1 & m_2 & m_3 \end{pmatrix}. \tag{27.49}$$

Problem 27.1 Prove Properties (1) to (4) stated above for the 3-j symbols.

Problem 27.2 Recall Problem 26.1, which dealt with the coupling of 3 angular momenta j_1, j_2 and j_3. We saw that in general

$$|\, j_1 j_2 (j_{12}) j_3;\, j\,\rangle \neq |\, j_2 j_3 (j_{23}) j_1;\, j\,\rangle\,.$$

In fact, one can expand each element of one set in terms of the other set by

$$|\, j_2 j_3 (j_{23}) j_1;\, j\,\rangle = \sum_{j_{12}} |\, j_1 j_2 (j_{12}) j_3;\, j\,\rangle \langle\, j_1 j_2 (j_{12}) j_3;\, j\,|\, j_2 j_3 (j_{23}) j_1;\, j\,\rangle\,.$$

The coupling coefficients are usually written as

$$\langle\, j_1 j_2 (j_{12}) j_3;\, j\,|\, j_2 j_3 (j_{23}) j_1;\, j\,\rangle \equiv (-1)^{j_1+j_2+j_3+j} \sqrt{(2j_{12}+1)(2j_{23}+1)} \begin{Bmatrix} j_1 & j_2 & j_{12} \\ j_3 & j & j_{23} \end{Bmatrix}$$

The quantity in braces is called a 6-j **symbol**, and is given in general by

$$\begin{Bmatrix} j_1 & j_2 & j_3 \\ l_1 & l_2 & l_3 \end{Bmatrix} = (-1)^{j_1+j_2+l_1+l_2}\, \Delta(j_1 j_2 j_3)\Delta(l_1 l_2 j_3)\Delta(l_1 j_2 l_3)\Delta(j_1 l_2 l_3)$$

$$\times \sum_k [\,(-1)^k (j_1+j_2+l_1+l_2-k)!/\{(j_1+j_2-j_3-k)!(l_1+l_2-j_3-k)!$$

$$(j_1+l_2-l_3-k)!(l_1+j_2-l_3-k)!(-j_1-l_1+j_3+l_3+k)!(-j_2-l_2+j_3+l_3+k)!\}\,]\,,$$

where

$$\Delta(abc) \equiv \sqrt{\frac{(a+b-c)!(a-b+c)!(b+c-a)!}{(a+b+c+1)!}}\,.$$

Verify the above results for the special case $j_1 = 1$, $j_2 = 1/2$ and $j_3 = 1/2$, and for $j = m = 1$, by expanding all eigenfunctions in terms of the appropriate $|\, j_1, m_1 \rangle$, $|\, j_2, m_2 \rangle$ and $|\, j_3, m_3 \rangle$ states, and by making use of the CG coefficients derived in this chapter.

Problem 27.3 Suppose a system with angular momentum $\boldsymbol{J}_1$ is coupled to another with angular momentum $\boldsymbol{J}_2$. Let $T^{(k)}$ be an irreducible tensor operator that acts only on the first system. Prove the following relationship between reduced matrix elements:

$$\langle \tau', j_1' j_2';\, j' \,\|\, T^{(k)} \,\|\, \tau, j_1, j_2;\, j \,\rangle$$

$$= (-1)^{j_1'+j_2'+j'+k}\, \delta_{j_2' j_2}\, \sqrt{(2j'+1)(2j+1)} \begin{Bmatrix} j' & k & j \\ j_1 & j_2' & j_1' \end{Bmatrix} \langle \tau',\, j_1' \,\|\, T^{(k)} \,\|\, \tau,\, j_1 \,\rangle\,.$$

Chapter 28

Applications of the Wigner-Eckart Theorem

We first recall the definition of an irreducible tensor operator $T_q^{(k)}$ as given in Definition 25.1. These operators transform under the rotation group $SO(3)$ as follows:

$$U(g)T_q^{(k)}U^{-1}(g) = D^{(k)}(g)_q^{q'}T_{q'}^{(k)} , \quad g \in SO(3) . \tag{28.1}$$

For infinitesimal rotations,

$$U(g) \approx 1 - i\varepsilon\, \boldsymbol{u} \cdot \boldsymbol{J} , \tag{28.2}$$

where $\boldsymbol{u}$ is the unit vector along the axis of rotation specified by g and ε is the infinitesimal angle of rotation. Hence

$$(1 - i\varepsilon\, \boldsymbol{u} \cdot \boldsymbol{J})\, T_q^{(k)}\, (1 + i\varepsilon\, \boldsymbol{u} \cdot \boldsymbol{J}) = \sum_{q'} (\delta_q^{q'} - i\varepsilon \,\langle\, kq' \,|\, \boldsymbol{u} \cdot \boldsymbol{J} \,|\, kq \,\rangle)\, T_{q'}^{(k)} . \tag{28.3}$$

Thus, to first order in ε,

$$[\, \boldsymbol{u} \cdot \boldsymbol{J} \,,\, T_q^{(k)} \,] = \sum_{q'} T_{q'}^{(k)} \,\langle\, kq' \,|\, \boldsymbol{u} \cdot \boldsymbol{J} \,|\, kq \,\rangle . \tag{28.4}$$

Setting $\boldsymbol{u} = \boldsymbol{x}, \boldsymbol{y}, \boldsymbol{z}$, we have

$$[\, J_x \,,\, T_q^{(k)} \,] = \sum_{q'} T_{q'}^{(k)} \,\langle\, kq' \,|\, J_x \,|\, kq \,\rangle , \tag{28.5a}$$

$$[\, J_y \,,\, T_q^{(k)} \,] = \sum_{q'} T_{q'}^{(k)} \,\langle\, kq' \,|\, J_y \,|\, kq \,\rangle , \tag{28.5b}$$

$$[\, J_z \,,\, T_q^{(k)} \,] = \sum_{q'} T_{q'}^{(k)} \,\langle\, kq' \,|\, J_z \,|\, kq \,\rangle = qT_q^{(k)} . \tag{28.5c}$$

Recalling [cf. (20.4), (20.37) and (20.40)] that

$$J_{\pm} = J_x \pm iJ_y \,, \tag{28.6}$$

and

$$J_{\pm} \,|\, kq \,\rangle = \sqrt{k(k+1) - q(q \pm 1)}\,|\, k, q \pm 1 \,\rangle = \sqrt{(k \mp q)(k \pm q + 1)}\,|\, k, q \pm 1 \,\rangle \,, \tag{28.7}$$

the equations (28.5) can be written

$$[\, J_+ \,,\, T_q^{(k)} \,] = \sqrt{(k-q)(k+q+1)}\; T_{q+1}^{(k)} \,, \tag{28.8a}$$

$$[\, J_- \,,\, T_q^{(k)} \,] = \sqrt{(k+q)(k-q+1)}\; T_{q-1}^{(k)} \,, \tag{28.8b}$$

$$[\, J_z \,,\, T_q^{(k)} \,] = qT_q^{(k)} \,. \tag{28.8c}$$

The above set of equations can be considered as alternative definitions for the spherical components of irreducible tensors of rank k, instead of (28.1).

Consider the vector operator $T_q^{(1)} = J_q^{(1)}$ corresponding to the angular momentum operator $\boldsymbol{J}$. The equations (28.8) imply that

$$[\, J_+ \,,\, J_{-1}^{(1)} \,] = \sqrt{2}\, J_0^{(1)} \,, \tag{28.9a}$$

$$[\, J_- \,,\, J_1^{(1)} \,] = \sqrt{2}\, J_0^{(1)} \,, \tag{28.9b}$$

$$[\, J_z \,,\, J_0^{(1)} \,] = 0 \,. \tag{28.9c}$$

Comparing these equations with [cf. (20.7)]

$$[\, J_+ \,,\, J_- \,] = 2J_z \,, \tag{28.10}$$

we see that

$$J_{-1}^{(1)} = \frac{J_-}{\sqrt{2}} = \frac{1}{\sqrt{2}}\,(J_x - iJ_y) \,, \tag{28.11a}$$

$$J_1^{(1)} = -\frac{J_+}{\sqrt{2}} = -\frac{1}{\sqrt{2}}\,(J_x + iJ_y) \,, \tag{28.11b}$$

$$J_0^{(1)} = J_z \,. \tag{28.11c}$$

Thus for any general vector operator $\boldsymbol{V}$, that is, an irreducible tensor operator of rank 1, the spherical components V_1, V_0, V_{-1} are expressed in terms of the Cartesian components V_x, V_y, V_z by

$$V_1 = -\frac{1}{\sqrt{2}}\,(V_x + iV_y) \,, \tag{28.12a}$$

$$V_{-1} = \frac{1}{\sqrt{2}}\,(V_x - iV_y) \,, \tag{28.12b}$$

$$V_0 = V_z \,. \tag{28.12c}$$

A **scalar operator** is an irreducible tensor operator of rank zero, and is invariant under $U(g)$:

$$U(g)SU^{-1}(g) = S\ . \tag{28.13}$$

The simplest example of the Wigner-Eckart Theorem is its application to a scalar operator. We have, from the statement of the theorem given by (25.11),

$$\langle\, \tau', j', m' \,|\, S \,|\, \tau, j, m\,\rangle = \langle\, \tau', j' \,\|\, S \,\|\, \tau, j\,\rangle \langle\, 0, j\,;\, 0, m \,|\, j', m'\,\rangle\ . \tag{28.14}$$

But the CG coefficient on the right-hand side of this equation is given by

$$\langle\, 0, j\,;\, 0, m \,|\, j', m'\,\rangle = \delta_{jj'}\delta_{mm'}\ . \tag{28.15}$$

Hence

$$\langle\, \tau', j', m' \,|\, S \,|\, \tau, j, m\,\rangle = \langle\, \tau', j' \,\|\, S \,\|\, \tau, j\,\rangle\, \delta_{jj'}\delta_{mm'}\ . \tag{28.16}$$

This matrix element is diagonal in the angular momentum indices and is independent of the magnetic quantum number $m = m'$.

Next consider matrix elements of spherical components of the angular momentum operator, which, as Eq. (28.9) shows, is a vector operator. The Wigner-Eckart Theorem implies that

$$\langle\, \tau', j', m' \,|\, J_q^{(1)} \,|\, \tau, j, m\,\rangle = \langle\, \tau', j' \,\|\, J^{(1)} \,\|\, \tau, j\,\rangle \langle\, 1, j\,;\, q, m \,|\, j', m'\,\rangle\ . \tag{28.17}$$

Although the reduced matrix element can be calculated by its defining equation (25.10), in practice it is much more convenient to evaluate the LHS of the above equation for a special case. Thus

$$\langle\, \tau', j', m' \,|\, J_0^{(1)} \,|\, \tau, j, m\,\rangle = \langle\, \tau', j', m' \,|\, J_z \,|\, \tau, j, m\,\rangle = m\, \delta_{\tau'\tau}\delta_{j'j}\delta_{m'm}\ . \tag{28.18}$$

So it suffices to calculate the matrix element in (28.17) for the case $j' = j = m = m'$. We then have

$$\langle\, \tau', j, j \,|\, J_0^{(1)} \,|\, \tau, j, j\,\rangle = \langle\, \tau', j \,\|\, J^{(1)} \,\|\, \tau, j\,\rangle \langle\, 1, j\,;\, 0, j \,|\, j, j\,\rangle = j\, \delta_{\tau'\tau}\ , \tag{28.19}$$

where in the second equality we have used the following value for the CG coefficient:

$$\langle\, 1, j\,;\, 0, j \,|\, j, j\,\rangle = \sqrt{\frac{j}{j+1}}\ , \tag{28.20}$$

which can be calculated by the methods discussed in the last chapter, or obtained in standard references. It follows that

$$\langle\, \tau', j \,\|\, J^{(1)} \,\|\, \tau, j\,\rangle = \delta_{\tau'\tau}\, \sqrt{j(j+1)}\ . \tag{28.21}$$

Now (28.17) and (28.18) imply

$$\langle\, \tau', j' \,\|\, J^{(1)} \,\|\, \tau, j\,\rangle \langle\, 1, j\,;\, 0, m \,|\, j', m\,\rangle = m\, \delta_{\tau'\tau}\delta_{j'j}\ . \tag{28.22}$$

The CG coefficient in the LHS of the above equation need not vanish for $j \neq j'$. Hence the reduced matrix element must vanish for $j \neq j'$. Equation (28.21) then implies

$$\boxed{\langle \tau', j' \,\|\, J^{(1)} \,\|\, \tau, j \rangle = \delta_{\tau'\tau}\delta_{j'j}\sqrt{j(j+1)}} \,. \tag{28.23}$$

We now consider products of irreducible tensor operators, of the form $X_q^{(k)} Z_m^{(l)}$. As in product representations of groups, these products are in general reducible. But we can show that the following tensor operator of rank L is irreducible:

$$T_M^{(L)} \equiv \sum_{q,m} \langle k, l;\, q, m \,|\, L, M \rangle \, X_q^{(k)} Z_m^{(l)} \,. \tag{28.24}$$

This fact can be demonstrated by proving that $T_M^{(L)}$ as defined above transforms under $g \in SO(3)$ according to (28.1). Indeed, for arbitrary $g \in SO(3)$, we can transform $T_M^{(L)}$ by the direct product representation $U(g) = D^{(k)}(g) \otimes D^{(l)}(g)$ as follows:

$$\begin{aligned}
&U(g) T_M^{(L)} U^{-1}(g) \\
&= \sum_{qm} \langle k, l;\, q, m \,|\, L, M \rangle \, U(g) X_q^{(k)} U^{-1}(g)\, U(g) Z_m^{(l)} U^{-1}(g) \\
&= \sum_{qm} \sum_{q'm'} \langle k, l;\, q, m \,|\, L, M \rangle \, (D^{(k)}(g))_q^{q'} \, (D^{(l)}(g))_m^{m'} \, X_{q'}^{(k)} Z_{m'}^{(l)} \\
&= \sum_{qm} \sum_{q'm'} \sum_{JQQ'} \langle k, l;\, q, m \,|\, L, M \rangle \\
&\qquad \times \langle k, l;\, q', m' \,|\, J, Q' \rangle \, (D^{(J)}(g))_Q^{Q'} \, \langle J, Q \,|\, k, l;\, q, m \rangle \, X_{q'}^{(k)} Z_{m'}^{(l)} \\
&= \sum_{q'm'} \sum_{JQQ'} \langle k, l\,,\, q', m' \,|\, J, Q' \rangle \langle J, Q \,|\, \sum_{qm} |\, k, l;\, q, m \rangle \langle k, l;\, q, m \,|\, L, M \rangle \\
&\qquad \times (D^{(J)}(g))_Q^{Q'} \, X_{q'}^{(k)} Z_{m'}^{(l)} \\
&= \sum_{q'm'} \sum_{JQQ'} \langle k, l\,,\, q', m' \,|\, J, Q' \rangle \langle J, Q \,|\, L, M \rangle \, (D^{(J)}(g))_Q^{Q'} \, X_{q'}^{(k)} Z_{m'}^{(l)} \\
&= \sum_{Q'} (D^{(L)}(g))_M^{Q'} \sum_{q'm'} \langle k, l;\, q', m' \,|\, L, Q' \rangle \, X_{q'}^{(k)} Z_{m'}^{(l)} \\
&= \sum_{Q'} (D^{(L)}(g))_M^{Q'} \, T_{Q'}^{(L)} \,,
\end{aligned} \tag{28.25}$$

where, in the second equality we have used (28.1), in the third equality the Clebsch-Gordon series as given by (23.16), in the fifth equality the completeness relation $\sum_{qm} |\, k, l;\, q, m \rangle \langle k, l;\, q, m \,| = 1$, and in the sixth equality the orthonormality condition $\langle J, Q \,|\, L, M \rangle = \delta_{JL}\delta_{QM}$.

Consider the ordinary scalar product of two vector operators

$$\boldsymbol{U} \cdot \boldsymbol{V} = U_x V_x + U_y V_y + U_z V_z \,. \tag{28.26}$$

In terms of the spherical components of vector operators [(28.12)] it can be readily verified that

$$\boxed{\boldsymbol{U} \cdot \boldsymbol{V} = \textstyle\sum_m (-1)^m \, U_m^{(1)} V_{-m}^{(1)} \,; \qquad m = 1, 0, -1 \,.} \tag{28.27}$$

On the other hand, (28.24) implies

$$T_0^{(0)} = \sum_{qm} \langle\, 1,1\,;\, q, m \,|\, 0,0 \,\rangle\, U_q^{(1)} V_m^{(1)} = \sum_{m=1,0,-1} \langle\, 1,1\,;\, -m, m \,|\, 0,0 \,\rangle\, U_{-m}^{(1)} V_m^{(1)} \,. \tag{28.28}$$

The three CG coefficients appearing in the above equation:

$$a = \langle\, 1,1\,;\, 1,-1 \,|\, 0,0 \,\rangle \,, \tag{28.29a}$$

$$b = \langle\, 1,1\,;\, -1,1 \,|\, 0,0 \,\rangle \,, \tag{28.29b}$$

$$c = \langle\, 1,1\,;\, 0,0 \,|\, 0,0 \,\rangle \,, \tag{28.29c}$$

can be easily determined. Consider the following equality

$$|\, 0,0 \,\rangle = \sum_m |\, 1,1\,;\, -m, m \,\rangle \langle\, 1,1\,;\, -m, m \,|\, 0,0 \,\rangle \,, \tag{28.30}$$

in which all kets are assumed to be normalized. We then have

$$|a|^2 + |b|^2 + |c|^2 = 1 \,. \tag{28.31}$$

On setting $j_1 = j_2 = 1$, $j = m = m_1 = 0$ and $m_2 = 1$ in the recursion relation (27.19), we obtain

$$\langle\, 1,1\,;\, -1,1 \,|\, 0,0 \rangle = -\langle\, 1,1\,;\, 0,0 \,|\, 0,0 \,\rangle \,. \tag{28.32}$$

The same equation [(27.19)] also yields, on setting $j_1 = j_2 = 1$, $j = m = m_2 = 0$ and $m_1 = 1$,

$$\langle\, 1,1\,;\, 1,-1 \,|\, 0,0 \,\rangle = -\langle\, 1,1\,;\, 0,0 \,|\, 0,0 \,\rangle \,. \tag{28.33}$$

The phase requirement (R) mentioned in the first paragraph of Chapter 27 dictates that

$$\langle\, j_1, j_2\,;\, j_1, j - j_1 \,|\, j, j \,\rangle = \langle\, 1,1\,;\, 1,-1 \,|\, 0,0 \,\rangle \tag{28.34}$$

be real and positive. Thus (28.31) entails

$$\langle\, 1,1\,;\, 1,-1 \,|\, 0,0 \,\rangle = \langle\, 1,1\,;\, -1,1 \,|\, 0,0 \,\rangle = \frac{1}{\sqrt{3}} \,, \tag{28.35a}$$

$$\langle\, 1,1\,;\, 0,0 \,|\, 0,0 \,\rangle = -\frac{1}{\sqrt{3}} \,. \tag{28.35b}$$

Equations (28.27) and (28.28) then imply

$$\boldsymbol{U}\cdot\boldsymbol{V} = -\sqrt{3}\,T_0^{(0)}\,. \tag{28.36}$$

This confirms the fact that $\boldsymbol{U}\cdot\boldsymbol{V}$ is indeed a scalar operator.

As a special example of a scalar operator, consider the operator $\boldsymbol{J}\cdot\boldsymbol{V}$, where $\boldsymbol{J}$ is the angular momentum operator and $\boldsymbol{V}$ is an arbitrary vector operator. Equation (28.16) implies that the matrix with elements $\langle\,\tau',j',m'\,|\,\boldsymbol{J}\cdot\boldsymbol{V}\,|\,\tau,j,m\,\rangle$ is diagonal with respect to the angular momentum indices. Thus we need only consider the diagonal elements $\langle\,\tau',j,m\,|\,\boldsymbol{J}\cdot\boldsymbol{V}\,|\,\tau,j,m\,\rangle$. We have

$$\begin{aligned}
\langle\,\tau',j,m\,|\,\boldsymbol{J}\cdot\boldsymbol{V}\,|\,\tau,j,m\,\rangle &= \sum_M \langle\,\tau',j,m\,|\,(-1)^M J^{(1)}_{-M}V^{(1)}_M\,|\,\tau,j,m\,\rangle \\
&= \sum_{\tau'',j'',m''}\sum_M (-1)^M\langle\,\tau',j,m\,|\,J^{(1)}_{-M}\,|\,\tau'',j'',m''\,\rangle\langle\,\tau'',j'',m''\,|\,V^{(1)}_M\,|\,\tau,j,m\,\rangle \\
&= \sum_{m''}\sum_M (-1)^M\langle\,\tau',j,m\,|\,J^{(1)}_{-M}\,|\,\tau',j,m''\,\rangle\langle\,\tau',j,m''\,|\,V^{(1)}_M\,|\,\tau,j,m\,\rangle \\
&= \langle\,\tau',j\,\|\,V^{(1)}\,\|\,\tau,j\,\rangle\sum_{m''}\sum_M (-1)^M\langle\,j,m\,|\,J^{(1)}_{-M}\,|\,j,m''\,\rangle\langle\,1,j\,;\,M,m\,|\,j,m''\,\rangle \\
&\equiv C_{jm}\,\langle\,\tau',j\,\|\,V^{(1)}\,\|\,\tau,j\,\rangle\,,
\end{aligned} \tag{28.37}$$

where

$$C_{jm} \equiv \sum_{m',M}(-1)^M\langle\,j,m\,|\,J^{(1)}_{(-M)}\,|\,j,m'\,\rangle\langle\,1,j\,;\,M,m\,|\,j,m'\,\rangle\,. \tag{28.38}$$

The following justifications have been used in (28.37): In the first equality we have used (28.27); in the second equality the completeness of the $|\,\tau,j,m\,\rangle$ states; in the third equality the fact that the matrix elements of $J^{(1)}_{-M}$ are diagonal in the indices τ and j [recall (28.17) and (28.23)]; and in the fourth equality we have applied the Wigner-Eckart Theorem [(25.11)] to the matrix elements of $V^{(1)}_M$. Since the coefficients C_{jm} are independent of τ',τ and $\boldsymbol{V}$, they can be evaluated by setting $\boldsymbol{V}=\boldsymbol{J}$ in (28.37). Thus

$$\langle\,\tau,j,m\,|\,\boldsymbol{J}\cdot\boldsymbol{J}\,|\,\tau,j,m\,\rangle = C_{jm}\langle\,\tau,j\,\|\,J^{(1)}\,\|\,\tau,j\,\rangle = C_{jm}\,\sqrt{j(j+1)} = j(j+1)\,, \tag{28.39}$$

where the second equality follows from (28.23) and the last from (20.11). We then have

$$C_{jm} = \sqrt{j(j+1)}\,, \tag{28.40}$$

and so (28.37) implies

$$\langle\,\tau',j,m\,|\,\boldsymbol{J}\cdot\boldsymbol{V}\,|\,\tau,j,m\,\rangle = \sqrt{j(j+1)}\,\langle\,\tau',j\,\|\,V^{(1)}\,\|\,\tau,j\,\rangle\,, \tag{28.41}$$

or

$$\boxed{\langle\,\tau',j\,\|\,V^{(1)}\,\|\,\tau,j\,\rangle = \frac{\langle\,\tau',j,m\,|\,\boldsymbol{J}\cdot\boldsymbol{V}\,|\,\tau,j,m\,\rangle}{\sqrt{j(j+1)}}}\,. \tag{28.42}$$

The last equation expresses the reduced matrix element of a vector operator $\boldsymbol{V}$ in terms of a matrix element of the scalar operator $\boldsymbol{J}\cdot\boldsymbol{V}$, which is often easier to calculate.

The Wigner-Eckart Theorem, applied separately to the vector operators $\boldsymbol{V}$ and $\boldsymbol{J}$, yields

$$\langle\,\tau',j',m'\,|\,V_q^{(1)}\,|\,\tau,j,m\,\rangle = \langle\,\tau',j'\,\|\,V^{(1)}\,\|\,\tau,j\,\rangle\langle\,1,j\,;\,q,m\,|\,j',m'\,\rangle\,, \tag{28.43}$$

$$\langle\,\tau',j',m'\,|\,J_q^{(1)}\,|\,\tau,j,m\,\rangle = \delta_{\tau'\tau}\delta_{j'j}\,\langle\,\tau,j\,\|\,J^{(1)}\,\|\,\tau,j\,\rangle\langle\,1,j\,;\,q,m\,|\,j',m'\,\rangle\,. \tag{28.44}$$

For the case $j'=j$, the above two equations, together with (28.42) and (28.23), lead to

$$\boxed{\langle\,\tau',j,m'\,|\,V_q^{(1)}\,|\,\tau,j,m\,\rangle = \frac{\langle\,\tau',j,m\,|\,\boldsymbol{J}\cdot\boldsymbol{V}\,|\,\tau,j,m\,\rangle\langle\,\tau,j,m'\,|\,J_q^{(1)}\,|\,\tau,j,m\,\rangle}{j(j+1)}}\,. \tag{28.45}$$

This important result is known as the **Projection Theorem for Vector Operators**.

As a final exercise in this chapter we will apply this theorem to the calculation of a diagonal matrix element of the **magnetic moment** $\boldsymbol{\mu}$ of an atom, given by

$$\boldsymbol{\mu} = -\frac{e}{2m_e c}\,(g_L\boldsymbol{L} + g_S\boldsymbol{S})\,, \tag{28.46}$$

where $e(>0)$ is the charge of the electron, m_e is the mass of the electron, c is the speed of light, $\boldsymbol{L}$ is the total orbital angular momentum, S is the total spin angular momentum, and $g_L \approx 1\,,\, g_S \approx 2$.

Using (28.45) we can write

$$\langle\,\tau,j,m'\,|\,\boldsymbol{\mu}\,|\,\tau,j,m\,\rangle = -\frac{e}{2m_e c}g\,\langle\,j,m'\,|\,\boldsymbol{J}\,|\,j,m\,\rangle\,, \tag{28.47}$$

where

$$-\frac{e}{2m_e c}g \equiv \frac{\langle\,\tau,j,m\,|\,\boldsymbol{\mu}\cdot\boldsymbol{J}\,|\,\tau,j,m\,\rangle}{j(j+1)}\,. \tag{28.48}$$

The dimensionless numerical factor g is known as the **Lande g-factor**. From (28.46) it is given by

$$g = \frac{\langle\,\tau,j,m\,|\,g_L\boldsymbol{L}\cdot\boldsymbol{J} + g_S\boldsymbol{S}\cdot\boldsymbol{J}\,|\,\tau,j,m\,\rangle}{j(j+1)}\,. \tag{28.49}$$

Now

$$\boldsymbol{L}\cdot\boldsymbol{J} = \boldsymbol{L}\cdot(\boldsymbol{L}+\boldsymbol{S}) = L^2 + \boldsymbol{L}\cdot\boldsymbol{S} = \frac{1}{2}(J^2 + L^2 - S^2)\,, \tag{28.50}$$

and similarly

$$\boldsymbol{S}\cdot\boldsymbol{J} = \boldsymbol{S}\cdot(\boldsymbol{L}+\boldsymbol{S}) = S^2 + \boldsymbol{L}\cdot\boldsymbol{S} = \frac{1}{2}(J^2 + S^2 - L^2)\,. \tag{28.51}$$

Hence

$$g = \frac{\langle \tau, j, m \,|\, (g_L + g_S)J^2 + (g_L - g_S)(L^2 - S^2) \,|\, \tau, j, m \rangle}{2j(j+1)} \,. \tag{28.52}$$

Using $g_L = 1$ and $g_S = 2$, this reduces to

$$\boxed{g = 1 + \frac{j(j+1) - L(L+1) + S(S+1)}{2j(j+1)}} \,. \tag{28.53}$$

This formula is important in the interpretation of the **Zeeman split spectra** of atoms for which the **Russell-Saunders** (LS) **coupling** scheme is valid.

Problem 28.1 Consider the electric dipole couplings [refer to (34.37)] between the ground (g) manifold ${}^1S(6s^2)$ and the first excited (e) manifold ${}^1P(6s6p)$ of atomic states in the barium isotopes ${}^{137}Ba$ or ${}^{138}Ba$, with **hyperfine** magnetic sublevels labeled by

$$\begin{aligned} |\, g \,\rangle &= |\, \alpha = [(6s^2), L = 0, S = 0];\ (J = 0, I);\ (F, m_F) \,\rangle \\ \text{and} \quad |\, e \,\rangle &= |\, \alpha' = [(6s, 6p), L' = 1, S' = 0];\ (J' = 1, I);\ (F', m_{F'}) \,\rangle \,, \end{aligned}$$

respectively, where I denotes the nuclear spin and F the total angular momentum. For ${}^{138}Ba$, $I = 0$, so the ground manifold consists of only one magnetic sublevel $|\, \alpha;\ (0,0);\ (0,0) \,\rangle$ while the excited manifold consists of the three Zeeman-split magnetic sublevels $|\, \alpha';\ (1,0);\ (1, m_{F'} = 0, \pm 1) \,\rangle$. For ${}^{137}Ba$ ($I = 3/2$), the ground manifold consists of the four magnetic sublevels $|\, \alpha;\ (0, 3/2);\ (3/2, m_{F'} = \pm 3/2, \pm 1/2) \,\rangle$, and the excited manifold consists of the twelve Zeeman-split hyperfine magnetic sublevels

$$\begin{aligned} &|\, \alpha';\ (1, 3/2);\ (F' = 5/2,\ m_{F'} = \pm 5/2, \pm 3/2, \pm 1/2) \,\rangle \,, \\ &|\, \alpha';\ (1, 3/2);\ (F' = 3/2,\ m_{F'} = \pm 3/2, \pm 1/2) \,\rangle \,, \\ &|\, \alpha';\ (1, 3/2);\ (F' = 1/2,\ m_{F'} = \pm 1/2) \,\rangle \,. \end{aligned}$$

Use the Wigner-Eckart Theorem and the result of Problem 27.3 to show that the electric dipole matrix elements D_{eg} for all cases are given by

$$\begin{aligned} &\langle\, [(6s, 6p), L'S'];\ J'I;\ F'm_{F'} \,|\, D^{(1)}_{\Delta m_F} \,|\, [(6s^2), L, S];\ JI;\ Fm_F \,\rangle \\ &= (-1)^{F' - m_{F'} + J' + I + F + L' + S' + J} \left[(2F' + 1)(2F + 1)(2J' + 1)(2J + 1)\right]^{1/2} \delta_{S'S} \\ &\times \begin{pmatrix} F' & 1 & F \\ -m_{F'} & \Delta m & m_F \end{pmatrix} \begin{Bmatrix} F' & 1 & F \\ J & I & J' \end{Bmatrix} \begin{Bmatrix} J' & 1 & J \\ L & S' & L' \end{Bmatrix} \langle\, (6s, 6p)L' \,||\, D^{(1)} \,||\, (6s^2)L \,\rangle \,, \end{aligned}$$

where $\Delta m_F \equiv m_{F'} - m_F$. Hence show that

$$D_{eg} = \begin{cases} \dfrac{M}{\sqrt{3}} \,, & {}^{138}Ba \\ (-1)^{2F' - m_{F'} + 3/2} \sqrt{\dfrac{(2F' + 1)}{3}} \begin{pmatrix} F' & 1 & 3/2 \\ -m_{F'} & \Delta m & m_F \end{pmatrix} M \,, & {}^{137}Ba \end{cases} \,,$$

where $M = \langle\, (6s, 6p)L' = 1 \,||\, D^{(1)} \,||\, (6s^2)L = 0 \,\rangle$.

Chapter 29

The Symmetric Groups

The symmetric group S_n is the group of permutations of n objects. For a particular positive integer n, S_n is a finite group with $n!$ elements. The importance of these groups arises from the fact that, in the non-relativistic quantum mechanical description of a system of n identical particles (for example, n electrons), the particles are considered to be *indistinguishable* and so the Hamiltonian H is invariant under any permutation of the particles. Mathematically this condition is given by [cf. Theorem 17.1]

$$(r^{\mathcal{V}_n}(\sigma))^{-1}H(r^{\mathcal{V}_n}(\sigma)) = H\ , \quad \text{for all } \sigma \in S_n\ , \tag{29.1}$$

where $r^{\mathcal{V}_n}$ is any representation of S_n on a vector space $\mathcal{V}_n$ of n-particle state vectors. Thus, according to the discussion in Chapter 18, knowledge of the irreducible representation spaces of S_n within $\mathcal{V}_n$ will be most useful in the determination of the spectrum of H. Physically this corresponds to the fact that the energy eigenspaces of an n-particle Hamiltonian will each consist of states having certain symmetry properties under permutations of the n particles, and *no dynamical effects arising from the Hamiltonian will couple states of different permutational symmetries.* (To avoid cumbersome notation, we will henceforth write σ instead of $r^{\mathcal{V}_n}(\sigma)$ in this chapter, when it is clear that the symbol is used to stand for some representation.)

The situation most frequently encountered is that the unperturbed Hamiltonian is also rotationally invariant (under the central field approximation), so that the single-particle eigenstates of H belong to certain irreducible representations of $SO(3)$. Working within a particular irreducible space V of $SO(3)$ [or more generally of $SU(2)$] for single-particle states, $\mathcal{V}_n$ can be taken to be the tensor product of n copies of V:

$$\mathcal{V}_n = T_nV \equiv \underbrace{V \otimes \cdots \otimes V}_{n \text{ times}}\ . \tag{29.2}$$

If $dimV = m$, then $dim\mathcal{V}_n = m^n$. For example, consider a 3-electron system with respect to the $D^{(1)}$ $(j = 1)$ representation of $SO(3)$. Then $m = n = 3$, so V is 3-dimensional, and $\mathcal{V}_n$ is 27-dimensional.

If $dimV = m$, then V is certainly an irreducible representation space of $GL(m)$ (the **general linear group** of $m \times m$ invertible matrices). The tensor space T_nV, however, is in general *reducible* under both $GL(m)$ and the symmetric group S_n. It is a remarkable mathematical fact that, the solution to the problem of finding the irreducible representation spaces of S_n within T_nV ($dimV = m$) will at the same time furnish a solution to the problem of finding the irreducible representation spaces of $GL(m)$ within the same tensor space. Since $SU(2) \subset GL(2)$, the irreducible spaces within T_nV under $GL(m)$ may then be further broken down into irreducible spaces under $SU(2)$ if $m \geq 2$. The present chapter will be devoted to illustrating this procedure, and to discussing its application to some specific examples in atomic spectroscopy. Mathematical proofs will for the most part be omitted.

We will begin with a general discussion of the symmetric groups. An element $\sigma \in S_n$ is usually represented by

$$\begin{pmatrix} 1 & 2 & 3 & \cdots & n \\ \sigma(1) & \sigma(2) & \sigma(3) & \cdots & \sigma(n) \end{pmatrix} . \tag{29.3}$$

For example, in S_4, we have the following permutation:

$$\sigma = \begin{pmatrix} 1 & 2 & 3 & 4 \\ 4 & 2 & 1 & 3 \end{pmatrix} \in S_4 \quad . \tag{29.4}$$

The identity $e \in S_4$ is, of course,

$$\begin{pmatrix} 1 & 2 & 3 & 4 \\ 1 & 2 & 3 & 4 \end{pmatrix} .$$

Group multiplication of two elements $\sigma, \rho \in S_n$ is defined as the composition

$$(\sigma\rho)(i) = (\sigma \circ \rho)(i) = \sigma(\rho(i)), \quad i = 1, \ldots, n \quad . \tag{29.5}$$

For instance, if σ is given by (29.4) and

$$\rho = \begin{pmatrix} 1 & 2 & 3 & 4 \\ 1 & 3 & 2 & 4 \end{pmatrix} , \tag{29.6}$$

then

$$\sigma\rho = \begin{pmatrix} 1 & 2 & 3 & 4 \\ 4 & 1 & 2 & 3 \end{pmatrix} . \tag{29.7}$$

The inverse of σ, σ^{-1}, is simply the permutation which undoes the effects of σ. For σ given by (29.4), σ^{-1} would be

$$\sigma^{-1} = \begin{pmatrix} 1 & 2 & 3 & 4 \\ 3 & 2 & 4 & 1 \end{pmatrix} . \tag{29.8}$$

We note that S_n is a non-abelian group for any $n > 2$.

The action of S_n on the tensor space T_nV ($dimV = m$) is defined by its action on a **monomial** in the tensor space. Suppose $t \in T_nV$ is a monomial given by

$$t = v_1 \otimes \cdots \otimes v_n \; ; \quad v_1, \ldots, v_n \in V , \tag{29.9}$$

then, for any $\sigma \in S_n$,

$$\boxed{\sigma t = v_{\sigma^{-1}(1)} \otimes \cdots \otimes v_{\sigma^{-1}(n)}} . \tag{29.10}$$

As noted at the end of the first paragraph of this chapter, in the above equation we have written σ for the representation instead of $r^{\mathcal{V}_n}(\sigma)$. The exact meaning of the symbol (whether it refers to a group element or the representation of that element by a linear operator) should be clear from the context. The extension of the action of σ to the whole tensor space is then determined by the imposition of the property of **linearity**:

$$\sigma \sum_\alpha c_\alpha t_\alpha = \sum_\alpha c_\alpha \, \sigma t_\alpha , \tag{29.11}$$

where the c_α's are arbitrary complex numbers and the t_α's are arbitrary monomials in T_nV. Note that in (29.10), the *positions* of the vectors $v_1, \ldots, v_n$ are permuted. Thus, with σ given by (29.4),

$$\sigma(v_1 \otimes v_2 \otimes v_3 \otimes v_4) = v_3 \otimes v_2 \otimes v_4 \otimes v_1 . \tag{29.12}$$

The above definition of the action of σ, in which the positions of the vectors in monomials are permuted, is equivalent to the following definition, in which the indices of the components of a tensor are permuted. For an arbitrary tensor $t \in T_nV$, we first express it in terms of components as

$$t = \sum_{i_1,\ldots,i_n} t^{i_1 \ldots i_n} \, e_{i_1} \otimes \cdots \otimes e_{i_n} \; ; \quad 1 \le i_1, \ldots, i_n \le n , \tag{29.13}$$

where $\{e_1, \ldots, e_n\}$ is a basis of V. Then

$$\boxed{\sigma t = \textstyle\sum_{i_1,\ldots,i_n} t^{i_{\sigma(1)} \cdots i_{\sigma(n)}} \, e_{i_1} \otimes \cdots \otimes e_{i_n} \; ; \quad 1 \le i_1, \ldots, i_n \le n} . \tag{29.14}$$

Any permutation $\sigma \in S_n$ can be expressed as the product of a certain number p of pair interchanges. If p is even (odd), then σ is said to be an **even (odd) permutation.**

Definition 29.1. *Suppose $t \in T_n V$. If for any $\sigma \in S_n$ we have*

$$\sigma t = t \quad , \tag{29.15}$$

then t is called a **symmetric tensor of rank** *n. If for any $\sigma \in S_n$ we have*

$$\sigma t = sgn\,\sigma \cdot t \quad , \tag{29.16}$$

where $sgn\,\sigma$ denotes the sign of the permutation σ:

$$sgn\,\sigma = \begin{cases} +1 & \text{if } \sigma \text{ is an even permutation,} \\ -1 & \text{if } \sigma \text{ is an odd permutation,} \end{cases} \tag{29.17}$$

then t is called an **anti-symmetric** *(or* **alternating***)* **tensor of rank** *n.*

It is clear that *a tensor is symmetric (anti-symmetric) if and only if all its components are symmetric (anti-symmetric).* If t is anti-symmetric, then the component $t^{i_1 \dots i_n} = 0$ whenever any pair of indices have the same value. Denote the subset of all symmetric $t \in T_n V$ by $P_n V$, and the subset of all anti-symmetric $t \in T_n V$ by $\Lambda_n V$. Then both $P_n V$ and $\Lambda_n V$ are vector subspaces of $T_n V$. In fact, they can be obtained from $T_n V$ by the so-called **symmetrizing map** $\mathcal{S}_n$ and the **anti-symmetrizing map** $\mathcal{A}_n$, respectively:

$$P_n V = \mathcal{S}_n\,(T_n V)\,, \quad \Lambda_n V = \mathcal{A}_n\,(T_n V)\,, \tag{29.18}$$

where

$$\mathcal{S}_n(t) \equiv \frac{1}{n!} \sum_{\sigma \in S_n} \sigma t\,, \tag{29.19}$$

$$\mathcal{A}_n(t) \equiv \frac{1}{n!} \sum_{\sigma \in S_n} (sgn\,\sigma)\,\sigma t\,, \tag{29.20}$$

for all $t \in T_n V$. Each sum in the above two equations runs through all permutations in S_n.

As a simple but important example of the concepts introduced above, consider the spin states of a two-electron system, with V being the two-dimensional irreducible space of the $D^{(1/2)}$ representation of $SU(2)$. For a chosen basis $\{e_1, e_2\}$ of V, an arbitrary tensor $t \in T_2 V$ can be expressed in components as

$$\begin{aligned} t &= \sum_{i_1, i_2} t^{i_1 i_2}\, e_{i_1} \otimes e_{i_2} \\ &= t^{11}\, e_1 \otimes e_1 + t^{12}\, e_1 \otimes e_2 + t^{21}\, e_2 \otimes e_1 + t^{22}\, e_2 \otimes e_2\,, \end{aligned} \tag{29.21}$$

where, in general, the four components t^{11}, t^{12}, t^{21} and t^{22} of t do not bear any relationship to each other. Now the symmetry group S_2 has just two elements: the identity

$$e = \begin{pmatrix} 1 & 2 \\ 1 & 2 \end{pmatrix}\,,$$

and the single interchange

$$\sigma = \begin{pmatrix} 1 & 2 \\ 2 & 1 \end{pmatrix} .$$

The inverse of the latter, σ^{-1}, is again equal to σ. Applying (29.10), we then have

$$\mathcal{S}_2(e_1 \otimes e_1) = \frac{1}{2}(e_1 \otimes e_1 + e_1 \otimes e_1) = e_1 \otimes e_1 \quad , \tag{29.22}$$

$$\mathcal{S}_2(e_1 \otimes e_2) = \frac{1}{2}(e_1 \otimes e_2 + e_2 \otimes e_1) = \mathcal{S}_2(e_2 \otimes e_1) \quad , \tag{29.23}$$

$$\mathcal{S}_2(e_2 \otimes e_2) = e_2 \otimes e_2 \quad , \tag{29.24}$$

$$\mathcal{A}_2(e_1 \otimes e_1) = \frac{1}{2}(e_1 \otimes e_1 - e_1 \otimes e_1) = 0 = \mathcal{A}_2(e_2 \otimes e_2) \quad , \tag{29.25}$$

$$\mathcal{A}_2(e_1 \otimes e_2) = \frac{1}{2}(e_1 \otimes e_2 - e_2 \otimes e_1) \quad , \tag{29.26}$$

$$\mathcal{A}_2(e_2 \otimes e_1) = \frac{1}{2}(e_2 \otimes e_1 - e_1 \otimes e_2) = -\mathcal{A}_2(e_1 \otimes e_2) \quad . \tag{29.27}$$

Hence,

$$\mathcal{S}_2(t) = t^{11}\, e_1 \otimes e_1 + t^{22}\, e_2 \otimes e_2 + \frac{1}{2}(t^{12} + t^{21})(e_1 \otimes e_2 + e_2 \otimes e_1) \,, \tag{29.28}$$

$$\mathcal{A}_2(t) = \frac{1}{2}(t^{12} - t^{21})(e_1 \otimes e_2 - e_2 \otimes e_1) \,. \tag{29.29}$$

Thus we see that P_2V is 3-dimensional (with a possible basis set $\{e_1 \otimes e_1,\, e_2 \otimes e_2,\, e_1 \otimes e_2 + e_2 \otimes e_1\}$), and $\Lambda_2 V$ is 1-dimensional (with a possible basis $\{e_1 \otimes e_2 - e_2 \otimes e_1\}$). In this case

$$T_2V = P_2V \oplus \Lambda_2 V \quad , \tag{29.30}$$

that is, the 4-dimensional T_2V is the direct sum of the 3-dimensional P_2V and the 1-dimensional Λ_2V. But in general, as we shall see below, T_nV cannot be decomposed into a direct sum of just totally symmetric and totally anti-symmetric subspaces [similar to (29.30)] when $n > 2$, but the decomposition will consist of subspaces of other symmetry types also.

Returning to the $n = 2$ case, if we choose

$$e_1 = |\, j = 1/2\,, m = 1/2\,\rangle \equiv |\uparrow\rangle\,, \quad e_2 = |\, j = 1/2\,, m = -1/2\,\rangle \equiv |\downarrow\rangle\,, \tag{29.31}$$

we can use the notation introduced at the beginning of Chapter 26 and write

$$e_1 \otimes e_1 = |\uparrow\uparrow\rangle \quad , \tag{29.32}$$

$$e_2 \otimes e_2 = |\downarrow\downarrow\rangle \quad , \tag{29.33}$$

$$e_1 \otimes e_2 + e_2 \otimes e_1 = |\uparrow\downarrow\rangle + |\downarrow\uparrow\rangle \quad , \tag{29.34}$$

$$e_1 \otimes e_2 - e_2 \otimes e_1 = |\uparrow\downarrow\rangle - |\downarrow\uparrow\rangle \quad . \tag{29.35}$$

The set $\{|\uparrow\uparrow\rangle, |\downarrow\downarrow\rangle, |\uparrow\downarrow\rangle + |\downarrow\uparrow\rangle\}$ is a basis of the 3-dimensioanl space of symmetric two-particle states (under exchange of the particles), while the single state $\{|\uparrow\downarrow\rangle - |\downarrow\uparrow\rangle\}$ forms a basis of the 1-dimensional space of antisymmetric two-particle states (under exchange of the particles). The one-particle states $|\uparrow\rangle$ and $|\downarrow\rangle$ are usually assumed to be normalized so that

$$\langle\uparrow|\uparrow\rangle = \langle\downarrow|\downarrow\rangle = 1 \quad . \tag{29.36}$$

Then the **triplet** of normalized symmetric states which span P_2V are

$$|\uparrow\uparrow\rangle \quad , \quad |\downarrow\downarrow\rangle \quad , \quad \frac{1}{\sqrt{2}}(|\uparrow\downarrow\rangle + |\downarrow\uparrow\rangle)\,; \tag{29.37}$$

while the **singlet** of normalized antisymmetric state which spans $\Lambda_2 V$ is

$$\frac{1}{\sqrt{2}}(|\uparrow\downarrow\rangle - |\downarrow\uparrow\rangle) \quad . \tag{29.38}$$

Each of the triplet states can serve as the basis vector for a one-dimensional (irreducible) representation of the symmetric group S_2, and together they yield three equivalent one-dimensional representations. The singlet state also gives a one-dimensional representation of S_2, but this representation is not equivalent to the ones given by the triplet states. On the other hand, the spaces P_2V and Λ_2V are each invariant subspaces of T_2V under the rotation group $SO(3)$ [as well as $SU(2)$], as discussed in Chapter 26. In other words, they provide irreducible representation spaces for $SO(3)$ and $SU(2)$. This gives an example of the relationship between the irreducible representations of S_n and $SU(m)$ [$GL(m)$] mentioned above. Generalizations to $n > 2$ and $m > 2$ will constitute the main development in the remainder of this chapter.

Let us return to a review of some basic group-theoretic facts of S_n. Each element $\sigma \in S_n$ can be written as a **cycle** or a product of cycles. A cycle of length r, or an r-cycle, is written $(i_1\, i_2 \dots i_r)$, where $i_j \in \{1, 2, \dots, n\}, j = 1, \dots, r$. Its action is given by

$$i_1 \to i_2 \to i_3 \to \cdots \to i_{r-1} \to i_r \to i_1 \, .$$

Thus, for S_3, the 2-cycle (13) gives the permutation

$$\begin{pmatrix} 1 & 2 & 3 \\ 3 & 2 & 1 \end{pmatrix},$$

while the 3-cycle (123) gives

$$\begin{pmatrix} 1 & 2 & 3 \\ 2 & 3 & 1 \end{pmatrix}.$$

Note that all 1-cycles are equal to the identity $e \in S_n$. *All $\sigma \in S_n$ except the identity e can be written as a product of 2-cycles (pair exchanges), or* ***transpositions****.* In fact, an even permutation σ (with $sgn\,\sigma = 1$) is one which can be written as a product of an even number of transpositions, while an odd permutation ($sgn\,\sigma = -1$) is one expressible as the product of an odd number of transpositions. Thus

$$(123) = (13)(12) = (23)(13) \tag{29.39}$$

is even while

$$(13) = (12)(23)(12) = (12)(13)(23)\,, \tag{29.40}$$

is odd. Note that *the decomposition into transpositions is not unique.*

As for every group, elements of S_n can be partitioned into disjoint **conjugacy classes**, a basic group-theoretic concept defined as follows.

Definition 29.2. *Two elements g_1 and g_2 in a group G are said to be* ***conjugate*** *to each other if there exists some element $p \in G$ such that $g_2 = p\,g_1 p^{-1}$. Elements of a group which are conjugate to each other are said to form a* ***conjugacy class****.*

For example, S_3 can be partitioned into the three classes

$$C_1 = \{e\}\,,\; C_2 = \{(12), (13), (23)\}\,,\; C_3 = \{(123), (321)\}\,,$$

since

$$(13) = (23)(12)(23)^{-1} = (12)(23)(12)^{-1}\,, \tag{29.41}$$

and

$$(321) = (12)(123)(12)^{-1}\,. \tag{29.42}$$

The representation theory of S_n rests on the following theorem, which will be stated without proof.

Theorem 29.1. *Each conjugacy class C_i of the symmetric group S_n is characterized by the common cycle-structure of elements $g \in C_i$. By a cycle-structure of an element $\sigma \in S_n$ we mean an ordered set of non-negative integers $\{\alpha_1, \alpha_2, \ldots, \alpha_n\}$ such that σ can be expressed as a product of α_1 1-cycles, α_2 2-cycles, ... , and α_n n-cycles; and*

$$\alpha_1 + 2\alpha_2 + \cdots + n\alpha_n = n\,. \tag{29.43}$$

The set $\{\alpha_1, \ldots, \alpha_n\}$ satisfying (29.43) is called a **partition** of the integer n. For example, in S_3, the three classes C_1, C_2 and C_3 have the following cycle-structures:

$$\begin{aligned}
&C_1 = \{e\} = \{(1)(2)(3)\}\,, && \alpha_1 = 3\,, \alpha_2 = 0\,, \alpha_3 = 0\,,\\
&C_2 = \{(3)(12), (2)(13), (1)(23)\}\,, && \alpha_1 = 1\,, \alpha_2 = 1\,, \alpha_3 = 0\,,\\
&C_3 = \{(123), (321)\}\,, && \alpha_1 = 0\,, \alpha_2 = 0\,, \alpha_3 = 1\,.
\end{aligned}$$

Note that if an element has more than one type of cycle (such as those in C_2), the numbers that occur in one type must all be different from those that occur in another type.

Using (29.43), the partition $\{\alpha_1, \ldots, \alpha_n\}$ can also be labeled by the set of integers $\{\mu_1, \ldots, \mu_n\}$ given by

$$\begin{aligned} \alpha_1 + \alpha_2 + \alpha_3 + \cdots + \alpha_n &= \mu_1 , \\ \alpha_2 + \alpha_3 + \cdots + \alpha_n &= \mu_2 , \\ \alpha_3 + \cdots + \alpha_n &= \mu_3 , \\ &\vdots \\ \alpha_n &= \mu_n . \end{aligned} \tag{29.44}$$

It is clear that

$$\mu_1 + \mu_2 + \cdots + \mu_n = n , \tag{29.45}$$

and

$$\mu_1 \geq \mu_2 \geq \cdots \geq \mu_n \geq 0 . \tag{29.46}$$

The set $\{\mu_1, \ldots, \mu_n\}$ can also be used to label a partition, and thus a conjugacy class. Using the μ-labeling for the conjugacy classes of S_3, we have

$$C_1 : \mu_1 = 3, \mu_2 = \mu_3 = 0 \, , \; C_2 : \mu_1 = 2, \mu_2 = 1, \mu_3 = 0 \, , \; C_3 : \mu_1 = \mu_2 = \mu_3 = 1 \, .$$

Each partition $\{\boldsymbol{\mu}\} = \{\mu_1, \ldots, \mu_n\}$ of n can be conveniently represented by what is called a **Young diagram**, each of which is an array of exactly n boxes, with at most n rows. The first row contains μ_1 boxes, the second row μ_2 ($\leq \mu_1$) boxes, and so on. Thus for S_3 there are three Young diagrams:

$$\{\boldsymbol{\mu}\} = \{3, 0, 0\} \quad \square\square\square \; ,$$

$$\{\boldsymbol{\mu}\} = \{2, 1, 0\} \quad \begin{array}{l} \square\square \\ \square \end{array} \; ,$$

$$\{\boldsymbol{\mu}\} = \{1, 1, 1\} \quad \begin{array}{l} \square \\ \square \\ \square \end{array} \; .$$

For S_2 there are only two Young diagrams:

$$\{\boldsymbol{\mu}\} = \{2, 0\} \quad \square\square \; , \qquad \{\boldsymbol{\mu}\} = \{1, 1\} \quad \begin{array}{l} \square \\ \square \end{array} \; ;$$

while for S_4 there are five:

$$\{\boldsymbol{\mu}\} = \{4,0,0,0\} \quad , \qquad \{\boldsymbol{\mu}\} = \{3,1,0,0\} \quad ,$$

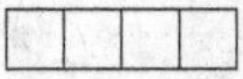

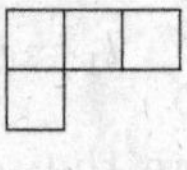

$$\{\boldsymbol{\mu}\} = \{2,2,0,0\} \quad , \qquad \{\boldsymbol{\mu}\} = \{2,1,1,0\} \quad ,$$

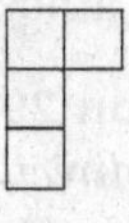

$$\{\boldsymbol{\mu}\} = \{1,1,1,1\} \quad .$$

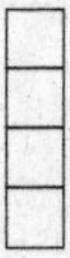

To link Theorem 29.1 to the irreducible representations of S_n we invoke the following general theorem, stated again without proof:

Theorem 29.2. *The number of inequivalent irreducible representations of any finite group G is equal to the number of distinct conjugacy classes of G.*

Thus *the inequivalent irreducible representations of S_n are also characterized by the Young diagrams.* Based on the above enumeration of Young diagrams, we conclude that S_2 has two, S_3 has three, and S_4 has five inequivalent irreducible representations. As a further illustration, we show below the seven Young diagrams of S_5:

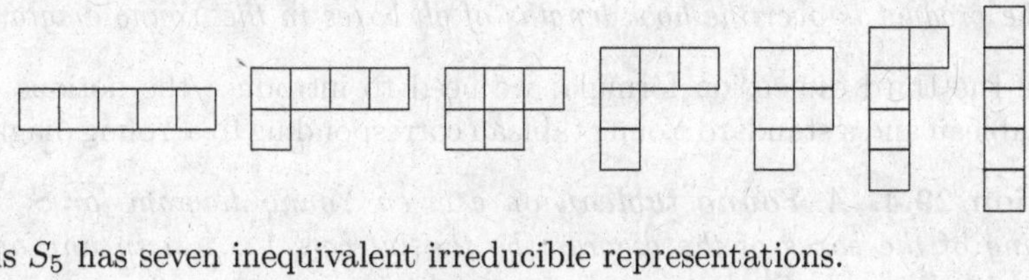

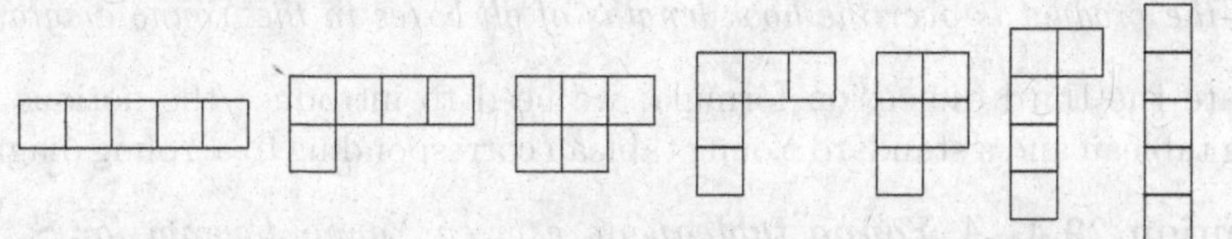

Thus S_5 has seven inequivalent irreducible representations.

We will now present three useful general formulas for the calculation of the dimension of the irreducible representation space $\mathbb{V}_{\{\mu\}}$ of S_n characterized by a particular Young diagram. The first formula is given by the following theorem.

Theorem 29.3 (Frobenius Dimension Formula). *For the irreducible representation of S_n characterized by the Young diagram $\{\boldsymbol{\mu}\} = \{\mu_1, \dots, \mu_n\}$, with $\mu_1 \geq \mu_2 \geq \dots \geq \mu_p > 0$, $\mu_{p+1} = \mu_{p+2} = \dots = \mu_n = 0$, the dimension of the representation space $\mathbb{V}_{\{\mu\}}$ is given by*

$$\boxed{dim\, \mathbb{V}_{\{\mu\}} = \frac{n!\, \Delta(l_1, \dots, l_p)}{l_1!\, l_2! \dots l_p!}} \quad , \tag{29.47}$$

where

$$\Delta(l_1, \dots, l_p) \equiv \prod_{i<k}^{p} (l_i - l_k) \; , \tag{29.48}$$

and the p *integers* $l_1, \ldots, l_p$ *are defined by*

$$l_1 \equiv \mu_1 + p - 1\,, \quad l_2 \equiv \mu_2 + p - 2\,, \quad \ldots\,, \quad l_p \equiv \mu_p\,. \tag{29.49}$$

To state the second dimension formula we need to define the so-called hook length of a box in a Young diagram.

Definition 29.3. *The* ***hook length*** *of a box in a Young diagram is the number of boxes directly to the right and directly below, including the box itself.*

In the following Young diagram, for example, each box is labeled by its hooked length.

6	4	1
4	2	
3	1	
1		

The second dimension formula is then given by the following theorem.

Theorem 29.4 (Hook Length Dimension Formula). *The dimension of the (irreducible) representation space* $\mathbb{V}_{\{\mu\}}$ *corresponding to a given Young diagram of* S_n *is given by*

$$\boxed{\, dim\, \mathbb{V}_{\{\mu\}} = \frac{n!}{\prod(\textit{hook lengths})} \,}\,, \tag{29.50}$$

where the product is over the hook lengths of all boxes in the Young diagram.

To state the third dimension formula, we need to introduce the notions of a Young tableau and a standard Young tableau corresponding to a Young diagram.

Definition 29.4. *A* ***Young tableau*** *on a given Young diagram for* S_n *is a numbering of the boxes of the diagram by the integers* $1, \ldots, n$ *in any order, each number being used only once. For example,*

7	8	2
6	4	
5	1	
3		

Definition 29.5. *A* ***standard Young tableau*** *is a Young tableau in which the numbers in each row increase to the right (not necessarily in strict order) and those in each column increase to the bottom. For example,*

1	3	4
2	5	
6	8	
7		

The third dimension formula is given by the following theorem.

Theorem 29.5. *The dimension of the irreducible representation space* $\mathbb{V}_{\{\mu\}}$ *corresponding to the Young diagram with partition* $\{\boldsymbol{\mu}\}$ *is given by*

$$\boxed{dim\,\mathbb{V}_{\{\mu\}} = \textit{number of distinct standard Young tableau for } \{\mu\}} \,. \tag{29.51}$$

As an illustration of the above theorems consider the Young diagram characterized by the partition $\{\mu\} = \{2, 1, 1, 1, 0\}$ for S_5. To apply Theorem 29.3 we note that $n = 5$, $\mu_1 = 2$, $\mu_2 = \mu_3 = \mu_4 = 1$, $\mu_5 = 0$. Hence $p = 4$ and by (29.49), $l_1 = 5$, $l_2 = 3$, $l_3 = 2$, $l_4 = 1$. Thus, by (29.48),

$$\begin{aligned}\Delta(l_1, l_2, l_3, l_4) &= (l_1 - l_2)(l_1 - l_3)(l_1 - l_4)(l_2 - l_3)(l_2 - l_4)(l_3 - l_4) \\ &= 2 \cdot 3 \cdot 4 \cdot 1 \cdot 2 \cdot 1 = 48 \,.\end{aligned}$$

It follows from (29.47) that

$$dim\,\mathbb{V}_{\{\mu\}} = \frac{5!\,48}{5!\,3!\,2!\,1!} = 4 \,.$$

To apply Theorem 29.4, we see that the tableau with hook lengths filled in is

$$\begin{array}{|c|c|} \hline 5 & 1 \\ \hline 3 & \multicolumn{1}{c}{} \\ \cline{1-1} 2 & \multicolumn{1}{c}{} \\ \cline{1-1} 1 & \multicolumn{1}{c}{} \\ \cline{1-1} \end{array} \,.$$

Thus, by (29.50),

$$dim\,\mathbb{V}_{\{\mu\}} = \frac{5!}{5.3.2.1.1} = 4 \,.$$

To apply Theorem 29.5 we note that all the possible standard Young tableau are the following four:

$$\begin{array}{|c|c|} \hline 1 & 2 \\ \hline 3 & \multicolumn{1}{c}{} \\ \cline{1-1} 4 & \multicolumn{1}{c}{} \\ \cline{1-1} 5 & \multicolumn{1}{c}{} \\ \cline{1-1} \end{array} \quad \begin{array}{|c|c|} \hline 1 & 3 \\ \hline 2 & \multicolumn{1}{c}{} \\ \cline{1-1} 4 & \multicolumn{1}{c}{} \\ \cline{1-1} 5 & \multicolumn{1}{c}{} \\ \cline{1-1} \end{array} \quad \begin{array}{|c|c|} \hline 1 & 4 \\ \hline 2 & \multicolumn{1}{c}{} \\ \cline{1-1} 3 & \multicolumn{1}{c}{} \\ \cline{1-1} 5 & \multicolumn{1}{c}{} \\ \cline{1-1} \end{array} \quad \begin{array}{|c|c|} \hline 1 & 5 \\ \hline 2 & \multicolumn{1}{c}{} \\ \cline{1-1} 3 & \multicolumn{1}{c}{} \\ \cline{1-1} 4 & \multicolumn{1}{c}{} \\ \cline{1-1} \end{array} \,.$$

Hence, by (29.51), $dim\,\mathbb{V}_{\{\mu\}} = 4$.

We will now construct explicitly the irreducible subspaces of the tensor space $T_n V$ under the action of the symmetric group S_n. Recall that for $\sigma \in S_n$ this action is given by [cf. (29.10) and (29.14)]

$$\begin{aligned}\sigma \sum_{i_1,\dots,i_n} t^{i_1\dots i_n}\, e_{i_1} \otimes \cdots \otimes e_{i_n} &= \sum_{i_1,\dots,i_n} t^{i_1\dots i_n}\, e_{i_{\sigma^{-1}(1)}} \otimes \cdots \otimes e_{i_{\sigma^{-1}(n)}} \\ &= \sum_{i_1,\dots,i_n} t^{i_{\sigma(1)}\dots i_{\sigma(n)}}\, e_{i_1} \otimes \cdots \otimes e_{i_n} \,.\end{aligned} \tag{29.52}$$

The action gives rise to the so-called group algebra $A(S_n)$ of the group S_n. The general notion of a group algebra arises as follows. In a group G, only group multiplication of group elements is defined, so that $g_1 g_2 \in G$ if $g_1 \in G$ and $g_2 \in G$. Neither addition, $g_1 + g_2$, nor scalar multiplication, αg (where $\alpha \in \mathbb{C}$ is some complex number and $g \in G$), is defined. For a given representation $D(G)$, however, both $D(g_1) + D(g_2)$ and $\alpha D(g)$, as well as $D(g_1)D(g_2)$, are well defined, since by definition $D(g)$ is a linear operator on a certain vector space for all $g \in G$. *The set of all linear superpositions and products of all elements in $D(G)$, with the usual distributive and associative laws, is called the* ***group algebra*** *of the group G with respect to the representation D.*

It is clear that the group algebra $A(S_n)$ can also be regarded as a vector space whose dimension is equal to $n!$ (the number of elements in the group S_n). In fact, it is a (reducible) representation space for S_n. To find the irreducible subspaces, we introduce special elements in $A(S_n)$, called Young symmetrizers. Corresponding to each standard Young tableau we define a **Young symmetrizer** as follows:

$$\boxed{\mathcal{Y}_n = \frac{N_{\{\mu\}}}{n!} \sum_{\sigma_{(c)}} (sgn\, \sigma_{(c)})\, \sigma_{(c)} \sum_{\sigma_{(r)}} \sigma_{(r)}} \,, \tag{29.53}$$

where $N_{\{\mu\}}$ = dimension of the irreducible representation labeled by the partition $\{\boldsymbol{\mu}\}$ corresponding to the Young diagram of the given Young tableau, $\sigma_{(r)} \in S_n$ is a permutation of the numbers appearing in a row, and $\sigma_{(c)} \in S_n$ is a permutation of the numbers appearing in a column. We have seen two important Young symmetrizers already [cf. (29.19) and (29.20)]:

$$\mathcal{Y}_n\left(\boxed{1}\boxed{2}\boxed{3}\boxed{\cdot}\boxed{\cdot}\boxed{\cdot}\boxed{n}\right) = \text{ the symmetrizing map } \mathcal{S}_n \,, \tag{29.54a}$$

$$\mathcal{Y}_n\begin{pmatrix}\boxed{1}\\\boxed{2}\\\boxed{3}\\\boxed{\cdot}\\\boxed{\cdot}\\\boxed{\cdot}\\\boxed{n}\end{pmatrix} = \text{ the alternating map } \mathcal{A}_n \,. \tag{29.54b}$$

As examples of Young symmetrizers corresponding to standard Young tableau of mixed symmetry, consider the standard tableau $\begin{matrix}\boxed{1}\boxed{2}\\\boxed{3}\phantom{\boxed{2}}\end{matrix}$ and $\begin{matrix}\boxed{1}\boxed{3}\\\boxed{2}\phantom{\boxed{3}}\end{matrix}$. We have

$$\mathcal{Y}_3\left(\begin{matrix}\boxed{1}\boxed{2}\\\boxed{3}\phantom{\boxed{2}}\end{matrix}\right) = \frac{2}{3!}(e-(13))(e+(12)) = \frac{1}{3}(e-(13)+(12)-(123))\,, \tag{29.55}$$

$$\mathcal{Y}_3\left(\begin{matrix}\boxed{1}\boxed{3}\\\boxed{2}\phantom{\boxed{3}}\end{matrix}\right) = \frac{2}{3!}(e-(12))(e+(13)) = \frac{1}{3}(e+(13)-(12)-(321))\,. \tag{29.56}$$

The importance of Young symmetrizers comes from the following theorem.

Theorem 29.6. *All Young symmetrizers* $\mathcal{Y}_n$ *are* ***primitive idempotents*** *of the group algebra* $A(S_n)$. *Idempotency of an element* s *in the group algebra means that* $s^2 = s$, *and primitivity of* s *means that if* $s = s_1 + s_2$, *where* s_1 *and* s_2 *are both idempotent and* $s_1 s_2 = s_2 s_1 = 0$, *then either* $s_1 = 0$ *or* $s_2 = 0$.

To proceed we need the following definition and theorem.

Definition 29.6. *A subalgebra* A' *of an algebra* A *is a* ***left ideal*** *of* A *if, for all* $s \in A$, $sA' \subset A'$. *Similarly,* A' *is a* ***right ideal*** *of* A *if, for all* $s \in A$, $A's \subset A'$.

Theorem 29.7. *Let* $\mathcal{Y}_n^{(i)}$ *be all the Young symmetrizers corresponding to the standard Young tableau of the symmetric group* S_n. *Each* $\mathcal{Y}_n^{(i)}$ *generates a left ideal* $L^{(i)}$ *of the group algebra* $A(S_n)$ *given by*

$$L^{(i)} = \{\, s\mathcal{Y}_n^{(i)} \mid s \in A(S_n) \,\}\,, \tag{29.57}$$

and $A(S_n)$, *as a vector space, can be decomposed as the direct sum*

$$A(S_n) = \sum_{\oplus i} L^{(i)}\,. \tag{29.58}$$

Each $L^{(i)}$ *is an irreducible representation space of* S_n, *some of which may be equivalent.*

For example, $A(S_3)$ is a $6\,(= 3!)$-dimensional vector space which can be decomposed as

$$A(S_3) = L^{(1)} \oplus L^{(2)} \oplus L^{(3)} \oplus L^{(4)}\,, \tag{29.59}$$

where

$$L^{(1)} \equiv L\left(\,\boxed{1\,|\,2\,|\,3}\,\right), \tag{29.60a}$$

$$L^{(2)} \equiv L\left(\begin{array}{|c|}\hline 1\\ \hline 2\\ \hline 3\\ \hline\end{array}\right), \tag{29.60b}$$

$$L^{(3)} \equiv L\left(\begin{array}{|c|c|}\hline 1 & 2\\ \hline 3 \\ \cline{1-1}\end{array}\right), \tag{29.60c}$$

$$L^{(4)} \equiv L\left(\begin{array}{|c|c|}\hline 1 & 3\\ \hline 2 \\ \cline{1-1}\end{array}\right). \tag{29.60d}$$

As irreducible representation spaces of S_n, $L^{(1)}$ and $L^{(2)}$ are clearly both one-dimensional, while $L^{(3)}$ and $L^{(4)}$ are both two-dimensional. The latter two give two equivalent two-dimensional representations, while the first two give two inequivalent one-dimensional representations. In particular the first one gives the trivial representation.

To show that $L^{(3)}$ and $L^{(4)}$ are both two-dimensional irreducible representation spaces for S_3, we will construct a basis set for each. We claim that $\{\epsilon_1, \epsilon_2\}$

is a basis for $L^{(3)}$, where

$$\epsilon_1 \equiv \mathcal{Y}_3\left(\begin{array}{|c|c|}\hline 1 & 2 \\ \hline 3 \\ \cline{1-1}\end{array}\right) = \frac{1}{3}(e + (12) - (13) - (123)) \,, \tag{29.61a}$$

$$\epsilon_2 \equiv (12)\mathcal{Y}_3\left(\begin{array}{|c|c|}\hline 1 & 2 \\ \hline 3 \\ \cline{1-1}\end{array}\right) = \frac{1}{3}(e + (12) - (23) - (321)) \,. \tag{29.61b}$$

It can be readily verified that

$$(12)\epsilon_1 = \epsilon_2 \,, \qquad (12)\epsilon_2 = \epsilon_1 \,, \tag{29.62}$$

$$(13)\epsilon_1 = -\epsilon_1 \,, \qquad (13)\epsilon_2 = -\epsilon_1 + \epsilon_2 \,, \tag{29.63}$$

$$(23)\epsilon_1 = \epsilon_1 - \epsilon_2 \,, \qquad (23)\epsilon_2 = -\epsilon_2 \,. \tag{29.64}$$

Since all elements of S_3 can be written as products of $e, (12), (13)$, and (23), we have a two-dimensional irreducible representation of S_3 with the representation space spanned by ϵ_1 and ϵ_2. With respect to these basis vectors, the matrix representations of (12), (13) and (23) are given by [cf. (18.8) and the discussion following]

$$(12)_\epsilon = \begin{pmatrix} 0 & 1 \\ 1 & 0 \end{pmatrix} \,, \quad (13)_\epsilon = \begin{pmatrix} -1 & -1 \\ 0 & 1 \end{pmatrix} \,, \quad (23)_\epsilon = \begin{pmatrix} 1 & 0 \\ -1 & -1 \end{pmatrix} \,. \tag{29.65}$$

Similarly, a basis $\{\varepsilon_1, \varepsilon_2\}$ for $L^{(4)}$ is given by

$$\varepsilon_1 \equiv \mathcal{Y}_3\left(\begin{array}{|c|c|}\hline 1 & 3 \\ \hline 2 \\ \cline{1-1}\end{array}\right) = \frac{1}{3}(e + (13) - (12) - (321)) \,, \tag{29.66a}$$

$$\varepsilon_2 \equiv (13)\mathcal{Y}_3\left(\begin{array}{|c|c|}\hline 1 & 3 \\ \hline 2 \\ \cline{1-1}\end{array}\right) = \frac{1}{3}(e + (13) - (23) - (123)) \,, \tag{29.66b}$$

with the following results analogous to Eqs. (29.62), (29.63) and (29.64):

$$(12)\varepsilon_1 = -\varepsilon_1 \,, \qquad (12)\varepsilon_2 = -\varepsilon_1 + \varepsilon_2 \,, \tag{29.67}$$

$$(13)\varepsilon_1 = \varepsilon_2 \,, \qquad (13)\varepsilon_2 = \varepsilon_1 \,, \tag{29.68}$$

$$(23)\varepsilon_1 = \varepsilon_1 - \varepsilon_2 \,, \qquad (23)\varepsilon_2 = -\varepsilon_2 \,. \tag{29.69}$$

With respect to the basis $\{\varepsilon_1, \varepsilon_2\}$ the matrix representations of $(12), (13)$ and (23) are given by

$$(12)_\varepsilon = \begin{pmatrix} -1 & -1 \\ 0 & 1 \end{pmatrix} \,, \quad (13)_\varepsilon = \begin{pmatrix} 0 & 1 \\ 1 & 0 \end{pmatrix} \,, \quad (23)_\varepsilon = \begin{pmatrix} 1 & 0 \\ -1 & -1 \end{pmatrix} \,. \tag{29.70}$$

The two representations in $L^{(3)}$ and $L^{(4)}$ are equivalent since there exists the invertible transformation

$$S = \begin{pmatrix} 1 & 0 \\ -1 & -1 \end{pmatrix}$$

from $L^{(3)}$ to $L^{(4)}$ such that, for all $\sigma \in S_3$, the condition $\sigma_\varepsilon S = S\sigma_\epsilon$ holds, where σ_ε and σ_ϵ are the matrix representations of σ with respect to the bases $\{\varepsilon_1, \varepsilon_2\}$ and $\{\epsilon_1, \epsilon_2\}$, respectively [cf. 18.6)]. Note that as a special case of the decomposition of the group algebra $A(S_3)$ given by (29.59), the identity element $e \in S_3$ is given by

$$e = \mathcal{S}_3 + \mathcal{A}_3 + \epsilon_1 + \varepsilon_1 \,, \tag{29.71}$$

where $\mathcal{S}_3$ and $\mathcal{A}_3$ are the symmetrizing and anti-symmetrizing maps for S_3, respectively [cf. (29.54)].

We will now proceed to the representation of S_3 on the tensor space T_3V. Since $\mathcal{Y}_3\left(\begin{array}{|c|c|c|}\hline 1&2&3\\\hline\end{array}\right)$ is the symmetrizing map and $\mathcal{Y}_3\left(\begin{array}{|c|}\hline 1\\\hline 2\\\hline 3\\\hline\end{array}\right)$ the anti-symmetrizing map, Eq. (29.18) implies that

$$P_3(V) = \mathcal{Y}_3\left(\begin{array}{|c|c|c|}\hline 1&2&3\\\hline\end{array}\right)(T_3V)\,, \quad \Lambda_3(V) = \mathcal{Y}_3\left(\begin{array}{|c|}\hline 1\\\hline 2\\\hline 3\\\hline\end{array}\right)(T_3V)\,, \tag{29.72}$$

where $P_3(V)$ $[\Lambda_3(V)]$ is the subspace of T_3V consisting of all totally symmetric (totally anti-symmetric) tensors in T_3V. Note that if $dim\, V < 3$, $\Lambda_3(V) = 0$. To find the explicit forms of tensors belonging to the various symmetry classes we consider general basis vectors in T_3V of the form

$$\varphi_{\mu\nu\lambda} \equiv e_\mu \otimes e_\nu \otimes e_\lambda \,, \tag{29.73}$$

where $\mu, \nu, \lambda = 1, \dots, dim(V)$, and $\{e_1, \dots, e_{dim(V)}\}$ is a basis of V, and allow the Young symmetrizers to act on them. We have, for example

$$\begin{aligned}\mathcal{Y}_3\left(\begin{array}{|c|c|c|}\hline 1&2&3\\\hline\end{array}\right)\varphi_{\mu\nu\lambda} &\equiv S_{\mu\nu\lambda} \equiv \begin{array}{|c|c|c|}\hline \mu&\nu&\lambda\\\hline\end{array} \\ &= \frac{1}{6}\left(\varphi_{\mu\nu\lambda} + \varphi_{\nu\lambda\mu} + \varphi_{\lambda\mu\nu} + \varphi_{\mu\lambda\nu} + \varphi_{\lambda\nu\mu} + \varphi_{\nu\mu\lambda}\right)\,,\end{aligned} \tag{29.74}$$

$$\begin{aligned}\mathcal{Y}_3\left(\begin{array}{|c|}\hline 1\\\hline 2\\\hline 3\\\hline\end{array}\right)\varphi_{\mu\nu\lambda} &\equiv A_{\mu\nu\lambda} \equiv \begin{array}{|c|}\hline \mu\\\hline \nu\\\hline \lambda\\\hline\end{array} \\ &= \frac{1}{6}\left(\varphi_{\mu\nu\lambda} + \varphi_{\nu\lambda\mu} + \varphi_{\lambda\mu\nu} - \varphi_{\mu\lambda\nu} - \varphi_{\lambda\nu\mu} - \varphi_{\nu\mu\lambda}\right)\,,\end{aligned} \tag{29.75}$$

where $S_{\mu\nu\lambda}$ $[\in P_3(V)]$ is a totally symmetric, and $A_{\mu\nu\lambda}$ $[\in \Lambda_3(V)]$ is a totally anti-symmetric tensor. One-dimensional subspaces generated by any vector in either $P_3(V)$ or $\Lambda_3(V)$ clearly form one-dimensional irreducible representations of S_3. As in the case of the group algebra representation, one-dimensional irreducible representations generated by symmetric tensors are inequivalent to one-dimensional representations generated by anti-symmetric tensors; but within the same symmetry class, all one-dimensional representations are equivalent.

Tensors of mixed symmetry (neither totally symmetric nor totally anti-symmetric) can be obtained by the actions of $\mathcal{Y}_3\left(\begin{array}{|c|c|}\hline 1&2\\\hline 3\\\cline{1-1}\end{array}\right)$ and $\mathcal{Y}_3\left(\begin{array}{|c|c|}\hline 1&3\\\hline 2\\\cline{1-1}\end{array}\right)$ on T_3V. We have, from (29.55) and (29.10),

$$\begin{aligned}M_{\mu\nu\lambda} &\equiv \mathcal{Y}_3\left(\begin{array}{|c|c|}\hline 1&2\\\hline 3\\\cline{1-1}\end{array}\right) e_\mu\otimes e_\nu\otimes e_\lambda \equiv \begin{array}{|c|c|}\hline \mu&\nu\\\hline \lambda\\\cline{1-1}\end{array}\\ &= \frac{1}{3}(e-(13))(e+(12))\,e_\mu\otimes e_\nu\otimes e_\lambda\\ &= \frac{1}{3}(e-(13))\,(e_\mu\otimes e_\nu\otimes e_\lambda + e_\nu\otimes e_\mu\otimes e_\lambda)\\ &= \frac{1}{3}(e_\mu\otimes e_\nu\otimes e_\lambda + e_\nu\otimes e_\mu\otimes e_\lambda - e_\lambda\otimes e_\nu\otimes e_\mu - e_\lambda\otimes e_\mu\otimes e_\nu)\,.\end{aligned} \tag{29.76}$$

Thus

$$M_{\mu\nu\lambda} = \frac{1}{3}(\varphi_{\mu\nu\lambda} + \varphi_{\nu\mu\lambda} - \varphi_{\lambda\nu\mu} - \varphi_{\lambda\mu\nu})\,. \tag{29.77}$$

Similarly, we can define

$$\tilde{M}_{\mu\nu\lambda} \equiv \mathcal{Y}_3\left(\begin{array}{|c|c|}\hline 1&3\\\hline 2\\\cline{1-1}\end{array}\right)\varphi_{\mu\nu\lambda} \equiv \begin{array}{|c|c|}\hline \mu&\lambda\\\hline \nu\\\cline{1-1}\end{array}\,, \tag{29.78}$$

and obtain

$$\tilde{M}_{\mu\nu\lambda} = \frac{1}{3}(\varphi_{\mu\nu\lambda} + \varphi_{\lambda\nu\mu} - \varphi_{\nu\mu\lambda} - \varphi_{\nu\lambda\mu})\,. \tag{29.79}$$

It is not hard to verify that the tensors $M_{\mu\nu\lambda}$ and $\tilde{M}_{\mu\nu\lambda}$ have the following symmetries:

$$M_{\mu\nu\lambda} = M_{\nu\mu\lambda}\,, \tag{29.80}$$

$$M_{\mu\nu\lambda} + M_{\nu\lambda\mu} + M_{\lambda\mu\nu} = 0\,, \tag{29.81}$$

$$\tilde{M}_{\mu\nu\lambda} = \tilde{M}_{\lambda\nu\mu}\,, \tag{29.82}$$

$$\tilde{M}_{\mu\nu\lambda} + \tilde{M}_{\nu\lambda\mu} + \tilde{M}_{\lambda\mu\nu} = 0\,. \tag{29.83}$$

Analogous to (29.71), it is readily seen that

$$\varphi_{\mu\nu\lambda} = S_{\mu\nu\lambda} + A_{\mu\nu\lambda} + M_{\mu\nu\lambda} + \tilde{M}_{\mu\nu\lambda}\,. \tag{29.84}$$

This shows explicitly that any tensor in T_3V can be written as a sum of a symmetric tensor, an anti-symmetric tensor, and two tensors of mixed symmetry.

Corresponding to the element ϵ_2 in the group algebra $A(S_3)$ we introduce the tensor

$$\xi_{\mu\nu\lambda} \equiv (1,2)\,M_{\mu\nu\lambda} = \frac{1}{3}(\varphi_{\nu\mu\lambda} + \varphi_{\mu\nu\lambda} - \varphi_{\nu\lambda\mu} - \varphi_{\mu\lambda\nu})\,. \tag{29.85}$$

Just as ϵ_1 and ϵ_2 [cf. (29.61)] in the group algebra of S_3 do, the tensors $M_{\mu\nu\lambda}$ and $\xi_{\mu\nu\lambda}$ (for fixed μ,ν and λ) in the tensor space T_3V constitute the basis of a two-dimensional irreducible space for S_3. The matrix representations for the elements in S_3 are also given by (29.65).

We will enumerate the tensors in T_3V belonging to the different symmetry classes when $dimV = 3$ (for the case of a three-electron system when each has orbital angular momentum $l = 1$, for example). In this case the indices μ, ν and λ can each assume the values 1, 2 and 3, and the space T_3V is 27-dimensional (with an obvious basis generated by the $\varphi_{\mu\nu\lambda}$). There are ten independent tensors generated by $S_{\mu\nu\lambda}$, given by

$$S_{111}, S_{222}, S_{333}, S_{112}, S_{113}, S_{221}, S_{223}, S_{331}, S_{332}, S_{123} \,. \tag{29.86}$$

Each of the above tensors generates a 1-dimensional representation space for S_3, and they are all equivalent to one another. The irreducible subspace generated by $A_{\mu\nu\lambda}$ is also one-dimensional, with A_{123} as the only independent basis tensor. Due to the symmetry conditions (29.80) and (29.81) for $M_{\mu\nu\lambda}$ (which are also shared by $\xi_{\mu\nu\lambda}$), we have the following independent tensors generated by $M_{\mu\nu\lambda}$ and $\xi_{\mu\nu\lambda}$:

$$M_{112}, M_{113}, M_{221}, M_{223}, M_{331}, M_{332}, M_{123}, M_{231} \,, \tag{29.87}$$

$$\xi_{112}, \xi_{113}, \xi_{221}, \xi_{223}, \xi_{331}, \xi_{332}, \xi_{123}, \xi_{231} \,. \tag{29.88}$$

As discussed above, each pair $\{M_{\mu\nu\lambda}, \xi_{\mu\nu\lambda}\}$ (with fixed μ, ν and λ), spans a two-dimensional irreducible representation space of S_3. These are all equivalent, so the eight pairs generate eight equivalent irreducible representations of S_3. All together we then have 27 independent tensors generated by the $S_{\mu\nu\lambda}, A_{\mu\nu\lambda}, M_{\mu\nu\lambda}$ and $\xi_{\mu\nu\lambda}$, which constitute a basis for the 27-dimensional T_3V. Note that instead of $M_{\mu\nu\lambda}$ and $\xi_{\mu\nu\lambda}$, we could also have chosen $\tilde{M}_{\mu\nu\lambda}$ [cf. (29.78)] and $\tilde{\xi}_{\mu\nu\lambda} \equiv (13)\tilde{M}_{\mu\nu\lambda}$ [cf. (29.66b)]. This decomposition of T_3V $(dim\, V = 3)$ is isomorphic to the following direct sum:

$$T_3V \sim (\mathbb{V}_{\square\square\square} \otimes \mathbb{C}^{10}) \oplus (\mathbb{V}_{\begin{smallmatrix}\square\\\square\\\square\end{smallmatrix}} \otimes \mathbb{C}) \oplus (\mathbb{V}_{\begin{smallmatrix}\square\square\\\square\end{smallmatrix}} \otimes \mathbb{C}^8) \quad (dim\, V = 3) \,, \tag{29.89}$$

where $\mathbb{V}_\mu$ is an irreducible representation space corresponding to the young diagram with partition $\{\boldsymbol{\mu}\}$, $\mathbb{C}$ is the field of complex numbers (considered as a one-dimensional complex vector space), and $\mathbb{C}^p$ is the (p-dimensional) complex vector space of p-tuples of complex numbers ($\mathbb{C}^1 = \mathbb{C}$). The right-hand side of the above equation can be interpreted as the direct sum of ten copies of $\mathbb{V}_{(3,0,0)}$, one copy of $\mathbb{V}_{(1,1,1)}$, and eight copies of $\mathbb{V}_{(2,1)}$. This equation implies the following one relating the dimensions of the various subspaces of T_3V $(dimV = 3)$:

$$3 \times 3 \times 3 = (1 \times 10) + (1 \times 1) + (2 \times 8) \,. \tag{29.90}$$

Now consider the action of the group $GL(3)$ (the **general linear group** of dimension 3, or the group of invertible 3×3 complex matrices) on the tensor space T_3V $(dimV = 3)$. Let $\{e_1, e_2, e_3\}$ be a basis of V. A general tensor in

T_3V can then be written as a linear combination of tensors of the form $\varphi_{\mu\nu\lambda}$ [cf. (29.73)]. Suppose $g \in GL(3)$ acts on e_μ by

$$ge_\mu = g^\alpha_\mu e_\alpha \,, \tag{29.91}$$

where g^α_μ is the $(\mu\alpha)$ element of the 3×3 matrix g. The induced action on T_3V is then given by

$$g\varphi_{\mu\nu\lambda} = g^\alpha_\mu g^\beta_\nu g^\gamma_\lambda \, \varphi_{\alpha\beta\gamma} \,. \tag{29.92}$$

From this, we can readily verify that, on recalling (29.74), (29.75), (29.77) and (29.85),

$$gS_{\mu\nu\lambda} = g^\alpha_\mu g^\beta_\nu g^\gamma_\lambda \, S_{\alpha\beta\gamma} \,, \tag{29.93}$$

$$gA_{\mu\nu\lambda} = g^\alpha_\mu g^\beta_\nu g^\gamma_\lambda \, A_{\alpha\beta\gamma} \,, \tag{29.94}$$

$$gM_{\mu\nu\lambda} = g^\alpha_\mu g^\beta_\nu g^\gamma_\lambda \, M_{\alpha\beta\gamma} \,, \tag{29.95}$$

$$g\xi_{\mu\nu\lambda} = g^\alpha_\mu g^\beta_\nu g^\gamma_\lambda \, \xi_{\alpha\beta\gamma} \,. \tag{29.96}$$

It follows that the set $\{S_{\mu\nu\lambda}\}$ in (29.86) forms the basis for a 10-dimensional irreducible representation of $GL(3)$, the single tensor A_{123} forms the basis for a 1-dimensional (irreducible) representation of $GL(3)$, and finally, the sets $\{M_{\mu\nu\lambda}\}$ [cf. (29.87)] and $\{\xi_{\mu\nu\lambda}\}$ [cf. (29.88)], each of eight independent tensors, form the bases for two equivalent 8-dimensional irreducible representations of $GL(3)$. We thus arrive at the remarkable fact that the decomposition of the tensor space T_3V ($dimV = 3$) according to (29.89) furnishes irreducible representations for both S_3 and $GL(3)$.

The above result can be generalized to the representations of $GL(m)$ and S_n on the tensor space T_nV ($dimV = m$). The central result is stated without proof in the following theorem (for a detailed proof, see, for example, Sternberg 1994).

Theorem 29.8. *The tensor space T_nV ($dimV = m$) decomposes into a direct sum of irreducible subspaces under both the symmetric group S_n and the general linear group $GL(m)$ as follows:*

$$T_nV \sim (\mathbb{V}_1 \otimes \mathbb{C}^{p_1}) \oplus \cdots \oplus (\mathbb{V}_q \otimes \mathbb{C}^{p_q}) \,, \tag{29.97}$$

where q is the total number of inequivalent irreducible representations of S_n, with $\mathbb{V}_1, \ldots, \mathbb{V}_q$ being the respective irreducible representation subspaces, p_i ($i = 1, \ldots, q$) is some non-negative integer, and $\mathbb{C}^{p_i}$ is the p_i-dimensional complex vector space of p_i-tuples of complex numbers. If $n > m$, the integers p_i vanish for all irreducible representations of S_n corresponding to Young diagrams with more than m rows. If $dim\mathbb{V}_i = d_i$, the above equation states that T_nV decomposes under S_n into a direct sum isomorphic to p_i copies of each of $\mathbb{V}_i$, and under $GL(m)$ into a direct sum isomorphic to d_i copies of each of $\mathbb{C}^{p_i}$.

We have discussed previously how to obtain the numbers q (cf. Theorem 29.2) and d_i [cf. (29.47), (29.50), and (29.51)]. We will now give a formula for calculating the numbers p_i.

Theorem 29.9. *The dimension of the irreducible representation of $GL(m)$ on T_nV ($dimV = m$) corresponding to an irreducible representation of S_n specified by a Young diagram with partition $\{\mu_1, \ldots, \mu_m\}$, $\mu_1 + \cdots + \mu_m = n$, is given by*

$$\boxed{d(\mu_1, \ldots, \mu_m) = \frac{\prod_{i<j}^{m} (l_i - l_j)}{1!\,2!\,3! \ldots (m-1)!}}\ , \tag{29.98}$$

where $l_1 > l_2 > \cdots > l_m$ are given by

$$l_1 \equiv \mu_1 + (m-1)\ ,\ l_2 \equiv \mu_2 + (m-2)\ , \ldots , l_j \equiv \mu_j + (m-j)\ , \ \ldots\ , l_m \equiv \mu_m\ .$$

It is easily checked that the above formula leads to the results obtained earlier for S_3 and $GL(3)$ represented on T_3V. As another illustration of the above two theorems, we consider the irreducible representations of S_3 and $GL(2)$ on T_3V ($dimV = 2$). In this case the analog of (29.89) is

$$T_3V \sim (\mathbb{V}_{\yng(3)} \otimes \mathbb{C}^4) \oplus (\mathbb{V}_{\yng(2,1)} \otimes \mathbb{C}^2) \qquad (dim\,V = 2)\ . \tag{29.99}$$

The corresponding equation relating the dimensions of the subspaces is

$$2 \times 2 \times 2 = (1 \times 4) + (2 \times 2)\ . \tag{29.100}$$

As noted in the beginning of this chapter, the subspaces of T_nV which are irreducible representation spaces of $GL(m), m \geq 2$, can further be decomposed into irreducible representation spaces of $SO(3)$ [or $SU(2)$]. These are the spaces of interest in atomic physics. Recall that the irreducible spaces of $SO(3)$ are each characterized by j (a positive integer or half-intreger), with corresponding dimension $2j + 1$. We will first illustrate how to determine these for the case of T_3V ($dim\,V = 3$) as the representation space of S_3 and $GL(3)$. First rewrite the tensors $S_{\mu\nu\lambda}$, $A_{\mu\nu\lambda}$, $M_{\mu\nu\lambda}$ and $\xi_{\mu\nu\lambda}$ using the Dirac notation. Let

$$S_{\mu\nu\lambda} \equiv |\,\yng(3)\,;\, d,\, p\,\rangle\ , \qquad (d = 1, p = 1, \ldots, 10)\ , \tag{29.101a}$$

$$A_{\mu\nu\lambda} \equiv |\,\yng(1,1,1)\,;\, d,\, p\,\rangle\ , \qquad (d = 1, p = 1)\ , \tag{29.101b}$$

$$M_{\mu\nu\lambda} \equiv |\,\yng(2,1)\,;\, d,\, p\,\rangle\ , \qquad (d = 1, p = 1, \ldots, 8)\ , \tag{29.101c}$$

$$\xi_{\mu\nu\lambda} \equiv |\,\yng(2,1)\,;\, d,\, p\,\rangle\ , \qquad (d = 2, p = 1, \ldots, 8)\ , \tag{29.101d}$$

where, with reference to (29.97), $d = 1, \ldots, d_i$, $p = 1, \ldots, p_i$ (for $p_i \neq 0$), with $i = 1, \ldots, q$ being an index specifying a particular irreducible representation of S_n. Now further let [cf. (29.73)]

$$\varphi_{m_1m_2m_3} = |\,j = 1,\ m_1\,\rangle \otimes |\,j = 1,\ m_2\,\rangle \otimes |\,j = 1,\ m_3\,\rangle\ , \tag{29.102}$$

where the states $|\,jm\,\rangle$ are the angular momentum states introduced in (20.11), and $m_1, m_2, m_3 = 1, 0, -1$. Making the correspondences $\mu = 1 \Leftrightarrow m = 1$, $\mu =$

$2 \Leftrightarrow m = 0$, $\mu = 3 \Leftrightarrow m = -1$, we can write [cf. (29.86)]

$$\begin{aligned}
S_{111} &\equiv S_{1,1,1} &&= |\, \yng(3)\, ; 1,1 ; m = 3 \,\rangle \,, && \text{(29.103a)}\\
S_{222} &\equiv S_{0,0,0} &&= |\, \yng(3)\, ; 1,2 ; m = 0 \,\rangle \,, && \text{(29.103b)}\\
S_{333} &\equiv S_{-1,-1,-1} &&= |\, \yng(3)\, ; 1,3 ; m = -3 \,\rangle \,, && \text{(29.103c)}\\
S_{112} &\equiv S_{1,1,0} &&= |\, \yng(3)\, ; 1,4 ; m = 2 \,\rangle \,, && \text{(29.103d)}\\
S_{113} &\equiv S_{1,1,-1} &&= |\, \yng(3)\, ; 1,5 ; m = 1 \,\rangle \,, && \text{(29.103e)}\\
S_{221} &\equiv S_{0,0,1} &&= |\, \yng(3)\, ; 1,6 ; m = 1 \,\rangle \,, && \text{(29.103f)}\\
S_{223} &\equiv S_{0,0,-1} &&= |\, \yng(3)\, ; 1,7 ; m = -1 \,\rangle \,, && \text{(29.103g)}\\
S_{331} &\equiv S_{-1,-1,1} &&= |\, \yng(3)\, ; 1,8 ; m = -1 \,\rangle \,, && \text{(29.103h)}\\
S_{332} &\equiv S_{-1,-1,0} &&= |\, \yng(3)\, ; 1,9 ; m = -2 \,\rangle \,, && \text{(29.103i)}\\
S_{123} &\equiv S_{1,0,-1} &&= |\, \yng(3)\, ; 1,10 ; m = 0 \,\rangle \,, && \text{(29.103j)}
\end{aligned}$$

where $m = m_1 + m_2 + m_3$ in each of the above totally symmetric states can be thought of as the z-component of the total orbital angular momentum $\boldsymbol{L}$ of a three-electron system, each in the p ($l = 1$) state. We see that the 10-dimensional subspace of T_3V spanned by the $S_{\mu\nu\lambda}$, which is an irreducible subspace under $GL(3)$, is decomposed into the direct sum of a 7-dimensional irreducible space ($L = 3$) and a 3-dimensional irreducible space ($L = 1$) under $SO(3)$:

$$10 = 7 + 3 \,, \qquad \text{where } 7 = (2 \times 3) + 1,\ 3 = (2 \times 1) + 1 \,. \tag{29.104}$$

Similarly, the totally antisymmetric state

$$A_{123} \equiv A_{1,0,-1} = |\, \yng(1,1,1)\, ; 1,1 ; m = 0 \,\rangle \tag{29.105}$$

spans a 1-dimensional irreducible space under $SO(3)$ corresponding to $L = 0$. We also have the following states of mixed symmetry [cf. (29.87) and (29.88)]

$$\begin{aligned}
M_{112} &\equiv M_{1,1,0} &&= |\, \yng(2,1)\, ; 1,1 ; m = 2 \,\rangle \,, && \text{(29.106a)}\\
M_{113} &\equiv M_{1,1,-1} &&= |\, \yng(2,1)\, ; 1,2 ; m = 1 \,\rangle \,, && \text{(29.106b)}\\
M_{221} &\equiv M_{0,0,1} &&= |\, \yng(2,1)\, ; 1,3 ; m = 1 \,\rangle \,, && \text{(29.106c)}\\
M_{223} &\equiv M_{0,0,-1} &&= |\, \yng(2,1)\, ; 1,4 ; m = -1 \,\rangle \,, && \text{(29.106d)}\\
M_{331} &\equiv M_{-1,-1,1} &&= |\, \yng(2,1)\, ; 1,5 ; m = -1 \,\rangle \,, && \text{(29.106e)}\\
M_{332} &\equiv M_{-1,-1,0} &&= |\, \yng(2,1)\, ; 1,6 ; m = -2 \,\rangle \,, && \text{(29.106f)}\\
M_{123} &\equiv M_{1,0,-1} &&= |\, \yng(2,1)\, ; 1,7 ; m = 0 \,\rangle \,, && \text{(29.106g)}\\
M_{231} &\equiv M_{0,-1,1} &&= |\, \yng(2,1)\, ; 1,8 ; m = 0 \,\rangle, && \text{(29.106h)}
\end{aligned}$$

and

$$
\begin{aligned}
\xi_{112} &\equiv \xi_{1,1,0} &&= |\,\yng(2,1)\,;\,2,1\,;\,m=2\,\rangle\,, &&\text{(29.107a)}\\
\xi_{113} &\equiv \xi_{1,1,-1} &&= |\,\yng(2,1)\,;\,2,2\,;\,m=1\,\rangle\,, &&\text{(29.107b)}\\
\xi_{221} &\equiv \xi_{0,0,1} &&= |\,\yng(2,1)\,;\,2,3\,;\,m=1\,\rangle\,, &&\text{(29.107c)}\\
\xi_{223} &\equiv \xi_{0,0,-1} &&= |\,\yng(2,1)\,;\,2,4\,;\,m=-1\,\rangle\,, &&\text{(29.107d)}\\
\xi_{331} &\equiv \xi_{-1,-1,1} &&= |\,\yng(2,1)\,;\,2,5\,;\,m=-1\,\rangle\,, &&\text{(29.107e)}\\
\xi_{332} &\equiv \xi_{-1,-1,0} &&= |\,\yng(2,1)\,;\,2,6\,;\,m=-2\,\rangle\,, &&\text{(29.107f)}\\
\xi_{123} &\equiv \xi_{1,0,-1} &&= |\,\yng(2,1)\,;\,2,7\,;\,m=0\,\rangle\,, &&\text{(29.107g)}\\
\xi_{231} &\equiv \xi_{0,-1,1} &&= |\,\yng(2,1)\,;\,2,8\,;\,m=0\,\rangle\,. &&\text{(29.107h)}
\end{aligned}
$$

It is clear that, by looking at the m values of the above two groups of states, the decomposition of the 8-dimensional subspaces spanned by $M_{\mu\nu\lambda}$ and $\xi_{\mu\nu\lambda}$ under $SO(3)$ is according to

$$
8 = 5 + 3\,, \qquad \text{where } 5 = (2\times 2)+1,\; 3 = (2\times 1)+1\,. \tag{29.108}
$$

Thus the irreducible subspaces spanned by the states in (29.106) [and by the equivalent representation of (29.107)] under $SO(3)$ correspond to $L=2$ and $L=1$.

To summarize, the eigenstates of a Hamiltonian describing a system of three p-electrons (for example, the 3 valence electrons of the nitrogen atom N) can be grouped according to symmetry under permutations of the electrons and the orbital angular momentum quantum numbers as follows:

$$
\begin{aligned}
&|\,\yng(3)\,;\,L=3,m_L\,\rangle\,,\quad |\,\yng(3)\,;\,L=1,m_L\,\rangle\,,\\
&|\,\yng(1,1,1)\,;\,L=0,m_L=0\,\rangle\,,\\
&|\,\yng(2,1)\,;\,1\,;\,L=2,m_L\,\rangle\,,\quad |\,\yng(2,1)\,;\,1\,;\,L=1,m_L\,\rangle\,,\\
&|\,\yng(2,1)\,;\,2\,;\,L=2,m_L\,\rangle\,,\quad |\,\yng(2,1)\,;\,2\,;\,L=1,m_L\,\rangle\,,
\end{aligned}
\tag{29.109}
$$

where the last two rows of states represent two equivalent representations under $SO(3)$.

For applications of group theory to atomic spectroscopy, one has to consider, in addition to the orbital angular momentum of the electrons, their spin angular momentum. In the cases where the so-called **Russell-Saunders (LS) coupling** is valid (such as in the case of the nitrogen atom), the total orbital momentum L of the electrons can be coupled to the total spin angular momentum S. Since the total spin angular momentum of each electron is $S=1/2$, each electron can be in one of two spin states: spin up ($m_S=1/2$) or spin down ($m_S=-1/2$) [cf. (29.31)]. For an n-electron system the appropriate tensor space describing the spin states is then T_nV, $dim\,V=2$. It is clear then that the Young diagrams describing the symmetry classes of the spin states cannot have more than 2 rows. (Diagrams with more than two rows would involve antisymmetrization with respect to more than two indices each of which can only assume two values. Hence the corresponding tensors would automatically vanish.) For example, for $n = 3$, the only possible Young diagrams for the spin states are

$$\begin{array}{|c|c|c|}\hline & & \\ \hline \end{array}\ , \quad \begin{array}{|c|c|}\hline & \\ \hline & \multicolumn{1}{c}{} \\ \cline{1-1} \end{array}\ .$$

The decomposition of the tensor space T_3V under S_3 and $GL(2)$ has been given by (29.99) and (29.100). The possible spin states for a 3-electron system are then given by

$$\begin{aligned} &|\,\square\square\square\,;\, S = 3/2, m_S\,\rangle\,, \\ &|\,\begin{smallmatrix}\square\square\\ \square\end{smallmatrix}\,;\,1\,;\, S = 1/2, m_S\,\rangle\,, \quad |\,\begin{smallmatrix}\square\square\\ \square\end{smallmatrix}\,;\,2\,;\, S = 1/2, m_S\,\rangle\,. \end{aligned} \tag{29.110}$$

The two sets of states labeled by 1 and 2 in the above equation again constitute equivalent representations of $SO(3)$.

The total wave function of the electrons is the direct product of the orbital and spin wave functions. Even in non-relativistic quantum theory, the fundamental requirement of the Lorentz invariance of relativistic wave equations intervenes to impose a strong restriction on this wave function, in the form of the **Pauli Exclusion Principle**, which states that *the product state representing the total wave function of an n-electron system must be totally antisymmetric with respect to any interchange of electrons.* This is a consequence of *the connection between spin and statistics* first demonstrated by Pauli: **fermions** (particles of half-integral spin) obey the **Fermi-Dirac statistics**, while **bosons** (particles of integral spin) obey the **Bose-Einstein statistics**. That special relativity imprints in such a profound fashion on non-relativistic quantum mechanics must be regarded as a testament to the fundamental inseparability of relativity and and quantum theory. Unfortunately it is beyond the scope of this book to delve into these issues more deeply.

To probe the consequences of the Exclusion Principle, we first introduce a mathematical concept and then state a theorem related to the concept (without proof).

Definition 29.7. *Corresponding to each irreducible representation of the symmetric group S_n is an* ***associated representation*** *of the same dimension obtained from the original one as follows: In the associated representation the matrix representation of an even permutation is identical to the original one, while the matrix representation of an odd permutation is the negative of the original one.*

Thus the direct product of an irreducible representation of S_n with its associated representation must be equivalent to the completely antisymmetric (one-dimensional) representation.

Theorem 29.10. *Irreducible representations of a symmetric group that are associated with each other are represented by Young diagrams whose rows and columns are interchanged. Such pairs of diagrams are referred to as being* ***conjugate*** *to each other. The only direct products of two irreducible representations* *equivalent to the completely antisymmetric representation are those between associated representations.*

Note that since there are many Young diagrams which remain unchanged when the rows and columns are interchanged, the corresponding irreducible representations of S_n are self-associative, that is, the associated representation is the same as the original one.

Applied to atomic spectroscopy, the Pauli Exclusion Principle and the above theorem imply that the orbital state and the spin state of an n-electron system must be characterized by associate irreducible representations of S_n, that is, by conjugate Young diagrams. Returning to the example of three p-electrons (the $(2p)^3$ configuration for the N atom), we have the following possible associations between the spin and orbital states:

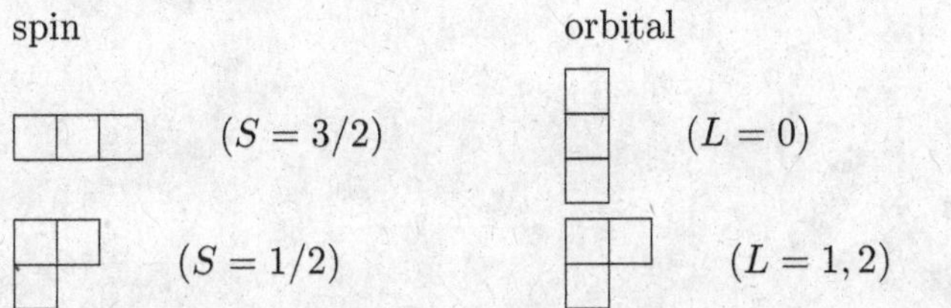

By the rules of angular momentum coupling, the possible states are given by

$$\begin{aligned}
&|\, L=0, S=3/2\,;\, J=3/2, m_J\,\rangle\,, \\
&|\, L=2, S=1/2\,;\, J=5/2, m_J\,\rangle\,, \\
&|\, L=2, S=1/2\,;\, J=3/2, m_J\,\rangle\,, \\
&|\, L=1, S=1/2\,;\, J=3/2, m_J\,\rangle\,, \\
&|\, L=1, S=1/2\,;\, J=1/2, m_J\,\rangle\,.
\end{aligned} \tag{29.111}$$

Using the spectroscopic notation ${}^{2S+1}L_J$, the above are designated as

$${}^4S_{3/2}\,,\ {}^2D_{5/2}\,,\ {}^2D_{3/2}\,,\ {}^2P_{3/2}\,,\ {}^2P_{1/2}\,,$$

a total of 20 states, where, for the orbital quantum numbers, S means $L = 0$, P means $L = 1$, and D means $L = 2$. The number 20 can be obtained by an elementary application of Pauli's Exclusion principle, which implies that each of the three $2p$ one-electron energy levels can accommodate at most two electrons, one spin-up and the other spin-down. Twenty is exactly the number of ways in which one can put three arrows [pointing up or down, with upward pointing (downward pointing) arrows representing spin-up (spin-down) electrons] in three boxes, such that no box can contain more than two arrows, and such that when a box does contain two arrows, one has to be upward-pointing and the other downward-pointing. Note that without the Pauli Exclusion Principle, a total of $216\,(= 27 \times 8)$ states would have been possible!

Problem 29.1 Verify that the possible number of spectroscopic terms for the ground state of atomic nitrogen is 20 by the elementary application of Pauli's exclusion principle mentioned in the last paragraph of this chapter.

Problem 29.2 Work out the possible spectroscopic terms for the ground state of atomic carbon with the electronic configuration $1s^2\,2s^2\,2p^2$ under the LS coupling scheme, using the group theoretic methods discussed in this chapter.

Chapter 30

The Lie Algebra of $SO(4)$ and the Hydrogen Atom

In Chapter 20 we have given an account of the properties of the angular momentum operators, which essentially arise from a study of the **Lie algebra** $\mathcal{SO}(3)$ of the group $SO(3)$ (recall the discussion in the first paragraph of that chapter). We saw that L_x, L_y, L_z, the three spatial components of the vector angular momentum operator $\boldsymbol{L}$ that were shown to be **infinitesimal generators** of spatial rotation in Chapter 17, are essentially a choice of basis vectors of the Lie algebra, which can be considered as a vector space. In this chapter we will study a remarkable application of the Lie algebra of the group $SO(4)$ to the hydrogen atom problem. This problem is traditionally treated using the Schrödinger wave equation, as discussed in the beginning of Chapter 16. We will not present the details of this partial-differential-equation approach here, but will instead focus on the group-theoretic approach. We shall see that the familiar spectrum of quantized energy levels (with their degeneracies) of the hydrogen atom is basically a consequence not only of rotational symmetry [symmetry under $SO(3)$], but of a larger symmetry [symmetry under $SO(4)$] – sometimes called **dynamical symmetry** – of the electronic Hamiltonian. First we need to review some basic facts of Lie algebra theory.

The idea of a Lie group as a differentiable manifold was discussed in Chapter 19, for the examples of $SU(2)$ and $SO(3)$. *The* ***Lie algebra*** $\mathcal{G}$ *of a Lie group* G *is an algebra of operators isomorphic to the tangent space of* G *at the identity, when* G *is considered as a differentiable manifold.* The dimension of $\mathcal{G}$ (as a vector space) is thus the same as the manifold-dimension of G. The multiplication rule for $\mathcal{G}$ is given by the **Lie bracket** (or commutator):

$$[A, B] = AB - BA\,, \tag{30.1}$$

for all $A, B \in \mathcal{G}$.

We will not give the technical definition of the notion of the tangent space here. For our purposes it suffices to consider an intuitive picture, such as the tangent plane to a spherical surface at a certain point. As we have seen in Chapter 19, because of the double-connectedness of the group manifold of $SO(3)$, the "shape" of $SO(3)$ is hard to visualize. Indeed, the Lie algebra, viewed as the tangent space (a linear vector space) at the identity, provides us with a convenient tool for the study of the properties of the group manifold near the identity element.

The Lie bracket multiplication rule obviously satisfies the following:

$$[A, B] = -[B, A], \quad \text{(antisymmetry)}, \tag{30.2}$$

$$[\alpha A + \beta B, C] = \alpha[A, C] + \beta[B, C], \text{ for all } \alpha, \beta \in \mathbb{R}, \quad \text{(associativity)}, \tag{30.3}$$

$$[A, [B, C]] + [B, [C, A]] + [C, [A, B]] = 0, \quad \text{(Jacobi identity)}. \tag{30.4}$$

The structure of a Lie algebra is completely determined by the so-called structure constants, defined below:

Definition 30.1. *Let $(e_1, \dots, e_n)$ be a basis of the Lie algebra $\mathcal{G}$ of a Lie group G. The **structure constants** $c^k{}_{ij} \in \mathbb{R}$ are defined by*

$$[e_i, e_j] = c^k{}_{ij}\, e_k . \tag{30.5}$$

The Jacobi identity [(30.4)] implies that the structure constants $c^k{}_{ij}$ satisfy the following identity:

$$c^h{}_{im}\, c^m{}_{jk} + c^h{}_{km}\, c^m{}_{ij} + c^h{}_{jm}\, c^m{}_{ki} = 0\,, \tag{30.6}$$

for all i, j, h, k. This equation is also called the Jacobi identity.

Problem 30.1 Verify the Jacobi identity given by (30.6).

Returning to $SO(n)$ we can identify a basis for its Lie algebra as follows. Let $g \in SO(n)$ be an element near the identity. Then we can write

$$a = 1 + \epsilon + O(\epsilon^2)\,, \tag{30.7}$$

where 1 is the $n \times n$ identity matrix and ϵ is an $n \times n$ real matrix of infinitesimal quantities. Since, by virtue of the orthogonality property of $SO(n)$, $aa^T = a^T a = 1$, we have

$$(1+\epsilon)(1+\epsilon^T) \approx 1\,. \tag{30.8}$$

Thus, to first order in ϵ,

$$\epsilon^T = -\epsilon\,, \tag{30.9}$$

which says that ϵ is an antisymmetric matrix, with $d = n(n-1)/2$ independent quantities. The integer d is thus the dimension of the group manifold of $SO(n)$, as well as the dimension of the Lie algebra $\mathcal{SO}(n)$. Now let $U(g)$ be an N-dimensional representation of g. $U(g)$ is thus an $N \times N$ matrix, where, in general, $N \neq n$. Analogous to (17.57), we can write

$$U(g) = 1 - \frac{1}{2}\, J_i^j \epsilon_j^i + O(\epsilon^2)\,, \tag{30.10}$$

where each J_i^j is an $N \times N$ matrix, with the property

$$J_i^j = -J_j^i\,. \tag{30.11}$$

The factor 1/2 is introduced to take care of the antisymmetric nature of ϵ_i^j and J_i^j. The $n(n-1)/2$ matrices J_i^j $(i < j)$ are then elements of the Lie algebra $\mathcal{SO}(n)$, and in fact constitute a basis of $\mathcal{SO}(n)$. Although we will not present the proof here (for a proof, see, for example, Lam 2003), it can be shown quite straightforwardly that the structure constants of $SO(n)$ are given by the following commutation relations:

$$\boxed{[J_i^j, J_k^l] = \delta_k^j J_i^l - \delta_i^l J_k^j + \delta_i^k J_l^j - \delta_l^j J_i^k \quad ; \quad i,j,k,l = 1,\dots,n \quad .} \tag{30.12}$$

For $SO(3)$ there are only three independent matrices J_i^j. One choice of the three is (J_1^2, J_3^1, J_2^3). According to (30.12),

$$[\,J_2^3,\, J_3^1\,] = J_2^1\,, \quad [\,J_3^1,\, J_1^2\,] = J_3^2\,, \quad [\,J_1^2,\, J_2^3\,] = J_1^3\,. \tag{30.13}$$

Making the identifications

$$\hbar J_1^2 = iL^3\,, \quad \hbar J_3^1 = iL^2\,, \quad \hbar J_2^3 = iL^1\,, \tag{30.14}$$

the above equations become

$$[\,L^1,\, L^2\,] = i\hbar L^3\,, \quad [\,L^2,\, L^3\,] = i\hbar L^1\,, \quad [\,L^3,\, L^1\,] = i\hbar L^2\,, \tag{30.15}$$

which are precisely the commutation relations for the orbital angular momentum operators given by (17.62).

With the above mathematical background we will now consider the simplest model of a hydrogen-like atom, in which the electronic Hamiltonian only includes the Coulomb interaction between the single electron and the nucleus. This Hamiltonian is given by

$$H = \frac{p^2}{2m} - \frac{Ze^2}{r}\,, \quad \left(p_j = -i\hbar \frac{\partial}{\partial x^j} \right)\,, \tag{30.16}$$

where m is the reduced mass of the electron, e is the magnitude of the electronic charge, Z is the charge number of the nucleus, and $r = \sqrt{(x^1)^2 + (x^2)^2 + (x^3)^2}$ is the magnitude of the position vector $\boldsymbol{r}$ of the electron relative to the nucleus.

Before we embark on the quantum mechanical solution of this problem, it is useful to consider an elegant classical mechanical solution, starting with Hamilton's equations of motion. These are given by [cf. (2.8)]

$$\frac{dq^i}{dt} = \frac{\partial H}{\partial p_i}, \quad \frac{dp_i}{dt} = -\frac{\partial H}{\partial q^i}, \tag{30.17}$$

where q^i and p_i are the canonical coordinates and momenta, respectively. With $q^i = x^i$ and the Hamiltonian in (30.16), Hamilton's equations yield

$$\frac{d\boldsymbol{r}}{dt} = \frac{\boldsymbol{p}}{m}, \quad \frac{d\boldsymbol{p}}{dt} = -\frac{Ze^2\boldsymbol{r}}{r^3}. \tag{30.18}$$

The second of the above equations is just Newton's second law applied to the present situation. The objective is to find all constants of motion, quantities whose total time derivatives vanish. One easily sees that the orbital angular momentum

$$\boldsymbol{L} = \boldsymbol{r} \times \boldsymbol{p} \tag{30.19}$$

is one such quantity, since by (30.18)

$$\frac{d\boldsymbol{L}}{dt} = 0. \tag{30.20}$$

Another (much less obvious) one is the so-called **Runge-Lenz vector**, defined by

$$\boldsymbol{M} \equiv \frac{\boldsymbol{p} \times \boldsymbol{L}}{m} - \frac{Ze^2\boldsymbol{r}}{r}, \tag{30.21}$$

which also satisfies

$$\frac{d\boldsymbol{M}}{dt} = 0. \tag{30.22}$$

Problem 30.2 Prove (30.22) within the context of classical mechanics.

It is also clear that

$$\boldsymbol{L} \cdot \boldsymbol{r} = \boldsymbol{L} \cdot \boldsymbol{M} = 0. \tag{30.23}$$

Thus $\boldsymbol{L}$ is perpendicular to the plane of the orbit and $\boldsymbol{M}$ lies on that plane. The latter fact immediately yields an expression for the orbit, since (30.21) gives

$$\boldsymbol{M} \cdot \boldsymbol{r} = \frac{L^2}{m} - Ze^2 r. \tag{30.24}$$

This, in fact, is the equation of a conic section. To see this we only have to let the x-y plane coincide with the orbital plane and align the positive x-axis along the direction of the constant vector $\boldsymbol{M}$. Then (30.24) can be written in polar coordinates as

$$r = \frac{L^2}{mZe^2}\left(1 + \frac{M}{Ze^2}\cos\theta\right)^{-1}, \tag{30.25}$$

where θ is the polar angle between $\boldsymbol{r}$ and the x-axis (the direction of $\boldsymbol{M}$), and $M/(Ze^2)$ is the eccentricity of the conic. A bit more vector algebra will reveal that the constant magnitude of $\boldsymbol{M}$ is given by

$$M^2 = \frac{2HL^2}{m} + Z^2e^4\,, \tag{30.26}$$

which implies that the conserved total energy is

$$H = \frac{m}{2L^2}\left(M^2 - Z^2e^4\right). \tag{30.27}$$

In the quantum mechanical formulation of the problem, $\boldsymbol{r}, \boldsymbol{p}$ and $\boldsymbol{L}$ are all operators in the Hilbert space of quantum states of the electron and do not commute with each other. Indeed, the non-commutativity of $\boldsymbol{r}$ and $\boldsymbol{p}$ [cf. (1.37)] is the mathematical basis of the Uncertainty Principle, as discussed in Chapter 11. In particular, as operators, $\boldsymbol{p}\times\boldsymbol{L} \neq -\boldsymbol{L}\times\boldsymbol{p}$, and we have to redefine a quantum mechanical Runge-Lenz vector by antisymmetrizing the product $\boldsymbol{p}\times\boldsymbol{L}$:

$$\boldsymbol{M} \equiv \frac{1}{2m}\left(\boldsymbol{p}\times\boldsymbol{L} - \boldsymbol{L}\times\boldsymbol{p}\right) - \frac{Ze^2\boldsymbol{r}}{r}\,. \tag{30.28}$$

Now our classical mechanical analysis above showed that both $\boldsymbol{L}$ and $\boldsymbol{M}$ are constants of motion. In quantum mechanics, this has to be checked by using Heisenberg's equation of motion [cf. (2.13)]

$$\frac{dA}{dt} = \frac{[A,\,H]}{i\hbar} + \frac{\partial A}{\partial t} \tag{30.29}$$

for any observable A. It is obvious from the definitions of $\boldsymbol{L}$ and $\boldsymbol{M}$ [given by (30.19) and (30.28) respectively] that

$$\frac{\partial \boldsymbol{L}}{\partial t} = \frac{\partial \boldsymbol{M}}{\partial t} = 0\,. \tag{30.30}$$

Thus it remains to compute the commutators $[\,\boldsymbol{L},\,H\,]$ and $[\,\boldsymbol{M},H\,]$. Using the commutation relations

$$[\,x^l,\,p_m\,] = i\hbar\delta^l_m\,, \tag{30.31}$$

or from (16.2), (16.3), and (17.63), it is relatively easy to establish that

$$[\,\boldsymbol{L},\,H\,] = 0\,. \tag{30.32}$$

A somewhat more tedious exercise, using the commutation relation [cf. (12.5)]

$$[\,f(\boldsymbol{r}),\,\boldsymbol{p}\,] = i\hbar\nabla f(\boldsymbol{r})\,, \tag{30.33}$$

where $f(\boldsymbol{r})$ is any analytic function of the components of $\boldsymbol{r}$, will also establish the fact that

$$[\boldsymbol{M}, H] = 0 . \tag{30.34}$$

Thus, quantum mechanically, $\boldsymbol{L}$ and $\boldsymbol{M}$ are also constants of motion.

Problem 30.3 Prove (30.34) with the help of (30.33).

The fact that M^1, M^2, M^3 are also constants of motion in addition to L^1, L^2 and L^3 suggests that the symmetry group here is larger than $SO(3)$. In fact we can establish that M^i and L^j form the basis of a six-dimensional Lie algebra by working out all the commutators of these objects and show that they satisfy (30.5). The following results (whose tedious but straightforward demonstrations will not be presented here) are obtained.

$$[L^i, L^j] = i\hbar\, \varepsilon^{ij}{}_k L^k , \tag{30.35a}$$

$$[L^i, M^j] = i\hbar\, \varepsilon^{ij}{}_k M^k , \tag{30.35b}$$

$$[M^i, M^j] = -\frac{2i\hbar}{m} H\, \varepsilon^{ij}{}_k L^k , \tag{30.35c}$$

where $\varepsilon^{ij}{}_k$ is the (completely antisymmetric) Levi-Civita tensor.

Problem 30.4 Verify the commutation relations (30.35) directly, using the definition of the quantum mechanical Runge-Lenz vector given by (30.28).

Actually the last equation seems to spoil things a bit since H occurs on the right-hand side. But since H commutes with both $\boldsymbol{M}$ and $\boldsymbol{L}$, $H = EI$ within an irreducible representation of our group (on a subspace of the Hilbert space of electronic states), where $E \in \mathbb{R}$ and I is the identity operator. If we work within an irreducible representation, and view all of the L^i, M^j as linear operators on the representation space, then H on the RHS of (30.37) can be replaced by a real number E, and the closure requirement of (30.5) will be satisfied. Within an irreducible representation let us then define

$$\boldsymbol{M}' \equiv \sqrt{-\frac{m}{2E}}\,\boldsymbol{M} , \tag{30.36}$$

where $E < 0$ (bound states of electrons). Note that $\boldsymbol{M}'$ has the same physical dimensions as the angular momentum $\boldsymbol{L}$. The commutation relations (30.35) can then be written as

$$[L^i, L^j\,] = i\hbar\,\varepsilon^{ij}{}_k L^k\,, \tag{30.37a}$$
$$[L^i, (M')^j\,] = i\hbar\,\varepsilon^{ij}{}_k (M')^k\,, \tag{30.37b}$$
$$[(M')^i, (M')^j\,] = i\hbar\,\varepsilon^{ij}{}_k L^k\,. \tag{30.37c}$$

The 3-dimensional algebra $\mathcal{SO}(3)$ generated by the L^i is clearly a subalgebra of our 6-dimensional algebra. The six operators L^i and $(M')^j$ can be viewed as the six independent components of a 4×4 antisymmetric matrix $-iJ$ (of operators) as follows:

$$\begin{aligned} &-i\hbar J \\ &= -i\hbar \begin{pmatrix} 0 & J_1^2 & J_1^3 & J_1^4 \\ -J_1^2 & 0 & J_2^3 & J_2^4 \\ -J_1^3 & -J_2^3 & 0 & J_3^4 \\ -J_1^4 & -J_2^4 & -J_3^4 & 0 \end{pmatrix} \equiv \begin{pmatrix} 0 & L^3 & -L^2 & (M')^1 \\ -L^3 & 0 & L^1 & (M')^2 \\ L^2 & -L^1 & 0 & (M')^3 \\ -(M')^1 & -(M')^2 & -(M')^3 & 0 \end{pmatrix}. \end{aligned} \tag{30.38}$$

In terms of the J_i^j, then, the commutation relations (30.37) appear as the single equation

$$[J_i^j, J_k^l\,] = \delta_k^j J_i^l - \delta_i^l J_k^j + \delta_i^k J_l^j - \delta_l^j J_i^k\,, \quad i,j,k,l = 1,2,3,4\,. \tag{30.39}$$

This equation is of exactly the same form as (30.12), and thus the commutation relations given by (30.37) are precisely those for the Lie algebra $\mathcal{SO}(4)$. We can conclude, then, that the relevant symmetry group for the hydrogen-atom problem (with only the Coulomb interaction) is $SO(4)$.

The Lie algebra $\mathcal{SO}(4)$ can in fact be decomposed into a direct sum of two copies of $\mathcal{SO}(3)$:

$$\mathcal{SO}(4) = \mathcal{SO}(3) \oplus \mathcal{SO}(3)\,. \tag{30.40}$$

This can be achieved by a change of basis from $\{\boldsymbol{L}, \boldsymbol{M}'\}$ to

$$\boldsymbol{L}^{\pm} \equiv \frac{1}{2}(\boldsymbol{L} \pm \boldsymbol{M}')\,. \tag{30.41}$$

Using the commutation relations between the L^i's and the $(M')^i$'s [(30.37)] it can be shown that the structure constants with respect to this new basis are given by

$$[(L^{\pm})^i, (L^{\pm})^j\,] = i\hbar\,\varepsilon^{ij}{}_k (L^{\pm})^k\,, \tag{30.42a}$$
$$[(L^+)^i, (L^-)^j\,] = 0\,, \tag{30.42b}$$

as expected from the result (30.40). It is also obvious from (30.32) and (30.34) that

$$[\boldsymbol{L}^+, H] = [\boldsymbol{L}^-, H] = 0\,. \tag{30.43}$$

Thus $\boldsymbol{L}^+$ and $\boldsymbol{L}^-$ are both constants of motion.

Problem 30.5 Verify the commutation relations (30.42) by using those in (30.37).

Each copy of $\mathcal{SO}(3)$ in $\mathcal{SO}(4)$ has a Casimir operator [recall discussion following (20.3)]. These are $\boldsymbol{L}^+ \cdot \boldsymbol{L}^+$ and $\boldsymbol{L}^- \cdot \boldsymbol{L}^-$, with eigenvalues $l^+(l^+ + 1)$ and $l^-(l^- + 1)$, respectively; and so the Casimir operator for $\mathcal{SO}(4)$ is simply

$$C \equiv \boldsymbol{L}^+ \cdot \boldsymbol{L}^+ + \boldsymbol{L}^- \cdot \boldsymbol{L}^- \,, \tag{30.44}$$

with

$$[C, (L^{\pm})^i\,] = 0 \,. \tag{30.45}$$

Within the same irreducible representation of $SO(4)$ [or $\mathcal{SO}(4)$], $\boldsymbol{L}^+ \cdot \boldsymbol{L}^+$ and $\boldsymbol{L}^- \cdot \boldsymbol{L}^-$ have the same eigenvalues, namely,

$$l^+(l^+ + 1) = l^-(l^- + 1) \,. \tag{30.46}$$

We thus have

$$l^+ = l^- = 0, 1/2, 1, 3/2, 2, \dots \,. \tag{30.47}$$

From the fundamental commutation relations

$$[\, x^l, p_m\,] = i\hbar\delta^l_m \,, \quad [\, x^i, x^j\,] = [\, p_i, p_j\,] = 0 \,, \quad [\, f(\boldsymbol{r}), p_k\,] = i\hbar\,\frac{\partial f}{\partial x^k} \,, \tag{30.48}$$

we can verify the following

$$[\, x^l, L^m\,] = i\hbar\,\epsilon^{lm}{}_k x^k \,, \quad [\, p_l, L^m\,] = i\hbar\,\epsilon_l{}^{mk} p_k \,; \tag{30.49}$$

and thus

$$(\boldsymbol{r} \times \boldsymbol{L}) + (\boldsymbol{L} \times \boldsymbol{r}) = 2i\hbar\boldsymbol{r} \,, \quad (\boldsymbol{p} \times \boldsymbol{L} + \boldsymbol{L} \times \boldsymbol{p}) = 2i\hbar\boldsymbol{p} \,. \tag{30.50}$$

Problem 30.6 Verify (30.50) by using the commutation relations in (30.49).

The above two equations can be used to establish the following operator relations:

$$\boldsymbol{r} \cdot \boldsymbol{L} = \boldsymbol{L} \cdot \boldsymbol{r} = 0 \,, \tag{30.51}$$

$$\boldsymbol{p} \cdot \boldsymbol{L} = \boldsymbol{L} \cdot \boldsymbol{p} = 0 \,, \tag{30.52}$$

$$\boldsymbol{M} \cdot \boldsymbol{L} = \boldsymbol{L} \cdot \boldsymbol{M} = 0 \,. \tag{30.53}$$

The classical analogs of these have already been shown previously. The last equation immediately leads to

$$\boldsymbol{L}\cdot\boldsymbol{M}'=\boldsymbol{M}'\cdot\boldsymbol{L}=0\,, \tag{30.54}$$

which implies

$$\boldsymbol{L}^+\cdot\boldsymbol{L}^+-\boldsymbol{L}^-\cdot\boldsymbol{L}^-=0\,, \tag{30.55}$$

corroborating (30.46). Through a somewhat tedious exercise using (30.48) to (30.50), we can also show the following quantum mechanical analog of (30.26):

$$\boldsymbol{M}\cdot\boldsymbol{M}=\frac{2E}{m}(\boldsymbol{L}\cdot\boldsymbol{L}+\hbar^2)+Z^2e^4\,, \tag{30.56}$$

within a certain eigen-subspace of H (where $H=EI$). we then have

$$C=\frac{1}{2}\left(\boldsymbol{L}\cdot\boldsymbol{L}+\boldsymbol{M}'\cdot\boldsymbol{M}'\right)=-\left(\frac{\hbar^2}{2}+\frac{mZ^2e^4}{4E}\right)\,. \tag{30.57}$$

Problem 30.7 | Verify (30.56) using (30.48) to (30.50).

Replacing C by $2l^+(l^++1)$ [cf. (30.46)] one obtains

$$E=-\frac{mZ^2e^4}{2\hbar^2(2l^++1)^2}\,,\quad l^+=0,1/2,1,3/2,2,\ldots \tag{30.58}$$

Now, from (30.41), the orbital angular momentum is given by

$$\boldsymbol{L}=\boldsymbol{L}^++\boldsymbol{L}^-\,, \tag{30.59}$$

with the eigenvalues of $\boldsymbol{L}\cdot\boldsymbol{L}$ being $l(l+1)\hbar^2$, where $l=0,1,2,3,\ldots$. The theory of the addition of angular momentum [or, equivalently, the theory of product representations of irreducible representations of $SO(3)$] [cf. Chapter 26] then implies that l can assume the values

$$l=\begin{cases} l^++l^-=2l^+\,,\\ l^++l^--1\,,\\ \ldots\,,\\ \ldots\,,\\ l^+-l^-=0\,. \end{cases} \tag{30.60}$$

Since $2l^+$ is always integral, we can replace $2l^++1$ by n, a positive integer. Finally we have the following expression for the energy eigenvalues.

$$\boxed{E=-\frac{mZ^2e^4}{2\hbar^2n^2}\,,\quad n=1,2,3,\ldots} \tag{30.61}$$

In the above equation n is identified as the **principal quantum number**. For a given n, (30.60) implies that the orbital angular momentum quantum number l can assume values from 0 to $n-1$, in unit intervals. Also, for a given $l^+ = l^-$, each of $(L^+)^3$ and $(L^-)^3$ has $2l^+ + 1 = n$ values. Thus every energy level is n^2-fold degenerate (ignoring spin). The actual degeneracy is $2n^2$, with spin.

The main outlines of the algebraic approach presented in this chapter for solving the hydrogen-atom problem was first worked out by W. Pauli (Pauli 1926, an English translation can be found in Van der Waerden 1968), although without making explicit use of group theory. This was done slightly before Schrödinger used the Schrödinger wave equation to solve the problem. Pauli's historical landmark paper convinced most physicists that quantum mechanics was correct.

Chapter 31

Stationary Perturbations

Generally the method of approximation known as **stationary perturbation theory** is useful when the full Hamiltonian H is independent of time. It starts with the exact solution of the eigenvalue problem for an unperturbed Hamiltonian H_0 that in some sense differs only slightly from H. H_0 is usually taken to have certain symmetries described by certain groups so that group representation theory can be exploited for the solution of the unperturbed problem. The perturbation Hamiltonian $H - H_0$ can then be interpreted as that part of the Hamiltonian which breaks the symmetries possessed by H_0. The eigenstates and eigenvalues of H are expressed as power series in $H - H_0$, which is usually characterized by a small parameter, denoted below by λ. We will first develop this method for the case of bound states. Problems involving continuous spectra (collision states) will be treated in Chapters 35, 36, and 37.

We set

$$H = H_0 + \lambda V \,, \tag{31.1}$$

where $\lambda \in \mathbb{R}$ is a small, real parameter, and both H_0 and V are time-independent Hermitian operators. Assume that the eigenvalue problem for H_0 is solved, and write

$$H_0 \,|\, \epsilon_i,\, \alpha \,\rangle = \epsilon_i \,|\, \epsilon_i,\, \alpha \,\rangle \,, \tag{31.2}$$

where the quantum numbers α distinguish between eigenvectors belonging to a degenerate eigenvalue. The method is particularly simple when the energy eigenvalue ϵ_i is **non-degenerate**. We will first treat this case.

The spectrum of H is assumed to vary continuously with the parameter λ such that, as $\lambda \to 0$, it coincides with that of H_0. Our problem can be stated as follows. Given a *discrete* eigenvalue ϵ_a of H_0, we wish to calculate the eigenvalue(s) of H which approach ϵ_a as $\lambda \to 0$, and to determine the corresponding eigenstate(s). In this chapter we will also assume that the entire spectrum of H_0 is discrete (so as to simplify the development), although the results remain valid even if part of that spectrum is continuous (as qualitatively illustrated in Fig. 31.1), as long as ϵ_a belongs to the discrete part.

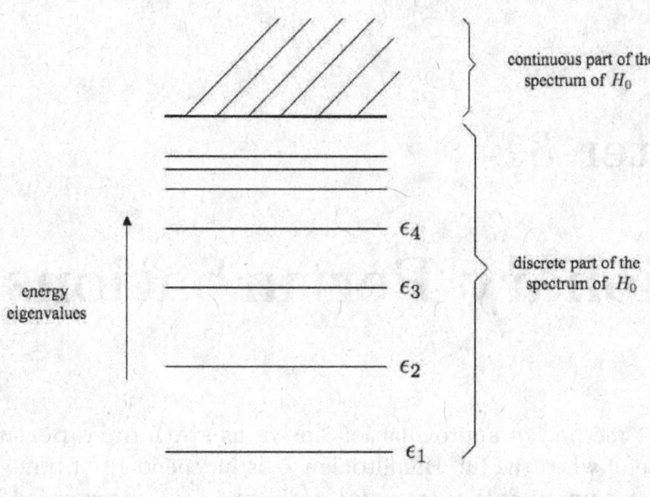

Fig. 31.1

Suppose E is an eigenvalue of H such that

$$\lim_{\lambda \to 0} E = \epsilon_a \,. \tag{31.3}$$

It is non-degenerate if ϵ_a is. The corresponding eigenvector $|\,\psi\,\rangle$, which statisfies

$$H\,|\,\psi\,\rangle = E\,|\,\psi\,\rangle \,, \tag{31.4}$$

will be normalized in the somewhat unusual way as follows:

$$\langle\,0\,|\,\psi\,\rangle = \langle\,0\,|\,0\,\rangle = 1 \,, \tag{31.5}$$

where we have written $|\,\epsilon_a\,\rangle$ as $|\,0\,\rangle$. This convention is adopted so that

$$\lim_{\lambda \to 0} \,|\,\psi\,\rangle = |\,0\,\rangle \,. \tag{31.6}$$

We now expand E and $|\,\psi\,\rangle$ as power series in λ:

$$E = \epsilon_a + \lambda\varepsilon_1 + \lambda^2\varepsilon_2 + \cdots + \lambda^n\varepsilon_n + \ldots \,, \tag{31.7}$$

$$|\,\psi\,\rangle = |\,0\,\rangle + \lambda\,|\,1\,\rangle + \lambda^2\,|\,2\,\rangle + \cdots + \lambda^n\,|\,n\,\rangle + \ldots \tag{31.8}$$

Plugging (31.1), (31.7) and (31.8) in (31.4), we have

$$\begin{aligned} &(H_0 + \lambda V)(\,|\,0\,\rangle + \lambda\,|\,1\,\rangle + \lambda^2\,|\,2\,\rangle + \cdots + \lambda^n\,|\,n\,\rangle + \ldots) \\ &= (\epsilon_a + \lambda\,\varepsilon_1 + \lambda^2\,\varepsilon_2 + \cdots + \lambda^n\,\varepsilon_n + \ldots)\times \\ &\quad (\,|\,0\,\rangle + \lambda\,|\,1\,\rangle + \lambda^2\,|\,2\,\rangle + \cdots + \lambda^n\,|\,n\,\rangle + \ldots)\,. \end{aligned} \tag{31.9}$$

Equating coefficients of each of the first few powers of λ we obtain

$$H_0\,|\,0\,\rangle = \epsilon_a\,|\,0\,\rangle\,, \tag{31.10a}$$

$$(H_0 - \epsilon_a)\,|\,1\,\rangle + (V - \varepsilon_1)\,|\,0\,\rangle = 0\,, \tag{31.10b}$$

$$(H_0 - \epsilon_a)\,|\,2\,\rangle + (V - \varepsilon_1)\,|\,1\,\rangle - \varepsilon_2\,|\,0\,\rangle = 0\,. \tag{31.10c}$$

In general,

$$(H_0 - \epsilon_a)\,|\,n\,\rangle + (V - \varepsilon_1)\,|\,n-1\,\rangle - \varepsilon_2\,|\,n-2\,\rangle - \cdots - \varepsilon_n\,|\,0\,\rangle = 0\,. \tag{31.11}$$

We note that the normalization condition (31.5) implies

$$\langle\,0\,|\,1\,\rangle = \langle\,0\,|\,2\,\rangle = \cdots = \langle\,0\,|\,n\,\rangle = \cdots = 0\,. \tag{31.12}$$

Projecting (31.10b) onto the subspace spanned by $|\,0\,\rangle$ [that is, forming the inner product of (31.10b) with $\langle\,0\,|$] we have the first order correction to the energy:

$$\boxed{\varepsilon_1 = \langle\,0\,|\,V\,|\,0\,\rangle}\,. \tag{31.13}$$

Similarly, the second-order correction is given by projecting (31.10c) onto the subspace spanned by $|\,0\,\rangle$:

$$\boxed{\varepsilon_2 = \langle\,0\,|\,V\,|\,1\,\rangle}\,. \tag{31.14}$$

In general, the n-th order correction is given by projecting (31.11) onto the subspace spanned by $|\,0\,\rangle$:

$$\boxed{\varepsilon_n = \langle\,0\,|\,V\,|\,n-1\,\rangle}\,. \tag{31.15}$$

We assume that the eigenstates of H_0 form a complete set and that $|\,\epsilon_a\,\rangle \equiv |\,0\,\rangle$ is non-degenerate. Hence we have the completeness relation [cf. (7.11)]

$$|\,0\,\rangle\langle\,0\,| + \sum_{b\neq a}\sum_{\alpha}|\,\epsilon_b,\,\alpha\,\rangle\langle\,\epsilon_b,\,\alpha\,| = 1\,. \tag{31.16}$$

Define the projection operator onto the orthogonal complement of the subspace spanned by $|\,0\,\rangle$:

$$Q_0 \equiv 1 - |\,0\,\rangle\langle\,0\,| = \sum_{b\neq a}\sum_{\alpha}|\,\epsilon_b,\,\alpha\,\rangle\langle\,\epsilon_b,\,\alpha\,|\,. \tag{31.17}$$

Then, since $\langle\,0\,|\,n\,\rangle = 0$ [(34.12)],

$$|\,n\,\rangle = \sum_{b\neq a}\sum_{\alpha}|\,\epsilon_b,\,\alpha\,\rangle\langle\,\epsilon_b,\,\alpha\,|\,n\,\rangle = Q_0\,|\,n\,\rangle\,. \tag{31.18}$$

Taking the inner product of (31.11) with $\langle\, \epsilon_b,\, \alpha \,|$, we obtain

$$\begin{aligned}&\langle\, \epsilon_b,\, \alpha \,|\, H_0 - \epsilon_a \,|\, n \,\rangle = (\epsilon_b - \epsilon_a)\,\langle\, \epsilon_b,\, \alpha \,|\, n \,\rangle \\ &= -\langle\, \epsilon_b,\, \alpha \,|\, V - \varepsilon_1 \,|\, n-1 \,\rangle + \varepsilon_2\,\langle\, \epsilon_b,\, \alpha \,|\, n-2 \,\rangle + \ldots\ldots \\ &\quad + \varepsilon_{n-1}\,\langle\, \epsilon_b,\, \alpha \,|\, 1 \,\rangle + \varepsilon_n\,\langle\, \epsilon_b,\, \alpha \,|\, 0 \,\rangle\,.\end{aligned} \tag{31.19}$$

The last term on the right-hand side vanishes since $\langle\, \epsilon_b,\, \alpha \,|\, 0 \,\rangle = 0$ (cf. Theorem 8.3). Hence

$$\begin{aligned}\langle\, \epsilon_b,\, \alpha \,|\, n \,\rangle = \frac{1}{\epsilon_a - \epsilon_b}\,\{\,&\langle\, \epsilon_b,\, \alpha \,|\, V - \varepsilon_1 \,|\, n-1 \,\rangle - \varepsilon_2\,\langle\, \epsilon_b,\, \alpha \,|\, n-2 \,\rangle - \ldots\ldots \\ &- \varepsilon_{n-1}\,\langle\, \epsilon_b,\, \alpha \,|\, 1 \,\rangle\,\}\,,\end{aligned} \tag{31.20}$$

and, from (31.18),

$$\begin{aligned}|\, n \,\rangle = &\left(\sum_{b\neq a}\sum_{\alpha} \frac{|\, \epsilon_b,\, \alpha \,\rangle\langle\, \epsilon_b,\, \alpha \,|}{\epsilon_a - \epsilon_b}\right) \\ &\{\,(V - \varepsilon_1)\,|\, n-1 \,\rangle - \varepsilon_2\,|\, n-2 \,\rangle - \ldots\ldots - \varepsilon_{n-1}\,|\, 1 \,\rangle\,\}\,.\end{aligned} \tag{31.21}$$

Using the completeness relation (31.16), the operator (within the parentheses) in the above equation can be rewritten as

$$\begin{aligned}\sum_{b\neq a}\sum_{\alpha} \frac{|\, \epsilon_b,\, \alpha \,\rangle\langle\, \epsilon_b,\, \alpha \,|}{\epsilon_a - \epsilon_b} &= \left(\,|\, 0 \,\rangle\langle\, 0 \,| + \sum_{a\neq b',\,\beta} \,|\, \epsilon_{b'},\, \beta \,\rangle\langle\, \epsilon_{b'},\, \beta \,|\,\right) \\ &\qquad \sum_{b\neq a}\sum_{\alpha} \frac{|\, \epsilon_b,\, \alpha \,\rangle\langle\, \epsilon_b,\, \alpha \,|}{\epsilon_a - \epsilon_b} \\ &= \sum_{b'\neq a}\sum_{\beta} |\, \epsilon_{b'},\, \beta \,\rangle\langle\, \epsilon_{b'},\, \beta \,|\, \frac{1}{\epsilon_a - H_0} \sum_{b\neq a}\sum_{\alpha} |\, \epsilon_b,\, \alpha \,\rangle\langle\, \epsilon_b,\, \alpha \,| \\ &= Q_0\,\frac{1}{\epsilon_a - H_0}\,Q_0\,.\end{aligned} \tag{31.22}$$

Thus (31.21) can be rewritten as

$$\boxed{|\, n \,\rangle = \left(Q_0 \frac{1}{\epsilon_a - H_0} Q_0\right)\{(V - \varepsilon_1)\,|\, n-1 \,\rangle - \varepsilon_2\,|\, n-2 \,\rangle - \ldots - \varepsilon_{n-1}\,|\, 1 \,\rangle\}}\,. \tag{31.23}$$

Equations (31.7), (31.15) and (31.23) give the complete perturbation solution for the energy eigenvalues of the non-degenerate problem.

Equation (31.23) immediately implies

$$|\, 1 \,\rangle = Q_0\,\frac{1}{\epsilon_a - H_0}\,Q_0(V - \varepsilon_1)\,|\, 0 \,\rangle\,; \tag{31.24}$$

or, since $Q_0\,|\,0\,\rangle = 0$,

$$|\,1\,\rangle = Q_0\,\frac{1}{\epsilon_a - H_0}\,Q_0V|\,0\,\rangle\,. \tag{31.25}$$

Thus, by (31.8), the exact eigenstate can be written as

$$\boxed{|\,\psi\,\rangle = \left(1 + \lambda Q_0\,\frac{1}{\epsilon_a - H_0}\,Q_0V\right)|\,0\,\rangle + O(\lambda^2)}\quad. \tag{31.26}$$

The norm of $|\,\psi\,\rangle$ is given by

$$\langle\,\psi\,|\,\psi\,\rangle = 1 + \lambda^2\langle\,1\,|\,1\,\rangle + \lambda^4\langle\,2\,|\,2\,\rangle + \dots \quad. \tag{31.27}$$

The components of $|\,1\,\rangle$ along $|\,\epsilon_b,\alpha\,\rangle$, $b \neq a$, are, from (31.22) and (31.25),

$$\begin{aligned}\lambda\,\langle\,\epsilon_b,\alpha\,|\,1\,\rangle &= \langle\,\epsilon_b,\alpha\,|\,Q_0\,\frac{1}{\epsilon_a - H_0}\,Q_0(\lambda V)\,|\,0\,\rangle \\ &= \frac{\langle\,\epsilon_b,\alpha\,|\,(\lambda V)\,|\,0\,\rangle}{\epsilon_a - \epsilon_b}\,,\quad \epsilon_b \neq \epsilon_a\,.\end{aligned} \tag{31.28}$$

The smallness of these quantities gives a measure of how rapidly the perturbation series converges.

We will now consider explicitly higher-order corrections than the first. From (31.14) and (31.25) the second-order correction to the energy eigenvalue is

$$\boxed{\varepsilon_2 = \langle\,0\,|\,VQ_0\,\frac{1}{\epsilon_a - H_0}\,Q_0V\,|\,0\,\rangle}\quad; \tag{31.29}$$

while from (31.23) and (31.24), the second-order correction to the energy eigenstate is

$$\begin{aligned}|\,2\,\rangle &= Q_0\,\frac{1}{\epsilon_a - H_0}\,Q_0(V - \varepsilon_1)\,|\,1\,\rangle \\ &= Q_0\,\frac{1}{\epsilon_a - H_0}\,Q_0(V - \varepsilon_1)Q_0\,\frac{1}{\epsilon_a - H_0}\,Q_0V\,|\,0\,\rangle\,,\end{aligned} \tag{31.30}$$

or,

$$\begin{aligned}|\,2\,\rangle = \Big[\,& Q_0\,\frac{1}{\epsilon_a - H_0}\,Q_0VQ_0\,\frac{1}{\epsilon_a - H_0}\,Q_0 \\ &- Q_0\,\frac{1}{\epsilon_a - H_0}\,Q_0\,\frac{1}{\epsilon_a - H_0}\,Q_0V\,|\,0\,\rangle\langle\,0\,|\,\Big]\,V\,|\,0\,\rangle\,,\end{aligned} \tag{31.31}$$

where, in the second term in the right-hand side, we have used the projection operator property of Q_0, namely, $Q_0Q_0 = Q_0$.

Higher order corrections can be similarly and straightforwardly written down, but the expressions are much more involved. However, the n-th order corrections will simplify considerably in the special case that $\varepsilon_1 = \varepsilon_2 = \cdots = \varepsilon_{n-1} = 0$. In this case, (21.23) immediately implies

$$\begin{aligned} |\,n\,\rangle &= Q_0\,\frac{1}{\epsilon_a - H_0}\,Q_0 V\,|\,n-1\,\rangle \\ &= \left(Q_0\,\frac{1}{\epsilon_a - H_0}\,Q_0 V\right)^2 |\,n-2\,\rangle \\ &= \left(Q_0\,\frac{1}{\epsilon_a - H_0}\,Q_0 V\right)^n |\,0\,\rangle\,. \end{aligned} \tag{31.32}$$

Then (31.15) yields

$$\varepsilon_n = \langle\,0\,|\,V\left(Q_0\,\frac{1}{\epsilon_a - H_0}\,Q_0 V\right)^{n-1} |\,0\,\rangle\,; \tag{31.33}$$

and the vanishing of the perturbation energies to the $(n-1)$-th order is equivalent to the following result:

$$\begin{aligned} \langle\,0\,|\,V\,|\,0\,\rangle &= \langle\,0\,|\,V\left(Q_0\,\frac{1}{\epsilon_a - H_0}\,Q_0 V\right)|\,0\,\rangle \\ &= \dots\;\dots \\ &= \langle\,0\,|\,V\left(Q_0\,\frac{1}{\epsilon_a - H_0}\,Q_0 V\right)^{n-2} |\,0\,\rangle = 0\,. \end{aligned} \tag{31.34}$$

Returning to (31.29) the second-order correction to the energy eigenvalue, ε_2, can be written in terms of matrix elements of V when the expression (31.17) for Q_0 is used:

$$\begin{aligned} \varepsilon_2 &= \langle\,0\,|\,VQ_0\,\frac{1}{\epsilon_a - H_0}\,Q_0 V\,|\,0\,\rangle \\ &= \sum_{b\neq a,b'\neq a}\,\sum_{\alpha,\alpha'} \langle\,0\,|\,V\,|\,\epsilon_b,\alpha\,\rangle\,\langle\,\epsilon_b,\alpha\,|\,\frac{1}{\epsilon_a - H_0}\,|\,\epsilon_{b'},\alpha'\,\rangle\langle\,\epsilon_{b'},\alpha'\,|\,V\,|\,0\,\rangle\,. \end{aligned} \tag{31.35}$$

Noting that

$$\langle\,\epsilon_b,\alpha\,|\,\frac{1}{\epsilon_a - H_0}\,|\,\epsilon_{b'},\alpha'\,\rangle = \delta_{bb'}\delta_{\alpha\alpha'}\,\frac{1}{\epsilon_a - \epsilon_b}\,, \tag{31.36}$$

we have

$$\boxed{\varepsilon_2 = \sum\nolimits_{b\neq a} \frac{\sum_\alpha |\,\langle\,0\,|\,V\,|\,\epsilon_b,\alpha\,\rangle\,|^2}{\epsilon_a - \epsilon_b}}\,. \tag{31.37}$$

While the first-order correction in energy [(31.13)] involves the calculation of only one matrix element, the above expression for the second-order correction is in general an infinite series.

Let us now consider the degenerate case, when the energy eigenvalue ϵ_a of the unperturbed Hamiltonian H_0 is g_a-fold degenerate. Let $\mathcal{E}_a$ be the subspace (of the Hilbert space $\mathcal{H}$ describing the physical system under study) of eigenvectors of H_0 with the eigenvalue ϵ_a, and let P_0 be the projection operator onto $\mathcal{E}_a$. More than one eigenvalue of the total Hamiltonian H will now tend to ϵ_a as $\lambda \to 0$. Let these be $E_1, \ldots, E_n$; their orders of degeneracy be $g_1, \ldots, g_n$; and their respective subspaces be $\mathcal{E}_1, \ldots, \mathcal{E}_n$. We thus have

$$g_1 + g_2 + \cdots + g_n = g_a \,, \tag{31.38}$$

$$\lim_{\lambda \to 0} \mathcal{E}_1 \oplus \mathcal{E}_2 \oplus \cdots \oplus \mathcal{E}_n = \mathcal{E}_a \,. \tag{31.39}$$

If P is the projection operator onto $\mathcal{E}_1 \oplus \cdots \oplus \mathcal{E}_n$, then we also have

$$\lim_{\lambda \to 0} P = P_0 = \sum_\alpha |\, \epsilon_a, \alpha \,\rangle\langle\, \epsilon_a, \alpha \,| \,. \tag{31.40}$$

Suppose E is one of the $E_1, \ldots, E_n$, and $|\, \psi \,\rangle$ is an eigenvector corresponding to E. In the limit $\lambda \to 0$, $|\, \psi \,\rangle$ tends to a certain $|\, 0 \,\rangle \in \mathcal{E}_a$. Expansions (31.7) and (31.8) for E and $|\, \psi \,\rangle$ as power series in λ are still assumed to be valid, as is the normalization condition (31.5). Equations (31.10) and (31.11) will then also determine the perturbation corrections to the energy eigenvalues and state vectors in the degenerate case.

First, since $|\, 0 \,\rangle \in \mathcal{E}_a$,

$$P_0 \,|\, 0 \,\rangle = |\, 0 \,\rangle \,. \tag{31.41}$$

Next, we project (31.10b) onto $\mathcal{E}_a$, to obtain

$$\sum_\alpha |\, \epsilon_a, \alpha \,\rangle\langle\, \epsilon_a, \alpha \,|\, H_0 - \epsilon_a \,|\, 1 \,\rangle + P_0(V - \varepsilon_1) \,|\, 0 \,\rangle = 0 \,. \tag{31.42}$$

Since $\langle\, \epsilon_a, \alpha \,|\, H_0 - \epsilon_a \,|\, 1 \,\rangle = (\epsilon_a - \epsilon_a) \langle\, \epsilon_a, \alpha \,|\, 1 \,\rangle = 0$, the above equation implies

$$P_0(V - \varepsilon_1) \,|\, 0 \,\rangle = 0 \,. \tag{31.43}$$

The complementary projection operator to P_0, Q_0, is given by

$$\begin{aligned} Q_0 = 1 - P_0 &= 1 - \sum_\alpha |\, \epsilon_a, \alpha \,\rangle\langle\, \epsilon_a, \alpha \,| \\ &= \sum_{b \neq a} \sum_\beta |\, \epsilon_b, \beta \,\rangle\langle\, \epsilon_b, \beta \,| \,. \end{aligned} \tag{31.44}$$

Application of Q_0 to (31.10b) then yields, since $Q_0 \,|\, 0 \,\rangle = 0$,

$$Q_0(H_0 - \epsilon_a) \,|\, 1 \,\rangle + Q_0 V \,|\, 0 \,\rangle = 0 \,. \tag{31.45}$$

Also, since

$$[\, Q_0, H_0 \,] = 0 \,, \tag{31.46}$$

Eq. (31.45) implies

$$Q_0 \,|\, 1 \,\rangle = \frac{1}{\epsilon_a - H_0} Q_0 V \,|\, 0 \,\rangle \,. \tag{31.47}$$

Application of Q_0 again on both sides yields

$$Q_0 \,|\, 1 \,\rangle = Q_0 \frac{1}{\epsilon_a - H_0} Q_0 V \,|\, 0 \,\rangle \,, \tag{31.48}$$

which should be compared with (31.25).

Equation (31.43) is an eigenvalue equation in $\mathcal{E}_a$ and provides the solution for ε_1. Indeed, written as

$$P_0 V \,|\, 0 \,\rangle = \varepsilon_1 P_0 \,|\, 0 \,\rangle \,, \tag{31.49}$$

and remembering that $P_0 \,|\, 0 \,\rangle = |\, 0 \,\rangle$, it is equivalent to

$$P_0 V P_0 \,|\, 0 \,\rangle = \varepsilon_1 \,|\, 0 \,\rangle \,. \tag{31.50}$$

Thus $|\, 0 \,\rangle$ is an eigenvector of $P_0 V P_0$ with eigenvalue ε_1. Projecting (31.50) onto a particular $|\, \epsilon_a, \alpha \,\rangle$ we have, since $\langle\, \epsilon_a, \alpha \,|\, P_0 = \langle\, \epsilon_a, \alpha \,|$,

$$\langle\, \epsilon_a, \alpha \,|\, V P_0 \,|\, 0 \,\rangle = \varepsilon_1 \,\langle\, \epsilon_a, \alpha \,|\, 0 \,\rangle \,, \tag{31.51}$$

or,

$$\boxed{\textstyle\sum_{\alpha'} \langle\, \epsilon_a, \alpha \,|\, V \,|\, \epsilon_a, \alpha' \,\rangle \langle\, \epsilon_a, \alpha' \,|\, 0 \,\rangle = \varepsilon_1 \,\langle\, \epsilon_a, \alpha \,|\, 0 \,\rangle} \quad . \tag{31.52}$$

This equation indicates that ε_1 is obtained by diagonalizing the $g_a \times g_a$ matrix with elements $\langle\, \epsilon_a, \alpha \,|\, V \,|\, \epsilon_q, \alpha' \,\rangle$, that is, ε_1 are the eigenvalues of this matrix. These eigenvalues may be either non-degenerate or degenerate. If all the eigenvalues are distinct, they are all non-degenerate. The degeneracy of ϵ_a is then said to be completely removed (or lifted). For a particular non-degenerate ε_1, the corresponding zeroth-order eigenstate $|\, 0 \,\rangle$ is determined up to a normalization constant by (31.52). Equation (31.48) gives the projection of the first-order correction for $|\, \psi \,\rangle$ onto the complementary subspace of $\mathcal{E}_a$. Its projection onto $\mathcal{E}_a$ remains undetermined, except for the normalization condition (31.5). If ε_1 is g_1-fold degenerate, then (31.50) [or (31.52)] only shows that $|\, 0 \,\rangle$ belongs to the corresponding g_1-dimensional subspace $\mathcal{E}^{(1)}$. In this case, a more precise determination of $|\, 0 \,\rangle$ depends on higher-order calculations.

Let $P^{(1)}$ be the projection operator onto $\mathcal{E}^{(1)}$, that is, the eigen-subspace of $P_0 V P_0$ corresponding to the eigenvalue ε_1. Clearly, $P^{(1)}$ is a subspace of $\mathcal{E}_a$. Thus we can write

$$P_0 = P^{(1)} + P' \,, \tag{31.53}$$

$$P^{(1)} + P' + Q_0 = 1 \,. \tag{31.54}$$

Then,

$$\begin{aligned} P^{(1)} H_0 &= P^{(1)}(P_0 + Q_0) H_0 = P^{(1)} P_0 H_0 + P^{(1)} Q_0 H_0 \\ &= P^{(1)} P_0 H_0 = P^{(1)} \sum_{\alpha} |\, \epsilon_a, \alpha \,\rangle \langle\, \epsilon_a, \alpha \,|\, H_0 \\ &= \epsilon_a P^{(1)} P_0 = \epsilon_a P^{(1)} (P^{(1)} + P') \\ &= \epsilon_a P^{(1)} \,, \end{aligned} \tag{31.55}$$

and

$$\begin{aligned} P^{(1)}V &= P^{(1)}V(P^{(1)} + P' + Q_0) \\ &= \varepsilon_1 P^{(1)} + P^{(1)}VQ_0 \,. \end{aligned} \tag{31.56}$$

The last equality in the above equation can be demonstrated as follows. We have

$$P^{(1)} = \sum_{\gamma=1}^{g_1} |\, \varepsilon_1, \gamma \,\rangle\langle\, \varepsilon_1, \gamma \,| \,, \tag{31.57}$$

where

$$V \,|\, \varepsilon_1, \gamma \,\rangle = \varepsilon_1 \,|\, \varepsilon_1, \gamma \,\rangle \,, \tag{31.58}$$

and all the $|\, \varepsilon_1, \gamma \,\rangle$ can be assumed to be orthonormalized. Thus

$$\begin{aligned} P^{(1)}VP^{(1)} &= \sum_{\gamma,\gamma'} |\, \varepsilon_1, \gamma \,\rangle\langle\, \varepsilon_1, \gamma \,|\, V \,|\, \varepsilon_1, \gamma' \,\rangle\langle\, \varepsilon_1, \gamma' \,| \\ &= \sum_{\gamma,\gamma'} |\, \varepsilon_1, \gamma \,\rangle\, \varepsilon_1 \delta_{\gamma,\gamma'} \langle\, \varepsilon_1, \gamma' \,| \\ &= \varepsilon_1 P^{(1)} \,, \end{aligned} \tag{31.59}$$

and

$$P^{(1)}VP' = \sum_{\gamma} |\, \varepsilon_1, \gamma \,\rangle\langle\, \varepsilon_1, \gamma \,|\, VP' = \varepsilon_1 P^{(1)} P' = 0 \,. \tag{31.60}$$

Recalling (31.10c), and projecting that equation onto $\mathcal{E}^{(1)}$ by $P^{(1)}$, we have

$$P^{(1)}(H_0 - \epsilon_a) \,|\, 2 \,\rangle + P^{(1)}(V - \varepsilon_1) \,|\, 1 \,\rangle - \epsilon_2 P^{(1)} \,|\, 0 \,\rangle = 0 \,. \tag{31.61}$$

Equations (31.55) and (31.56) can then be directly applied to this equation to yield

$$P^{(1)}VQ_0 \,|\, 1 \,\rangle - \varepsilon_2 P^{(1)} \,|\, 0 \,\rangle = 0 \,, \tag{31.62}$$

whence, with the aid of (31.48),

$$P^{(1)} \left[\left(VQ_0 \, \frac{1}{\epsilon_a - H_0} \, Q_0 V \right) - \varepsilon_2 \right] |\, 0 \,\rangle = 0 \,. \tag{31.63}$$

Just as (31.50) is an eigenvalue equation in $\mathcal{E}_a$, (31.63) is an eigenvalue equation in $\mathcal{E}^{(1)} \subset \mathcal{E}_a$. Using $P^{(1)} \,|\, 0 \,\rangle = |\, 0 \,\rangle$, the latter can be recast as

$$P^{(1)} \left(VQ_0 \, \frac{1}{\epsilon_a - H_0} \, Q_0 V \right) P^{(1)} \,|\, 0 \,\rangle = \varepsilon_2 \,|\, 0 \,\rangle \,. \tag{31.64}$$

If ε_1 is non-degenerate, that is, $g_1 = 1$, $|\, 0 \,\rangle$ is already determined up to a constant by (31.50) [or (31.52)], and can be rendered unique by requiring it to be normalized. In this case we recover the result (31.29) for ε_2. If $g_1 > 1$, ε_2 is found by the diagonalization of a $g_1 \times g_1$ matrix [as seen by (31.64)]. If the eigenvalues are all distinct, then we say that the degeneracy of ϵ_a is completely removed in the second order. If not, one may proceed to higher orders. However, it sometimes happens that a degeneracy persists to all orders.

Since the existence of degenerate eigenvalues can usually be attributed to symmetries of the Hamiltonian, knowledge of the invariance groups of both H_0 and H can be useful in predicting within what limits the degeneracy of a level ϵ_a can be removed by the perturbation. In the next chapter we will study an interesting application of degenerate perturbation theory to atomic physics.

Problem 31.1 Consider the hydrogen atom problem. Let us assume that we do not know the exact solution to the radial equation, and we want to solve it approximately. As a guess for the ground state wave function, use the Gaussian orbital

$$\psi_{nlm}(\boldsymbol{r}) = \langle\, \boldsymbol{r} \,|\, nlm \,\rangle = N_{nlm}\, r^{n-1} e^{-\alpha r^2}\, Y_l^m(\theta,\varphi)\,,$$

where N_{nlm} is a normalization constant and α is a free parameter to be varied. Using the single orbital $|\,100\,\rangle$ for the ground state obtain an expression for $E(\alpha)$ for the ground state energy, and then vary α so as to minimize the energy. How does this variational energy compare to the exact one? The following integrals will be useful:

$$\int_0^\infty e^{-\alpha r^2}\, dr = \frac{1}{2}\sqrt{\frac{\pi}{\alpha}}\,,$$

$$\int_0^\infty r^{2m} e^{-\alpha r^2}\, dr = \frac{1\cdot 3\cdot 5\cdot\cdots(2m-1)}{2^{m+1}\alpha^m}\sqrt{\frac{\pi}{\alpha}}\,,$$

$$\int_0^\infty r^{2m+1} e^{-\alpha r^2}\, dr = \frac{m!}{2\alpha^{m+1}}\,.$$

Problem 31.2 Consider the hydrogen atom problem again, and the Gaussian orbitals introduced in the last problem. Use the value of α obtained there to find an improved upper bound for the ground state energy by mixing the orbitals $|\,100\,\rangle$ and $|\,200\,\rangle$. This mixing is called **configuration interaction**. Proceed with the following steps:

1. Compute normalization constants.
2. Evaluate matrix elements of the Hamiltonian H.
3. Since $|\,200\,\rangle$ is not orthogonal to $|\,100\,\rangle$, Schmidt orthogonalize to find a suitable basis.
4. Express matrix elements with respect to this new basis.
5. Diagonalize the Hamiltonian matrix.

Problem 31.3 Recall Problem 14.1, which dealt with a one-dimensional simple harmonic oscillator perturbed by a constant force. Use second-order perturbation theory to calculate the changes in energy and compare with the exact results obtained from the earlier problem.

Problem 31.4 Recall Problem 16.4, which dealt with a particle confined in a two-dimensional square box. Suppose now a perturbation $V(x,y) = \alpha xy$ (α = constant) is introduced. Calculate the energy changes of the ground state and the first excited state, to the lowest nonvanishing order, and construct the zeroth order wave functions for the perturbed problem for the first excited state. [Hints: Use degenerate perturbation theory for the first excited state.]

Problem 31.5 Consider a general $n \times n$ matrix A. The eigenvalues λ of the matrix are determined by the n roots of the secular equation

$$\det(A - \lambda \mathbb{I}) = 0 ,$$

where $\mathbb{I}$ is the $n \times n$ identity matrix. Obtain an approximation to the k-th eigenvalue by the following steps:

(a) In the diagonal elements of determinant of the secular equation, replace λ by A_{kk}, except for the element $A_{kk} - \lambda$.

(b) Set all the nondiagonal elements equal to zero, except for the A_{ij} when *either* i or j is equal to k (in other words, except for those in either the k-th row or k-th column).

(c) Expand the determinant thus obtained (which will be a first order polynomial in λ) and solve for λ.

Compare the result of this approximate method to that obtained from second order perturbation theory.

Chapter 32

The Fine Structure of Hydrogen: Application of Degenerate Perturbation Theory

The zeroth-order (non-relativistic) Hamiltonian H_0 for an electron in the field of a proton is

$$H_0 = \frac{p^2}{2\mu} - \frac{e^2}{r} \,. \tag{32.1}$$

As given by (30.61), the corresponding unperturbed energy eigenvalues are

$$\epsilon_n = -\frac{\alpha^2 \mu c^2}{2n^2} \,, \quad (n = 1, 2, \dots) \,, \tag{32.2}$$

where

$$\alpha = \frac{e^2}{\hbar c} = \frac{1}{137.04} \tag{32.3}$$

is the **fine structure constant** (a fundamental constant measuring the strength of the electromagnetic interaction), and the reduced mass of the electron-proton system μ equals $0.9995 m_e$ (with m_e being the mass of the electron). The spectrum of H_0 is highly degenerate: the degree of degeneracy for each ϵ_n is $2n^2$.

The exact Hamiltonian of the hydrogen atom contains many corrections to (32.1). Ultimately the problem has to be treated within the framework of **quantum electrodynamics** (QED), which is a quantum field theory. Here we will only consider the most important perturbations to H_0 arising from the effects of relativistic kinematics and the electron spin; the latter, as we shall see, leads to the so-called spin-orbit coupling.

The relativistic kinetic energy is given by

$$\begin{aligned}\sqrt{p^2c^2+\mu^2c^4}-\mu c^2 &= \mu c^2\left(1+\frac{p^2}{\mu^2c^2}\right)^{\frac{1}{2}}-\mu c^2\\ &\approx \mu c^2\left(1+\frac{p^2}{2\mu^2c^2}-\frac{1}{8}\left(\frac{p^2}{\mu^2c^2}\right)^2+\ldots\right)-\mu c^2 \qquad (32.4)\\ &\approx \frac{p^2}{2\mu}-\frac{1}{2}\left(\frac{p^2}{2\mu}\right)^2\frac{1}{\mu c^2}+\ldots\ .\end{aligned}$$

The spin-orbit coupling also arises from relativistic effects. The electron, in its rest frame, "sees" a magnetic field $\boldsymbol{\mathcal{H}}$ arising from the relative motion of the proton given, to lowest order in v/c, by

$$\boldsymbol{\mathcal{H}}=-\frac{1}{c}\,\boldsymbol{v}\times\boldsymbol{\mathcal{E}}\,, \qquad (32.5)$$

where $\boldsymbol{v}$ is the velocity of the electron relative to the proton, and $\boldsymbol{\mathcal{E}}$ is the electric field at the position of the electron in the rest frame of the proton. Equation (32.5) is the result of a Lorentz transformation of field quantities from the proton's rest frame to the electron's rest frame. Since the electron also possesses a spin magnetic moment given by

$$\boldsymbol{\mu}=\frac{e\hbar}{2m_ec}\,\boldsymbol{\sigma}\,, \qquad (32.6)$$

where $\boldsymbol{\sigma}$ is the Pauli spin vector, it interacts with $\boldsymbol{\mathcal{H}}$ with an energy $-\boldsymbol{\mu}\cdot\boldsymbol{\mathcal{H}}$. For the Coulomb potential energy $V=-e^2/r$, we know that

$$\boldsymbol{\mathcal{E}}=-\frac{1}{e}\,\nabla V=-\frac{1}{e}\hat{\boldsymbol{r}}\frac{dV}{dr}\,. \qquad (32.7)$$

Thus

$$\begin{aligned}\boldsymbol{\mathcal{H}} &= \frac{1}{ec}\,\boldsymbol{v}\times\hat{\boldsymbol{r}}\,\frac{dV}{dr}=\frac{1}{ec}\,\boldsymbol{v}\times\boldsymbol{r}\,\frac{1}{r}\frac{dV}{dr}\\ &= -\frac{1}{em_ec}\,(\boldsymbol{r}\times m_e\boldsymbol{v})\,\frac{1}{r}\frac{dV}{dr} \qquad (32.8)\\ &= -\frac{\hbar}{em_ec}\,\boldsymbol{L}\,\frac{1}{r}\frac{dV}{dr}\,,\end{aligned}$$

where $\boldsymbol{L}$ is the orbital angular momentum of the electron. Thus the spin-orbit coupling interaction Hamiltonian is expected to be

$$\frac{1}{2}\left(\frac{\hbar}{m_ec}\right)^2\boldsymbol{L}\cdot\boldsymbol{\sigma}\,\frac{1}{r}\frac{dV}{dr}\,.$$

This formula is actually incorrect! There should be an extra factor of $1/2$ due to the so-called **Thomas precession**, which arises from the fact that the electron rest frame is a rotating frame with respect to the proton (see Jackson 1966, p. 364). Since $\boldsymbol{S} = \boldsymbol{\sigma}/2$ [cf. (22.11)], where we denote by $\boldsymbol{S}$ the spin angular momentum of the electron, we finally have

$$H = H_0 + H' , \tag{32.9}$$

where the perturbation Hamiltonian H' is given by

$$H' = H_{kin} + H_{LS} , \tag{32.10}$$

with [cf. (32.4)]

$$H_{kin} = -\frac{p^4}{8\mu^3 c^2} , \tag{32.11}$$

and the spin-orbit coupling Hamiltonian given by

$$\boxed{H_{LS} = \frac{1}{2}\left(\frac{\hbar}{m_e c}\right)^2 \boldsymbol{L}\cdot\boldsymbol{S}\,\frac{1}{r}\frac{dV}{dr}} \quad . \tag{32.12}$$

A clean derivation of this term will require the (relativistic) Dirac theory of the electron.

We can estimate the relative strengths of H_{kin} and H_{LS} as follows:

$$\frac{H_{kin}}{H_0} \approx \frac{p^4}{\mu^3 c^2}\frac{\mu}{p^2} = \frac{p^2}{\mu^2 c^2} \approx \frac{H_0}{\mu c^2} \approx \frac{\alpha^2 \mu c^2}{\mu c^2} = \alpha^2 , \tag{32.13}$$

where the third approximate equality follows from (32.2). On the other hand,

$$\frac{H_{LS}}{H_0} \approx \left(\frac{\hbar}{m_e c}\right)^2 \frac{e^2}{r^3}\frac{r}{e^2} \approx \left(\frac{\hbar}{m_e c a_0}\right)^2 , \tag{32.14}$$

where a_0 is the **Bohr radius**. Using the fact that

$$a_0 \equiv \frac{\hbar^2}{\mu e^2} \quad (= 0.5292 \times 10^{-8}\, cm) , \tag{32.15}$$

we conclude that

$$\frac{H_{LS}}{H_0} \approx \left(\frac{\hbar}{m_e c}\,\frac{m_e e^2}{\hbar^2}\right)^2 = \left(\frac{e^2}{\hbar c}\right)^2 = \alpha^2 . \tag{32.16}$$

H_{kin} and H_{LS} are therefore of the same order of magnitude, both being approximately $\alpha^2 H_0$, or $5.32 \times 10^{-5} H_0$ [cf. (32.3)]. QED shows that all other corrections to H_0 are smaller by higher powers of α. Since, as will be seen below, H' leads to the fine structure splitting, α is called the fine structure constant.

According to (31.52) we need to diagonalize H' in each degenerate subspace belonging to a definite value ϵ_n [cf. (32.2)]. One can either work with the basis

$\{\,|\,n,l,1/2\,;m_l,m_s\,\rangle\,\}$ or $\{\,|\,n,l,j,m\,\rangle = |\,nl\,\rangle|\,ljm\,\rangle\,\}$, where $l = 0,\ldots,n-1$, $m_l = l, l-1,\ldots,-l$, $m_s = \pm 1/2$, $j = l+1/2,\ldots,|\,l-1/2\,|$, and $m = j, j-1,\ldots,-j$. For $n = 2$, for example, the degenerate subspace corresponding to ϵ_2 is eight-dimensional. H_{kin} is diagonal in both basis sets; but H_{LS} is only so in the basis set $\{\,|\,nljm\,\rangle\,\}$ (for fixed n). Thus, in the words of Sakurai (Sakurai 1985), "one would either have to be a fool or a masochist to choose the basis set $\{\,|\,n,l,1/2\,;m_l,m_s\,\rangle\,\}$".

From (32.11) we have

$$\begin{aligned}
\langle\,nljm\,|\,H_{kin}\,|\,nljm\,\rangle &= \langle\,nljm\,|\,-\frac{1}{2\mu c^2}\left(\frac{p^2}{2\mu}\right)^2\,|\,nljm\,\rangle \\
&= -\frac{1}{2\mu c^2}\,\langle\,nljm\,|\,\left(H_0+\frac{e^2}{r}\right)\left(H_0+\frac{e^2}{r}\right)|\,nljm\,\rangle \\
&= -\frac{1}{2\mu c^2}\left[\epsilon_n^2 + 2e^2\epsilon_n\,\langle\,nl\,|\,\frac{1}{r}\,|\,nl\,\rangle + e^4\langle\,nl\,|\,\frac{1}{r^2}\,|\,nl\,\rangle\right].
\end{aligned} \tag{32.17}$$

Recalling that $\boldsymbol{L}\cdot\boldsymbol{S} = (J^2 - L^2 - S^2)/2$ and $V = -e^2/r$, we have

$$\begin{aligned}
&\langle\,nljm\,|\,H_{LS}\,|\,nljm\,\rangle \\
&= \frac{e^2}{2}\left(\frac{\hbar}{m_e c}\right)^2\,\langle\,nl\,|\,\frac{1}{r^3}\,|\,nl\,\rangle\langle\,ljm\,|\,\boldsymbol{L}\cdot\boldsymbol{S}\,|\,ljm\,\rangle \\
&= \frac{e^2}{2}\left(\frac{\hbar}{m_e c}\right)^2\,\langle\,nl\,|\,\frac{1}{r^3}\,|\,nl\,\rangle\,\frac{1}{2}\,[j(j+1)-l(l+1)-3/4] \\
&= \frac{e^2}{2}\left(\frac{\hbar}{m_e c}\right)^2\,\langle\,nl\,|\,\frac{1}{r^3}\,|\,nl\,\rangle \times \begin{cases} l/2\,, & j = l+1/2\,, \\ -(l+1)/2\,, & j = l-1/2\,. \end{cases}
\end{aligned} \tag{32.18}$$

We will need the following radial integrals for hydrogen wave functions (Condon and Shortley 1970, p. 117):

$$\langle\,nl\,|\,\frac{1}{r}\,|\,nl\,\rangle = \frac{1}{a_0 n^2}\,, \tag{32.19}$$

$$\langle\,nl\,|\,\frac{1}{r^2}\,|\,nl\,\rangle = \frac{1}{a_0^2 n^3(l+1/2)}\,, \tag{32.20}$$

$$\langle\,nl\,|\,\frac{1}{r^3}\,|\,nl\,\rangle = \frac{1}{a_0^3 n^3(l+1)(l+1/2)l}\,, \tag{32.21}$$

where

$$\langle\,nl\,|\,f(r)\,|\,nl\,\rangle = \int_0^\infty dr\,r^2 R_{nl}^2(r) f(r)\,, \tag{32.22}$$

and the radial wave functions as given by

$$R_{nl}(r) = \langle\,r\,|\,nl\,\rangle \tag{32.23}$$

are normalized according to

$$\int_0^\infty dr\,r^2 R_{nl}(r) R_{n'l'}(r) = \delta_{nn'}\delta_{ll'}\,. \tag{32.24}$$

Making use of (32.19) to (32.21) in (32.17) and (32.18) we have

$$\langle\, nljm \,|\, H_{kin} \,|\, nljm \,\rangle = \frac{\epsilon_n \alpha^2}{n^2} \left(\frac{n}{l+1/2} - \frac{3}{4} \right) , \tag{32.25}$$

and

$$\langle\, nljm \,|\, H_{LS} \,|\, nljm \,\rangle = -\frac{\epsilon_n \alpha^2}{n(2l+1)} \times \begin{cases} 1/(l+1)\,, & j = l+1/2\,, \\ -1/l\,, & j = l-1/2\,, \quad l \neq 0 \end{cases} . \tag{32.26}$$

Note that for $l=0$, $\langle\, \boldsymbol{L}\cdot\boldsymbol{S} \,\rangle = 0$, but according to (32.21), $\langle\, 1/r^3 \,\rangle \sim 1/l$ diverges. In this case we use $\langle\, \boldsymbol{L}\cdot\boldsymbol{S} \,\rangle_{j=l+1/2} = l/2$ [cf. (32.18)], which cancels the l factor in the denominator of (32.21), and so the expression for $j = l+1/2$ for $\langle\, H_{LS} \,\rangle$ in (32.26) still applies.

Combining (32.25) and (32.26), we have, for all values of $j\,(= l \pm 1/2)$, the following expression for the fine structure energies of the hydrogen atom to first order:

$$\boxed{E_{nlj} = \epsilon_n \left\{ 1 + \frac{\alpha^2}{n^2} \left(\frac{n}{j+1/2} - \frac{3}{4} \right) \right\}} \; . \tag{32.27}$$

For $n=2$, the (nlj) configurations are $(2,0,1/2)$, $(2,1,3/2)$, and $(2,1,1/2)$. In conventional spectroscopic notation, these are labeled $2s_{1/2}$, $2p_{3/2}$, and $2p_{1/2}$, respectively, where $s,p,d,f,\ldots$ stand for $l-0,1,2,3,\ldots$. The above equations then give

$$E_{2,l,3/2} = -\frac{\alpha^2 \mu c^2}{8} \left(1 + \frac{\alpha^2}{16} \right) , \tag{32.28a}$$

$$E_{2,l,1/2} = -\frac{\alpha^2 \mu c^2}{8} \left(1 + \frac{5\alpha^2}{16} \right) . \tag{32.28b}$$

Figure 32.1 shows the hydrogen fine structure energy level diagram for $n=2$. The numerical values shown in the figure can be calculated by using $\alpha = 1/137.04$, $\alpha^2 \mu c^2/2 = 13.6\, eV$, $1\, eV = 2.418 \times 10^{14}\, sec^{-1}$.

The $2p_{3/2}$ configuration is still 4-fold degenerate. The degeneracy of the $2p_{1/2}$ and $2s_{1/2}$ configurations actually still holds in the relativistic Dirac theory of the hydrogen atom to *all* orders of α^2. In reality, however, these levels are split by **radiative corrections** by the tiny amount (in the radio frequency range) of about $1040\, MHz$ (see Fig. 32.1). This effect, known as the **Lamb shift**, can only be properly accounted for by QED. In practice, the so-called 1040 line is of great importance in radio astronomy.

Problem 32.1 The magnetic moment operator $\boldsymbol{\mu}$ for a nucleon of mass μ_{nc} can be written as

$$\boldsymbol{\mu} = \frac{e}{2\mu_{nc} c} (g_l \boldsymbol{L} + g_s \boldsymbol{S}) ,$$

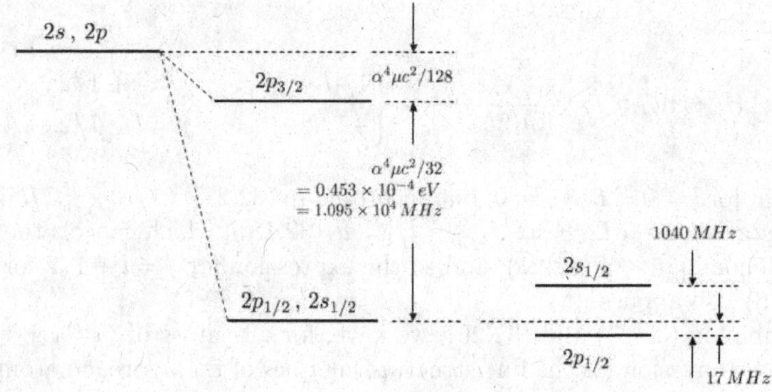

Fig. 32.1

where $g_l = 1$, $g_s = 5.587$ for a proton, and $g_l = 0$, $g_s = -3.826$ for a neutron. Assume that a nucleon is moving in a central field with an additional spin-orbit coupling so that the orbital shells are characterized by the quantum numbers l and $j = l \pm 1/2$. Calculate, for both the proton and the neutron, the magnetic moment as a function of j, distinguishing between the cases $j = l + 1/2$ and $j = l - 1/2$.

Problem 32.2 The Hamiltonian of the positronium atom (a bound electron-positron system) in the $1S$ state and in the presence of a uniform magnetic field B along the z-direction is, to a good approximation, given by

$$H = A\boldsymbol{S}_1 \cdot \boldsymbol{S}_2 + \frac{eB}{\mu c}[(S_1)_z - (S_2)_z] , \quad (\mu = \text{reduced mass}) ,$$

if all higher energy states are neglected. In the above expression, the electron is labeled as particle 1 and the positron as particle 2.

(a) Using the representation in which $\boldsymbol{S}^2 = (\boldsymbol{S}_1 + \boldsymbol{S}_2)^2$ and $S_z = (S_1)_z + (S_2)_z$ are diagonal, calculate the energy eigenvalues (by diagonalization of H) and the corresponding eigenstates. Draw a diagram showing the splitting of the energy levels. [Hints: $\boldsymbol{S}_1 \cdot \boldsymbol{S}_2 = (1/2)[\boldsymbol{S}^2 - (\boldsymbol{S}_1)^2 - (\boldsymbol{S}_2)^2]$ is diagonal in this representation, but H is not.]

(b) Experimentally it is known that for $B = 0$ the frequency of the $1^3S \to 1^1S$ transition is $2.0338 \times 10^{11}\ s^{-1}$, and that the mean lifetimes (against pair-annihilation) of the singlet state and the triplet state are $10^{-10}\ s$ (two-photon decay) and $10^{-7}\ s$ (three-photon decay), respectively. Estimate the magnetic field strength B which will cause the lifetime of the longer-lived $S_z = 0$ state to be reduced from $10^{-7}\,s$ to $10^{-8}\,s$. [Hints: When $B \neq 0$, the longer-lived eigenstate can be written as the following linear combination of triplet and singlet states:

$$|-\rangle = a_- \,|\, S=1, S_z=0\,\rangle + b_- \,|\, S=0, S_z=0\,\rangle \,.$$

Then the corresponding lifetimes are related by

$$\tau_{|-\rangle} = |a_-|^2\,\tau(^3S) + |b_-|^2\,\tau(^1S) \,.]$$

Chapter 33

Time-Dependent Perturbation Theory

Let us consider a Hamiltonian of the form

$$H = H_0 + V(t) , \tag{33.1}$$

where the Schrödinger equation for the unperturbed Hamiltonian H_0,

$$H_0 \,|\, \epsilon_i, \alpha \,\rangle = \epsilon_i \,|\, \epsilon_i, \alpha \,\rangle , \tag{33.2}$$

is assumed to be solved, and the perturbation Hamiltonian $V(t)$ may be explicitly time-dependent. We are mostly interested in the following situation: At $t = 0$, the system is in a particular eigenstate of H_0, say, $|\, \epsilon_i, \alpha \,\rangle$, for particular values of i and α. For $t > 0$, $V(t)$ can then cause transitions to other eigenstates of H_0. The basic problem of time-dependent perturbation theory is to calculate **transition probabilities** as functions of time from the initial state to other states.

Suppose at $t = 0$, the state of the system is given by

$$|\, \psi(0) \,\rangle = \sum_{i,\alpha} c_{i,\alpha}(0) \,|\, \epsilon_i, \alpha \,\rangle . \tag{33.3}$$

We assume all the basis states in the set $\{|\, \epsilon, \alpha \,\rangle\}$ are orthonormalized and relabel the index set (i, α) by the single index n. Thus

$$|\, \psi(0) \,\rangle = \sum_n c_n(0) \,|\, n \,\rangle , \tag{33.4}$$

where the set $\{|\, n \,\rangle , \, n = 1, 2, 3, \dots\}$ span the Hilbert space of the system. We wish to find the quantities $c_n(t)$, $t > 0$, such that

$$|\, \psi(t) \,\rangle = \sum_n c_n(t) \exp\left(-\frac{i\epsilon_n t}{\hbar} \right) |\, n \,\rangle , \tag{33.5}$$

and $|\,\psi(t)\,\rangle$ satisfies the time-dependent Schrödinger equation

$$i\hbar\frac{\partial}{\partial t}\,|\,\psi(t)\,\rangle = (H_0 + V(t))\,|\,\psi(t)\,\rangle\,. \tag{33.6}$$

Note that the exponential factor in (33.5) is exactly the time-evolution factor when $V = 0$ for a state which at $t = 0$ is one of the eigenstates of H_0 with energy eigenvalue ϵ_n. Thus, by introducing this exponential factor, we have relegated the time-dependence of c_n to the effects due solely to the perturbation Hamiltonian $V(t)$. As will be seen later, $c_n(t)$ satisfies a relatively simple differential equation. The probability of finding the system in the state $|\,n\,\rangle$ at time t is $|\,c_n(t)\,|^2 = |\,\langle\,n\,|\,\psi(t)\,\rangle\,|^2$.

To obtain the equation satisfied by $c_n(t)$ we introduce the so-called **interaction picture** (representation) of state vectors and operators in Hilbert space. To distinguish between this picture and the **Schrödinger picture** we will use the subscripts I and S, respectively, in the following. Thus, the interaction picture state vector is defined by

$$|\,\psi(t)\,\rangle_I = \exp\left(\frac{iH_0t}{\hbar}\right)\,|\,\psi(t)\,\rangle_S\;, \tag{33.7}$$

that is, $|\,\psi(t)\,\rangle_I$ is the unitary transformation of $|\,\psi(t)\,\rangle_S$ given by time-evolving the Schrödinger ket backwards (to $t = 0$) with only the unperturbed Hamiltonian H_0. The corresponding unitary transformation relating an operator $O = O_S$ in the two representations is

$$O_I = \exp\left(\frac{iH_0t}{\hbar}\right)\,O_S\,\exp\left(-\frac{iH_0t}{\hbar}\right)\,. \tag{33.8}$$

In particular

$$V_I = \exp\left(\frac{iH_0t}{\hbar}\right)\,V\,\exp\left(-\frac{iH_0t}{\hbar}\right)\,, \tag{33.9}$$

where V (without any subscript) is understood to be the perturbation potential in the Schrödinger picture. At this point, it is important to compare Eqs. (33.7) and (33.9) with the corresponding equations (for time-independent H) for state vectors and operators in the **Heisenberg picture**:

$$|\,\psi(t)\,\rangle_H = \exp\left(\frac{iHt}{\hbar}\right)\,|\,\psi(t)\,\rangle_S = |\,\psi(0)\,\rangle\,, \tag{33.10}$$

$$O_H = \exp\left(\frac{iHt}{\hbar}\right)\,O_S\,\exp\left(-\frac{iHt}{\hbar}\right)\,. \tag{33.11}$$

Taking the time derivative of (33.7) we have

$$i\hbar\,\frac{\partial|\,\psi(t)\,\rangle_I}{\partial t} = -H_0 \exp\left(\frac{iH_0 t}{\hbar}\right)|\,\psi(t)\,\rangle_S + \exp\left(\frac{iH_0 t}{\hbar}\right)(i\hbar)\frac{\partial|\,\psi(t)\,\rangle_S}{\partial t}$$

$$= -H_0 \exp\left(\frac{iH_0 t}{\hbar}\right)|\,\psi(t)\,\rangle_S + \exp\left(\frac{iH_0 t}{\hbar}\right)(H_0 + V)\,|\,\psi(t)\,\rangle_S$$

$$= \underbrace{\exp\left(\frac{iH_0 t}{\hbar}\right) V \exp\left(-\frac{iH_0 t}{\hbar}\right)}_{V_I}\;\underbrace{\exp\left(\frac{iH_0 t}{\hbar}\right)|\,\psi(t)\,\rangle_S}_{|\,\psi(t)\,\rangle_I}\;. \tag{33.12}$$

Thus

$$\boxed{i\hbar\,\frac{\partial|\,\psi(t)\,\rangle_I}{\partial t} = V_I\,|\,\psi(t)\,\rangle_I}\;, \tag{33.13}$$

which resembles the Schrödinger equation, but with H replaced by V_I. This equation shows explicitly that $|\,\psi(t)\,\rangle_I$ would be time-independent if $V = 0$. Differentiating (33.8) we can similarly show that

$$\boxed{i\hbar\,\frac{dO_I}{dt} = [\,O_I\,,\,H_0\,]}\;. \tag{33.14}$$

Problem 33.1 Prove Eq. (33.14).

The time evolution of the state vectors and operators in the three pictures discussed so far (Schrödinger, Heisenberg, and interaction) are summarized in the following table.

	Schrödinger picture	Heisenberg picture	interaction picture
state vector	time evolution determined by H	constant in time	time-evolution determined by V_I
operator	constant in time	time evolution determined by H	time evolution determined by H_0

Using (33.5) for $|\,\psi(t)\,\rangle_S$ in (33.7), we have

$$|\,\psi(t)\,\rangle_I = \sum_n c_n(t)\,|\,n\,\rangle\;. \tag{33.15}$$

Now multiply (33.13) on the left by $\langle n |$ to obtain

$$i\hbar \frac{\partial}{\partial t} \langle n | \psi(t) \rangle_I = \sum_m \langle n | V_I | m \rangle \langle m | \psi(t) \rangle_I \,, \tag{33.16}$$

where we have used the completeness condition

$$\sum_m | m \rangle \langle m | = 1 \,. \tag{33.17}$$

The matrix element appearing on the RHS of (33.16) can be written as

$$\begin{aligned} \langle n | V_I | m \rangle &= \langle n | \exp\left(\frac{iH_0 t}{\hbar}\right) V(t) \exp\left(\frac{-iH_0 t}{\hbar}\right) | m \rangle \\ &= \exp\left(\frac{i(\epsilon_n - \epsilon_m)t}{\hbar}\right) \langle n | V(t) | m \rangle = e^{i\omega_{nm} t} V_{nm}(t) \,, \end{aligned} \tag{33.18}$$

where

$$\omega_{nm} \equiv \frac{\epsilon_n - \epsilon_m}{\hbar} \,, \tag{33.19}$$

and

$$V_{nm}(t) \equiv \langle n | V(t) | m \rangle \,. \tag{33.20}$$

Since (33.15) implies

$$c_n(t) = \langle n | \psi(t) \rangle_I \,. \tag{33.21}$$

Equation (33.16) yields the following set of differential equations satisfied by the $c_n(t)$:

$$\boxed{i\hbar \frac{dc_n(t)}{dt} = \sum_m V_{nm}(t) e^{i\omega_{nm} t} c_m(t)} \,, \tag{33.22}$$

which is in general an infinite set of coupled first-order equations. This set can be displayed in the matrix form:

$$i\hbar \begin{pmatrix} \dot{c}_1 \\ \dot{c}_2 \\ \dot{c}_3 \\ \vdots \\ \vdots \end{pmatrix} = \begin{pmatrix} V_{11} & V_{12}\, e^{i\omega_{12} t} & \cdots & \cdots & \cdots \\ V_{21}\, e^{i\omega_{21} t} & V_{22} & \cdots & \cdots & \cdots \\ \cdots & \cdots & V_{33} & \cdots & \cdots \\ \cdots & \cdots & \cdots & \cdots & \cdots \\ \cdots & \cdots & \cdots & \cdots & \cdots \end{pmatrix} \begin{pmatrix} c_1 \\ c_2 \\ c_3 \\ \vdots \\ \vdots \end{pmatrix} \,. \tag{33.23}$$

Approximate solutions to $c_n(t)$ can be obtained by an iterative procedure, which was in fact used by Dirac in his development of time-dependent perturbation theory. Here, however, we will work directly with the time-evolution operator in the interaction picture, $U_I(t, t_0)$, and then relate $c_n(t)$ to matrix elements of this unitary operator. This powerful formalism has the advantage that it generalizes to quantum field theory and the applications thereof to many-body problems. The time-evolution operator is defined by the following condition:

$$|\,\psi(t)\,\rangle_I = U_I(t,t_0)\,|\,\psi(t_0)\,\rangle_I \;. \tag{33.24}$$

Equation (33.13) then implies that

$$i\hbar\frac{dU_I(t,t_0)}{dt} = V_I(t)\,U_I(t,t_0)\;, \tag{33.25}$$

with the initial condition

$$U_I(t_0,t_0) = 1\;. \tag{33.26}$$

The operator differential equation (33.25), with the initial condition (33.26), is equivalent to the following integral equation:

$$U_I(t,t_0) = 1 - \frac{i}{\hbar}\int_{t_0}^{t} V_I(t')U_I(t',t_0)\,dt'\;. \tag{33.27}$$

A formal solution to the above equation can be obtained by iteration:

$$\begin{aligned}
U_I(t,t_0) &= 1 - \frac{i}{\hbar}\int_{t_0}^{t} V_I(t')\left[1-\frac{i}{\hbar}\int_{t_0}^{t'} V_I(t'')U_I(t'',t_0)\,dt''\right]dt' \\
&= 1 - \frac{i}{\hbar}\int_{t_0}^{t} V_I(t')dt' + \left(-\frac{i}{\hbar}\right)^2\int_{t_0}^{t}dt'\int_{t_0}^{t'}dt''\,V_I(t')V_I(t'') \\
&\quad + \dots \\
&\quad + \left(-\frac{i}{\hbar}\right)^n\int_{t_0}^{t}dt'\int_{t_0}^{t'}dt''\dots\int_{t_0}^{t^{(n-1)}}dt^{(n)}\,V_I(t')V_I(t'')\dots V_I(t^{(n)}) \\
&\quad + \dots\;.
\end{aligned} \tag{33.28}$$

This infinite series is known as the **Dyson series**. Note that the operators V_I with different time-arguments in the integrand do not in general commute, and thus the order in which they appear has to be preserved.

According to the definition (33.24), we have

$$|\,\psi(t)\,\rangle_I = U_I(t,t_0)\,|\,i\,\rangle\;, \tag{33.29}$$

if, at $t=t_0$, the initial state is an eigenstate $|\,i\,\rangle$ of H_0. Thus

$$|\,\psi(t)\,\rangle_I = \sum_n |\,n\,\rangle\langle\,n\,|\,U_I(t,t_0)\,|\,i\,\rangle\;, \tag{33.30}$$

where $\{|\,n\,\rangle\}$ is the complete set of eigenstates of H_0. Comparing this with (33.15) we see that

$$c_n(t) = \langle\,n\,|\,U_I(t,t_0)\,|\,i\,\rangle\;. \tag{33.31}$$

$U_I(t, t_0)$ can be related to the Schrödinger picture time-evolution operator $U(t, t_0)$ as follows. We have

$$\begin{aligned} |\,\psi(t)\,\rangle_I &= e^{iH_0t/\hbar}\,|\,\psi(t)\,\rangle_S = e^{iH_0t/\hbar}\,U(t,t_0)\,|\,\psi(t_0)\,\rangle_S \\ &= e^{iH_0t/\hbar}\,U(t,t_0)\,e^{-iH_0t/\hbar}\;e^{iH_0t/\hbar}\,|\,\psi(t_0)\,\rangle_S \\ &= e^{iH_0t/\hbar}\,U(t,t_0)\,e^{-iH_0t/\hbar}\;|\,\psi(t_0)\,\rangle_I\;, \end{aligned} \tag{33.32}$$

where the last equality follows from (33.7). Equation (33.24) then implies that

$$U_I(t,t_0) = \exp\left(\frac{iH_0t}{\hbar}\right)\,U(t,t_0)\,\exp\left(-\frac{iH_0t_0}{\hbar}\right)\,. \tag{33.33}$$

[Note carefully the difference between this equation and (33.8).] Thus

$$\langle\, n\,|\,U_I(t,t_0)\,|\,i\,\rangle = e^{i(\epsilon_n t-\epsilon_i t_0)/\hbar}\,\langle\, n\,|\,U(t,t_0)\,|\,i\,\rangle\,. \tag{33.34}$$

The matrix element $\langle\, n\,|\,U(t,t_0)\,|\,i\,\rangle$ is called the **transition amplitude** between the states $|\,i\,\rangle$ and $|\,n\,\rangle$. If these are both eigenstates of H_0 (as assumed to be the case in the present discussion), Eq. (33.34) implies that the transition probability $|\,c_n(t)\,|^2$ can be calculated with either U_I or U, that is,

$$|\langle\, n\,|\,U_I(t,t_0)\,|\,i\,\rangle|^2 = |\langle\, n\,|\,U(t,t_0)\,|\,i\,\rangle|^2\,. \tag{33.35}$$

On expanding $c_n(t)$ according to

$$c_n(t) = c_n^{(0)}(t) + c_n^{(1)}(t) + c_n^{(2)}(t) + \dots\,, \tag{33.36}$$

where $c_n^{(i)}(t)$ is of the i-th order in V_I, the Dyson series (33.28) implies that

$$c_n^{(0)}(t) = \delta_{ni} \qquad \text{(independent of } t)\,, \tag{33.37a}$$

$$c_n^{(1)}(t) = -\frac{i}{\hbar}\int_{t_0}^{t}\langle\, n\,|\,V_I(t')\,|\,i\,\rangle\,dt' = -\frac{i}{\hbar}\int_{t_0}^{t} e^{i\omega_{ni}t'}\,V_{ni}(t')\,dt'\,, \tag{33.37b}$$

$$c_n^{(2)}(t) = \left(-\frac{i}{\hbar}\right)^2\sum_m\int_{t_0}^{t}dt'\int_{t_0}^{t'}dt''\,e^{i\omega_{nm}t'}\,V_{nm}(t')\,e^{i\omega_{mi}t''}\,V_{mi}(t'')\,, \tag{33.37c}$$

$$\vdots\;\;;$$

where the matrix elements $V_{nm}(t)$ on the right-hand sides are those of the Schrödinger picture interaction potential.

Now we consider the most important application of (33.37): the case of time-constant V given by

$$V(t) = \begin{cases} 0 & ;\, t<0\,, \\ V & ;\, t\geq 0 \quad \text{(independent of time)}\,. \end{cases} \tag{33.38}$$

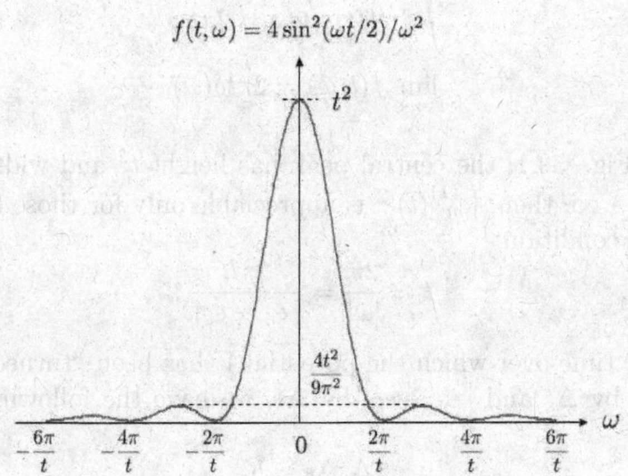

Fig. 33.1

Let $|\psi(0)\rangle = |i\rangle$, an eigenstate of H_0. Then, setting $t_0 = 0$ in (33.37), we obtain

$$c_n^{(0)}(t) = c_n^{(0)}(0) = \delta_{ni} , \tag{33.39}$$

$$c_n^{(1)}(t) = -\frac{i}{\hbar} V_{ni} \int_0^t e^{i\omega_{ni}t'}\, dt' = \frac{V_{ni}}{(\epsilon_n - \epsilon_i)} (1 - e^{i\omega_{ni}t}) . \tag{33.40}$$

Hence

$$|c_n^{(1)}(t)|^2 = \frac{|V_{ni}|^2}{(\epsilon_n - \epsilon_i)^2} (1 - e^{i\omega_{ni}t})(1 - e^{-i\omega_{ni}t}) = \frac{2|V_{ni}|^2}{(\epsilon_n - \epsilon_i)^2} (1 - \cos\omega_{ni}t) , \tag{33.41}$$

or

$$\boxed{|c_n^{(1)}(t)|^2 = \frac{4|V_{ni}|^2}{(\epsilon_n - \epsilon_i)^2} \sin^2\left\{\frac{(\epsilon_n - \epsilon_i)t}{2\hbar}\right\}} . \tag{33.42}$$

Thus the first-order transition probability not only depends on the strength of the interaction potential (as measured by $|V_{ni}|^2$), but also critically on the energy difference $\epsilon_n - \epsilon_i$. It is then of interest to examine the behavior of

$$f(t,\omega) \equiv \frac{4\sin^2(\omega t/2)}{\omega^2}$$

as a function of ω for constant t, where $\omega = \omega_{ni} = (\epsilon_n - \epsilon_i)/\hbar$. A qualitative plot of this function is shown in Fig. 33.1.

We note the following properties of the function $f(t,\omega)$:

$$\int_{-\infty}^{\infty} f(t,\omega)d\omega = 2\pi t \,, \tag{33.43}$$

$$\lim_{t\to\infty} f(t,\omega) = 2\pi t\delta(\omega) \,. \tag{33.44}$$

Referring to Fig. 33.1, the central peak has height t^2 and width proportional to $1/t$. As $t \to \infty$, then, $|c_n^{(1)}(t)|^2$ is appreciable only for those final states $|\, n \,\rangle$ satisfying the condition

$$t \approx \frac{2\pi}{|\omega|} = \frac{2\pi\hbar}{|\,\epsilon_n - \epsilon_i\,|} \,, \tag{33.45}$$

where t is the time over which the potential V has been "turned on". Relabeling this time by Δt and $|\,\epsilon_n - \epsilon_i\,|$ by $\Delta\epsilon$, we have the following "uncertainty relationship":

$$\Delta t \Delta\epsilon \approx h \,. \tag{33.46}$$

This equation has very important physical consequences, and needs to be interpreted properly. Its meaning is that for very short times Δt, quantum mechanical processes that do not conserve energy can occur with significant probability. As Δt increases, so does the requirement for energy conservation.

A situation that is frequently encountered is when a discrete initial energy level ϵ_i is embedded in a continuous band of final energy levels ϵ_n (we will retain the discrete index n for the time being for convenience). In such a case, we are interested in the total transition probability to all final states with $\epsilon_n \approx \epsilon_i$, given by the quantity

$$\sum_{n,\epsilon\approx\epsilon_i} |c_n^{(1)}(t)|^2 \,.$$

This sum can be converted to an integral over a continuous range of energy values by the introduction of an energy density of final states $\rho(\epsilon)$, which is defined in terms of a projection operator, $P_{\Delta\epsilon}$, onto an energy interval $\Delta\epsilon$:

$$P_{\Delta\epsilon} = \int_{\Delta\epsilon} d\epsilon\, \rho(\epsilon) |\, \epsilon\,\rangle\langle\, \epsilon\,| \,. \tag{33.47}$$

The continuum states $|\,\epsilon\,\rangle$ are required to satisfy the normalization condition

$$\langle\, \epsilon \,|\, \epsilon' \,\rangle = \frac{\delta(\epsilon - \epsilon')}{\rho(\epsilon)} \,; \tag{33.48}$$

so that, indeed,

$$\begin{aligned} P_{\Delta\epsilon}\,|\,\epsilon'\,\rangle &= \int_{\Delta\epsilon} d\epsilon\, \rho(\epsilon)\,|\,\epsilon\,\rangle\langle\,\epsilon\,|\,\epsilon'\,\rangle \\ &= \int_{\Delta\epsilon} d\epsilon\, \rho(\epsilon) |\,\epsilon\,\rangle\, \frac{\delta(\epsilon-\epsilon')}{\rho(\epsilon)} = \begin{cases} 0\,, & \text{if } \epsilon' \notin \Delta\epsilon\,, \\ |\,\epsilon'\,\rangle\,, & \text{if } \epsilon' \in \Delta\epsilon\,. \end{cases} \end{aligned} \tag{33.49}$$

Then, by (33.42),

$$\begin{aligned}\sum_{n,\epsilon_n\approx\epsilon_i} |c_n^{(1)}(t)|^2 &\longrightarrow \int d\epsilon_n\, \rho(\epsilon_n)|c_n^{(1)}|^2 \\ &= 4\int d\epsilon_n\, \rho(\epsilon_n)\,\frac{|V_{ni}|^2}{(\epsilon_n-\epsilon_i)^2}\,\sin^2\left\{\frac{(\epsilon_n-\epsilon_i)t}{2\hbar}\right\}.\end{aligned} \tag{33.50}$$

For $t\to\infty$, we can use the representation of the Dirac delta function given by (33.44) to obtain

$$\lim_{t\to\infty}\frac{1}{(\epsilon_n-\epsilon_i)^2}\sin^2\left\{\frac{(\epsilon_n-\epsilon_i)t}{2\hbar}\right\} = \frac{\pi t}{2\hbar}\,\delta(\epsilon_n-\epsilon_i)\,. \tag{33.51}$$

It follows that

$$\sum_{n,\epsilon_n\approx\epsilon_i} |c_n^{(1)}(t)|^2 \xrightarrow[t\to\infty]{} \frac{2\pi t}{\hbar}\int d\epsilon_n\, \rho(\epsilon_n)\delta(\epsilon_n-\epsilon_i)\,|V_{ni}|^2\,. \tag{33.52}$$

The interaction matrix element V_{ni} depends on ϵ_n through the index n. In the event that all the states with roughly the same ϵ_n also have roughly the same V_{ni}, we can replace $|V_{ni}|^2$ in the integrand by its average value $\overline{|V_{ni}|^2}$, which can then be taken outside of the integral. Then the integration in (33.52) can be performed easily due to the presence of the Dirac delta function. Denoting the **transition probability per unit time** (or **transition rate**) from an initial state i to a final group of states $\{n\}$ with $\epsilon_n\approx\epsilon_i$ by $w_{i\to\{n\}}$, we have

$$\boxed{w_{i\to\{n\}} = \frac{2\pi}{\hbar}\,\overline{|V_{ni}|^2}\,\rho(\epsilon_n)\big|_{\epsilon_n\approx\epsilon_i}}\;. \tag{33.53}$$

This formula, called **Fermi's Golden Rule**, is one of the most useful in quantum mechanics. It was actually due to Dirac (but perhaps made famous and rendered immortal by Fermi). Equivalently, the transition rate $w_{i\to n}$ to a particular state n can be written as

$$\boxed{w_{i\to n} = \frac{2\pi}{\hbar}\,|V_{ni}|^2\,\delta(\epsilon_n-\epsilon_i)}\;. \tag{33.54}$$

If the V_{ni} differ significantly among themselves for $\epsilon_n\approx\epsilon_i$, then in practice one just considers the **differential transition rates** to states which not only conserve energy approximately, but also with similar values for V_{ni}. For example, in elastic scattering, the final states are labeled by the momentum $\boldsymbol{k}$ of the scattered particle, but the energy only depends on $|\boldsymbol{k}|$. In this case, V_{ni} depends strongly on the direction of the momentum as well as its magnitude.

Let us now consider the second-order perturbation term [(33.37c)], still with a time-independent V given by (33.38). Equation (33.37c) gives

$$\begin{aligned} c_n^{(2)}(t) &= \left(-\frac{i}{\hbar}\right)^2 \sum_m V_{nm} V_{mi} \int_0^t dt'\, e^{i\omega_{nm}t'} \int_0^{t'} dt''\, e^{i\omega_{mi}t''} \\ &= \frac{i}{\hbar} \sum_m \frac{V_{nm}V_{mi}}{\epsilon_m - \epsilon_i} \int_0^t dt' \left(e^{i\omega_{ni}t'} - e^{i\omega_{nm}t'}\right) . \end{aligned} \tag{33.55}$$

Thus, using (33.36) and (33.40), the transition amplitude is given to second order by

$$\begin{aligned} c_n(t) &\approx c_n^{(1)}(t) + c_n^{(2)}(t) \\ &= \left(-\frac{i}{\hbar}\right) \left[V_{ni} - \sum_m \frac{V_{nm}V_{mi}}{\epsilon_m - \epsilon_i} \right] \int_0^t dt'\, e^{i\omega_{ni}t'} \\ &\quad - \frac{i}{\hbar} \sum_m \frac{V_{nm}V_{mi}}{\epsilon_m - \epsilon_i} \int_0^t dt'\, e^{i\omega_{nm}t'} , \quad n \neq i . \end{aligned} \tag{33.56}$$

The first term on the right-hand side has exactly the same time dependence as the first-order term [cf. (33.41)]; while the second one gives rise to rapid oscillations in time, and will not yield a significant contribution in comparison to the transition probability that grows linearly with time. Hence the second term can be ignored in the limit $t \to \infty$. Repeating the same argument as for the first-order term, we finally have

$$W_{i \to \{n\}} = \frac{2\pi}{\hbar} \left| V_{ni} + \sum_m \frac{V_{nm}V_{mi}}{\epsilon_i - \epsilon_m} \right|^2 \rho(\epsilon_n)|_{\epsilon \approx \epsilon_i} \quad \text{to second order} . \tag{33.57}$$

A special non-perturbative treatment is needed if $V_{nm}V_{mi} \neq 0$ and $\epsilon_i \approx \epsilon_m$ at the same time. We will pursue this problem in Chapter 35.

Problem 33.2 The observed lifetime for the **neutron decay** process

$$n \longrightarrow p + e + \nu$$

is $\tau \approx 15$ min. Use Fermi's Golden Rule to give an order-of-magnitude estimate for the dimensionless coupling constant g for the **weak interaction** responsible for this process. You should obtain

$$\frac{g^2}{4\pi} \approx 4 \times 10^{-14} .$$

Compared to the strength of the electromagnetic interaction characterized by the fine structure constant $\alpha = e^2/(\hbar c) \approx 1/137$, the weak interaction is indeed weak. Follow the steps below.

(a) Assume the neutron to be enclosed in a box of volume V, so that we can use **box quantization** (see the next chapter) and write normalized free-particle wave functions as

$$\phi_{\boldsymbol{k}}(\boldsymbol{x}) = \frac{1}{\sqrt{V}}\, e^{i\boldsymbol{k}\cdot\boldsymbol{x}} \,.$$

Show that

$$\langle\, pe\nu \,|\, H_{int} \,|\, N \,\rangle \approx \frac{g}{Vm_\pi^2}\, \delta_K(\boldsymbol{p}_p + \boldsymbol{p}_e + \boldsymbol{p}_\nu - \boldsymbol{p}_N) \,,$$

where H_{int} is some unknown interaction Hamiltonian describing the weak interaction, the $\boldsymbol{p}$'s are the respective momenta, and δ_K is the Kronecka delta function (and hence dimensionless). The choice of using the pion mass $m_\pi = 140\, MeV$ instead of any other mass is entirely arbitrary.

(b) Use Fermi's Golden Rule to obtain

$$\frac{1}{\tau} = \sum_{\boldsymbol{p}_p,\, \boldsymbol{p}_e,\, \boldsymbol{p}_\nu} \frac{2\pi g^2}{V^2 m_\pi^4}\, \delta(E_p + E_e + E_\nu - E_N)\, \delta_K(\boldsymbol{p}_p + \boldsymbol{p}_e + \boldsymbol{p}_\nu - \boldsymbol{p}_N) \,.$$

(In the above result we have ignored spin and set $\hbar = c = 1$.)

(c) Carry out the calculation in the neutron rest frame. Sum over $\boldsymbol{p}_p$ first. Make the following kinematical approximations.

$$\begin{aligned} &\boldsymbol{p}_N = 0 \,, && E_N = m_N = 940\, MeV \,; \\ &\boldsymbol{p}_p = -(\boldsymbol{p}_e + \boldsymbol{p}_\nu) \,, && E_p \approx m_p \,; \\ & && E_e \approx m_e + \frac{p_e^2}{2m_e} \,; \\ & && E_\nu = p_\nu \,; \\ &m_p = m_N - 1.3\, MeV \,; && \\ &m_e = 0.5\, MeV \,; && \\ &m_\nu = 0 \,. && \end{aligned}$$

In neglecting the term $p_p^2/(2m_p)$ in E_p we have made the approximation that the proton is at rest in the neutron rest frame, due to its large mass in comparison to the electron mass. Also, the non-relativistic approximation for the electron energy E_e is warranted because we are only interested in an order-of-magnitude estimate for g.

(d) Next perform the sum over $\boldsymbol{p}_e$ and $\boldsymbol{p}_\nu$, and pass to the limit $V \longrightarrow \infty$, to obtain

$$\frac{1}{\tau} = \frac{g^2}{m_\pi^4}\, \frac{(2\pi)(4\pi)^2}{(2\pi)^6} \int_0^\infty dp_e\, p_e^2 \int_0^\infty dp_\nu\, p_\nu^2\, \delta\left(\frac{p_e^2}{2m_e} + p_\nu - \Delta\right) ,$$

where $\Delta \equiv m_N - m_p - m_e = 0.8\, MeV$, and $\Delta/m_e = 8/5$.

(e) Do the p_ν integral to get

$$\begin{aligned} \frac{1}{\tau} &= \frac{g^2}{m_\pi^4}\, \frac{(2\pi)(4\pi)^2}{(2\pi)^6}\, \Delta^2 \int_0^{\sqrt{2m_e\Delta}} dp_e\, p_e^2 \left(1 - \frac{p_e^2}{2m_e\Delta}\right)^2 \\ &= m_\pi \left(\frac{m_e}{m_\pi}\right)^5 g^2 \left[\frac{8(2)^{3/2}}{105\pi^3} \left(\frac{8}{5}\right)^{7/2}\right] . \end{aligned}$$

(f) Finally restore units. Let

$$t_0 \equiv \frac{\hbar}{m_\pi c^2} = 0.5 \times 10^{-23}\, sec\,.$$

Show that

$$\frac{t_0}{\tau} = \left(\frac{m_e}{m_\pi}\right)^5 \frac{g^2}{2}\left[\frac{8(2)^{3/2}}{105\pi^3}\left(\frac{8}{5}\right)^{7/2}\right],$$

and solve for g^2 with $\tau \approx 15\, min$.

Chapter 34

Interaction of Matter with the Classical Radiation Field: Application of Time-Dependent Perturbation Theory

The rigorous theory to treat the interaction of charged particles with the electromagnetic field is quantum electrodynamics (QED), where relativistic effects and the quantized nature of the electromagnetic field are fully accounted for. As an introduction to the subject, we will consider neither of these complications, and study only the non-relativistic Schrödinger equation describing the interaction between an electron (in an atom) subject to a static Coulomb potential ϕ and a classical radiation field. The latter is described by a vector potential $\boldsymbol{A}(\boldsymbol{x}, t)$.

One uses the so-called **minimal coupling** introduced by Dirac, where

$$\boldsymbol{p} \longrightarrow \boldsymbol{p} - \frac{e}{c}\boldsymbol{A} ,$$

and write the Schrödinger equation as

$$\left[\frac{1}{2m}\left(\boldsymbol{p} - \frac{e}{c}\boldsymbol{A}\right)^2 + e\phi\right]\psi = i\hbar\frac{\partial\psi}{\partial t} . \tag{34.1}$$

On replacing $\boldsymbol{p}$ by $-i\hbar\partial_\mu$, $\mu = 1, 2, 3$, Eq. (34.1) becomes

$$\left[-\frac{\hbar^2}{2m}\left(\partial_\mu - \frac{ie}{\hbar c}A_\mu\right)^2 + e\phi\right]\psi = i\hbar\frac{\partial\psi}{\partial t} . \tag{34.2}$$

This equation is *not* relativistically invariant (under the Lorentz transformation), but is invariant under a **gauge transformation**:

$$\boxed{A_\mu(\boldsymbol{x}) \longrightarrow A_\mu(\boldsymbol{x}) + \partial_\mu \lambda(\boldsymbol{x}) \; ; \quad \psi(\boldsymbol{x}) \longrightarrow \exp\left(\frac{ie}{\hbar c}\lambda(\boldsymbol{x})\right)\psi(\boldsymbol{x})} \quad , \tag{34.3}$$

where $\lambda(\boldsymbol{x})$ is an arbitrary function. The quantity

$$\boxed{D_\mu \psi \equiv \left(\partial_\mu - \frac{ie}{\hbar c}A_\mu\right)\psi} \tag{34.4}$$

is called the **covariant derivative** of the wave function ψ with respect to the **gauge potential** A_μ. These are differential geometrical concepts that we will explain more carefully in subsequent chapters. They will play a fundamental role in the application of differential geometry to quantum theory.

We demonstrate the gauge invariance of the Schrödinger equation (34.2) as follows. First we have

$$\begin{aligned} D_\mu \psi \longrightarrow \quad & D'_\mu \psi' \\ &= \left\{\partial_\mu - \frac{ie}{\hbar c}(A_\mu + \partial_\mu \lambda)\right\} \exp\left(\frac{ie}{\hbar c}\lambda\right)\psi \\ &= \frac{ie}{\hbar c} e^{\frac{ie}{\hbar c}\lambda}(\partial_\mu \lambda)\psi + e^{\frac{ie}{\hbar c}\lambda}\,\partial_\mu \psi - \frac{ie}{\hbar c}e^{\frac{ie}{\hbar c}\lambda}(\partial_\mu\lambda)\psi - \frac{ie}{\hbar c}e^{\frac{ie}{\hbar c}\lambda}\,A_\mu\psi \\ &= e^{\frac{ie}{\hbar c}\lambda}\left(\partial_\mu - \frac{ie}{\hbar c}A_\mu\right)\psi \\ &= e^{\frac{ie}{\hbar c}\lambda}\,D_\mu\psi\,. \end{aligned} \tag{34.5}$$

Hence

$$D_\mu^2 \psi \longrightarrow D'^{\,2}_\mu \psi' = D'_\mu(D'_\mu \psi') = e^{\frac{ie}{\hbar c}\lambda}\,D_\mu(D_\mu\psi) = e^{\frac{ie}{\hbar c}\lambda}\,D_\mu^2\psi\,. \tag{34.6}$$

Thus

$$\begin{aligned} &\left(-\frac{\hbar^2}{2m}D'^{\,2}_\mu + e\phi\right)\psi' = -\frac{\hbar^2}{2m}e^{\frac{ie}{\hbar c}\lambda}\,D_\mu^2\psi + e\phi\, e^{\frac{ie}{\hbar c}\lambda}\,\psi \\ &= e^{\frac{ie}{\hbar c}\lambda}\left(-\frac{\hbar^2}{2m}D_\mu^2 + e\phi\right)\psi = e^{\frac{ie}{\hbar c}\lambda}\,i\hbar\frac{\partial\psi}{\partial t} = i\hbar\frac{\partial}{\partial t}\left(e^{\frac{ie}{\hbar c}\lambda}\,\psi\right) \\ &= i\hbar\frac{\partial\psi'}{\partial t}\,. \end{aligned} \tag{34.7}$$

We will work with the so-called **Coulomb gauge**, in which

$$\nabla\cdot\boldsymbol{A} = \sum_\mu \partial_\mu A_\mu = \partial_\mu A^\mu = 0\,. \tag{34.8}$$

In this gauge,

$$
\begin{aligned}
&(\boldsymbol{p}\cdot\boldsymbol{A}+\boldsymbol{A}\cdot\boldsymbol{p})\psi \\
&= -i\hbar\{\partial_\mu(A^\mu\psi)+A^\mu\partial_\mu\psi\} = -i\hbar\{\overbrace{(\partial_\mu A^\mu)}^{=0}\psi+2A^\mu\partial_\mu\psi\} = -2i\hbar A^\mu\partial_\mu\psi \\
&= 2(\boldsymbol{A}\cdot\boldsymbol{p})\psi\,.
\end{aligned}
\tag{34.9}
$$

Thus we can write the Schrödinger equation (34.1) as

$$
\frac{1}{2m}\left(\boldsymbol{p}^2-\frac{2e}{c}\boldsymbol{A}\cdot\boldsymbol{p}+\frac{e^2}{c^2}\boldsymbol{A}^2\right)\psi+e\phi\,\psi=i\hbar\frac{\partial\psi}{\partial t}\,, \tag{34.10}
$$

or

$$
(H_0+V')\psi=i\hbar\frac{\partial\psi}{\partial t}\,, \tag{34.11}
$$

where

$$
H_0=\frac{\boldsymbol{p}^2}{2m}+e\phi\,, \tag{34.12}
$$

$$
V'=-\frac{e}{mc}\boldsymbol{A}\cdot\boldsymbol{p}+\frac{e^2}{2mc^2}\boldsymbol{A}^2\,. \tag{34.13}
$$

In perturbation theory, the $\boldsymbol{A}^2$ term is usually neglected, and we will focus on the following perturbation Hamiltonian:

$$
V=-\frac{e}{mc}\boldsymbol{A}\cdot\boldsymbol{p}\,. \tag{34.14}
$$

We consider the classical plane wave

$$
\boldsymbol{A}(\boldsymbol{x},t)=2A_0\hat{\boldsymbol{\varepsilon}}\cos(\boldsymbol{k}\cdot\boldsymbol{x}-\omega t)\,, \tag{34.15}
$$

where

$$
\omega=kc\,, \tag{34.16}
$$

and $\hat{\boldsymbol{\varepsilon}}$ is a **polarization vector**. The Coulomb gauge $\nabla\cdot\boldsymbol{A}=0$ requires that

$$
\hat{\boldsymbol{\varepsilon}}\cdot\boldsymbol{k}=0\,. \tag{34.17}
$$

Equations (34.15) and (34.17) give the familiar solutions to Maxwell's equations corresponding to plane waves. The perturbation Hamiltonian (34.14) can then be written as

$$
V=-\frac{e}{mc}A_0\hat{\boldsymbol{\varepsilon}}\cdot\boldsymbol{p}\left[e^{i(\boldsymbol{k}\cdot\boldsymbol{x}-\omega t)}+e^{-i(\boldsymbol{k}\cdot\boldsymbol{x}-\omega t)}\right]\,. \tag{34.18}
$$

This is a special case of the so-called **harmonic perturbations**, which can in general be written in the form

$$
V(t)=\mathcal{V}e^{i\omega t}+\mathcal{V}^\dagger e^{-i\omega t}\,, \tag{34.19}
$$

where the operator $\mathcal{V}$ may depend on $\boldsymbol{x}, \boldsymbol{p}$ and the spin $\boldsymbol{S}$. We will consider the general case (34.19) first and then apply the results to the radiative interaction (34.18). Assume that the perturbation $V(t)$ is turned on at $t = 0$; and for $t \geq 0$, it has the form (34.19). At $t = 0$, the system is in an eigenstate $|\, i\,\rangle$ of H_0. Equation (33.37b) then implies

$$\begin{aligned} c_n^{(1)}(t) &= -\frac{i}{\hbar}\int_0^t dt'\, e^{i\omega_{ni}t'}\left(\mathcal{V}_{ni}e^{i\omega t'} + \mathcal{V}_{ni}^{\dagger}e^{-i\omega t'}\right) \\ &= \frac{1}{\hbar}\left[\left(\frac{1-e^{i(\omega+\omega_{ni})t}}{\omega+\omega_{ni}}\right)\mathcal{V}_{ni} + \left(\frac{1-e^{i(\omega_{ni}-\omega)t}}{-\omega+\omega_{ni}}\right)\mathcal{V}_{ni}^{\dagger}\right], \end{aligned} \tag{34.20}$$

where $\mathcal{V}_{ni}^{\dagger} \equiv (\mathcal{V}^{\dagger})_{ni}$. This formula is the same as (33.40) for the case $V(t) =$ constant, provided we make the replacement $\omega_{ni} \to \omega_{ni} + \omega$.

Following the analysis of the last chapter, we see that as $t \to \infty$, $|\, c_n^{(1)}(t)\,|^2$ is appreciable only if

$$\omega_{ni} + \omega \approx 0\,, \quad \text{i.e.} \quad \epsilon_n \approx \epsilon_i - \hbar\omega\,, \tag{34.21a}$$

or

$$\omega_{ni} - \omega \approx 0\,, \quad \text{i.e.} \quad \epsilon_n \approx \epsilon_i + \hbar\omega\,. \tag{34.21b}$$

In the case when (34.19) represents the interaction with a radiation field, (34.21a) represents the process of **stimulated emission**, while (34.21b) represents the process of **absorption**, as shown in Fig. 34.1. Note that there is no energy conservation requirement (as $t \to \infty$) for the matter system alone. The time-dependent perturbation $V(t)$ can be regarded as an inexhaustible source or sink of energy. Thus we have, in analogy with (33.53), the following transition rates:

$$w_{i\to\{n\}} = \frac{2\pi}{\hbar}\overline{|\,\mathcal{V}_{ni}\,|^2}\;\rho(\epsilon_n)\big|_{\epsilon_n=\epsilon_i-\hbar\omega}\,, \quad \text{(stimulated emission)} \tag{34.22a}$$

$$w_{i\to\{n\}} = \frac{2\pi}{\hbar}\overline{|\,\mathcal{V}_{ni}^{\dagger}\,|^2}\;\rho(\epsilon_n)\big|_{\epsilon_n=\epsilon_i+\hbar\omega}\,. \quad \text{(absorption)} \tag{34.22b}$$

We have two separate equations since (34.21a) and (34.21b) cannot be simultaneously satisfied for finite ω. The corresponding equations analogous to (33.54) are

$$w_{i\to n} = \frac{2\pi}{\hbar}\,|\mathcal{V}_{ni}|^2\,\delta(\epsilon_n - \epsilon_i + \hbar\omega)\,, \tag{34.23a}$$

$$w_{i\to n} = \frac{2\pi}{\hbar}\left|\mathcal{V}_{ni}^{\dagger}\right|^2\delta(\epsilon_n - \epsilon_i - \hbar\omega)\,. \tag{34.23b}$$

Since $\langle\, n\,|\,\mathcal{V}^{\dagger}\,|\,i\,\rangle = \langle\, i\,|\,\mathcal{V}\,|\,n\,\rangle^*$, we have

$$|\,\mathcal{V}_{ni}\,|^2 = |\,\mathcal{V}_{in}^{\dagger}\,|^2\,. \tag{34.24}$$

This fact, together with (34.22a) and (34.22b), imply that

$$\frac{\text{stimulated emission rate for } i \to \{n\}}{\text{density of final states for } \{n\}} = \frac{\text{absorption rate for } n \to \{i\}}{\text{density of final states for } \{i\}}\,. \tag{34.25}$$

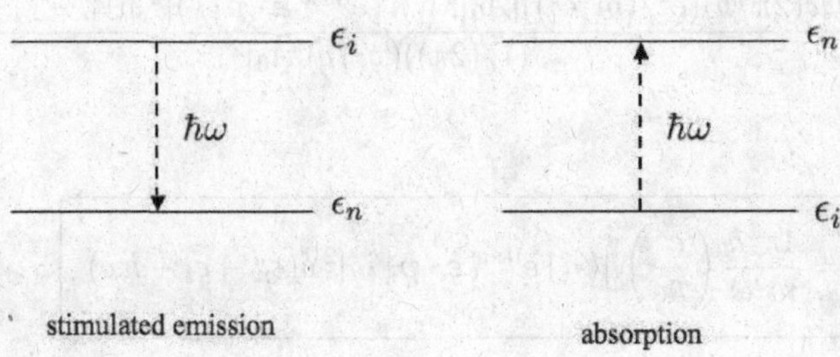

Fig. 34.1

The above statement, known as the **principle of detailed balancing**, expresses the reciprocity between the radiative processes of absorption and stimulated emission.

We will now return to the case of (34.28), and treat the absorption case in detail. We have

$$\mathcal{V}_{ni}^{\dagger} = -\frac{eA_0}{mc}(e^{i\boldsymbol{k}\cdot\boldsymbol{x}}\,\hat{\boldsymbol{\varepsilon}}\cdot\boldsymbol{p})_{ni}\,. \tag{34.26}$$

Thus, the absorption transition probability per unit time (transition rate) is given by

$$\boxed{w_{i\to n} = \frac{2\pi}{\hbar}\frac{e^2}{m^2c^2}|A_0|^2\,|\langle\, n\,|\, e^{i\boldsymbol{k}\cdot\boldsymbol{x}}\,\hat{\boldsymbol{\varepsilon}}\cdot\boldsymbol{p}\,|\, i\,\rangle|^2\,\delta(\epsilon_n-\epsilon_i-\hbar\omega)}\quad . \tag{34.27}$$

The **absorption cross section**, σ_{abs}, on the other hand, is defined by

$$\sigma_{abs} = \frac{\text{energy per unit time absorbed } (i\to n)}{\text{intensity (energy flux) of radiation field}}\,, \tag{34.28}$$

where the intensity $[I = energy/(area \times time)]$ is given, according to classical electrodynamics, by

$$I = cU = \frac{\omega^2}{2\pi c}\,|A_0|^2\,. \tag{34.29}$$

In the above equation, the **energy density** U is in turn given by

$$U = \frac{1}{2}\left(\frac{E_{max}^2}{8\pi}+\frac{B_{max}^2}{8\pi}\right)\,, \tag{34.30}$$

with

$$\boldsymbol{E} = -\frac{1}{c}\frac{\partial \boldsymbol{A}}{\partial t}\,, \qquad \boldsymbol{B} = \nabla \times \boldsymbol{A}\,. \tag{34.31}$$

Putting everything together in (34.28) we have

$$\sigma_{abs} = \frac{\hbar\omega(2\pi/\hbar)(e^2/(m^2c^2))|A_0|^2\,|\langle\, n\,|\, e^{i\boldsymbol{k}\cdot\boldsymbol{x}}\,\hat{\boldsymbol{\varepsilon}}\cdot\boldsymbol{p}\,|\, i\,\rangle|^2\,\delta(\epsilon_n - \epsilon_i - \hbar\omega)}{(1/(2\pi))(\omega^2/c)|A_0|^2}\,, \tag{34.32}$$

or, finally,

$$\boxed{\sigma_{abs} = \frac{4\pi^2\hbar}{m^2\omega}\left(\frac{e^2}{\hbar c}\right)|\langle\, n\,|\, e^{i\boldsymbol{k}\cdot\boldsymbol{x}}\,\hat{\boldsymbol{\varepsilon}}\cdot\boldsymbol{p}\,|\, i\,\rangle|^2\,\delta(\epsilon_n - \epsilon_i - \hbar\omega)}\,. \tag{34.33}$$

Note that the RHS of the above equation indeed has the dimensions of an area.

The so-called **electric dipole** (E1) **approximation** is based on maintaining only the leading term, 1, in the following expansion:

$$e^{i\boldsymbol{k}\cdot\boldsymbol{x}} = 1 + i\frac{\omega}{c}\,\hat{\boldsymbol{k}}\cdot\boldsymbol{x} + \dots\;. \tag{34.34}$$

This approximation is justified when the wavelength of the radiation field is much larger than atomic dimensions. Indeed, for radiative frequencies ω corresponding to electronic transitions in atoms [cf. (30.61)],

$$\hbar\omega \approx \frac{Ze^2}{(a_0/Z)} \approx \frac{Ze^2}{R_{atom}}\,, \tag{34.35}$$

where a_0 is the Bohr radius [(32.15)] and Z is the atomic number. Hence

$$\frac{\omega}{c}\,R_{atom} \approx Z\left(\frac{e^2}{\hbar c}\right) \approx \frac{Z}{137}\,, \tag{34.36}$$

where, in the last approximation, we have made use of the value of the fine structure constant given by (32.3). Now consider the matrix element in (34.33), in the E1 approximation:

$$\langle\, n\,|\, e^{i\boldsymbol{k}\cdot\boldsymbol{x}}\,\hat{\boldsymbol{\varepsilon}}\cdot\boldsymbol{p}\,|\, i\,\rangle \longrightarrow \langle\, n\,|\,\hat{\boldsymbol{\varepsilon}}\cdot\boldsymbol{p}\,|\, i\,\rangle = \hat{\boldsymbol{\varepsilon}}\cdot\langle\, n\,|\,\boldsymbol{p}\,|\, i\,\rangle\,. \tag{34.37}$$

Suppose the radiation field is linearly polarized, with $\hat{\boldsymbol{\epsilon}}$ along the x-axis and $\boldsymbol{k}$ along the z-axis. Then we have to calculate $\langle\, n\,|\, p_x\,|\, i\,\rangle$. Recalling that

$$H_0 = \frac{p^2}{2m} + V(r)\,,$$

where $V(r)$ is a central potential, we have

$$[\,x\,,\,H_0\,] = [\,x\,,\,\frac{p^2}{2m}\,] = \frac{i\hbar}{2m}\frac{\partial(p^2)}{\partial p_x} = \frac{i\hbar}{m}\,p_x\,, \tag{34.38}$$

where in the second equality we have made use of the commutation relation (12.4). Thus

$$\begin{aligned}\langle n \,|\, p_x \,|\, i \rangle &= \frac{m}{i\hbar} \langle n \,|\, [\, x \,,\, H_0 \,] \,|\, i \rangle = \frac{m}{i\hbar} (\epsilon_i - \epsilon_n) \langle n \,|\, x \,|\, i \rangle \\ &= im\omega_{ni} \langle n \,|\, x \,|\, i \rangle \,.\end{aligned} \tag{34.39}$$

Now x is an irreducible tensor operator of rank one, of the form $T_q^{(1)}$ with $q = \pm 1$ [cf. (28.12)]. Thus the Wigner-Eckart Theorem gives the following **selection rules** for the **electric dipole transition** $|\, i \,\rangle = |\, jm \,\rangle \longrightarrow |\, n \,\rangle = |\, j'm' \,\rangle$:

$$m' - m = \pm 1 \,, \quad |\, j' - j \,| = 0, 1 \,, \quad (j = 0 \to j' = 0 \text{ not allowed}) \,. \tag{34.40}$$

In the dipole approximation, the absorption coefficient given by (34.33) now takes the form

$$\sigma_{abs} = 4\pi^2 \left(\frac{e^2}{\hbar c} \right) \omega_{ni} \,|\langle\, n \,|\, x \,|\, i \,\rangle|^2 \,\delta(\omega - \omega_{ni}) \,. \tag{34.41}$$

As an example we will consider the **photoelectric effect**: the ejection of a bound electron when an atom is exposed to a radiation field. The final state $|\, n \,\rangle$ is a free electron state with energy $\epsilon > 0$ in a continuum, which is assumed to be a plane wave state: $|\, n \,\rangle = |\, \boldsymbol{k}_f \,\rangle$. As we shall see, the matrix element $\langle\, n \,|\, x \,|\, i \,\rangle$ depends strongly not only on the magnitude of $\boldsymbol{k}_f$, but also its direction, that is, the direction of motion of the ejected electron. Thus we have to consider a differential cross section $d\sigma/d\Omega$, where $d\Omega$ is a solid angle element around the direction $\boldsymbol{k}_f$.

To facilitate the counting of plane-wave states, we use the so-called **box quantization** procedure, where the free electron is imagined to be confined within a large cubic box of volume $V = L^3$. Periodic boundary conditions are then imposed on the faces of the cube to obtain the possible quantized values for k_i, $(i = x, y, z)$. Thus

$$e^{ik_x x} = e^{ik_x (x+L)} \,, \tag{34.42}$$

with similar conditions for k_y and k_z. It follows that

$$k_i = \frac{2n\pi}{L} \,, \quad n = 0, \pm 1, \pm 2, \dots \,, \tag{34.43}$$

and the density of states in $\boldsymbol{k}$-space is

$$\rho(\boldsymbol{k}) = \frac{V}{(2\pi)^3} \,. \tag{34.44}$$

The number of states in an infinitesimal volume d^3k in $\boldsymbol{k}$-space is then

$$\rho(\boldsymbol{k})\, d^3k = \frac{V}{(2\pi)^3}\, k^2 dk \, d\Omega \,. \tag{34.45}$$

This result and (34.33) then imply

$$\begin{aligned} d\sigma_{abs} = & \frac{V}{(2\pi)^3} k^2 \frac{dk}{d\epsilon} \, d\epsilon \, d\Omega \\ & \times \frac{4\pi^2 \hbar}{m^2 \omega} \left(\frac{e^2}{\hbar c} \right) |\langle n | e^{i\boldsymbol{k}\cdot\boldsymbol{x}} \,\hat{\boldsymbol{\varepsilon}} \cdot \boldsymbol{p} \,| i \rangle|^2 \, \delta(\epsilon_n - \epsilon_i - \hbar\omega) \,. \end{aligned} \tag{34.46}$$

Now for a particle of mass m, $\epsilon = \hbar^2 k^2/(2m)$, so that $d\epsilon/dk = \hbar^2 k/m$. Putting this last result in (34.46) we have

$$\boxed{\frac{d\sigma_{abs}}{d\Omega} = \frac{4\pi^2 \hbar}{m^2 \omega} \left(\frac{e^2}{\hbar c} \right) \frac{V}{(2\pi^3)} \frac{m k_f}{\hbar^2} |\langle \boldsymbol{k}_f | e^{i\boldsymbol{k}\cdot\boldsymbol{x}} \,\hat{\boldsymbol{\varepsilon}} \cdot \boldsymbol{p} \,| i \rangle|^2} \,, \tag{34.47}$$

where the integration over ϵ in (34.36) has already been carried out using the δ-function, and, in the above equation, it is understood that k_f satisfies the energy-shell condition

$$\frac{\hbar^2 k_f^2}{2m} - \epsilon_i = \hbar\omega \,. \tag{34.48}$$

In the box-quantization scheme, the free-electron wave function is given by

$$\langle \boldsymbol{x} | \boldsymbol{k}_f \rangle = \frac{e^{i\boldsymbol{k}_f\cdot\boldsymbol{x}}}{\sqrt{V}} \,. \tag{34.49}$$

To be specific, we consider the ejection of a K-shell electron of a hydrogen-like atom ($n = 1, l = 0$) by absorption of a photon of frequency ω. The initial state wave function is thus the ground state hydrogen atom wave function with a_0 (Bohr radius) replaced by a_0/Z, where Z is the atomic number of the atom:

$$\psi_i(\boldsymbol{x}) = \left(\frac{Z}{a_0} \right)^{3/2} e^{-Zr/a_0} \,. \tag{34.50}$$

Thus the matrix element in (34.47) is given by

$$\langle \boldsymbol{k}_f | e^{i\boldsymbol{k}\cdot\boldsymbol{x}} \,\hat{\boldsymbol{\varepsilon}} \cdot \boldsymbol{p} \,| i \rangle = \frac{1}{\sqrt{V}} \hat{\boldsymbol{\varepsilon}} \cdot \int d^3 r \, e^{-i\boldsymbol{k}_f\cdot\boldsymbol{x} + i\boldsymbol{k}\cdot\boldsymbol{x}} (-i\hbar\nabla) \left\{ \left(\frac{Z}{a_0} \right)^{3/2} e^{-Zr/a_0} \right\} . \tag{34.51}$$

Integrating by parts and using the boundary condition

$$\lim_{r\to\infty} e^{-Z/a_0} = 0$$

for the bound state wave function, we have

$$\langle \boldsymbol{k}_f | e^{i\boldsymbol{k}\cdot\boldsymbol{x}} \,\hat{\boldsymbol{\varepsilon}} \cdot \boldsymbol{p} \,| i \rangle = \left(\frac{Z}{a_0} \right)^{3/2} \frac{i\hbar}{\sqrt{V}} \hat{\boldsymbol{\varepsilon}} \cdot \int d^3 r \, e^{-Z/a_0} \, \nabla \left(e^{i(\boldsymbol{k}-\boldsymbol{k}_f)\cdot\boldsymbol{x}} \right) . \tag{34.52}$$

The gradient term in the integrand can be expanded as follows:

$$\nabla\left(e^{i\boldsymbol{k}\cdot\boldsymbol{x}}e^{-i\boldsymbol{k}_f\cdot\boldsymbol{x}}\right)=e^{i\boldsymbol{k}\cdot\boldsymbol{x}}\nabla\left(e^{-i\boldsymbol{k}_f\cdot\boldsymbol{x}}\right)+\left(\nabla e^{i\boldsymbol{k}\cdot\boldsymbol{x}}\right)e^{-i\boldsymbol{k}_f\cdot\boldsymbol{x}}\,.$$

The second term does not contribute since in the Coulomb gauge $\hat{\boldsymbol{\varepsilon}}\cdot\boldsymbol{k}=0$. Thus

$$\langle\,\boldsymbol{k}_f\,|\,e^{i\boldsymbol{k}\cdot\boldsymbol{x}}\,\hat{\boldsymbol{\varepsilon}}\cdot\boldsymbol{p}\,|\,i\,\rangle=\left(\frac{Z}{a_0}\right)^{3/2}\frac{\hbar}{\sqrt{V}}\hat{\boldsymbol{\varepsilon}}\cdot\boldsymbol{k}_f\int d^3r\,e^{-i(\boldsymbol{k}_f-\boldsymbol{k})\cdot\boldsymbol{x}}\,e^{-Zr/a_0}\,. \tag{34.53}$$

The matrix element is seen to be proportional to the Fourier transform of the atomic wave function with respect to the **momentum transfer**

$$\boldsymbol{q}\equiv\boldsymbol{k}_f-\boldsymbol{k}\,. \tag{34.54}$$

The integral can be evaluated by letting the z-axis be along the direction of $\boldsymbol{q}$. Then, with $\alpha\equiv Z/a_0$,

$$\begin{aligned}
\int d^3r\,e^{-i\boldsymbol{q}\cdot\boldsymbol{x}}e^{-\alpha r}&=2\pi\int_0^\infty dr\,r^2e^{-\alpha r}\int_0^\pi d\theta\,\sin\theta e^{-iqr\cos\theta}\\
&=2\pi\int_0^\infty dr\,r^2e^{-\alpha r}\int_{-1}^1 dz\,e^{-iqrz}\qquad(z=\cos\theta)\\
&=2\pi\int_0^\infty dr\,r^2e^{-\alpha r}\left(\frac{e^{iqr}-e^{-iqr}}{iqr}\right)=\frac{4\pi}{q}\int_0^\infty dr\,re^{-\alpha r}\sin qr\\
&=\frac{8\pi\alpha}{(\alpha^2+q^2)^2}\,.
\end{aligned} \tag{34.55}$$

Finally, it follows from (34.53) and (34.47) that the absorption differential cross section is given by

$$\boxed{\frac{d\sigma_{abs}}{d\Omega}=\frac{32\pi e^2k_f}{mc\omega}\left(\frac{Z}{a_0}\right)^5(\hat{\boldsymbol{\varepsilon}}\cdot\boldsymbol{k}_f)^2\frac{1}{\left(\dfrac{Z^2}{a_0^2}+q^2\right)^4}}\,, \tag{34.56}$$

where m and e are the mass and charge of the electron, respectively, a_0 is the Bohr radius, ω is the frequency of the incident light, and, with respect to the geometry of the process depicted in Fig. 34.2,

$$(\hat{\boldsymbol{\varepsilon}}\cdot\boldsymbol{k}_f)^2=k_f^2\sin^2\theta\cos^2\phi\,, \tag{34.57}$$

$$q^2=k_f^2+k^2-2k_fk\cos\theta\,. \tag{34.58}$$

In Problem 37.1 we will examine a more complete non-relativistic, perturbative, quantum electrodynamical treatment of a two-state atom interacting with an external laser field, in the presence of radiative decay, but still within the dipole approximation. We will see that the Heisenberg equation approach (one that uses operators directly instead of wave functions) will offer certain advantages.

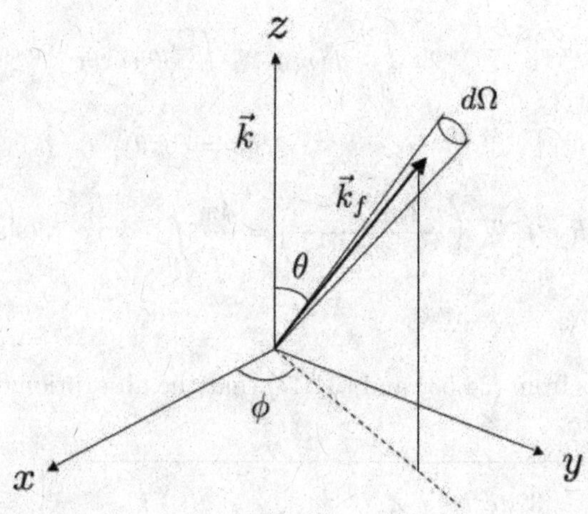

Fig. 34.2

Chapter 35

Potential Scattering Theory

Non-relativistic quantum potential scattering theory is mainly concerned with collisional situations where the interaction between the collision partners is described by a short-range potential (which is usually assumed to be time-independent), and where, in the remote past ($t \to -\infty$) and the remote future ($t \to \infty$), the state of the system asymptotically approaches certain solutions of the time-dependent Schrödinger equation for the unperturbed Hamiltonian. This is a large subject whose rigorous presentation would involve many technically sophisticated issues in Hilbert space theory and functional analysis, as well as the theory of analytic functions. Naturally we will not be able to dwell on many of these issues. In this chapter and the next we expand on some elementary notions already introduced in Chapter 16, and attempt to present in a more systematic but largely intuitive manner some of the most basic concepts and mathematical techniques that are fundamental to potential scattering theory, without any recourse to mathematical rigor. For further study of this rich and vital subject, the reader may consult, for example, Ross 1963, Goldberger and Watson 1975, and Taylor 1983 (for more physical approaches); and De Alfaro and Regge 1965, and Reed and Simon 1979 (for more mathematical approaches).

We assume that the Hamiltonian pertinent to the collision process can be split up according to

$$H = H_0 + V(\boldsymbol{r}) \,, \tag{35.1}$$

where H_0 is the unperturbed Hamiltonian describing the free motion of the collision partners, and the time-independent interaction potential $V(\boldsymbol{r})$ responsible for their interaction is **short-range** (approaches zero faster than r^{-3} as $r \to \infty$) and sufficiently "non-singular" for small r (less singular than r^{-2} as $r \to 0$). Note that these conditions immediately rule out the familiar Coulomb potential, which requires a special treatment, but are satisfied by a large number of problems in atomic, molecular, and nuclear physics. In general the Hilbert space $\mathcal{H}$ relevant to the problem can be written as a direct sum of a subspace $\mathcal{H}_d$ spanned by the discrete eigenstates of H, and its orthogonal complement $\mathcal{H}_c$

spanned by the continuum eigenstates of H – the so-called **scattering states:**

$$\mathcal{H} = \mathcal{H}_d \oplus \mathcal{H}_c \,. \tag{35.2}$$

A formal solution $|\,\psi\,\rangle \in \mathcal{H}$ of the time-dependent Schrödinger equation

$$H\,|\,\psi\,\rangle = i\hbar\frac{\partial|\,\psi\,\rangle}{\partial t} \tag{35.3}$$

is given by

$$|\,\psi(t)\,\rangle = U(t)\,|\,\psi\,\rangle\,, \tag{35.4}$$

where

$$U(t) = \exp\left(-iHt/\hbar\right) \tag{35.5}$$

is the unitary (Schrödinger picture) time-evolution operator [cf. (33.33)]. To set up the basic mathematical formalism of quantum scattering theory, it is useful to keep in mind the more intuitive picture of a classical scattering situation: Long before and long after the collision event, the actual orbit of a particle being scattered by a target will asymptotically approach free-particle orbits. Provided that the interaction potential $V(\boldsymbol{r})$ satisfies certain restrictions (referred to somewhat imprecisely above), this state of affairs can actually be stated as a mathematical fact in the quantum mechanical setting:

Suppose $|\,\psi_c\,\rangle \in \mathcal{H}_c$, then there exist two states $|\,\psi_{in}\,\rangle \in \mathcal{H}$ and $|\,\psi_{out}\,\rangle \in \mathcal{H}$, called the ***in-asymptote*** *and* ***out-asymptote****, respectively, such that*

$$\begin{aligned} U(t)\,|\,\psi_c\,\rangle &\xrightarrow[t\to-\infty]{} U_0(t)\,|\,\psi_{in}\,\rangle \\ &\xrightarrow[t\to+\infty]{} U_0(t)\,|\,\psi_{out}\,\rangle\,, \end{aligned} \tag{35.6}$$

where $U(t)$ is the "full" time-evolution operator given by (35.5) and

$$U_0(t) \equiv \exp\left(-iH_0t/\hbar\right) \tag{35.7}$$

is the time-evolution operator corresponding to the unperturbed Hamiltonian. Conversely, given any $|\,\phi\,\rangle \in \mathcal{H}$, there exist two continuum states $|\,\phi^{(+)}\,\rangle \in \mathcal{H}_c$ and $|\,\phi^{(-)}\,\rangle \in \mathcal{H}_c$, called the **outgoing state** *and* **incoming state**, *respectively, such that*

$$\begin{aligned} U(t)\,|\,\phi^{(+)}\,\rangle &\xrightarrow[t\to-\infty]{} U_0(t)\,|\,\phi\,\rangle\,, \\ U(t)\,|\,\phi^{(-)}\,\rangle &\xrightarrow[t\to+\infty]{} U_0(t)\,|\,\phi\,\rangle\,. \end{aligned} \tag{35.8}$$

This statement, sometimes referred to as the **Asymptotic-Condition Theorem**, is the basic mathematical result and starting point of nonrelativistic quantum scattering theory. In Eqs. (35.6) and (35.8), it is important to note that, whereas $|\,\psi_c\,\rangle$ and $|\,\phi^{(\pm)}\,\rangle$ are all vectors in $\mathcal{H}_c$, $|\,\psi_{in}\,\rangle, |\,\psi_{out}\,\rangle$ and $|\,\phi\,\rangle$

can either be in $\mathcal{H}_d$ or $\mathcal{H}_c$. Various proofs under diverse sets of restrictions for $V(\boldsymbol{r})$ have been given (see, for example, the article "*General Quantum Theory of Collision Processes*" by W. Brenig and R. Haag in Ross 1963, and Taylor 1983). Before we proceed we need to point out the precise mathematical sense in which the limits in (35.6) are understood. These are understood in the sense of **strong convergence**, defined as follows [cf. Def. 6.2]:

Definition 35.1. *The parametrized set of vectors* $|\,\psi(t)\,\rangle$ *in a Hilbert space* $\mathcal{H}$ *is said to* **converge strongly** *to a vector* $|\,\psi\,\rangle \in \mathcal{H}$ *as* $t \to \pm\infty$ *if*

$$\|\,\psi(t) - \psi\,\| \xrightarrow[t\to\pm\infty]{} 0\,, \tag{35.9}$$

where $\|\phi\|$ *denotes the norm of a vector* $|\,\phi\,\rangle \in \mathcal{H}$.

In comparison, we also have the notion of **weak convergence**:

Definition 35.2. *The parametrized set of vectors* $|\,\psi(t)\,\rangle$ *in a Hilbert space* $\mathcal{H}$ *is said to* **converge weakly** *to a vector* $|\,\psi\,\rangle \in \mathcal{H}$ *as* $t \to \pm\infty$ *if*

$$\langle\,\phi\,|\,\psi(t)\,\rangle \xrightarrow[t\to\pm\infty]{} 0 \tag{35.10}$$

for every (fixed) $|\,\phi\,\rangle \in \mathcal{H}$.

Strong convergence implies weak convergence, but the converse is not true in general.

Equations (35.6) and (35.8) allow us to define a pair of operators $\Omega_\pm : \mathcal{H} \longrightarrow \mathcal{H}_c$, called the **Møller operators**, as follows:

$$\boxed{\Omega_\pm \equiv \lim_{t\to\mp\infty} U(t)^\dagger U_0(t) = \lim_{t\to\mp\infty} \exp(iHt/\hbar)\exp(-iH_0t/\hbar)\,.} \tag{35.11}$$

We will also assume that the ranges of both Møller operators are identical, and are equal to the continuum subspace $\mathcal{H}_c$:

$$\Omega_+(\mathcal{H}) = \Omega_-(\mathcal{H}) = \mathcal{H}_c\,. \tag{35.12}$$

Indeed, it can be proved that $\Omega_\pm : \mathcal{H} \longrightarrow \mathcal{H}_c$ are both one-to-one (injective) and onto (surjective). Considered as maps from $\mathcal{H}$ to $\mathcal{H}$, however, they are injective but clearly not surjective. The operators $\Omega_\pm : \mathcal{H} \longrightarrow \mathcal{H}$, being limits of unitary operators, do satisfy the property of being **isometric**, that is, they preserve lengths of vectors in $\mathcal{H}$:

$$\|\Omega_\pm\phi\| = \|\phi\|\,, \quad \text{for all } \;|\,\phi\,\rangle \in \mathcal{H}\,. \tag{35.13}$$

The above equation implies that

$$\Omega_\pm^\dagger\Omega_\pm = 1\,, \quad \text{the identity map in } \;\mathcal{H}\,. \tag{35.14}$$

But clearly, $\Omega_\pm\Omega_\pm^\dagger$ is not the identity in $\mathcal{H}$. This last property prevents the Møller operators from being unitary operators in $\mathcal{H}$.

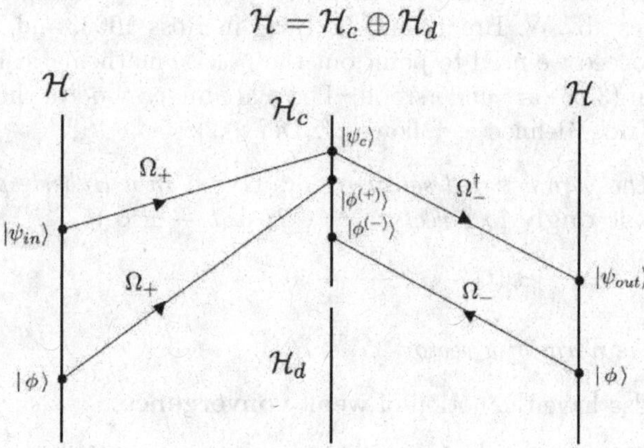

Fig. 35.1

In terms of the Møller operators we can restate the results contained in (35.6) and (35.8) as follows:

For any scattering state $|\,\psi_c\,\rangle \in \mathcal{H}_c$ *there exist asymptotic states* $|\,\psi_{in}\,\rangle \in \mathcal{H}$ *and* $|\,\psi_{out}\,\rangle \in \mathcal{H}$ *such that*

$$|\,\psi_{out}\,\rangle = \Omega_-^\dagger\,|\,\psi_c\,\rangle = \Omega_-^\dagger\Omega_+\,|\,\psi_{in}\,\rangle\,; \tag{35.15}$$

and for any $|\,\phi\,\rangle \in \mathcal{H}$,

$$\Omega_\pm\,|\,\phi\,\rangle = |\,\phi^{(\pm)}\,\rangle \in \mathcal{H}_c\,. \tag{35.16}$$

Figure 35.1 shows a schematic illustration of the action of $\Omega_\pm$ on $\mathcal{H}$.

We can now define the primary notion in scattering theory, the **scattering operator** S, a unitary operator acting on $\mathcal{H}$:

$$\boxed{S \equiv \Omega_-^\dagger\Omega_+}\;. \tag{35.17}$$

$S : \mathcal{H} \longrightarrow \mathcal{H}$ is unitary because it is a limit of unitary operators {since the $\Omega_\pm$ are themselves limits of unitary operators [cf. (35.11)]}, and because it is an onto map. The physical significance of the scattering operator rests on the following interpretation of (35.15) and (35.16). Suppose a system is prepared in the distant past in the state $\lim_{t\to-\infty} U_0\,|\,\phi\,\rangle$, then, according to (35.16), it will be in the scattering state $|\,\phi^{(+)}\,\rangle$ at $t = 0$. Now suppose, in the remote future, we are interested in finding the probability that the system will be in the state $\lim_{t\to+\infty} U_0 \,|\, \chi \,\rangle$, which evolved out of the state $|\, \chi^{(-)} \rangle$ at $t = 0$, then the transition probability amplitude of the scattering process $(\phi \to \chi)$ will be given by

$$A(\phi \to \chi) = \langle\, \chi^{(-)} \,|\, \phi^{(+)} \,\rangle = \langle\, \chi \,|\, \Omega_-^\dagger \Omega_+ \,|\, \phi \,\rangle = \langle\, \chi \,|\, S \,|\, \phi \,\rangle \, . \tag{35.18}$$

Thus *the transition probability amplitude between two asymptotic states is given by the S-matrix element corresponding to the two states.* The **S-matrix** is usually given as the matrix representation of the scattering operator S with respect to a basis of $\mathcal{H}$ consisting of eigenstates of the unperturbed Hamiltonian H_0. One can then show that S conserves energy between the asymptotic states $|\, \psi_{in} \,\rangle$ and $|\, \psi_{out} \,\rangle$. The proof of this fact can be facilitated by the so-called **intertwining relation** for the Møller operators:

$$H\Omega_\pm = \Omega_\pm H_0 \, . \tag{35.19}$$

Problem 35.1 Prove the intertwining relation (35.19) by showing that

$$e^{iH\tau/\hbar}\Omega_\pm = \Omega_\pm e^{iH_0\tau/\hbar} \, ,$$

and then differentiating the result with respect to τ and finally setting $\tau = 0$.

Using (35.19) twice we quickly see that S commutes with H_0:

$$SH_0 = \Omega_-^\dagger \Omega_+ H_0 = \Omega_-^\dagger H \Omega_+ = H_0 \Omega_-^\dagger \Omega_+ = H_0 S \, . \tag{35.20}$$

Thus, since $|\, \psi_{out} \,\rangle = S \,|\, \psi_{in} \,\rangle$,

$$\langle\, \psi_{out} \,|\, H_0 \,|\, \psi_{out} \,\rangle = \langle\, \psi_{in} \,|\, S^\dagger H_0 S \,|\, \psi_{in} \,\rangle = \langle\, \psi_{in} \,|\, H_0 \,|\, \psi_{in} \,\rangle \, . \tag{35.21}$$

In what follows we will assume that the free-particle Hamiltonian $H_0 = \boldsymbol{p}^2/2m$ is the unperturbed Hamiltonian, so that the eigenstates of H_0 can be chosen to be the (non-normalizable) plane-wave continuum states [cf. (10.2)]

$$\psi_{\boldsymbol{k}}(\boldsymbol{r}) = \langle\, \boldsymbol{r} \,|\, \boldsymbol{k} \,\rangle = \frac{1}{(2\pi)^{3/2}} \, e^{i\boldsymbol{k}\cdot\boldsymbol{r}} \, , \tag{35.22}$$

where $\boldsymbol{p} = \hbar \boldsymbol{k}$. Since energy is conserved by the scattering process, we can define a **transition operator**, T, in terms of the scattering operator S as follows:

$$\boxed{\langle\, \boldsymbol{k}' \,|\, S \,|\, \boldsymbol{k} \,\rangle = \delta^3(\boldsymbol{k}' - \boldsymbol{k}) - 2\pi i \delta(E_{k'} - E_k) \langle\, \boldsymbol{k}' \,|\, T \,|\, \boldsymbol{k} \,\rangle \, .} \tag{35.23}$$

The first term on the right-hand side of the above equation clearly represents the amplitude for no scattering, so that the second, involving the T-matrix elements, represents solely the effects of the scattering potential $V(\boldsymbol{r})$. Because of the energy-conserving condition, the quantities $\langle\, \boldsymbol{k}' \,|\, T \,|\, \boldsymbol{k} \,\rangle$ $(k' = k)$ are called **on-shell** T-matrix elements. Physically they are important because the so-called **scattering amplitude**, $f(\boldsymbol{k} \to \boldsymbol{k}')$, can be given in terms of them:

$$f(\boldsymbol{k} \to \boldsymbol{k}') = -(2\pi)^2 \frac{m}{\hbar^2} \langle\, \boldsymbol{k}' \,|\, T \,|\, \boldsymbol{k} \,\rangle\,, \quad (k' = k)\,. \tag{35.24}$$

The scattering amplitude is in turn related to the (directly physically observable) **differential scattering cross section** $d\sigma/d\Omega$ by [cf. (34.47)]

$$\frac{d\sigma}{d\Omega}(\boldsymbol{k} \to \boldsymbol{k}') = |\, f(\boldsymbol{k} \to \boldsymbol{k}')\,|^2\,, \tag{35.25}$$

where $d\Omega$ is the differential solid angle around the direction specified by the outgoing momentum $\boldsymbol{k}'$, into which the particle is scattered. We will not give the details of the derivation of (35.25). The reader should, however, check its dimensional consistency using (35.23) and (35.24).

We will now develop some methods for the calculation of the S or T matrices; and, in the process, arrive at a better understanding of the analytic properties of the scattering states as functions of their energy eigenvalues. Just as the plane wave states $|\, \boldsymbol{k} \,\rangle$ are eigenstates of H_0:

$$H_0 \,|\, \boldsymbol{k} \,\rangle = E_k \,|\, \boldsymbol{k} \,\rangle\,, \tag{35.26}$$

where $E_k = \hbar^2 k^2/(2m)$, the incoming and outgoing scattering states

$$|\, \boldsymbol{k}^{(\pm)} \,\rangle = \Omega_\pm \,|\, \boldsymbol{k} \,\rangle \tag{35.27}$$

are eigenstates of the full Hamiltonian H with the same energy eigenvalue. Indeed, using the intertwining relation (35.19) again, we see that

$$H \,|\, \boldsymbol{k}^{(\pm)} \,\rangle = H\Omega_\pm \,|\, \boldsymbol{k} \,\rangle = \Omega_\pm H_0 \,|\, \boldsymbol{k} \,\rangle = E_k \Omega_\pm \,|\, \boldsymbol{k} \,\rangle = E_k \,|\, \boldsymbol{k}^{(\pm)} \,\rangle\,. \tag{35.28}$$

So we have

$$(E_k - H)(|\, \boldsymbol{k}^{(\pm)} \,\rangle - |\, \boldsymbol{k} \,\rangle) = -(E_k - H)\,|\, \boldsymbol{k} \,\rangle = V \,|\, \boldsymbol{k} \,\rangle\,. \tag{35.29}$$

Formally we can then write

$$|\, \boldsymbol{k}^{(\pm)} \,\rangle - |\, \boldsymbol{k} \,\rangle = (E_k - H)^{-1} V \,|\, \boldsymbol{k} \,\rangle \quad \text{(tentative)}\,,$$

where $(E_k - H)^{-1}$ is understood to be a formal inverse of the operator $E_k - H$. This tentative equation, however, is incomplete, since the right-hand side does not offer a recipe to distinguish between the incoming and outgoing states. To remedy this situation we recall the definitions of the scattering states $|\, \boldsymbol{k}^{(\pm)} \,\rangle$ again [cf. (35.11)]:

$$|\, \boldsymbol{k}^{(\pm)} \,\rangle = \Omega_\pm \,|\, \boldsymbol{k} \,\rangle = \lim_{t \to \mp\infty} e^{iHt/\hbar} e^{-iH_0 t/\hbar} \,|\, \boldsymbol{k} \,\rangle\,. \tag{35.30}$$

Hence

$$\lim_{t\to\mp\infty} e^{i(H-E_k)t/\hbar}(|\,\boldsymbol{k}^{(\pm)}\,\rangle - |\,\boldsymbol{k}\,\rangle) = 0\,, \tag{35.31}$$

where the "zero" on the right-hand side is understood to be the zero-vector in $\mathcal{H}$. On taking the inner product with $\langle\,\boldsymbol{k}'^{(\pm)}\,|$ and using the above tentative equation, we arrive at another tentative equation:

$$-\lim_{t\to\mp\infty}\frac{e^{i(E_{k'}-E_k)t/\hbar}}{E_{k'}-E_k}\langle\,\boldsymbol{k}'^{(\pm)}\,|\,V\,|\,\boldsymbol{k}\,\rangle = 0 \quad \text{(tentative)}\,,$$

where the zero on the right-hand side is now the number zero. Since the matrix element of V is in general non-zero, we are forced to conclude that the limit factor must be zero. However, for real values of $E_{k'}-E_k$, this limit is undefined. The only way that the limit can be taken to be zero is to understand it in the following sense:

$$\int_{-\omega_0}^{\omega_0} d\omega\, f(\omega)\left(\lim_{t\to\mp\infty}\frac{e^{i\omega t}}{\omega\mp i\eta}\right) = 0\,, \quad \eta\to 0^+\,, \tag{35.32}$$

where $f(\omega)$ is any function analytic in a region around $\omega = 0$ in the complex ω-plane. The above result is usually presented in short form as

$$\lim_{t\to\mp\infty}\frac{e^{i\omega t}}{\omega\mp i\eta} = 0\,. \tag{35.33}$$

In fact we also have the similar results

$$\lim_{t\to+\infty}\frac{e^{i\omega t}}{\omega - i\eta} = -\lim_{t\to-\infty}\frac{e^{i\omega t}}{\omega + i\eta} = 2\pi i\delta(\omega)\,. \tag{35.34}$$

The above two (shorthand) equations should be compared with a similar (shorthand) equation, given by (5.29). Both (35.33) and (35.34) can be easily checked by using the theory of residues in contour integration [as can be done for (5.29), cf. Problem 5.2]. Using (35.33) the inverse operator in the first tentative equation above should be written as $-(H-E_k\mp i\eta)^{-1}$, and we then have the following equation for the scattering states:

$$\boxed{\,|\,\boldsymbol{k}^{(\pm)}\,\rangle = |\,\boldsymbol{k}\,\rangle + \frac{1}{E_k - H \pm i\eta}V\,|\,\boldsymbol{k}\,\rangle\,}\,. \tag{35.35}$$

This very important equation in scattering theory is known as the **Lippmann-Schwinger equation**. It can also be written as

$$|\,\boldsymbol{k}^{(\pm)}\,\rangle = |\,\boldsymbol{k}\,\rangle + G(E_k\pm i\eta)V\,|\,\boldsymbol{k}\,\rangle\,, \tag{35.36}$$

where

$$G(z) \equiv (z - H)^{-1}\,, \tag{35.37}$$

depending on the complex variable z, is called the **Green's operator** of the Hamiltonian H. Since it approaches two different values as z approaches the positive real axis from above and below, $G(z)$, as an operator-valued function of a complex variable z, has a square-root branch structure with a **branch cut** along the positive real axis and **branch point** at $z = 0$.

Problem 35.2 Prove Eqs. (35.33) and (35.34) (in the sense that the expressions should occur under an integral sign) by contour integration and using the residue theorem. Note the positions of the poles of the integrands. The contours should be closed in the appropriate half-planes (lower or upper) depending on the signs.

Now the incoming and outgoing scattering states are, like the plane-wave states $|\,\boldsymbol{k}\,\rangle$, orthonormal in the sense that

$$\langle\,\boldsymbol{k}'^{(\pm)}\,|\,\boldsymbol{k}^{(\pm)}\,\rangle = \delta^3(\boldsymbol{k}'-\boldsymbol{k})\,. \tag{35.38}$$

Hence, on taking the inner product of (35.35) with $\langle\,\boldsymbol{k}'^{(\pm)}\,|$, we have

$$\langle\,\boldsymbol{k}'^{(\pm)}\,|\,\boldsymbol{k}\,\rangle = \delta^3(\boldsymbol{k}'-\boldsymbol{k}) + \frac{1}{E_{k'}-E_k \mp i\eta}\,\langle\,\boldsymbol{k}'^{(\pm)}\,|\,V\,|\,\boldsymbol{k}\,\rangle\,. \tag{35.39}$$

The inner products on the left-hand side of the above equation will be useful for the calculation of S-matrix elements, since

$$\begin{aligned} S_{\boldsymbol{k}'\boldsymbol{k}} &= \langle\,\boldsymbol{k}'\,|\,S\,|\,\boldsymbol{k}\,\rangle = \langle\,\boldsymbol{k}'\,|\,\Omega_-^\dagger\Omega_+\,|\,\boldsymbol{k}\,\rangle = \langle\,\boldsymbol{k}'^{(-)}\,|\,\Omega_+\,|\,\boldsymbol{k}\,\rangle \\ &= \lim_{t\to-\infty}\langle\,\boldsymbol{k}'^{(-)}\,|\,e^{iHt/\hbar}e^{-iH_0t/\hbar}\,|\,\boldsymbol{k}\,\rangle \\ &= \lim_{t\to-\infty} e^{i(E_{k'}-E_k)t/\hbar}\,\langle\,\boldsymbol{k}'^{(-)}\,|\,\boldsymbol{k}\,\rangle\,. \end{aligned} \tag{35.40}$$

It follows from (35.39) that

$$S_{\boldsymbol{k}'\boldsymbol{k}} = \lim_{t\to-\infty} e^{i(E_{k'}-E_k)/\hbar}\,\delta^3(\boldsymbol{k}'-\boldsymbol{k}) + \lim_{t\to-\infty}\frac{e^{i(E_{k'}-E_k)t/\hbar}}{E_{k'}-E_k+i\eta}\,\langle\,\boldsymbol{k}'^{(-)}\,|\,V\,|\,\boldsymbol{k}\,\rangle\,. \tag{35.41}$$

In the first term on the right-hand side, $E_{k'}$ is constrained to be equal to E_k because of the delta function, so it just reduces to the delta function itself. This fact, together with (35.34), allow us to write

$$S_{\boldsymbol{k}'\boldsymbol{k}} = \delta^3(\boldsymbol{k}'-\boldsymbol{k}) - 2\pi i\delta(E_{k'}-E_k)\langle\,\boldsymbol{k}'^{(-)}\,|\,V\,|\,\boldsymbol{k}\,\rangle\,. \tag{35.42}$$

In an entirely similar manner, we can also show that

$$S_{\boldsymbol{k}'\boldsymbol{k}} = \delta^3(\boldsymbol{k}'-\boldsymbol{k}) - 2\pi i\delta(E_{k'}-E_k)\langle\,\boldsymbol{k}'\,|\,V\,|\,\boldsymbol{k}^{(+)}\,\rangle\,. \tag{35.43}$$

On comparison with (35.23) one concludes that the on-shell T-matrix elements are given by

$$\langle \boldsymbol{k}' \,|\, T \,|\, \boldsymbol{k} \rangle = \langle \boldsymbol{k}'^{(-)} \,|\, V \,|\, \boldsymbol{k} \rangle = \langle \boldsymbol{k}' \,|\, V \,|\, \boldsymbol{k}^{(+)} \rangle \quad (E_{k'} = E_k) \,. \tag{35.44}$$

Making use of the Lippmann-Schwinger equation (35.35) and the definition of the Green's operator given by (35.37), the second equality of the above equation implies

$$\begin{aligned}\langle \boldsymbol{k}' \,|\, T \,|\, \boldsymbol{k} \rangle &= \langle \boldsymbol{k}' \,|\, V \,|\, \boldsymbol{k} \rangle + \langle \boldsymbol{k}' \,|\, V \frac{1}{E_k - H + i\eta} V \,|\, \boldsymbol{k} \rangle \\ &= \langle \boldsymbol{k}' \,|\, V \,|\, \boldsymbol{k} \rangle + \langle \boldsymbol{k}' \,|\, V G(E_k + i\eta) V \,|\, \boldsymbol{k} \rangle \,.\end{aligned} \tag{35.45}$$

This equation suggests that one can define an operator-valued function of the complex variable z, the **transition operator** $T(z)$, by

$$T(z) = V + VG(z)V \,, \tag{35.46}$$

which has the same analytic (branch) structure as $G(z)$. The **off-shell** T-matrix elements $\langle \boldsymbol{k}' \,|\, T(z) \,|\, \boldsymbol{k} \rangle$ (where the energy conservation condition $E_{k'} = E_k$ is not required) then acquire a meaning independent of (35.44). The on- and off-shell T-matrix elements are then related by

$$\langle \boldsymbol{k}' \,|\, T \,|\, \boldsymbol{k} \rangle = \langle \boldsymbol{k}' \,|\, T(E_k + i\eta) \,|\, \boldsymbol{k} \rangle \,. \tag{35.47}$$

An iterative equation for $T(z)$, suitable for calculational purposes and completely equivalent to the Lippmann-Schwinger equation (35.5), can be deduced as follows. Analogous to the full $G(z)$ [(35.37)], we define a Green's operator for the unperturbed Hamiltonian H_0 by

$$G_0(z) = (z - H_0)^{-1} \,. \tag{35.48}$$

$G_0(z)$ has the advantage that its actions on the asymptotic (unperturbed) eigenstates $|\, \boldsymbol{k} \rangle$ are known. Based on the general (and quite obvious) operator identity

$$A^{-1} = B^{-1} + B^{-1}(B - A)A^{-1} \,, \tag{35.49}$$

which holds for any two invertible operators A and B, we immediately have, on setting $A = z - H$ and $B = z - H_0$,

$$G(z) = G_0(z) + G_0(z) V G(z) \,; \tag{35.50}$$

or equivalently, on setting $A = z - H_0$ and $B = z - H$,

$$G(z) = G_0(z) + G(z) V G_0(z) \,. \tag{35.51}$$

Multiplying the definition of $T(z)$ given by (35.46) on the left by G_0 we have

$$G_0 T = G_0 V + G_0 V G V = (G_0 + G_0 V G) V = G V \,, \tag{35.52}$$

where the last equality follows from (35.50). Thus we have

$$\boxed{T(z) = V + VG_0(z)T(z)} \quad , \tag{35.53}$$

which is the Lippmann-Schwinger equation for the T-operator. An iterative procedure then yields the so-called **Born series**, useful in perturbation calculations:

$$T = V + VG_0V + VG_0VG_0V + \dots \quad . \tag{35.54}$$

The **Born approximation** is obtained when only the first term is kept, that is, on setting $T \approx V$. We note that, since $H = H^\dagger$,

$$G(z^*) = (G(z))^\dagger \, . \tag{35.55}$$

This immediately implies that the transition operator satisfies the same analytic condition:

$$T(z^*) = (T(z))^\dagger \, . \tag{35.56}$$

To establish some contact between the abstract formalism introduced above and a more intuitive understanding, we will next study the asymptotic ($r \to \infty$) behavior of the coordinate representation of an outgoing scattering state, that is, the Schrödinger wave function $\psi_{\boldsymbol{k}}^{(\pm)} \equiv \langle \boldsymbol{r} \,|\, \boldsymbol{k}^{(\pm)} \rangle$. Recall from (35.52) that $GV = G_0T$. Also, from (35.44) and (35.56),

$$T(E_k \pm i\eta) \,|\, \boldsymbol{k} \,\rangle = V \,|\, \boldsymbol{k}^{(\pm)} \,\rangle \, . \tag{35.57}$$

The Lippmann-Schwinger equation (35.36) can then be rewritten as

$$\boxed{|\, \boldsymbol{k}^{(\pm)} \,\rangle = |\, \boldsymbol{k} \,\rangle + G_0(E_k \pm i\eta) V \,|\, \boldsymbol{k}^{(\pm)} \,\rangle = |\, \boldsymbol{k} \,\rangle + \frac{1}{E_k - H_0 \pm i\eta} V \,|\, \boldsymbol{k}^{(\pm)} \,\rangle} \quad . \tag{35.58}$$

This equation, more appropriate for perturbation calculations than (35.36), is also known as the Lippmann-Schwinger equation. Taking the inner product with $\langle \boldsymbol{r} \,|$ we have

$$\langle \boldsymbol{r} \,|\, \boldsymbol{k}^{(\pm)} \,\rangle = \langle \boldsymbol{r} \,|\, \boldsymbol{k} \,\rangle + \int d^3r' \, \langle \boldsymbol{r} \,|\, G_0(E_k \pm i\eta) \,|\, \boldsymbol{r}' \,\rangle \, V(\boldsymbol{r}') \, \langle \boldsymbol{r}' \,|\, \boldsymbol{k}^{\pm} \,\rangle \, , \tag{35.59}$$

or

$$\psi_{\boldsymbol{k}}^{(\pm)}(\boldsymbol{r}) = \psi_{\boldsymbol{k}}(\boldsymbol{r}) + \int d^3r' \, \langle \boldsymbol{r} \,|\, G_0(E_k \pm i\eta) \,|\, \boldsymbol{r}' \,\rangle \, V(\boldsymbol{r}') \, \psi_{\boldsymbol{k}}^{(\pm)}(\boldsymbol{r}') \, . \tag{35.60}$$

We will now carefully evaluate the free-particle Green's operator matrix element. Since $|\, \boldsymbol{k} \,\rangle$ is an eigenstate of $G_0(z)$ with eigenvalue $1/(z - E_k)$, we have, on recalling (35.22),

$$\begin{aligned}\langle \boldsymbol{r} \,|\, G_0(z) \,|\, \boldsymbol{r}' \rangle &= \int d^3k \, \langle \boldsymbol{r} \,|\, G_0(z) \,|\, \boldsymbol{k} \rangle \langle \boldsymbol{k} \,|\, \boldsymbol{r}' \rangle = \frac{1}{(2\pi)^3} \int d^3k \, \frac{e^{i\boldsymbol{k}\cdot(\boldsymbol{r}-\boldsymbol{r}')}}{z - E_k} \\ &= \frac{2\pi}{(2\pi)^3} \int_0^\infty dk \frac{k^2}{z - \frac{\hbar^2 k^2}{2m}} \int_0^\pi d\theta \, \sin\theta \, e^{ik|\boldsymbol{r}-\boldsymbol{r}'| \cos\theta} \\ &= -\frac{1}{2\pi^2} \int_0^\infty dk \, \frac{mk^2}{\hbar^2 k^2 - 2mz} \int_{-1}^1 dz \, e^{ik|\boldsymbol{r}-\boldsymbol{r}'|z} \\ &= \frac{im}{2\pi^2 |\boldsymbol{r}-\boldsymbol{r}'|} \int_0^\infty dk \, k \, \frac{e^{ik|\boldsymbol{r}-\boldsymbol{r}'|} - e^{-ik|\boldsymbol{r}-\boldsymbol{r}'|}}{\hbar^2 k^2 - 2mz} \\ &= \frac{im}{2\pi^2 |\boldsymbol{r}-\boldsymbol{r}'|} \int_{-\infty}^\infty dk \, \frac{k \, e^{ik|\boldsymbol{r}-\boldsymbol{r}'|}}{\hbar^2 k^2 - 2mz} \,. \end{aligned} \tag{35.61}$$

The last integral can be evaluated by contour integration (completing the closed contour by an infinite semicircle in the upper-half plane, so that it encloses the pole $k = +(2mz)^{1/2}/\hbar$). We then have

$$\langle \boldsymbol{r} \,|\, G_0(z) \,|\, \boldsymbol{r}' \rangle = -\frac{m}{2\pi\hbar^2} \frac{\exp\left(\frac{i}{\hbar} (2mz)^{1/2} |\boldsymbol{r}-\boldsymbol{r}'| \right)}{|\boldsymbol{r}-\boldsymbol{r}'|} \,. \tag{35.62}$$

Now, as $z \longrightarrow E_k \pm i\eta = \hbar^2 k^2/(2m) \pm i\eta$, $(2mz)^{1/2} \longrightarrow \pm\hbar k$. Thus we arrive at the following result for the free-particle **Green's function**:

$$\langle \boldsymbol{r} \,|\, G_0(E_k \pm i\eta) \,|\, \boldsymbol{r}' \rangle = -\frac{m}{2\pi\hbar^2} \frac{e^{\pm ik|\boldsymbol{r}-\boldsymbol{r}'|}}{|\boldsymbol{r}-\boldsymbol{r}'|} \,, \tag{35.63}$$

and the following Lippmann-Schwinger equation for the incoming and outgoing scattering-state Schrödinger wave functions:

$$\psi_{\boldsymbol{k}}^{(\pm)}(\boldsymbol{r}) = \frac{1}{(2\pi)^{3/2}} \, e^{i\boldsymbol{k}\cdot\boldsymbol{r}} - \frac{m}{2\pi\hbar^2} \int d^3r' \, \frac{e^{\pm ik|\boldsymbol{r}-\boldsymbol{r}'|}}{|\boldsymbol{r}-\boldsymbol{r}'|} \, V(\boldsymbol{r}') \psi_{\boldsymbol{k}}^{(\pm)}(\boldsymbol{r}') \,. \tag{35.64}$$

For $|\boldsymbol{r}-\boldsymbol{r}'| \gg 0$ we can make the expansion

$$|\boldsymbol{r}-\boldsymbol{r}'| \approx r \left(1 - \frac{\boldsymbol{r}\cdot\boldsymbol{r}'}{r^2} + O\left(\frac{r'}{r}\right)^2 \right) . \tag{35.65}$$

Then the asymptotic behavior of the scattering-state wave functions is given by

$$\psi_{\boldsymbol{k}}^{(\pm)}(\boldsymbol{r}) \xrightarrow[r\to\infty]{} \frac{1}{(2\pi)^{3/2}} \, e^{i\boldsymbol{k}\cdot\boldsymbol{r}} - \frac{m}{2\pi\hbar^2} \frac{e^{\pm ikr}}{r} \int d^3r' \, e^{\mp ik\hat{\boldsymbol{r}}\cdot\boldsymbol{r}'} \, V(\boldsymbol{r}') \, \psi_{\boldsymbol{k}}^{(\pm)}(\boldsymbol{r}') \,. \tag{35.66}$$

Recognizing that

$$
\begin{aligned}
&-\frac{m}{2\pi\hbar^2}\int d^3r'\, e^{-ik\hat{\boldsymbol{r}}\cdot\boldsymbol{r}'}\, V(\boldsymbol{r}')\,\psi_{\boldsymbol{k}}^{(+)}(\boldsymbol{r}') \\
&= -\frac{m}{2\pi\hbar^2}\int d^3r'\,(2\pi)^{3/2}\langle\, k\hat{\boldsymbol{r}}\,|\,\boldsymbol{r}'\,\rangle\, V(\boldsymbol{r}')\,\psi_{\boldsymbol{k}}^{(+)}(\boldsymbol{r}') = -\frac{m(2\pi)^{3/2}}{2\pi\hbar^2}\,\langle\, k\hat{\boldsymbol{r}}\,|\,V\,|\,\boldsymbol{k}^{(+)}\,\rangle \\
&= \frac{1}{(2\pi)^{3/2}}\, f(\boldsymbol{k}\to k\hat{\boldsymbol{r}})\,,
\end{aligned}
\tag{35.67}
$$

where $f(\boldsymbol{k} \to k\hat{\boldsymbol{r}})$ is the scattering amplitude introduced in (35.24), we can finally write

$$
\boxed{\psi_{\boldsymbol{k}}^{(+)}(\boldsymbol{r}) \xrightarrow[r\to\infty]{} \frac{1}{(2\pi)^{3/2}}\left(e^{i\boldsymbol{k}\cdot\boldsymbol{r}} + f(\boldsymbol{k}\to k\hat{\boldsymbol{r}})\,\frac{e^{ikr}}{r}\right)}\;. \tag{35.68}
$$

In this expression, the first term represents the "incident wave" while the second represents the "scattered wave". It justifies the terms "scattering amplitude" and "outgoing" wave. In fact we have already encountered a similar expression in (16.49).

We would now like to examine the use of the spherical basis $\{|\, E, l, m\,\rangle\}$ in scattering theory, which satisfies

$$
H_0\,|\, E,l,m\,\rangle = E\,|\, E,l,m\,\rangle\,, \tag{35.69}
$$

$$
L^2\,|\, E,l,m\,\rangle = l(l+1)\hbar^2\,|\, E,l,m\,\rangle\,, \tag{35.70}
$$

$$
L_z\,|\, E,l,m\,\rangle = m\hbar\,|\, E,l,m\,\rangle\,. \tag{35.71}
$$

These states satisfy the normalization condition

$$
\langle\, E',l',m'\,|\, E,l,m\,\rangle = \delta(E'-E)\delta_{ll'}\delta_{mm'}\,. \tag{35.72}
$$

In the coordinate representation we have the corresponding wave functions given by

$$
\psi_{Elm}(\boldsymbol{r}) = \langle\, \boldsymbol{r}\,|\, E,l,m\,\rangle\,, \tag{35.73}
$$

and the analog of (35.72) is

$$
\int d^3r\,\psi^*_{E'l'm'}(\boldsymbol{r})\psi_{Elm}(\boldsymbol{r}) = \delta(E'-E)\delta_{l'l}\delta_{m'm}\,. \tag{35.74}
$$

The wave functions $\psi_{Elm}(\boldsymbol{r})$ being solutions of the free-particle Schrödinger equation, it can be verified without too much difficulty that

$$
\psi_{Elm}(\boldsymbol{r}) = \langle\, \boldsymbol{r}\,|\, E,l,m\,\rangle = i^l\left(\frac{2m}{\pi\hbar^2 k}\right)^{1/2}\frac{1}{r}\,\hat{j}_l(kr)Y_l^m(\hat{\boldsymbol{r}})\,, \tag{35.75}
$$

where the normalization phase constant i^l is chosen for convenience, $Y_l^m(\hat{\boldsymbol{r}})$ are the spherical harmonics introduced in Chapter 21, and $\hat{j}_l(z)$ are the so-called **Riccati-Bessel functions** given by

$$\hat{j}_l(z) \equiv z j_l(z) = \sqrt{\frac{\pi z}{2}}\, J_{l+1/2}(z) = z^{l+1} \sum_{n=0}^{\infty} \frac{(-z^2/2)^n}{n!(2l+2n+1)!!} \,. \tag{35.76}$$

In the above equation, $j_l(z)$ are the ordinary **spherical Bessel functions** and $J_{l+1/2}(z)$ are the ordinary **Bessel functions**. The Riccati-Bessel functions appearing in (35.75) satisfy the normalization condition

$$\int_0^\infty dr\, \hat{j}_l(k'r)\hat{j}_l(kr) = \frac{\pi}{2}\delta(k'-k) \tag{35.77}$$

and the physical boundary condition that $\hat{j}_l(kr) \to 0$ as $r \to 0$.

One can also construct the free-particle momentum representation wave functions as follows:

$$\begin{aligned}\psi_{Elm}(\boldsymbol{k}) = \langle\, \boldsymbol{k} \,|\, E, l, m \,\rangle &= \int d^3r \langle\, \boldsymbol{k} \,|\, \boldsymbol{r} \,\rangle\langle\, \boldsymbol{r} \,|\, E, l, m \,\rangle \\ &= \int d^3r \, \frac{1}{(2\pi)^{3/2}} e^{-i\boldsymbol{k}\cdot\boldsymbol{r}} \,\langle\, \boldsymbol{r} \,|\, E, l, m \,\rangle \\ &= \sqrt{\frac{\hbar^2}{mk}}\, \delta(E_k - E) Y_l^m(\hat{\boldsymbol{k}}) \,.\end{aligned} \tag{35.78}$$

This result can be used to obtain an expansion of plane waves in terms of spherical waves. Indeed

$$\begin{aligned}|\, \boldsymbol{k} \,\rangle &= \int dE \sum_{lm} |\, E, l, m \,\rangle\langle\, E, l, m \,|\, \boldsymbol{k} \,\rangle \\ &= \sqrt{\frac{\hbar^2}{mk}} \int dE \sum_{lm} |\, E, l, m \,\rangle \delta(E_k - E)(Y_l^m(\hat{\boldsymbol{k}}))^* \\ &= \sqrt{\frac{\hbar^2}{mk}} \sum_{lm} (Y_l^m(\hat{\boldsymbol{k}}))^* \,|\, E_k, l, m \,\rangle \,.\end{aligned} \tag{35.79}$$

Hence

$$\begin{aligned}\langle\, \boldsymbol{r} \,|\, \boldsymbol{k} \,\rangle = \frac{1}{(2\pi)^{3/2}} e^{i\boldsymbol{k}\cdot\boldsymbol{r}} &= \sqrt{\frac{\hbar^2}{mk}} \sum_{lm} \langle\, \boldsymbol{r} \,|\, E_k, l, m \,\rangle (Y_l^m(\hat{\boldsymbol{k}}))^* \\ &= \sqrt{\frac{2}{\pi}}\, \frac{1}{kr} \sum_{lm} i^l \hat{j}_l(kr) Y_l^m(\hat{\boldsymbol{r}}) (Y_l^m(\hat{\boldsymbol{k}}))^* \,,\end{aligned} \tag{35.80}$$

where in the third equality on the right we have used (35.75). On choosing $\boldsymbol{k}$ to be along the z-axis, only the $m = 0$ term contributes to the above sum over

m [cf. (21.39)]. Then, with $\hat{\boldsymbol{r}} = (\theta, \varphi)$ and using (21.47), we have

$$\boxed{e^{i\boldsymbol{k}\cdot\boldsymbol{r}} = \frac{1}{kr}\sum_{l=0}^{\infty}(2l+1)\, i^l\, \hat{j}_l(kr) P_l(\hat{\boldsymbol{r}}\cdot\hat{\boldsymbol{k}})} \quad , \tag{35.81}$$

where $P_l(\cos\theta)$ are the Legendre polynomials given by (21.43). This is the expansion of plane waves in terms of the spherical waves $\hat{j}_l(kr)$.

Now the scattering operator S commutes with H_0 [cf. (35.20)] and the angular momentum $\boldsymbol{L}$ (and thus the raising and lowering operators $L_\pm$ also). We then have

$$S\,|\,E,l,m\,\rangle = S_l(E)\,|\,E,l,m\,\rangle\ , \tag{35.82}$$

where $S_l(E)$ is independent of m. Since S is unitary we can write

$$S_l(E) = e^{2i\delta_l(E)}\ . \tag{35.83}$$

Thus

$$\begin{aligned}
\langle\,\boldsymbol{k}'\,|\,S-1\,|\,\boldsymbol{k}\,\rangle &= \int dE \sum_{lm}\langle\,\boldsymbol{k}'\,|\,S-1\,|\,E,l,m\,\rangle\langle\,E,l,m\,|\,\boldsymbol{k}\,\rangle \\
&= \int dE \sum_{lm}(S_l(E)-1)\langle\,\boldsymbol{k}'\,|\,E,l,m\,\rangle\langle\,E,l,m\,|\,\boldsymbol{k}\,\rangle \\
&= \int dE \sum_{lm}(S_l(E)-1)\sqrt{\frac{\hbar^2}{mk}}\sqrt{\frac{\hbar^2}{mk'}}\,\delta(E_{k'}-E)\delta(E_k-E)\,Y_l^m(\hat{\boldsymbol{k}'})(Y_l^m(\hat{\boldsymbol{k}}))^* \\
&= \frac{\hbar^2}{mk}\,\delta(E_{k'}-E_k)\sum_{lm} Y_l^m(\hat{\boldsymbol{k}'})(Y_l^m(\hat{\boldsymbol{k}}))^*(S_l(E_k)-1)\ .
\end{aligned} \tag{35.84}$$

On the other hand, it follows from (35.23) and (35.24) that

$$\langle\,\boldsymbol{k}'\,|\,S-1\,|\,\boldsymbol{k}\,\rangle = \frac{i\hbar^2}{2\pi m}\delta(E_{k'}-E_k)\,f_{\boldsymbol{k}\to\boldsymbol{k}'}\ . \tag{35.85}$$

Comparison of the above two equations then yields

$$f_{\boldsymbol{k}\to\boldsymbol{k}'} = \frac{2\pi}{ik}\sum_{lm} Y_l^m(\hat{\boldsymbol{k}'})(Y_l^m(\hat{\boldsymbol{k}}))^*\,(S_l(E_k)-1)\ . \tag{35.86}$$

On choosing $\boldsymbol{k}$ to be along the z-axis as before and with $\hat{\boldsymbol{k}'} = (\theta,\varphi)$, we can rewrite the above expression for the scattering amplitude as

$$f(E_k,\theta) \equiv f_{\boldsymbol{k}\to\boldsymbol{k}'} = \frac{1}{2ik}\sum_{l=0}^{\infty}(2l+1)(S_l(E_k)-1)P_l(\cos\theta)\ . \tag{35.87}$$

Defining the **partial wave amplitude** by

$$f_l(E_k) \equiv \frac{S_l(E_k)-1}{2ik} = \frac{e^{2i\delta_l(E_k)}-1}{2ik} = \frac{e^{i\delta_l(E_k)}}{k}\,\sin\delta_l(E_k)\,, \tag{35.88}$$

we can then write

$$f(E_k,\theta)=\sum_{l=0}^{\infty}(2l+1)f_l(E_k)P_l(\cos\theta)\,. \tag{35.89}$$

This result is known as the **partial wave expansion** of the scattering amplitude.

Problem 35.3 Use (35.25) for $d\sigma/d\Omega(\boldsymbol{k}\to\boldsymbol{k}')$, (35.88) for $f_l(k)$, and (35.89) for the total scattering amplitude $f(\boldsymbol{k}\to\boldsymbol{k}')$ to show that, after integrating over Ω, that is, over all directions of $\boldsymbol{k}'$, the cross terms in the double sum arising from (35.89) all cancel out, and one obtains the following expression for the total scattering cross section σ:

$$\sigma=\sum_l \sigma_l(k)\,,\quad \sigma_l(k)=4\pi(2l+1)\frac{\sin^2\delta_l(k)}{k^2}\,.$$

Problem 35.4 Prove the **Optical Theorem** for the total scattering cross section:

$$\sigma=\frac{4\pi}{k}\,Im f(\theta=0)\,.$$

[Hints: Use the fact that the scattering operator S is unitary: $SS^\dagger=1$, and calculate a diagonal matrix element of this condition, with the help of (35.23) to (35.25).]

Let us now turn our attention to the radial wave equation (16.12), which will be reproduced here for convenience:

$$\frac{d^2\chi_{kl}}{dr^2}+\{k^2-U_l(r)\}\chi_{kl}=0\,,$$

$$U_l(r)\equiv\frac{l(l+1)}{r^2}+V'(r)\,,\qquad V'(r)\equiv\frac{2m}{\hbar^2}V(r)\,.$$

The free-particle wave equation is

$$\frac{d^2\chi_{kl}}{dr^2}+\left(k^2-\frac{l(l+1)}{r^2}\right)\chi_{kl}=0\,. \tag{35.90}$$

The solution to this equation that satisfies the physical boundary condition $\chi_{kl}\to 0$ as $r\to 0$ is precisely the Riccati-Bessel function $\hat{j}_l(kr)$ given by (35.76). The other solution, linearly independent of $\hat{j}_l(kr)$, behaves like r^{-l} as $r\to 0$, and is given by the so-called **Riccati-Neumann function**:

$$\hat{n}_l(z)\equiv z n_l(z)=(-1)^l\sqrt{\frac{\pi z}{2}}\,J_{-l-1/2}(z)=z^{-l}\sum_{n=0}^{\infty}\frac{(-z^2/2)^n(2l-2n-1)!!}{n!}$$
$$\xrightarrow[z\to 0]{} z^{-l}(2l-1)!![1+O(z^2)]\,, \tag{35.91}$$

where $n_l(z)$ are the ordinary spherical **Neumann functions** and $(-2n-1)!! \equiv (-1)^n/(2n-1)!!$ $[(-1)!! \equiv 1]$. As $r \to \infty$, $l(l+1)/r^2 \to 0$, and the solutions to (35.90) behave like $\chi_{kl}(r) \to e^{\pm ikr}$. The so-called **Riccati-Hankel functions** $\hat{h}_l^{\pm}$, which are linear combinations of $\hat{j}_l(z)$ and $\hat{n}_l(z)$, have precisely this property:

$$\hat{h}_l^{\pm}(z) \equiv \hat{n}_l(z) \pm i\hat{j}_l(z)\,, \tag{35.92}$$

$$\xrightarrow[z\to\infty]{} \exp\{\pm i(z - l\pi/2)\}[1 + O(z^{-1})]\,. \tag{35.93}$$

It is interesting to note that

$$\hat{j}_l(z) = \frac{\hat{h}_l^{+}(z) - \hat{h}_l^{-}(z)}{2i} \xrightarrow[z\to\infty]{} \sin(z - l\pi/2) + O(z^{-1})\,. \tag{35.94}$$

Consider the scattering states

$$|\, E, l, m^{(+)} \rangle = \Omega_+ \,|\, E, l, m \rangle\,. \tag{35.95}$$

Since $H\Omega_+ = \Omega_+ H_0$ and $[\,\Omega_+, \boldsymbol{L}\,] = 0$, these are eigenstates of H and L^2:

$$H \,|\, E, l, m^{(+)} \rangle = E \,|\, E, l, m^{(+)} \rangle\,, \tag{35.96}$$

$$L^2 \,|\, E, l, m^{(+)} \rangle = l(l+1)\hbar^2 \,|\, E, l, m^{(+)} \rangle\,; \tag{35.97}$$

with the normalization condition

$$\langle\, E', l', m'^{(+)} \,|\, E, l, m^{(+)} \rangle = \delta(E' - E)\delta_{ll'}\delta_{mm'}\,. \tag{35.98}$$

Analogous to (35.75), we can write

$$\langle\, \boldsymbol{r} \,|\, E, l, m^{(+)} \rangle = i^l \left(\frac{2m}{\pi\hbar^2 k}\right)^{1/2} \frac{1}{r}\, \psi_{l,k}(r) Y_l^m(\hat{\boldsymbol{r}})\,, \tag{35.99}$$

where $\psi_{l,k}(r)$ is a physical solution satisfying (16.12) with the boundary condition

$$\psi_{l,k}(r) \xrightarrow[r\to 0]{} 0\,. \tag{35.100}$$

Using (35.99) to expand the inner product on the left-hand side of (35.98), we have

$$\begin{aligned}
&\int d^3r \,\langle\, E', l', m'^{(+)} \,|\, \boldsymbol{r} \rangle\langle\, \boldsymbol{r} \,|\, E, l, m^{(+)} \rangle \\
&= \int_0^\infty dr\, r^2 \int d\Omega \left(\frac{2m}{\pi\hbar^2}\right) \frac{1}{\sqrt{k'k}\, r^2}\, \psi^*_{l',k'}(r)\psi_{lk}(r)(Y_{l'}^{m'}(\hat{\boldsymbol{r}}))^* Y_l^m(\hat{\boldsymbol{r}}) \\
&= \left(\frac{2m}{\pi\hbar^2}\right) \frac{1}{\sqrt{k'k}}\, \delta_{ll'}\delta_{mm'} \int_0^\infty dr\, \psi^*_{l',k'}(r)\psi_{l,k}(r) \\
&= \delta(E' - E)\delta_{ll'}\delta_{mm'} = \frac{m}{\hbar^2 k}\, \delta(k' - k)\delta_{ll'}\delta_{mm'}\,.
\end{aligned} \tag{35.101}$$

Thus we obtain the following normalization condition for the radial wave functions $\psi_{l,k}(r)$:

$$\int_0^\infty dr\, \psi^*_{l,k'}(r)\psi_{l,k}(r) = \frac{\pi}{2}\,\delta(k'-k)\,, \tag{35.102}$$

which is the same as that for the Riccati-Bessel functions $\hat{j}_l(kr)$ [cf. (35.77)]. The functions $\psi_{l,k}(r)$ are called the normalized radial wave functions.

The normalized radial wave functions play the same role in the coordinate representation of the scattering states $|\,\boldsymbol{k}^{(+)}\,\rangle$ as the Riccati-Bessel functions do in the plane-wave expansion (35.81). Analogous to that equation we have

$$\psi^{(+)}_{\boldsymbol{k}}(\boldsymbol{r}) = \langle\,\boldsymbol{r}\,|\,\boldsymbol{k}^{(+)}\,\rangle = \frac{1}{(2\pi)^{3/2}kr}\sum_{l=0}^{\infty}(2l+1)i^l\psi_{l.k}(r)P_l(\hat{\boldsymbol{r}}\cdot\hat{\boldsymbol{k}})\,. \tag{35.103}$$

On the other hand, it follows from the asymptotic condition for $\psi^{(+)}_{\boldsymbol{k}}(\boldsymbol{r})$ [(35.68)], the plane wave expansion [(35.81)], and the partial wave expansion for the scattering amplitude [(35.89)] that

$$\psi^{(+)}_{\boldsymbol{k}}(\boldsymbol{r}) \xrightarrow[r\to\infty]{} \frac{1}{(2\pi)^{3/2}kr}\sum_l (2l+1)\{i^l\hat{j}_l(kr) + kf_l(k)e^{ikr}\}P_l(\hat{\boldsymbol{r}}\cdot\hat{\boldsymbol{k}})\,, \tag{35.104}$$

where we have written $f_l(k)$ for $f_l(E_k)$. Comparing the above two equations we see that the asymptotic behavior of the normalized radial wave functions $\psi_{l,k}(r)$ is given by

$$\begin{aligned}\psi_{l,k}(r) &\xrightarrow[r\to\infty]{} \hat{j}_l(kr) + kf_l(k)\exp\{i(kr - l\pi/2)\}\\ &\xrightarrow[r\to\infty]{} \hat{j}_l(kr) + kf_l(k)\hat{h}^+_l(kr)\,,\end{aligned} \tag{35.105}$$

where the last limit follows from (35.93). Recalling (35.94) and (35.88), a few simple steps will lead to the result

$$\boxed{\psi_{l,k}(r) \xrightarrow[r\to\infty]{} e^{i\delta_l(k)}\sin(kr - l\pi/2 + \delta_l(k))}\;. \tag{35.106}$$

We have already seen a special case of this result (for $l = 0$) in Chapter 16 [Eq. (16.38)]. Again, the above equation justifies the name "**phase shift**" for the quantity $\delta_l(k)$. On using (35.94) for $\hat{j}_l(kr)$ and replacing $kf_l(k)$ by $(S_l(k)-1)/2i$ [cf. (35.88)], Eq. (35.105) can be rewritten as

$$\psi_{l,k}(r) \xrightarrow[r\to\infty]{} \frac{i}{2}\{\hat{h}^-_l(kr) - S_l(k)\hat{h}^+_l(kr)\}\,, \tag{35.107}$$

which is the l-th partial wave generalization of (16.19).

Analogous to (35.60), the Lippmann-Schwinger (integral) equation for $\psi_{l,k}(r)$ incorporating the appropriate boundary conditions is

$$\psi_{l,k}(r) = \hat{j}_l(kr) + \int_0^\infty dr'\, G^{l,k}_0(r,r')V'(r')\psi_{l,k}(r')\,, \tag{35.108}$$

where

$$G_0^{l,k}(r,r') = -\frac{1}{k}\hat{j}_l(kr_<)\hat{h}_l^+(kr_>)\,, \tag{35.109}$$

is the l-th partial wave spherical free-particle Green's function. In the above equation, $r_<$ ($r_>$) is the smaller (larger) of r and r'. On expanding the expression for the scattering amplitude given by (35.24) in terms of partial waves [and recalling (35.44) for the T-matrix element] we can also show that

$$f_l(k) = -\frac{1}{k^2}\int_0^\infty dr\,\hat{j}_l(kr)V'(r)\psi_{l,k}(r)\,. \tag{35.110}$$

Problem 35.5 Prove Eq. (35.109), using an approach similar to the derivation of Eq. (35.63).

Problem 35.6 Verify Eq. (35.110) for the l-th partial wave scattering amplitude.

As seen above, the normalized radial wave functions $\psi_{l,k}(r)$ are characterized by the small r behavior

$$\psi_{l,k}(r) \xrightarrow[r\to 0]{} \textit{constant} \times \hat{j}_l(kr)\,, \tag{35.111}$$

where *constant* means a constant independent of r. It turns out that it is immensely useful to consider the solution to the radial Schrödinger equation (16.12) specified by the single-point boundary condition

$$\phi_{l,k}(r) \xrightarrow[r\to 0]{} \hat{j}_l(kr)\,. \tag{35.112}$$

This is called the **regular solution**. It can be shown that it satisfies the following **Volterra**-type integral equation:

$$\phi_{l,k}(r) = \hat{j}_l(kr) + \int_0^r dr'\, g_0^{l,k}(r,r')V'(r')\phi_{l,k}(r')\,, \tag{35.113}$$

where $g_0^{l,k}(r,r')$, the corresponding free-particle Green's function, is given by

$$g_0^{l,k}(r,r') = \frac{1}{k}[\hat{j}_l(kr)\hat{n}_l(kr') - \hat{n}_l(kr)\hat{j}_l(kr')]\,. \tag{35.114}$$

(For a proof, the reader may consult, for example, Taylor 1983.) Since $\phi_{l,k}(r)$ is real we may write

$$\phi_{l,k}(r) \xrightarrow[r\to\infty]{} \frac{i}{2}\{\mathcal{F}_l(k)\hat{h}_l^-(kr) - (\mathcal{F}_l(k))^*\hat{h}_l^+(kr)\}\,. \tag{35.115}$$

Comparing this result with the asymptotic behavior for $\psi_{l,k}(r)$ given by (35.107), and realizing that $\phi_{l,k}(r) \propto \psi_{l,k}(r)$ for all r, we have the following important expression for the scattering matrix:

$$\boxed{S_l(k) = \frac{(\mathcal{F}_l(k))^*}{\mathcal{F}_l(k)}} \quad . \tag{35.116}$$

Furthermore,

$$\phi_{l,k}(r) = \mathcal{F}_l(k)\psi_{l,k}(r) \; . \tag{35.117}$$

The quantity $\mathcal{F}_l(k)$ is called the **Jost function**, and plays a central role in potential scattering theory. According to (35.113), (35.114), and (35.115), it is given by

$$\mathcal{F}_l(k) = 1 + \frac{1}{k}\int_0^\infty dr\, \hat{h}_l^+(kr)V'(r)\phi_{l,k}(r) \; . \tag{35.118}$$

In the next chapter we will explore some of the theoretical significance of this important function.

Problem 35.7 Verify (35.118) for the Jost function.

Problem 35.8 Consider the Schrödinger equation

$$-\frac{\hbar^2}{2\mu}\nabla^2\psi(\boldsymbol{r}) + \int d^3r'\, V(\boldsymbol{r},\boldsymbol{r}')\psi(\boldsymbol{r}') = E\psi(\boldsymbol{r}) \, ,$$

where the **nonlocal potential** has the following separable form:

$$V(\boldsymbol{r},\boldsymbol{r}') = -\frac{\hbar^2}{2\mu}\lambda u(|\boldsymbol{r}|)u(|\boldsymbol{r}'|) \, , \quad (\lambda > 0, \in \mathbb{R}) \; .$$

(a) Establish the integral equation that is equivalent to the Schrödinger equation with the boundary conditions for outgoing waves.

(b) Show that the integral equation obtained in (a) can be reduced to an algebraic equation. Solve this algebraic equation and show that the scattering amplitude for incident momentum $\hbar\boldsymbol{k}$ is given by

$$f(k) = \frac{4\pi\lambda|v(k)|^2}{1 + \frac{2\lambda}{\pi}\int d^3q\, \frac{|v(q)|^2}{k^2 - q^2 + i\epsilon^+}} \, ,$$

where

$$v(k) \equiv \int_0^\infty dr\, r^2 u(r)\frac{\sin kr}{kr} \; .$$

Problem 35.9 Suppose that in the nonlocal potential in the above problem, we have

$$u(r) = \frac{e^{-r/b}}{r}, \quad (b > 0) .$$

(a) Show that the scattering phase shift $\delta(k)$ is given by

$$k \cot \delta(k) = \frac{(k^2b^2 + 1)^2 + \xi(k^2b^2 - 1)}{2\xi b},$$

where $\xi \equiv 2\pi\lambda b^3$.

(b) For low energies ($k \to 0$), one can expand

$$k \cot \delta(k) = -\frac{1}{a} + O(k^2),$$

where a is called the **scattering length**. Evaluate a. Express the zero-energy limit of the total cross section σ in terms of a.

(c) Discuss the behavior of a as a function of ξ. Anticipating our discussion in the next chapter that *a bound state appears as a pole of the scattering amplitude on the positive imaginary axis when the latter is considered as a function of the complex variable* k, discuss how the occurrence of a bound state is reflected in the properties of the scattering length a and the total cross section σ.

Problem 35.10 Consider the **hard-sphere scattering** problem (recall Problem 16.2), where the potential is given by

$$V(r) = \begin{cases} \infty, & (r \leq a) \\ 0, & (r > a) . \end{cases}$$

(a) Show that

$$\tan \delta_l = \frac{j_l(ka)}{n_l(ka)}, \quad \text{where } k = \sqrt{\frac{2\mu E}{\hbar^2}} .$$

(b) Show that

$$\delta_l \xrightarrow[k\to 0]{} -\frac{(ka)^{2l+1}}{(2l+1)!!(2l-1)!!}, \quad \delta_l \xrightarrow[k\to\infty]{} -ka \ \text{(for fixed } l) .$$

(c) Expand the scattering amplitude $f(\theta)$ [as given by (35.89)] and the differential cross section $d\sigma/d\Omega$ in power series in ka, keeping only terms to order $(ka)^2$.

(d) Find the total cross section σ in the limit $k \to 0$ by using the Optical Theorem (cf. Problem 35.4) and by integrating $d\sigma/d\Omega$. Explain physically why the total cross section turns out to be considerably larger than the geometrical cross section of the hard sphere.

Problem 35.11 Consider the scattering of a spin 1/2 particle of mass m by a fixed scatterer. The interaction Hamiltonian operator can in general be written as

$$H' = \frac{\hbar^2}{2m}(U + \boldsymbol{V} \cdot \boldsymbol{\sigma}),$$

where U and $\boldsymbol{V}$ are (2×2) matrix-valued operators and $\boldsymbol{\sigma}$ represents the Pauli spin matrices.

(a) Show that in the first Born approximation the scattering amplitude (a 2×2 matrix) is given by

$$f^{(B)}(\boldsymbol{k}_i, \boldsymbol{k}_f) = -\frac{1}{4\pi} \int d^3r\, d^3r'\, e^{-i\boldsymbol{k}_f \cdot \boldsymbol{r}} \left(\langle \boldsymbol{r} \,|\, U \,|\, \boldsymbol{r}' \rangle + \langle \boldsymbol{r} \,|\, \boldsymbol{V} \,|\, \boldsymbol{r}' \rangle \cdot \boldsymbol{\sigma} \right) e^{i\boldsymbol{k}_i \cdot \boldsymbol{r}'} .$$

Explain the physical meanings of its elements. Note that in general the operators U and $\boldsymbol{V}$ lead to nonlocal potentials. [Hint: Use the Lippmann-Schwinger equation (35.58) and then insert the identity operator as given by the completeness relation (24.30).]

(b) Show that the differential cross section for scattering from a state with $s = 1/2$ along the z direction to a state with $s = -1/2$ is given by

$$\left| \frac{1}{4\pi} \int d^3r\, d^3r'\, e^{-i\boldsymbol{k}_f \cdot \boldsymbol{r}} \langle \boldsymbol{r} \,|\, V_x + iV_y \,|\, \boldsymbol{r}' \rangle\, e^{i\boldsymbol{k}_i \cdot \boldsymbol{r}'} \right|^2 .$$

(c) If $U = 0$ and $\boldsymbol{V} = \boldsymbol{p}W(r) + W(r)\boldsymbol{p}$, where $W(r)$ is a (matrix-valued) central field and $\boldsymbol{p}$ is the momentum operator, show that

$$f^{(B)} = -\frac{\hbar}{4\pi} \boldsymbol{q} \cdot \boldsymbol{\sigma} \tilde{W}(q) ,$$

where $\boldsymbol{q} \equiv \boldsymbol{k}_i - \boldsymbol{k}_f$ is the momentum transfer and $\tilde{W}(q)$ is the Fourier transform of $W(r)$.

(d) Show that for the case of spin-orbit coupling: $U = 0, \boldsymbol{V} = W(r)\boldsymbol{L}$, where $\boldsymbol{L}$ is the orbital angular momentum operator, $f^{(B)} = 0$. [Hint: Write $\boldsymbol{\sigma} \cdot \boldsymbol{L}$ as $(1/2)(\boldsymbol{\sigma} \cdot \boldsymbol{L} + \boldsymbol{\sigma} \cdot \boldsymbol{L})$ and make use of the fact that $\boldsymbol{L}$ is hermitian.]

Problem 35.12 A monoenergetic beam of particles is incident on a set of N static potentials. Let $\hbar\boldsymbol{q}$ be the momentum transfer in the collision.

(a) Show that the differential cross section in the lowest (Born) approximation is given by

$$\sigma(\boldsymbol{q}) = \sigma_0(\boldsymbol{q}) \left| \sum_{n=1}^{N} e^{i\boldsymbol{q} \cdot \boldsymbol{r}_n} \right|^2 \equiv \sigma_0(\boldsymbol{q}) S(\boldsymbol{q}) ,$$

where $\boldsymbol{r}_n$ is the center of the n-th potential. $S(\boldsymbol{q})$ and $\sigma_0(\boldsymbol{q})$ are referred to as the **structure factor** and the **form factor**, respectively. The structure factor accounts for the presence of diffraction effects.

(b) Assume that the scatterers are equally spaced on a straight line. Calculate the structure factor and plot it as a function of $\boldsymbol{q} \cdot \boldsymbol{a}$ (where $\boldsymbol{a}$ is the displacement vector from one scatterer to an adjacent one) for $N = 5, 10$ and 20. Deduce the positions of the diffraction peaks and the value of $S(\boldsymbol{q})$ at such peaks. Discuss how these peaks change as N increases.

(c) Derive an expression for $S(\boldsymbol{q})$ as $N \to \infty$.

(d) Consider the situation where the incident beam of particles is propagated along the positive y-direction and towards the origin of a coordinate frame, while the line of scatters is arranged along the z-direction and centered at the origin. A pair of photographic plates I and II are placed at some large distance away from the scatterers, with I perpendicular to the z-direction and II perpendicular to the y-direction. Assuming that the size of the scatterers is very much smaller than the distance between them, what kind of diffraction patterns would one expect to observe on plates I and II?

Chapter 36

Analytic Properties of the S-Matrix: Bound States and Resonances

Whereas the values of physical interest for the momentum variable k lie on the positive real axis $0 \leq k < \infty$, and those for the angular momentum l are the non-negative integers, there is much theoretical advantage in considering both k and l as complex variables in the Jost function $\mathcal{F}_l(k)$ [(35.117)] as well as the S-matrix element $S_l(k)$ [(35.116)]. Indeed, in the radial Schrödinger equation (16.12), k^2 and $l(l+1)$ can be considered to be complex; and by doing so, the mathematical (and frequently physical) properties of the solutions can be more readily explored and understood. This is a familiar theme in theoretical physics which plays out in many important topics. As the mathematician Hadamard put it succinctly : "*The shortest path between two truths on the real axis goes through the complex plane*"! In this chapter we will provide an introductory survey of some fascinating consequences of treating many relevant quantities in potential scattering theory as functions of the *complex* variable k only, and of the study their analytical properties. As in the last chapter, we will largely refrain from providing mathematical proofs. The focus will again be on their physical applications. However, the reader will need to be conversant with the basic notions of complex analysis.

We begin by stating a result in complex variable theory that is crucial in the study of the analytic properties of the S-matrix.

Theorem 36.1. ***Schwarz Reflection Principle.*** *If $f(z)$ is analytic in a region R of the complex z-plane that includes a segment of the real axis, and if $f(z)$ is real in this segment, then $f(z)$ can be analytically continued into the region R^*, and in $R \cup R^*$,*

$$f(z) = [f(z^*)]^* \,. \tag{36.1}$$

Under the assumptions for the asymptotic behavior of the interaction potential $V(r)$ as stated immediately following (35.1), hereafter referred to as the usual assumptions, the following key mathematical results can be established.

Theorem 36.2. *Under the usual assumptions for the potential, the regular solution $\phi_{l,k}(r)$ is an entire function of k, and*

$$\phi_{l,-k}(r) = (-1)^{l+1}\,\phi_{l,k}(r)\,. \tag{36.2}$$

Theorem 36.3. *Under the usual assumptions for the potential, the Jost function $\mathcal{F}_l(k)$ is analytic in $Im(k) > 0$ (the open upper-half k-plane), and continuous in $Im(k) \geq 0$ (the closed upper-half k-plane). Furthermore, for $Im(k) \geq 0$,*

$$\mathcal{F}_l(k) = [\mathcal{F}_l(-k^*)]^*\ ; \tag{36.3}$$

and $\mathcal{F}_l(k)$ is real for k lying on the positive imaginary axis.

The first of the above two theorems can be proved from the properties of the integral equation for $\phi_{l,k}(r)$ [(35.113)], while the second follows from the integral equation for $\mathcal{F}_l(k)$ [(35.118)] and the Schwarz reflection principle.

Now, for physical values of k (k real ≥ 0), $k = k^*$. For these values, then, $\mathcal{F}_l(k) = \mathcal{F}_l(k^*)$, and so from (35.116),

$$S_l(k) = \frac{(\mathcal{F}_l(k))^*}{\mathcal{F}_l(k)} = \frac{(\mathcal{F}_l(k^*))^*}{\mathcal{F}_l(k)} = \frac{\mathcal{F}_l(-k)}{\mathcal{F}_l(k)}\,, \quad (k\,\text{real} \geq 0)\,, \tag{36.4}$$

where the last equality follows from the Schwarz reflection principle. The mathematical problem is to analytically continue the function

$$S_l(k) = \frac{\mathcal{F}_l(-k)}{\mathcal{F}_l(k)}\,, \tag{36.5}$$

if possible, to the entire complex k-plane. The immediate difficulty is that, according to the result (2) above, while $\mathcal{F}_l(k)$ is analytic only in the upper-half k-plane, $\mathcal{F}_l(-k)$ is perforce analytic only in the lower-half k-plane, so that, a priori, there is no overlap in the regions of analyticity for the numerator and denominator of the expression for $S_l(k)$ as given in (36.5). The situation, however, can be remedied if we assume that $V(r)$, when viewed as a function of the complex variable r, is analytic for $Re(r) > 0$. Under this assumption, it can be shown that (see, for example, Taylor 1983) $\mathcal{F}_l(k)$ can be analytically continued to the entire complex k-plane except for the negative imaginary axis. Instead of (35.118), the analytically continued $\mathcal{F}_l(k)$ into the lower-half k-plane (except the negative imaginary axis) is given by the following expression:

$$\mathcal{F}_l(k) = 1 + \frac{e^{i\theta}}{k}\int_0^\infty d\rho\,\hat{h}_l^+(k\rho e^{i\theta})V'(\rho e^{i\theta})\phi_{l,k}(\rho e^{i\theta})\,, \tag{36.6}$$

where, for a given value of θ (within the allowable range $-\pi/2 < \theta < \pi/2$), the region of analyticity is $-\theta < arg(k) < \pi - \theta$. The nature of the singularities on the negative imaginary axis (the positions of poles and branch points) will

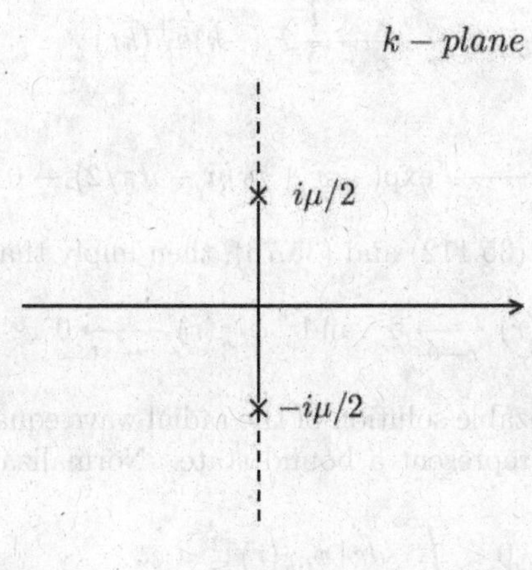

Fig. 36.1

depend sensitively on the nature of the potential $V(r)$ for large r. If, in addition to the usual conditions, the potential is exponentially bound as $r \to \infty$ [**Yukawa-type potentials**, for which $\int_0^\infty dr\, r|V(r)|e^{\nu r} < \infty$ for every ν less than a certain $\mu > 0$], then the region of non-analyticity of the continued $\mathcal{F}_l(k)$ will only be a subset of the negative imaginary axis: consisting of the part where $Im(k) < -i\mu/2$. In this case, the analytically continued $S_l(k)$ will be a *meromorphic function* (analytic except for poles) in the region shown in Fig. 36.1 (the whole k-plane excluding the dotted line segments on the imaginary axis). Furthermore, by result (2) above, $S_l(k)$ is real in the interval between $-i\mu/2$ and $i\mu/2$ on the imaginary axis. It should be mentioned that, in general, the analytic properties of the continued Jost function in the lower-half k-plane are highly unstable with respect to small changes in the potential at large distances. For this reason, the behavior of $\mathcal{F}_l(k)$ outside of the original region of analyticity (before continuation) is physically less relevant than its behavior inside that region.

We will now state the central result of this chapter, concerning the relationship between poles of $S_l(k)$ and bound states of the system.

Theorem 36.4. *Zeros of the Jost function $\mathcal{F}_l(k)$ in the upper-half k-plane ($Im(k) > 0$) are always purely imaginary simple zeroes ($\bar{k} = i\kappa$, κ real and > 0), and correspond to bound states. If $\mathcal{F}_l(k)$ can be analytically continued into the lower-half k-plane such that $\mathcal{F}_l(-k)$ is analytic at the values $\bar{k}$, then $\mathcal{F}_l(-\bar{k}) \neq 0$, and such values are simple poles of the S-matrix element $S_l(k)$ as well.*

Proof. Let $\bar{k} = \kappa_R + i\kappa$ (κ_R, κ real and $\kappa > 0$) be a zero of $\mathcal{F}_l(k)$, and assume that $\mathcal{F}_l(-k)$ is analytic at $\bar{k}$. Then $\mathcal{F}_l(-\bar{k})$ cannot vanish. For if it did, (35.115) would imply that the regular solution $\phi_{l,\bar{k}}(r)$ would vanish identically. Now (35.115) also implies that

$$\phi_{l,\bar{k}}(r) \xrightarrow[r\to\infty]{} -\frac{i}{2}\mathcal{F}_l(-\bar{k})\hat{h}_l^+(\bar{k}r)\,. \tag{36.7}$$

By (35.93)

$$\hat{h}_l^+(\bar{k}r) \xrightarrow[r\to\infty]{} \exp(-\kappa r + i\kappa_R r - il\pi/2) \to 0\,. \tag{36.8}$$

This fact, together with (35.112) and (35.76), then imply that

$$\phi_{l,\bar{k}}(r) \xrightarrow[r\to 0]{} 0 \quad \text{and} \quad \phi_{l,\bar{k}}(r) \xrightarrow[r\to\infty]{} 0\,. \tag{36.9}$$

Thus $\phi_{l,\bar{k}}(r)$ is a normalizable solution of the radial wave equation with angular momentum l and must represent a bound state. Normalizability in this case means

$$0 < \int_0^\infty dr\,|\,\phi_{l,\bar{k}}(r)\,|^2 < \infty\,. \tag{36.10}$$

Since $\phi_{l,\bar{k}}(r)$ satisfies the radial Schrödinger equation (16.12), we have

$$\frac{d^2\phi_{l,\bar{k}}(r)}{dr^2} + \left[\bar{k}^2 - \frac{l(l+1)}{r^2} - V'(r)\right]\phi_{l,\bar{k}}(r) = 0\,. \tag{36.11}$$

Taking the complex conjugate of this equation, we obtain, on assuming l to be real and that $V'(r)$ is real for positive real values of r,

$$\frac{d^2\phi^*_{l,\bar{k}}(r)}{dr^2} + \left[(\bar{k}^2)^* - \frac{l(l+1)}{r^2} - V'(r)\right]\phi^*_{l,\bar{k}}(r) = 0\,. \tag{36.12}$$

Multiplying (36.11) by $\phi^*_{l,\bar{k}}(r)$ and (36.12) by $\phi_{l,\bar{k}}(r)$, and subtracting, we get

$$\frac{d}{dr}\left(\phi^*_{l,\bar{k}}\frac{d\phi_{l,\bar{k}}}{dr} - \frac{d\phi^*_{l,\bar{k}}}{dr}\phi_{l,\bar{k}}\right) + 2i\,Im(\bar{k}^2)\,|\phi_{l,\bar{k}}|^2 = 0\,. \tag{36.13}$$

Upon integrating both sides over r from 0 to ∞ and recalling the boundary conditions (36.9) for $\phi_{l,\bar{k}}(r)$, we see that

$$2i\,Im(\bar{k}^2)\int_0^\infty dr\,|\phi_{l,\bar{k}}(r)|^2 = 0\,. \tag{36.14}$$

It then follows from (36.10) that $Im(\bar{k}^2) = 0$, which in turn implies that $\kappa_R\kappa = 0$. But since $\kappa > 0$, we conclude that κ_R must vanish. Thus any zero of $\mathcal{F}_l(k)$ in the upper-half plane must lie on the positive imaginary axis. Each such zero, $\bar{k} = i\kappa\,(\kappa > 0)$, then corresponds to a bound state with angular momentum l and energy $E = -\hbar^2\kappa^2/(2m)$. We will not present here the proof that all the zeros $\bar{k}$ are simple zeros. The reader may consult, for example, Taylor (1983). □

We note two circumstances which may cause confusion with respect to the above theorem. First, Yukawa-type (exponentially bounded) potentials in general lead to the existence of branch points (with corresponding branch cuts called **dynamical cuts**) for $\mathcal{F}_l(k)$ in the lower-half imaginary k-axis, and consequently for $\mathcal{F}_l(-k)$ in the upper-half imaginary k-axis. For pure exponential potentials, these cuts degenerate into a series of poles called **false poles** for $S_l(k)$ in the upper-half imaginary axis – so called because they do not correspond to true bound states. Second, zeros of $\mathcal{F}_l(k)$ in the negative imaginary k-axis lead to solutions $\psi_{l,k}(r)$ which are clearly non-normalizable in the sense (35.102). These poles of $S_l(k)$ correspond to what are sometimes referred to as **anti-bound states**. (For more extensive discussions of these two situations, see De Alfaro and Regge 1965.)

It is naturally of interest to determine the number of bound states for a particular potential satisfying the usual assumptions. For this purpose we will state two more theorems (without proof) for the high-energy ($|k| \to \infty$) behavior and the threshold ($k = 0$) behavior of the Jost function, respectively:

Theorem 36.5. *The Jost function $\mathcal{F}_l(k)$ cannot have any zeros on the real k-axis, except possibly at $k = 0$. It also satisfies the following asymptotic condition in the closed upper-half k-plane:*

$$\mathcal{F}_l(k) - 1 = O\left(\frac{1}{k^\eta}\right) \qquad (\textit{for some } \eta \geq 0) \,, \tag{36.15}$$

as $|k| \to \infty$ in the upper-half k-plane, the stated convergence being a uniform one.

Theorem 36.6. *For $l > 0$, $\mathcal{F}_l(0) = 0$ if and only if $k = 0$ corresponds to a bound state. If $k = 0$ is a zero of $\mathcal{F}_l(k)$ for $l > 0$, then it is a simple zero. If $k = 0$ is a zero of $\mathcal{F}_0(k)$, then it is a double zero.*

It follows immediately from Theorem 36.5 that there is a finite radius $|k| = \rho$ beyond which $\mathcal{F}_l(k)$ has no zeros in the upper-half plane. We can then conclude that there are only a finite number of bound states of angular momentum l. The following classical theorem in scattering theory determines this number in terms of the scattering phase shift $\delta_l(k)$, a measurable quantity.

Theorem 36.7. ***Levinson's Theorem****. For a spherically symmetric potential satisfying the usual assumptions, the partial-wave phase shift $\delta_l(k)$ satisfies the condition*

$$\boxed{\delta_l(0) - \delta_l(\infty) = n_l \pi \qquad (\mathcal{F}_0(0) \neq 0)} \,, \tag{36.16}$$

where n_l is the number of bound states of angular momentum l. In the case $\mathcal{F}_0(0) = 0$, the above result still holds for $l > 0$. But for $l = 0$,

$$\delta_0(0) - \delta_0(\infty) = (n_0 + 1/2)\pi \qquad (\mathcal{F}_0(0) = 0) \,. \tag{36.17}$$

Problem 36.1 Prove Levinson's Theorem by evaluating the contour integral

$$I = \oint_C dk \, \frac{d\mathcal{F}_l(k)/dk}{\mathcal{F}_l(k)} ,$$

where the closed contour C is the infinite semicircle in the upper-half k-plane for the case $\mathcal{F}_l(0) \neq 0$, and the same contour with the infinitesimal real segment $-\epsilon < k < \epsilon$ ($\epsilon \to 0$) replaced by the infinitesimal semicircle of radius ϵ in the upper-half plane for the case $\mathcal{F}_l(0) = 0$. The following result in complex function theory will be useful: If $f(z)$ is analytic inside and on a closed contour C, then

$$\frac{1}{2\pi i} \oint_C dz \, \frac{f'(z)}{f(z)} = N - P ,$$

where N = the number of zeros of $f(z)$ inside the contour C (a zero of order m being counted m times), and P = the number of poles of $f(z)$ inside the contour C (a pole of order m being counted m times).

So far we have only considered zeros of $\mathcal{F}_l(k)$ (or poles of $S_l(k)$) in the upper-half plane corresponding to bound states (all lying on the positive imaginary axis); the "false poles" of $S_l(k)$ on the positive imaginary axis (not corresponding to bound states); and the zeros of $\mathcal{F}_l(k)$ in the negative imaginary axis (corresponding to "anti-bound" states). We will now consider zeros of $\mathcal{F}_l(k)$ in the lower-half imaginary plane which do not lie on the negative imaginary axis, that is, zeros which can be written in the form $k = h + ib \, (h \neq 0, \, b < 0)$. First we note that for such k's, $\phi_{l,k}(r)$ is not normalizable [cf. (36.8)] and thus cannot represent bound states. On the other hand, by (36.3), if $k = h + ib$ is a zero, $-k^* = -h + ib$ is a zero also. Let us assume that $k = h + ib$ is a simple zero of $\mathcal{F}_l(k)$ in the lower-half plane; so we can write, in a region near $h + ib$,

$$\mathcal{F}_l(k) \approx C^*(k - h - ib) , \tag{36.18}$$

where $C = \rho e^{i\theta}$ is a (complex) constant. Thus, by (36.3) again,

$$\mathcal{F}_l(-k) \approx C(k - h + ib) . \tag{36.19}$$

Near the value $k = h + ib$, then,

$$S_l(k) = \frac{\mathcal{F}_l(-k)}{\mathcal{F}_l(k)} \approx \frac{C}{C^*} \frac{k - h + ib}{k - h - ib} = e^{2i\theta} \, \frac{k - h + ib}{k - h - ib} . \tag{36.20}$$

We will assume that $\theta = 0$. It follows from (35.83) that

$$e^{2i\delta_l(k)} \approx \frac{k - h + ib}{k - h - ib} . \tag{36.21}$$

A simple calculation then shows that, if k is real,

$$\sin^2 \delta_l(k) \approx \frac{b^2}{(k-h)^2+b^2} \,. \tag{36.22}$$

The result from Problem 35.3 clearly indicates that when $k = h > 0$, the l-th partial wave scattering cross section $\sigma_l(k)$ peaks in comparison to its values at neighboring real values of k, and the quantity $|b|$ gives a measure of the width of the peak. For this reason, the zeros $k = h + ib$ are said to correspond to **resonances** (or metastable states). Analysis of the expression on the right-hand side of (36.21) shows that, for $b \ll |h|$, as k ranges across the real axis from $k < h$ to $k > h$, the scattering phase shift $\delta_l(k)$ changes sharply from roughly 0 to roughly π, going through the value $\pi/2$ at precisely $k = h$. This is the key feature of any resonance.

Going over to the energy variable and setting $E = \hbar^2k^2/(2m)$ (k real), $E_R \equiv \hbar^2h^2/(2m)$, and $\Gamma \equiv 2h|b|(\hbar^2/(2m))$, (36.22) implies

$$\boxed{\sin^2 \delta_l(E) = \frac{\Gamma^2/4}{(E-E_0)^2+\Gamma^2/4}} \,. \tag{36.23}$$

This is known as the **Breit-Wigner formula** for a resonance, with Γ as the width. Since $E \propto k^2$, when E is treated as a complex variable in $\mathcal{F}_l(k(E))$ or $S_l(k(E))$, the Riemann surface of these functions is a two-sheeted surface cut along the positive real axis. The first sheet, corresponding to $Im(k) > 0$, is called the "physical sheet". The bound states are given by zeros of $\mathcal{F}_l(k(E))$ on the negative real axis of the "physical sheet", while the resonances are given by zeros of the form $E = E_R - i\Gamma/2$ ($E_R > 0, \Gamma > 0$), close to and below the positive real axis on the second sheet.

Problem 36.2 Verify (36.22) and (36.23).

Problem 36.3 Consider the resonance scattering of a particle of mass μ by the following "hollow-shell" potential:

$$V(r) = \begin{cases} 0 & (0 \le r < R_1) \\ \dfrac{\hbar^2\kappa_0^2}{2\mu} & (R_1 \le r \le R) \\ 0 & (r > R) \,. \end{cases}$$

We will see that for certain energies $E = \hbar^2k^2/(2\mu)$ so that $k^2 < \kappa_0^2$, the particle gets trapped inside the shell for a long time, and thus forms a resonance. For simplicity, consider only S-wave ($l = 0$) scattering. The radial Schrödinger equation (16.12) becomes

$$\left[\frac{d^2}{dr^2} - U_0(r) + k^2\right] \chi_0 = 0 \,,$$

where $U_0(r) = (2\mu)/\hbar^2)V(r)$. The outgoing solution $\chi_0^{(+)}$, which approaches $\exp(ikr)$ as $r \to \infty$, can be written in all regions as

$$\chi_0^{(+)}(r) = \begin{cases} e^{ikr} & (r > R)\,, \\ \alpha e^{\kappa r} + \beta e^{-\kappa r} & (R_1 \le r \le R)\,, \\ ae^{ikr} + be^{-ikr} & (r < R_1)\,, \end{cases}$$

where $\kappa \equiv \sqrt{\kappa_0^2 - k^2}$. The incoming wave $\chi_0^{(-)}$ is just the complex conjugate of $\chi_0^{(+)}$.

(a) By matching boundary conditions show that the constants a, b, α and β in the above equation are given by

$$\alpha = \frac{1}{2}\left(1 + i\frac{k}{\kappa}\right) e^{ikR-\kappa R}\,, \quad \beta = \frac{1}{2}\left(1 - i\frac{k}{\kappa}\right) e^{ikR+\kappa R}\,,$$

$$a = \frac{\kappa}{4ik}\, e^{ik(R-R_1)+\kappa\rho} \left\{ \left(1 + i\frac{k}{\kappa}\right)^2 e^{-2\kappa\rho} - \left(1 - i\frac{k}{\kappa}\right)^2 \right\}\,,$$

$$b = \frac{\kappa}{4ik}\, e^{ik(R+R_1)+\kappa\rho} \left\{ \left(1 - e^{-2\kappa\rho}\right)\left(1 + \frac{k^2}{\kappa^2}\right) \right\}\,,$$

where $\rho \equiv R - R_1$.

(b) Recalling (16.50) and (16.51), and requiring that $\chi_0 \to 0$ as $r \to 0$, show that the scattering matrix S can be expressed as

$$S(k) = \frac{a^* + b^*}{a + b} = e^{-2ikR}\left(\frac{\kappa - ik}{\kappa + ik}\right) \left[\frac{e^{-2\kappa\rho} + \zeta(k)\left(\dfrac{\kappa + ik}{\kappa - ik}\right)}{e^{-2\kappa\rho} + \zeta(k)\left(\dfrac{\kappa - ik}{\kappa + ik}\right)} \right].$$

where

$$\zeta(k) \equiv \frac{k\cot(kR_1) + \kappa}{k\cot(kR_1) - \kappa}\,.$$

(c) Under the condition that the potential barrier is sufficiently high and wide so that $\kappa\rho \gg 1$, show that

$$S(k) \approx e^{-2ikR}\left(\frac{\kappa + ik}{\kappa - ik}\right)$$

provided $\zeta(k)$ is not close to zero. Show that in the region of k in which the above approximation holds, $S(k)$ has no poles.

(d) When $\zeta(k)$ is close to zero, the above approximation for $S(k)$ no longer holds. Denote the roots of $\zeta(k) = 0$ by k_n, and expand $\zeta(k)$ in powers of $(k - k_n)$ in the neighborhood of a particular k_n. Show that

$$\zeta(k) = \frac{(k - k_n)}{2k_n}\left[1 + \left(\frac{k_n}{\kappa_n}\right)^2\right](1 + \kappa_n R_1)\,,$$

where $\kappa_n \equiv \sqrt{\kappa_0^2 - k_n^2}$.

(e) Show that there is a pole of $S(k)$ at

$$k_0 = k_n - e^{-2\kappa\rho}\,\frac{2k_n\kappa_n^2(\kappa_n + ik_n)^2}{(\kappa_n^2 + k_n^2)^2(1 + \kappa_n R_1)} \equiv k_1 - ik_2\,.$$

Note that k_0 is in the lower half k-plane.

(f) Show that near the pole k_0, $S(k)$ has the form

$$S(k) \approx e^{-2ikR} \frac{(\kappa + ik)(k - k_0^*)}{(\kappa - ik)(k - k_0)} .$$

(g) Observe that in the immediate neighborhood of the pole k_0, the factor $\exp(-2ikR)(\kappa + ik)/(\kappa - ik)$ in the above expression can be approximated by a constant. Defining the constant

$$e^{2i\phi} \equiv e^{-2ikR} \left(\frac{\kappa_1 + ik_1}{\kappa_1 - ik_1} \right) , \quad (\kappa_1 \equiv \sqrt{\kappa_0^2 - k_1^2}) ,$$

we then have

$$S(k) \approx e^{2i\phi} \frac{k - k_0^*}{k - k_0} .$$

Show that when S is expressed as a function of the complex energy E the above expression is equivalent to [compare with (36.20)]

$$S(E) = e^{2i\phi} \frac{E - E_0 - i\Gamma/2}{E - E_0 + i\Gamma/2} ,$$

where $E_0 \equiv \hbar^2 k_n^2/(2\mu)$ is the resonance energy and

$$\Gamma \equiv 16 E_0 e^{-2\kappa\rho} \frac{\kappa_n^3 k_n}{(\kappa_n^2 + k_n^2)^2 (1 + \kappa_n R_1)}$$

is the width of the resonance.

Chapter 37

Non-Perturbative Bound-State and Scattering-State Solutions: Radiation-Induced Bound-Continuum Interactions

In this chapter we will examine a problem for which perturbation treatment (either time-independent or time-dependent) fails because of vanishing energy differences in the denominators of higher-order perturbative solutions [cf. (31.37) and (33.57)]. The model presented here has an exact solution, and is based on the so-called **Lee model** of unstable particles (see, for example, Glaser and Källén 1956; Araki, Munakata and Kawaguchi 1957), which is originally introduced to treat the problem of the renormalizability of field theories (Lee 1954). It consists of a discrete state embedded in a continuum, and can be used to describe, for instance, the physical processes of single-photon **radiative dissociation** or **recombination** of molecular complexes, and radiation-induced **collisional ionization** of atomic-molecular systems. In all these physical processes the key feature is that the spectrum of the unperturbed Hamiltonian (the part excluding the radiative interaction) involves discrete (bound) states as well as continuum (free) states. If the energy ($\hbar\omega$) of an incident photon is such that some discrete state (whose energy originally lies beneath the continuum threshold) is brought into resonance with a level within the continuum, the above-mentioned processes may occur. It is beyond the scope of this book to treat the various kinds of atomic-molecular dynamics in detail. We will instead

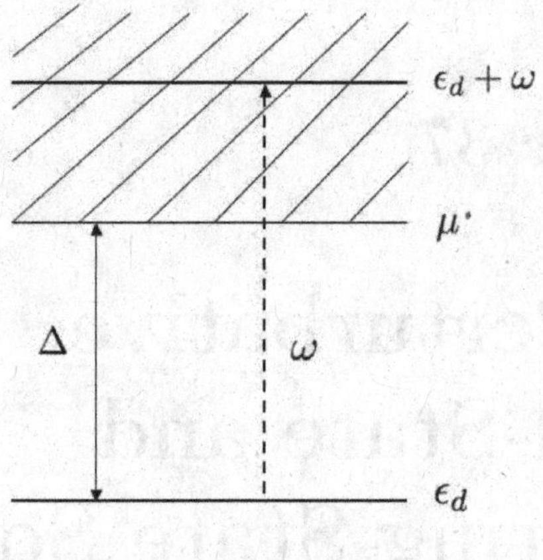

Fig. 37.1

focus on the formal aspects and some of the main predictions of our phenomenological model, as an example of the use of non-perturbative techniques in the solution of the Schrödinger equation, and the introductory ideas of **scattering theory** presented in Chapters 16 and 35. For more details see Lam and George 1986.

Figure 37.1 depicts schematically the energy spectrum of the model. The single discrete (bound) state, of energy ϵ_d, lies completely below the continuum band, with a finite gap Δ separating ϵ_d and the threshold μ of the band. The gap is presumed sufficiently large so that there is no non-radiative bound-continuum coupling. A radiative bound-continuum coupling, however, can be induced by a photon with frequency $\omega > \Delta/\hbar$. We can write the Hamiltonian as

$$H = H_0 + V , \tag{37.1}$$

where the unperturbed Hamiltonian H_0 is given by (on setting $\hbar = 1$)

$$H_0 = \epsilon_d \, c_d^\dagger c_d + \sum_\epsilon \epsilon \, c_\epsilon^\dagger c_\epsilon + \omega \, a_\omega^\dagger a_\omega , \tag{37.2}$$

and the interaction Hamiltonian V by

$$V = \sum_\epsilon (a_\omega^\dagger + a_\omega)(V_{\epsilon\omega} \, c_d^\dagger c_\epsilon + V_{\epsilon\omega}^* \, c_d c_\epsilon^\dagger) . \tag{37.3}$$

The above expressions for the Hamiltonian are given in what is called the **second-quantized** form. These second-quantized operators act on a Hilbert space consisting of various **occupation-number (Fock) states** [cf. Chapter 14]. The operators $c_d\,(c_d^\dagger)$, $c_\epsilon\,(c_\epsilon^\dagger)$, and $a_\omega\,(a_\omega^\dagger)$ are the **annihilation (creation) operators** for the discrete state, the continuum state of energy ϵ, and the photon state of frequency ω, respectively, and satisfy (bosonic) commutation relations analogous to (14.8). $V_{\epsilon\omega}$ (in general a complex quantity) is the radiative bound-continuum coupling for energy ϵ. In addition to the assumption of a single (frequency)-mode radiation field, the spin of the photon (which may lead to polarization effects) has also been neglected. Note that both H_0 and V are hermitian, as required by first principles. The physical interpretation of H is quite obvious. Since $c_d^\dagger c_d$, $c_\epsilon^\dagger c_\epsilon$ and $a_\omega^\dagger a_\omega$ all represent **number operators** [cf. Chapter 14], the three terms in H_0 just represent the total energies of the three different kinds of bosons. These bosons can all be created or annihilated within the model, as described by the different terms in V.

Denoting by n_d and n_ϵ the number of bosons in the discrete state (with energy ϵ_d) and the continuum state with energy ϵ, respectively, and by n_ω the number of photons with frequency ω, the set of (number) eigenstates $\{|\,n_d, n_\epsilon, n_\omega\,\rangle\}$ of H_0 forms a complete set of states in the Hilbert space of the model. It can immediately be seen that, for any ϵ within the continuum, the operator $c_d^\dagger c_d + c_\epsilon^\dagger c_\epsilon$ commutes with H, so that $n_d + n_\epsilon$ is a constant of motion. We will work in the invariant subspace with $n_d + n_\epsilon = 1$, with the completeness relation

$$\sum_n [\,|\,d_n\,\rangle\langle\,d_n\,| + \sum_\epsilon |\,\epsilon_n\,\rangle\langle\,\epsilon_n\,|\,] = 1\,, \tag{37.4}$$

where

$$|\,d_n\,\rangle \equiv |\,1, 0, n\,\rangle\,, \tag{37.5}$$

$$|\,\epsilon_n\,\rangle \equiv |\,0, 1, n\,\rangle\,, \tag{37.6}$$

and n is any positive integer or zero. In this representation all matrix elements of the form $\langle\,d_n\,|\,H\,|\,\epsilon_{n\pm 1}\,\rangle$ are nonvanishing. Thus the entire subspace spanned by $\{|\,d_n\,\rangle, |\,\epsilon_n\,\rangle\}$ is in principle required for the description of the radiative interaction. This subspace, however, further partitions into a direct sum of an infinity of subspaces (one for each n), each invariant under H, if the so-called **rotating wave approximation** (RWA) is made. The RWA states that all antiresonant (energy non-conserving) processes are strictly forbidden. [Recall the discussion following Eq. (33.46).] In the present context, this means that excitation cannot be accompanied by emission of a photon, and conversely, de-excitation cannot be accompanied by absorption of a photon. It implies that

$$\langle\,\epsilon_{n+1}\,|\,H\,|\,d_n\,\rangle = 0 \tag{37.7}$$

for all n. Under the RWA, then, each irreducible subspace (labeled by a fixed n) has a completeness relation of the form

$$|\,d_{n+1}\,\rangle\langle\,d_{n+1}\,| + \sum_\epsilon |\,\epsilon_n\,\rangle\langle\,\epsilon_n\,| = 1\,. \tag{37.8}$$

In this chapter we will assume the validity of the RWA and work exclusively in the subspace where $n = N$ and $n_d + n_\epsilon = 1$ (for each ϵ within the continuum band). The fixed integer N will be taken to be the number of photons in the external radiation field, and gives a measure of the latter's intensity.

We will now proceed to solve the eigenvalue problem for H in the invariant subspace specified by (37.8) above. Let $|\, E(N)\,\rangle$ be an eigenstate of H, so that

$$H|\, E(N)\,\rangle = E\,|\, E(N)\,\rangle \,. \tag{37.9}$$

It follows from (37.8) that $|\, E(N)\,\rangle$ can be expanded as

$$|\, E(N)\,\rangle = \beta_E\,|\, d_{N+1}\,\rangle + \sum_\epsilon \chi_{E\epsilon}\,|\,\epsilon_N\,\rangle \,. \tag{37.10}$$

Using this equation and (37.1) through (37.3) in the Schrödinger equation (37.9) [and recalling (14.23) through (14.26)], it is straightforward to show that

$$\begin{aligned} &E\beta_E\,|\, d_{N+1}\,\rangle + \sum_\epsilon \chi_{E\epsilon} E\,|\,\epsilon_N\,\rangle \\ &= \beta_E\left[\{\epsilon_d + (N+1)\omega\}\,|\, d_{N+1}\,\rangle + \sqrt{N+1}\sum_\epsilon V^*_{\epsilon\omega}\,|\,\epsilon_N\,\rangle\right] \\ &\quad + \sum_\epsilon \chi_{E\epsilon}\left[(\epsilon + N\omega)\,|\,\epsilon_N\,\rangle + \sqrt{N+1}\,V_{\epsilon\omega}\,|\, d_{N+1}\,\rangle\right] \,. \end{aligned} \tag{37.11}$$

Taking the inner product of the above equation with $\langle\, d_{N+1}\,|$ and $\langle\,\epsilon'_N\,|$ leads, respectively, to the equations

$$E\beta_E = \beta_E[\epsilon_d + (N+1)\omega] + \sum_\epsilon \chi_{E\epsilon}\sqrt{N+1}\,V_{\epsilon\omega} \tag{37.12}$$

and

$$\chi_{E\epsilon'}E = \beta_E\sqrt{N+1}\,V^*_{\epsilon'\omega} + \chi_{E\epsilon'}(\epsilon' + N\omega) \,, \tag{37.13}$$

from which β_E and $\chi_{E\epsilon}$ can be eliminated to obtain

$$E - [\epsilon_d + (N+1)\omega] + \sum_\epsilon \frac{(N+1)|V_{\epsilon\omega}|^2}{\epsilon + N\omega - E} = 0 \,. \tag{37.14}$$

Defining an energy scale such that $N\omega = 0$ and converting the sum in (37.14) to an integral, we have

$$E - (\epsilon_d + \omega) + g^2\int_\mu^\infty d\epsilon\,\frac{\rho(\epsilon)|V_{\epsilon\omega}|^2}{\epsilon - E} = 0 \,, \tag{37.15}$$

where $g^2 \equiv N + 1$ designates a dimensionless radiative coupling strength that is proportional to the intensity of the external radiation field, and $\rho(\epsilon)$ is a density of states appropriate to the continuum. This integral equation is the eigenvalue equation for H. In what follows we will demonstrate that its solution implies the existence of both bound and metastable states.

A bound state exists if (37.15) has a real solution $E = \epsilon_b < \mu$. Since

$$\phi(E) \equiv g^2 \int_\mu^\infty d\epsilon \, \frac{\rho(\epsilon)|V_{\epsilon\omega}|^2}{\epsilon - E} \tag{37.16}$$

is a monotonically increasing function of E for $E < \mu$, we infer that there exists one and only one bound state solution for (37.15) if

$$\phi(\mu) > \omega - \Delta \, ; \tag{37.17}$$

and no bound state solution can exist if the coupling strength g^2 falls below the critical value given by

$$g^2_{crit} = (\omega - \Delta) \Big/ \int_\mu^\infty d\epsilon \, \frac{\rho(\epsilon)|V_{\epsilon\omega}|^2}{\epsilon - \mu} \, . \tag{37.18}$$

It is also clear that the energy ϵ_b of the bound state decreases as g^2 is increased and/or ω is decreased, provided $\omega > \Delta$. We thus have the following interesting conclusions: As the intensity of the external radiation field is increased beyond the critical value given by (37.18), the latter has the effect of "pulling down" the energy level of the induced bound state; while, with a tunable laser, for example, the same result can be achieved by lowering the field frequency. Hence, there exists the possibility of "fine-tuning" the energy of the radiation-induced bound state by manipulating the field strength and frequency independently or simultaneously.

An explicit expression for the normalized bound state can be given as follows:

$$|\, b_N \,\rangle = \beta_b \left[\, |\, d_{N+1} \,\rangle - g \int_\mu^\infty d\epsilon \, \frac{\rho(\epsilon) V^*_{\epsilon\omega}}{\epsilon - \epsilon_b} \, |\, \epsilon_N \,\rangle \, \right] \, , \tag{37.19}$$

where

$$\beta_b = \left[1 + g^2 \int_\mu^\infty d\epsilon \, \frac{\rho(\epsilon)|V_{\epsilon\omega}|^2}{(\epsilon - \epsilon_b)^2} \right]^{-1/2} \, , \tag{37.20}$$

and $\epsilon < \mu$. This result follows from Eqs. (37.10) and (37.13), and the normalization requirement

$$\langle\, b_N \,|\, b_N \,\rangle = 1 \tag{37.21}$$

for the bound state. Thus a system initially in the discrete state $|\, d_{N+1} \,\rangle$ has a probability

$$P_{bd} = |\, \langle\, b_N \,|\, d_{N+1}, t \,\rangle \,|^2 = \beta_b^2 \tag{37.22}$$

of being in the tunable bound state $|\, b_N \,\rangle$ at later times t; while the probability of bound-state formation from a continuous initial state $|\, \epsilon_N \,\rangle$, $P_{b\epsilon}$, is given by

$$P_{b\epsilon} = |\, \langle\, b_N \,|\, \epsilon_N, t \,\rangle \,|^2 = \frac{\beta_b^2 g^2 |V_{\epsilon\omega}|^2}{(\epsilon - \epsilon_b)^2} \, . \tag{37.23}$$

To investigate the possible existence of **metastable states**, we have to consider the continuum-to-continuum transitions specified by scattering processes of the type $|\,\epsilon_N\,\rangle \longrightarrow |\,\epsilon'_N\,\rangle$. For this purpose we introduce the **incoming** and **outgoing scattering states**, denoted by $|\,E^-(N)\,\rangle$ and $|\,E^+(N)\,\rangle$ respectively, which are exact continuum eigenstates of the full Hamiltonian H (recall the introduction of these concepts in Chapter 16 and their further elaboration in Chapter 35):

$$H\,|\,E^\pm(N)\,\rangle = E\,|\,E^\pm(N)\,\rangle\,, \quad \mu < E < \infty\,. \tag{37.24}$$

Analogous to (37.10) we can expand the scattering states as

$$|\,E^\pm(N)\,\rangle = |\,E_N\,\rangle + \sum_\epsilon \chi^\pm_{E\epsilon}\,|\,\epsilon_N\,\rangle + \beta^\pm_E\,|\,d_{N+1}\,\rangle\,. \tag{37.25}$$

Note the difference between $|\,E(N)\,\rangle$ and $|\,E_N\,\rangle$: the former is an exact eigenstate of H, while the latter is an eigenstate of H_0. In the above equation, $|\,E_N\,\rangle$ can be interpreted as the incident wave of the scattering process, while $\sum_\epsilon \chi^\pm_{E\epsilon}\,|\,\epsilon_N\,\rangle$ can be interpreted as the scattered wave. We will state without proof that the scattering states $|\,E^\pm(N)\,\rangle$ satisfy the following completeness relations (for a proof see Lam and George 1986):

$$\sum_E |\,E^\pm(N)\,\rangle\langle\,E^\pm(N)\,| = 1\,, \quad g < g_{crit}\,, \tag{37.26}$$

$$\sum_E |\,E^\pm(N)\,\rangle\langle\,E^\pm(N)\,| + |\,b_N\,\rangle\langle\,b_N\,| = 1\,, \quad g > g_{crit}\,, \tag{37.27}$$

where g_{crit} is given by (37.18).

The expansion coefficients $\chi^\pm_{E\epsilon}$ and $\beta^\pm_E$ in (37.25) can be determined unambiguously from (37.24):

$$\beta^\pm_E = \frac{gV_{E\omega}}{h(E \pm i\eta)}\,, \quad \eta \longrightarrow 0^+\,, \tag{37.28}$$

$$\chi^\pm_{E\epsilon} = \frac{g\beta^\pm_E V^*_{\epsilon\omega}}{E - \epsilon \pm i\eta}\,, \quad \eta \longrightarrow 0^+\,, \tag{37.29}$$

where $h(z)$ is the analytic function of a complex variable z given by

$$h(z) \equiv z - (\epsilon_d + \omega) + \phi(z)\,, \tag{37.30}$$

and $\phi(z)$ is the **Cauchy integral**

$$\phi(z) \equiv g^2 \int_\mu^\infty d\epsilon\, \frac{\rho(\epsilon)|V_{\epsilon\omega}|^2}{\epsilon - z} \tag{37.31}$$

[cf. (37.16)]. The function $\phi(z)$ has a square-root branch structure with **branch point** at $z = \mu$ and a discontinuity of $2i\pi g^2 \rho(\epsilon)|V_{\epsilon\omega}|^2$ across the **branch cut** (μ, ∞) at $z = \epsilon > \mu$. This double-valuedness differentiates between the incoming and outgoing scattering-state solutions.

Using (37.25), (37.28), (37.29), (35.43) and (35.44), we can write the following exact expression for the S-matrix for the scattering process $|\,\epsilon_N\,\rangle \to |\,\epsilon'_N\,\rangle$:

$$\begin{aligned} S_{\epsilon'\epsilon} &= \delta_{\epsilon'\epsilon} - 2\pi i\delta(\epsilon'-\epsilon)\,T_{\epsilon'\epsilon} \\ &= \delta_{\epsilon'\epsilon} - 2\pi i\delta(\epsilon'-\epsilon)\,\langle\,\epsilon'_N\,|\,V\,|\,\epsilon_N^+\,\rangle \\ &= \delta_{\epsilon'\epsilon} - 2\pi i\delta(\epsilon'-\epsilon)\,\frac{g^2|\,V_{\epsilon\omega}\,|^2}{h(\epsilon+i\eta)} \\ &= \delta_{\epsilon'\epsilon}\left(1 - \frac{i\Gamma_0(\epsilon)}{\epsilon-(\epsilon_d+\omega)+g^2P(\epsilon)+i\Gamma_0(\epsilon)/2}\right), \end{aligned} \tag{37.32}$$

where

$$\Gamma_0(\epsilon) \equiv 2\pi g^2\rho(\epsilon)|\,V_{\epsilon\omega}\,|^2 \tag{37.33}$$

and

$$P(\epsilon) \equiv \mathrm{P}\int_\mu^\infty d\epsilon'\,\frac{\rho(\epsilon')|\,V_{\epsilon'\omega}\,|^2}{\epsilon'-\epsilon}\,, \tag{37.34}$$

with P denoting the principal value of the integral. Noting that [cf. (35.83)]

$$S_{\epsilon\epsilon} = e^{2i\delta(\epsilon)}\,, \tag{37.35}$$

where $\delta(\epsilon)$ is the scattering phase shift, the latter is given by

$$\cot\delta(\epsilon) = -\frac{[\epsilon-(\epsilon_d+\omega)+g^2P(\epsilon)]}{\Gamma_0(\epsilon)/2}\,. \tag{37.36}$$

Recalling from the last chapter that resonances occur at energies satisfying $\delta(\epsilon)=\pi/2$, the observed energies of the metastable states must satisfy approximately the equation

$$\epsilon-(\epsilon_d+\omega)+g^2P(\epsilon) = 0\,. \tag{37.37}$$

This equation is to be compared with (37.15), whose solution ($\epsilon=\epsilon_b<\mu$) for $g>g_{crit}$ is a pole of the S matrix $S_{\epsilon\epsilon}$ on the real axis and corresponds to a laser-induced bound state.

We can proceed further to study the roots of (37.37) by making a specific choice for $\rho(\epsilon)$ and certain general assumptions for $|\,V_{\epsilon\omega}\,|^2$. A convenient choice for $\rho(\epsilon)$ is the free-electron energy density of states of the form

$$\rho(\epsilon) = \alpha\sqrt{\epsilon-\mu}\,, \tag{37.38}$$

where α is a real constant. We will also assume that

$$f(\epsilon) \equiv |\,V_{\epsilon\omega}\,|^2 \tag{37.39}$$

is an analytic function of the complex variable ϵ which is real and single-valued on the segment of the real axis (μ,∞), and has the property that

$$\rho(\epsilon)f(\epsilon) \xrightarrow[|\epsilon|\to\infty]{} 0\,. \tag{37.40}$$

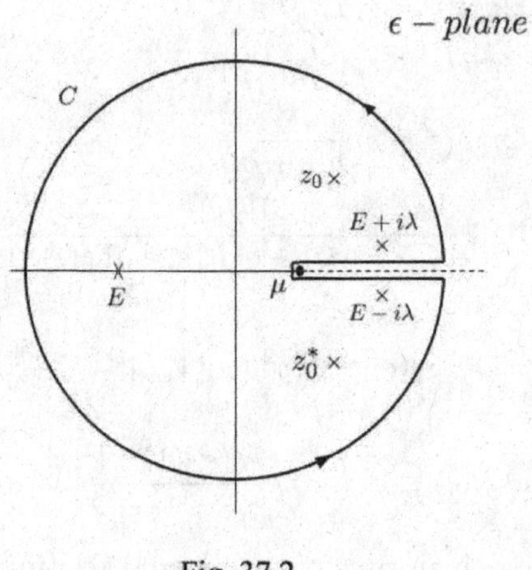

Fig. 37.2

The principal-value integral in (37.34) can then be evaluated readily by contour integration, without as yet specifying the form of $f(\epsilon)$. Thus

$$\begin{aligned} P(E) &\equiv \mathrm{P}\int_{\mu}^{\infty} d\epsilon\, \frac{\alpha\sqrt{\epsilon-\mu}\, f(\epsilon)}{\epsilon - E} \\ &= \lim_{\lambda\to 0} \alpha \int_{\mu}^{\infty} d\epsilon\, \frac{\sqrt{\epsilon-\mu}\, f(\epsilon)(\epsilon - E)}{(\epsilon - E)^2 + \lambda^2} \\ &= \lim_{\lambda\to 0} \frac{\alpha}{2} \int_C d\epsilon\, \frac{(\epsilon-\mu)^{1/2} f(\epsilon)(\epsilon - E)}{(\epsilon - E)^2 + \lambda^2}\,, \quad E \geq \mu\,, \end{aligned} \tag{37.41}$$

where the contour C is shown in Fig. 37.2. It follows from the residue theorem that

$$P(E) = i\pi\alpha \sum_{z_0} \frac{Res[f(z_0)](z_0-\mu)^{1/2}}{z_0 - E}\,, \quad E \geq \mu\,, \tag{37.42}$$

where the sum is over the poles z_0 of $f(z)$, and $Res[f(z_0)]$ denotes the residue of $f(z)$ at z_0.

According to the Schwarz's reflection principle (Theorem 36.1), the poles of $f(z)$, if they exist at all, must exist in conjugate pairs. This guarantees that $P(E)$ as given by the above equation is always real. We immediately note that $P(E)$ can be continued to the real values $E < \mu$ by

$$P(E) \equiv \int_{\mu}^{\infty} d\epsilon\, \frac{\alpha\sqrt{\epsilon-\mu}\, f(\epsilon)}{\epsilon - E} = \phi(E)/g^2\,, \quad E < \mu\,. \tag{37.43}$$

This integral (no longer a principal-value integral) can also be obtained by using the contour C (shown in Fig. 37.2) and the residue theorem, with the result

$$P(E) = i\pi\alpha \left((E-\mu)^{1/2} f(E) + \sum_{z_0} \frac{Res[f(z_0)](z_0-\mu)^{1/2}}{z_0 - E} \right) , \quad E < \mu . \tag{37.44}$$

It is clear from (37.42) and (37.44) that $P(E)$, as defined by the first equality of (37.41) and (37.43), is continuous at $E = \mu$.

Nothing more specific concerning the locations and properties of the metastable states can be said at this point without further specifying the form of the radiative coupling function $f(\epsilon)$. We will confine ourselves to the following general comments. In the usual perturbation treatments of resonance states, $P(\epsilon)$ and $\Gamma_0(\epsilon)$ are both regarded as small and weakly dependent on ϵ. For our model, there is then only one resonance (metastable) state, and it appears as an approximate pole, $\epsilon = \epsilon_d + \omega - g^2 P - i\Gamma_0/2$, of the S matrix slightly below the positive ϵ-axis on the second sheet (recall the discussion in the last chapter). The quantities $-g^2P$ and Γ_0 are interpreted as the shift and width of the state, respectively. This scheme, as embodied by the Breit-Wigner formula [(36.23)], becomes invalid when $P(\epsilon)$ and $\Gamma_0(\epsilon)$ vary significantly over a region ϵ of interest, a situation which may well result for many reasonable choices of $f(\epsilon)$. In this case, the solutions of (37.37) may bear no simple relationship to the "unperturbed" line center $\epsilon_d + \omega$. Furthermore, it may turn out that not all solutions actually correspond to stable metastable states, and the physical widths of these states may have to be determined by a renormalization procedure. (For more details see Lam and George 1986.)

To conclude this chapter we will consider the process of laser-induced dissociation in our model ($|\, d_{N+1}\rangle \to |\, \epsilon_N\rangle$), accompanied by the absorption of a photon; and the associated reverse process of recombination ($|\, \epsilon_N\rangle \to |\, d_{N+1}\rangle$), accompanied by the stimulated emission of a photon. By microreversibility, the probabilities for these processes are identical, both given by

$$P_{\epsilon d}(t) = |\, \langle\, \epsilon_N \,|\, d_{N+1}(t) \,\rangle\, |^2 , \tag{37.45}$$

where $|\, d_{N+1}(t)\,\rangle$ denotes the time-evolved state which at $t = 0$ is $|\, d_{N+1}\,\rangle$. The reverse process is to be distinguished from the related process $|\, \epsilon_N\,\rangle \to |\, b_N\,\rangle$, which is also recombination but does not involve the absorption or emission of photons, and for which the final state $|\, b_N\,\rangle$ only exists when the laser is kept on at long times. The probability for this last process has been given by (37.23).

The calculation of the matrix element appearing in (37.45) is most easily carried out by expanding $|\, d_{N+1}\,\rangle$ in terms of the complete set $\{|\, E^-(N)\,\rangle, |\, b_N\,\rangle\}$ of eigenstates of H [cf. (37.27)], assuming the general case where a bound state is present. The reason for using the incoming states will become apparent shortly.

Using (37.19), (37.25), and (37.27) through (37.29), we have

$$\begin{aligned} |\, d_{N+1}(t)\,\rangle &= \sum_E e^{-iEt}\,|\, E^-(N)\,\rangle\langle\, E^-(N)\,|\, d_{N+1}\,\rangle + e^{-i\epsilon_b t}\,|\, b_N\,\rangle\langle\, b_N\,|\, d_{N+1}\,\rangle \\ &= \sum_E (\beta_E^-)^*\, e^{-iEt}\,|\, E^-(N)\,\rangle + \beta_b\, e^{-i\epsilon_b t}\,|\, b_N\,\rangle\,, \end{aligned} \tag{37.46}$$

where β_E^- and β_b are given by (37.28) and (37.20), respectively. [Note that (37.20) implies that β_b is real.] Expanding $|\, E^-(N)\,\rangle$ and $|\, b_N\,\rangle$ again in terms of the complete set $\{|\, d_{N+1}\,\rangle, \,|\, \epsilon_N\,\rangle\}$ using (37.25) and (37.19), Eq. (37.46) yields

$$\langle\, \epsilon_N\,|\, d_{N+1}(t)\,\rangle = (\beta_E^-)^*\, e^{-i\epsilon t} - \frac{g\beta_b^2 V_{\epsilon\omega}^*\, e^{-i\epsilon_b t}}{\epsilon - \epsilon_b} - gV_{\epsilon\omega}^* \sum_E \frac{e^{-iEt}\,|\,\beta_E^-\,|^2}{E - \epsilon - i\eta}\,. \tag{37.47}$$

The last term represents the overlap between the scattered-wave part of $|\, E^-(N)\,\rangle$ and $|\, \epsilon_N\,\rangle$. It thus vanishes as $t \to \infty$, since in the distant future the scattered part of an incoming scattering state vanishes. The mathematical statement of this fact has already been given by (35.33), which also implies that the interference term due to the remaining two terms vanish (since ϵ is greater than μ and never equals ϵ_b). We finally have

$$P_{\epsilon d} = g^2 |V_{\epsilon\omega}|^2 \left[\frac{\beta_b^4}{(\epsilon - \epsilon_b)^2} + \frac{1}{\{\epsilon - (\epsilon_d + \omega) + g^2 P(\epsilon)\}^2 + \Gamma_0^2(\epsilon)/4} \right]. \tag{37.48}$$

The first term in this equation represents the effect of the bound state and vanishes if $g < g_{crit}$. It contributes only to the nonresonant background since ϵ is never equal to ϵ_b, and is most prominent near the dissociation threshold $\epsilon = \mu$. When metastable states exist, the second term gives rise to resonant peaks. In the absence of such states it also contributes to the nonresonant background.

Problem 37.1 In this problem we will consider in some detail a non-relativistic treatment of a two-level atom interacting with an external single-mode radiation field (for example, a laser) and capable of spontaneous radiative decay (**spontaneous emission** from the excited state). This is a typical problem in **non-relativistic quantum electrodynamics**. The ground state of the atom will be labeled $|\, g\,\rangle$ and the excited state labeled $|\, e\,\rangle$. We will use the Heisenberg equation of motion to describe the dynamics of the system. The total Hamiltonian of the system can be written as

$$H = H_{atom} + H_{field} + H_I\,,$$

where, in second-quantized notation,

$$\begin{aligned} H_{atom} &= \epsilon_g\, b_g^\dagger b_g + \epsilon_e\, b_e^\dagger b_e\,, \\ H_{field} &= \sum_k \hbar\omega_k\, a_k^\dagger a_k\,, \qquad k \equiv \{\boldsymbol{k}, \boldsymbol{\varepsilon_k}\}\,, \\ H_I &= -\boldsymbol{\mu}\cdot\boldsymbol{E} = \hbar \sum_k \left\{ g_{eg}^k\, b_e^\dagger b_g a_k + (g_{eg}^k)^*\, a_k^\dagger b_g^\dagger b_e \right\}, \end{aligned}$$

with the radiative coupling strengths given by

$$g^k_{eg} = -i\boldsymbol{\varepsilon}_{\boldsymbol{k}} \cdot \boldsymbol{D}_{eg} \sqrt{\frac{2\pi\omega_k}{\hbar V}} \,, \quad \boldsymbol{D}_{eg} \equiv \langle\, e \,|\, \boldsymbol{\mu} \,|\, g \,\rangle \,.$$

In the above equations, ϵ_g and ϵ_e denote the ground and excited atomic energy levels, respectively; $\boldsymbol{k}$ is the wave vector, $\boldsymbol{\varepsilon}_{\boldsymbol{k}}$ the associated unit polarization vectors, and ω_k is the frequency, for the k-th radiation (photon) mode (including the laser mode); $\boldsymbol{\mu}$ is the atomic dipole moment operator; $\boldsymbol{E}$ is the electric field operator (for the laser); and V is the volume of the system for the purpose of box quantization. The operators $a^\dagger_k(a_k)$ are the creation (annihilation) (bosonic) operators for photons of the k-th mode (satisfying the usual "*equal time*" commutation relations), while $b^\dagger_g(b_g)$ and $b^\dagger_e(b_e)$ are the creation (annihilation) (fermionic) operators for atomic excitations in the ground and excited states, respectively (satisfying the "*equal time*" anti-commutation relations given in Problem 14.2). The interaction Hamiltonian H_I as given satisfies the **rotating wave approximation**, in which atomic excitation is necessarily accompanied by absorption of a photon and de-excitation necessarily accompanied by emission of a photon. We will limit ourselves to the Hilbert subspace of the system for which $N_g + N_e = 1$, where N_g and N_e are the total number of atomic excitations in the ground and excited atomic states, respectively (the atom is either in the ground state or the excited state).

(a) Show that

$$[\, N_T,\, H\,] = 0 \,,$$

where $N_T = N_e + N_r$, $N_e = b^\dagger_e b_e$, $N_r = \sum_k a^\dagger_k a_k$ (number operator for the total number of photons). So $N_e + N_r$ is conserved. Alternatively, $N_+ + N_- + N_r$ is conserved, where $N_+ \equiv (N_e + N_g)/2$, $N_- \equiv (N_e - N_g)/2$.

(b) Define

$$\sigma_{ij} \equiv b^\dagger_i b_j \,.$$

Show that $\sigma_{ij}\sigma_{lm} = \sigma_{im}\delta_{jl}$.

(c) Recall the discussion of the density matrix ρ_{ij} in Problem 8.7. Show that

$$\rho_{ji}(t) = \langle\, \sigma_{ij}(t) \,\rangle \,,$$

where $\langle\, O(t) \,\rangle$, the average (expected) value of a dynamical variable O at time t, is given by $\langle\, O(t) \,\rangle = Tr(O(t)\rho(0)) = Tr(O\rho(t))$.

(d) Use Heisenberg's equation of motion $i\hbar\dot{O} = [\, O,\, H\,]$ to show that:

$$\dot{\sigma}_{eg} = -i(\omega_g - \omega_e)\sigma_{eg} - i\sum_k (g^k_{eg})^*\, a^\dagger_k \sigma_{ee} + i\sum_k (g^k_{eg})^*\, a^\dagger_k \sigma_{gg} \,,$$

$$\dot{\sigma}_{ge} = (\dot{\sigma}_{eg})^\dagger \,,$$

$$\dot{\sigma}_{ee} = -i\sum_k g^k_{eg}\, \sigma_{eg} a_k - i\sum_k (g^k_{eg})^*\, a^\dagger_k \sigma_{ge} \,,$$

$$\dot{\sigma}_{gg} = -i\sum_k (g^k_{eg})^*\, a^\dagger_k \sigma_{ge} + i\sum_k g^k_{eg}\, \sigma_{eg} a_k \,,$$

$$\dot{a}_k = -i\omega_k a_k - i(g^k_{eg})^*\, \sigma_{ge} \,,$$

where $\omega_g = \epsilon_g/\hbar$ and $\omega_e = \epsilon_e/\hbar$.

(e) Show that, formally, one can write the solution of the last equation in (d) [for the field operator $a_k(t)$] as

$$a_k(t) = a_k(0)e^{-i\omega_k t} - i(g_{eg}^k)^* \int_0^t dt' \, \sigma_{ge}(t') \exp(-i(\omega_k - i\epsilon)(t-t')) \,,$$

where the $-i\epsilon$ factor is inserted to ensure that we have the retarded solution [$a_k(t) = 0$ for $t < 0$].

(f) Define the **Liouville operator** L (acting on the space of observables) by

$$LO \equiv \frac{1}{\hbar} [H, O] \,,$$

so that Heisenberg's equation of motion $i\hbar\dot{O} = [O, H]$ can be written as

$$\dot{O} = iLO \,.$$

This then has the formal solution

$$O(t) = e^{iLt} O(0) \,.$$

So we can write

$$\sigma_{ge}(t) = e^{iL(t-t')} \sigma_{ge}(t') \,.$$

Inverting this equation and plugging $\sigma_{ge}(t')$ back into the one in (e), we have

$$a_k(t) = a_k(0) \, e^{-i\omega_k t} - i(g_{eg}^k)^* \int_0^t dt' \, e^{-i(\omega_k + L - i\epsilon)(t-t')} \, \sigma_{ge}(t) \,.$$

Now introduce the **Markoffian approximation**, by letting $t \to \infty$ in the above equation. This is justified when one is interested in large times t for which $(\omega_e - \omega_g)t \gg 1$. Show that in this approximation the above equation reduces to

$$a_k(t) = a_k(0) \, e^{-i\omega_k t} - (g_{eg}^k)^* \, \frac{1}{\omega_k + L - i\epsilon} \, \sigma_{ge}(t) \,.$$

Thus, under the Markoffian approximation, **memory effects** described by the time integral in $a_k(t)$ are eliminated.

(g) Write

$$L = L_0 + L_1 \,,$$

where

$$L_0 O \equiv [H_{atom} + H_{field}, O]/\hbar \,, \quad L_1 O \equiv [H_I, O]/\hbar \,.$$

Recalling the general operator identity relating two arbitrary invertible operators given by (35.49), prove the following **Lippmann-Schwinger equation** for L:

$$\frac{1}{\omega_k + L - i\epsilon} = \frac{1}{\omega_k + L_0 - i\epsilon} + \frac{1}{\omega_k + L_0 - i\epsilon} (-L_1) \, \frac{1}{\omega_k + L - i\epsilon} \,.$$

Thus, to lowest order in the interaction,

$$a_k(t) \approx a_k(0) \, e^{-i\omega_k t} - (g_{eg}^k)^* \, \frac{1}{\omega_k + L_0 - i\epsilon} \, \sigma_{ge}(t) \,.$$

(h) Show that

$$L_0\sigma_{ge} = \omega_{ge}\sigma_{ge} \ ,$$

where $\omega_{ge} \equiv \omega_g - \omega_e$. Then, to lowest order in the interaction,

$$a_k(t) \approx a_k(0)\, e^{-i\omega_k t} - (g^k_{eg})^* \frac{1}{\omega_k + \omega_{ge} - i\epsilon}\, \sigma_{ge}(t) \ .$$

(i) Use the above equation and its hermitian adjoint in the set of coupled equations in (d) to eliminate the radiation field operators $a_k(t)$ and $a^\dagger(t)$. Show that the following set of coupled equations for only the atomic excitation operators is obtained:

$$\dot\sigma_{eg} = -i\left(\omega_{ge} + \sum_k |g^k_{eg}|^2 \frac{1}{\omega_k + \omega_{ge} + i\epsilon}\right)\sigma_{eg} - i\left(\sum_k (g^k_{eg})^* e^{i\omega_k t} a^\dagger_k(0)\right)(\sigma_{ee} - \sigma_{gg}) \ ,$$

$$\dot\sigma_{ge} = (\dot\sigma_{eg})^\dagger \ ,$$

$$\dot\sigma_{ee} = -i\left(\sum_k g^k_{eg} e^{-i\omega_k t} a_k(0)\right)\sigma_{eg} - i\left(\sum_k |g^k_{eg}|^2 \frac{1}{\omega_k + \omega_{ge} + i\epsilon}\right)\sigma_{ee} + h.c. \ ,$$

$$\dot\sigma_{gg} = -i\left(\sum_k (g^k_{eg})^* a^\dagger_k(0) e^{i\omega_k t}\right)\sigma_{ge} - i\left(\sum_k |g^k_{eg}|^2 \frac{1}{\omega_k + \omega_{ge} - i\epsilon}\right)\sigma_{ee} + h.c. \ ,$$

where "*h.c.*" means hermitian conjugate.

(j) In order to relate the operators in the above set of equations to observable quantities we take expectation values [cf. result of (c)]. Assume that we have the following conditions of statistical independence:

$$\rho(0) = \rho_{atom}(0)\rho_{field}(0) \ ,$$
$$\langle\, a^\dagger_k(0)\sigma_{ij}(t)\,\rangle \approx \langle\, a^\dagger_k(0)\,\rangle_F \langle\, \sigma_{ij}(t)\,\rangle_A \ , \quad \langle\, \sigma_{ij}(t) a_k(0)\,\rangle \approx \langle\, a_k(0)\,\rangle_F \langle\, \sigma_{ij}(t)\,\rangle_A \ ,$$
$$\langle\, \sigma_{ij}(t)\,\rangle = \langle\, \sigma_{ij}\,\rangle_A \ ,$$

where the subscripts A and F mean atomic and field modes, respectively. We also assume that at $t = 0$ only the laser mode is present, so that

$$\rho_{field}(0) = |\, L\,\rangle\langle\, L\,| \ ,$$

where $|\, L\,\rangle$ is a pure state describing the laser mode. Thus

$$\langle\, a_k(0)\,\rangle_F = Tr_F\{a_k(0)\rho_{field}(0)\} = Tr_F\{a_k(0)|\, L\,\rangle\langle\, L\,|\} = \langle\, L\,|\, a_k(0)\,|\, L\,\rangle \ ,$$

where Tr_F means summing over the field modes only. We can assume that $|\, L\,\rangle$ only involves number states in a single $\{\boldsymbol{k}_L, \boldsymbol{\varepsilon}_{\boldsymbol{k}_L}\}$ (laser) mode (with $n_k = 0$ for $k \neq \{\boldsymbol{k}_L, \boldsymbol{\varepsilon}_{\boldsymbol{k}_L}\}$), so that in the sums over k involving $a_k(0)$ and $a^\dagger_k(0)$ in the equations in (i) only the term $k = \{\boldsymbol{k}_L, \boldsymbol{\varepsilon}_{\boldsymbol{k}_L}\}$ survives. This result suggests that it will be convenient to use a coherent state $|\, z_L\,\rangle\, (z_L \in \mathbb{C})$ to describe the laser field, since, according to the results in Problem 14.3(b) and 14.3(e), $\langle\, z_L\,|\, a_L(0)\,|\, z_L\,\rangle = z_L$, where a_L stands for $a_{k=\{\boldsymbol{k}_L, \boldsymbol{\varepsilon}_{\boldsymbol{k}_L}\}}$. On the other hand, z_L can be related to the expectation value of the electric field strength of the laser in the state $|\, z_L\,\rangle$ by

$$\langle\, z_L\,|\, \boldsymbol{E}_L(\boldsymbol{r}, t)\,|\, z_L\,\rangle = -2\sqrt{\frac{2\pi\hbar\omega_{\boldsymbol{k}_L}}{V}}\, |z_L| \sin(\boldsymbol{k}_L\cdot\boldsymbol{r} - \omega_{\boldsymbol{k}_L} t + \theta_L)\, \boldsymbol{\varepsilon}_{\boldsymbol{k}_L} \ ,$$

where $z_L = |z_L|e^{i\theta_L}$. Justify this equation by using the following:

$$\boldsymbol{A}_L(\boldsymbol{r}) = \sqrt{\frac{2\pi\hbar c^2}{V\omega_{\boldsymbol{k}_L}}}\,(a_L e^{i\boldsymbol{k}_L\cdot\boldsymbol{r}} + a_L^\dagger e^{-i\boldsymbol{k}_L\cdot\boldsymbol{r}})\,\boldsymbol{\varepsilon}_{\boldsymbol{k}_L}\,,$$

$$\boldsymbol{E}_L(\boldsymbol{r}) = -\frac{i}{\hbar c}\,[H_{field}, \boldsymbol{A}_L(\boldsymbol{r})]\,.$$

Define the quantity

$$\Omega \equiv g^L_{eg}\langle\, a_L(0)\,\rangle = -i\boldsymbol{\varepsilon}_{\boldsymbol{k}_L}\cdot\boldsymbol{D}_{eg}\, z_L\sqrt{\frac{2\pi\omega_L}{\hbar V}}\,.$$

On requiring $\boldsymbol{D}_{eg}$ to be real the phase of z_L can then be chosen so that Ω_{eg} is real, and one can then write

$$\Omega_{eg} = \Omega^*_{eg} = -\frac{1}{2}\boldsymbol{E}_L\cdot\boldsymbol{D}_{eg}/\hbar\,,$$

where $\boldsymbol{E}_L$ is the electric field strength of the pumping laser. $|\Omega_{eg}|$ is seen to be half the so-called **Rabi frequency**.

(k) On setting

$$\chi_{eg} = \sigma_{eg}\, e^{-i\omega_L t}\,, \quad \chi_{ee} = \sigma_{ee}\,, \quad \chi_{gg} = \sigma_{gg}\,,$$

the "fast" terms proportional to $\exp(\pm i\omega_k t)$ in the rate equations in (i) can be eliminated. Show that the set of coupled rate equations for the expectation values $\langle\,\chi_{eg}\,\rangle$, $\langle\,\chi_{ge}\,\rangle$, $\langle\,\chi_{ee}\,\rangle$ and $\langle\,\chi_{gg}\,\rangle$ is given by

$$\frac{d}{dt}\langle\,\chi_{eg}\,\rangle = -i\Delta_{eg}\langle\,\chi_{eg}\,\rangle + i\Omega^*_{eg}(\langle\,\chi_{gg}\,\rangle - \langle\,\chi_{ee}\,\rangle) - i\left(\sum_k \frac{|g^k_{eg}|^2}{\omega_k + \omega_{ge} + i\epsilon}\right)\langle\,\chi_{eg}\,\rangle$$

$$\frac{d}{dt}\langle\,\chi_{ge}\,\rangle = \left(\frac{d}{dt}\langle\,\chi_{eg}\,\rangle\right)^*\,,$$

$$\frac{d}{dt}\langle\,\chi_{ee}\,\rangle = -i\Omega_{eg}\langle\,\chi_{eg}\,\rangle + i\Omega^*_{eg}\langle\,\chi_{ge}\,\rangle + \left[i\left(\sum_k \frac{|g^k_{eg}|^2}{\omega_k + \omega_{ge} - i\epsilon}\right)\langle\,\chi_{ee}\,\rangle + c.c.\right]\,,$$

$$\frac{d}{dt}\langle\,\chi_{gg}\,\rangle = -i\Omega^*_{eg}\langle\,\chi_{ge}\,\rangle + i\Omega_{eg}\langle\,\chi_{eg}\,\rangle + \left[i\left(\sum_k \frac{|g^k_{eg}|^2}{\omega_k + \omega_{ge} + i\epsilon}\right)\langle\,\chi_{ee}\,\rangle + c.c.\right]\,,$$

where $\Delta_{eg} \equiv \omega_L - \omega_{eg}$ is called the **detuning**, and *c.c.* stands for the complex conjugate.

(l) Write

$$\sum_k \frac{|g^k_{eg}|^2}{\omega_k - \omega_{eg} + i\epsilon} = -i\frac{A}{2}\,,$$

where

$$\sum_k \longrightarrow \frac{V}{(2\pi)^3}\int d^3k \sum_{\boldsymbol{\varepsilon}_{\boldsymbol{k}}}$$

and we have assumed that the (real) principal part of the integral (which actually diverges) has been renormalized into the experimentally observed ω_{ge} [cf. (5.29)]. Show that

$$A = \frac{\omega^3_{eg}}{2\pi\hbar c^3}\,[4\pi\boldsymbol{D}_{ge}\cdot\boldsymbol{D}_{eg} - \int d\Omega_{\boldsymbol{k}}(\hat{\boldsymbol{k}}\cdot\boldsymbol{D}_{ge})(\hat{\boldsymbol{k}}\cdot\boldsymbol{D}_{eg}) = \frac{4|\boldsymbol{D}_{eg}|^2\omega^3_{eg}}{3\hbar c^3}\,.$$

This quantity is known as **Einstein's A coefficient**.

(m) Finally show that the set of coupled equations for a laser-pumped two-level atom capable of spontaneous radiative decay can be written as

$$\frac{d}{dt}\begin{pmatrix} \langle \chi_{gg} \rangle \\ \langle \chi_{ge} \rangle \\ \langle \chi_{eg} \rangle \\ \langle \chi_{ee} \rangle \end{pmatrix} = -i \begin{pmatrix} 0 & \Omega & -\Omega & iA \\ \Omega & -\Delta_{eg} - i\dfrac{A}{2} & 0 & -\Omega \\ -\Omega & 0 & \Delta_{eg} - i\dfrac{A}{2} & \Omega \\ 0 & -\Omega & \Omega & -iA \end{pmatrix} \begin{pmatrix} \langle \chi_{gg} \rangle \\ \langle \chi_{ge} \rangle \\ \langle \chi_{eg} \rangle \\ \langle \chi_{ee} \rangle \end{pmatrix}.$$

The quantities $\langle \chi_{gg} \rangle$ and $\langle \chi_{ee} \rangle$ are in fact equal to the density matrix elements $\langle \sigma_{gg} \rangle_A$ and $\langle \sigma_{ee} \rangle_A$, respectively, of the atomic system, and provide a measure of the time evolution of the populations in the ground and excited states.

The treatment given in this problem for a two-state system can be generalized to one involving manifolds of ground and excited states, such as that described in Problem 28.1.

Chapter 38

Geometric Phases: The Aharonov-Bohm Effect and the Magnetic Monopole

Our discussions thus far have dealt with the themes of dynamics and symmetry in quantum theory. Beginning with this chapter we will focus on the third main theme of this book – geometry – and explore the significance of some basic differential geometric notions in quantum theory. This will be done through specific examples, first described in conventional physics terminology, and then related to the appropriate mathematical notions. The relevant mathematics will be introduced as required, in a manner that favors physical clarity rather than mathematical rigor. We will demonstrate that a core group of geometric notions, namely, those surrounding the theory of **connections on fiber bundles**, will furnish a unifying mathematical description of diverse quantal (as well as classical) phenomena.

We start by considering again the interaction between an electron (of charge e and mass m) and an electromagnetic field, described by a vector potential $A_\mu(\boldsymbol{x})$ and a scalar potential $\phi(\boldsymbol{x})$ ($\mu = 1, 2, 3$ are spatial tensorial indices). Recalling our discussion at the beginning of Chapter 34, the (non-relativistic) Schrödinger equation with minimal coupling,

$$\left[-\frac{\hbar^2}{2m} \left(\partial_\mu - \frac{ie}{\hbar c} A_\mu \right)^2 + e\phi \right] \psi = i\hbar \frac{\partial \psi}{\partial t} \,, \tag{38.1}$$

is invariant under the **gauge transformation**

$$A_\mu(\boldsymbol{x}) \longrightarrow A_\mu(\boldsymbol{x}) + \partial_\mu \lambda(\boldsymbol{x}) \,; \quad \psi(\boldsymbol{x}) \longrightarrow \exp\left(\frac{ie}{\hbar c} \lambda(\boldsymbol{x}) \right) \psi(\boldsymbol{x}) \,, \tag{38.2}$$

where $\lambda(\boldsymbol{x})$ is an arbitrary function on the spatial domain in which $\psi(\boldsymbol{x})$ is defined. We will denote this domain by M. At a point $\boldsymbol{x} \in M$, the local value of $A_\mu(\boldsymbol{x})$ can be obtained by the gauge transformation

$$0 \longrightarrow \partial_\mu \lambda(\boldsymbol{x}) = A_\mu(\boldsymbol{x}) \,. \tag{38.3}$$

Hence we can write

$$\lambda(\boldsymbol{x}) = \int_{\boldsymbol{x}_0\,(C)}^{\boldsymbol{x}} \boldsymbol{A} \cdot d\boldsymbol{x} \,, \tag{38.4}$$

where the line integral is carried out over some path C from $\boldsymbol{x}_0$ to $\boldsymbol{x}$. Suppose that at $\boldsymbol{x}$, the solution to the Schrödinger equation (38.1) for $A_\mu = 0$ is $\psi'(\boldsymbol{x})$. Then the solution for A_μ as given by (38.3) is, according to the second equation in (38.2),

$$\psi(\boldsymbol{x}) = \psi'(\boldsymbol{x}) \exp\left(\frac{ie}{\hbar c} \int_{\boldsymbol{x}_0\,(C)}^{\boldsymbol{x}} \boldsymbol{A} \cdot d\boldsymbol{x} \right) \,. \tag{38.5}$$

An interesting situation arises if

$$\oint_{\partial S} \boldsymbol{A} \cdot d\boldsymbol{x} = \int_S (\nabla \times \boldsymbol{A}) \cdot d\boldsymbol{a} = \int_S \boldsymbol{B} \cdot d\boldsymbol{a} = \Phi \neq 0 \,, \tag{38.6}$$

where the closed loop ∂S is the boundary of some surface S and includes the path C, $\boldsymbol{B}$ is the magnetic field, Φ is the magnetic flux through the surface S, and the first equality follows from **Stokes' Theorem** in vector calculus. In this case the path integral from $\boldsymbol{x}_0$ to $\boldsymbol{x}$ in (38.5) depends on the path, and so does the value of the wave function $\psi(\boldsymbol{x})$. The path-dependent phase factor in (38.5) is referred to as a **geometric phase** and is said to be **nonintegrable**. This is precisely the situation in the **Aharonov-Bohm effect** (Aharonov and Bohm 1959). (For excellent reviews of the general subject of geometric phases, see Shapere and Wilczek 1989; Bohm, Mostafazadeh, Koizumi, Niu and Zwanziger 2003; Chruściński and Jamiołkowski 2004.)

This effect involves two paths followed by an electron, C_1 and C_2, joining two points $\boldsymbol{x}_0$ and $\boldsymbol{x}$, over which the magnetic field $\boldsymbol{B}$ vanishes, but such that the closed loop $C_2 - C_1$ is the boundary of a surface through which there is a magnetic flux Φ (Fig. 38.1). From (38.5) and (38.6), it follows then that the two wave functions

$$\psi_1(\boldsymbol{x}) \equiv \psi'(\boldsymbol{x}) \exp\left(\frac{ie}{\hbar c} \int_{\boldsymbol{x}_0\,(C_1)}^{\boldsymbol{x}} \boldsymbol{A} \cdot d\boldsymbol{x} \right) , \; \psi_2(\boldsymbol{x}) \equiv \psi'(\boldsymbol{x}) \exp\left(\frac{ie}{\hbar c} \int_{\boldsymbol{x}_0\,(C_2)}^{\boldsymbol{x}} \boldsymbol{A} \cdot d\boldsymbol{x} \right) \tag{38.7}$$

are not equal to each other. In fact, on propagating the wave function around the closed loop $C_1 - C_2$, it acquires a geometric phase as follows [cf. (38.6)]:

$$\psi(\boldsymbol{x}) \longrightarrow \psi(\boldsymbol{x}) \exp\left(\frac{ie\Phi}{\hbar c} \right) \,. \tag{38.8}$$

This geometric phase has readily measurable interference effects, which, it is important to note, have no classical analogs. Classically, the electron does not experience any magnetic force over the paths C_1 and C_2 that it traverses, since,

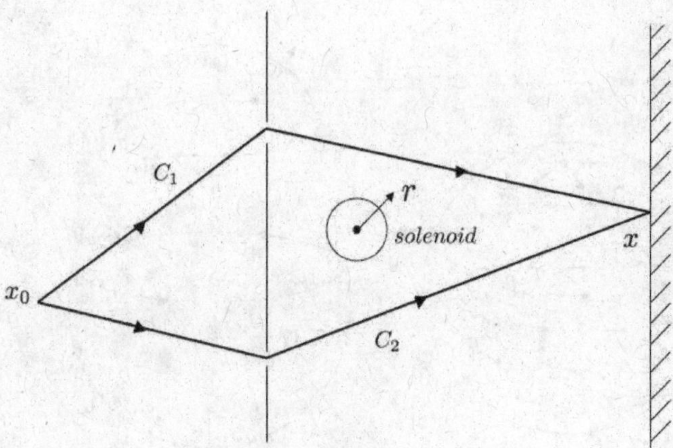

Fig. 38.1

by assumption, the magnetic field $\boldsymbol{B}$ vanishes over all points on these paths. The Aharonov-Bohm effect was historically the first indication that, in quantum mechanics, the vector potential $\boldsymbol{A}(\boldsymbol{x})$ is more basic than the electromagnetic field strength represented by $\boldsymbol{B}$. An example of a vector potential giving rise to the Aharonov-Bohm effect is (in Cartesian coordinates):

$$\boldsymbol{A}(\boldsymbol{x}) = \left(-\frac{y\Phi}{2\pi r^2}, \frac{x\Phi}{2\pi r^2}, 0 \right), \tag{38.9}$$

where Φ is a constant and $r^2 \equiv x^2 + y^2$. This vector potential describes a solendoidal flux along the z-axis and concentrated at the origin.

Problem 38.1 Show that, for the vector potential $\boldsymbol{A}(\boldsymbol{x})$ given by (38.9), $\boldsymbol{B} = \nabla \times \boldsymbol{A} = 0$ for $r \neq 0$, and

$$\oint_C \boldsymbol{A} \cdot d\boldsymbol{x} = \Phi,$$

for any closed counterclockwise loop C on the xy-plane encircling the origin. Draw a schematic plot of the vector field $\boldsymbol{A}$.

The next example that we will consider – the magnetic monopole – illustrates in an even more dramatic fashion the basic role played by the vector potential $\boldsymbol{A}$ (as opposed to the magnetic field $\boldsymbol{B}$) in quantum theory. Indeed, this example

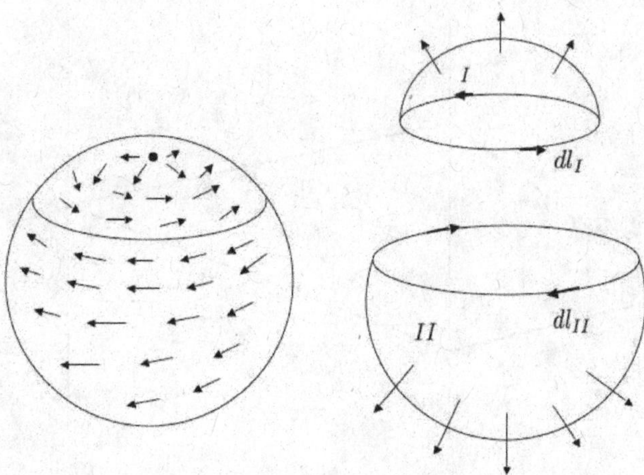

Fig. 38.2

can be used as a gateway to understand the intimate relationship between the physical concept of **gauge fields**, of which the electromagnetic vector potential $\boldsymbol{A}$ is an example, and the basic differential geometric one of **connections on fiber bundles**.

The magnetic field $\boldsymbol{B}$ due to a **magnetic monopole** of strength (**magnetic charge**) M is given, in analogy to Coulomb's law in electrostatics, by

$$\boldsymbol{B} = \frac{M\boldsymbol{r}}{r^3} , \tag{38.10}$$

where $\boldsymbol{r}$ is the radial vector from the position of the monopole to the field point. According the Gauss' law, the magnetic flux Φ through a spherical surface S^2 centered at the monopole is

$$\Phi = \oint_{S^2} \boldsymbol{B} \cdot d\boldsymbol{a} = 4\pi M . \tag{38.11}$$

Let us divide the 2-sphere S^2 into two regions, I and II, each including its boundaries, as shown in Fig. 38.2, and use Stokes' Theorem to convert the surface integral (38.11) to a pair of line integrals:

$$\begin{aligned}\oint_{S^2} \boldsymbol{B} \cdot d\boldsymbol{a} &= \int_I \nabla \times \boldsymbol{A} \cdot d\boldsymbol{a} + \int_{II} \nabla \times \boldsymbol{A} \cdot d\boldsymbol{a} \\ &= \oint \boldsymbol{A}_I \cdot d\boldsymbol{l}_I + \oint \boldsymbol{A}_{II} \cdot d\boldsymbol{l}_{II} = \oint (\boldsymbol{A}_I - \boldsymbol{A}_{II}) \cdot d\boldsymbol{l}_I ,\end{aligned} \tag{38.12}$$

where $\boldsymbol{A}_I$ and $\boldsymbol{A}_{II}$ are the vector potential fields defined on the boundaries of the top cap (region I) and the bottom cap (region II), respectively. (Note the opposite orientations of these boundaries in the integration.) Now these boundaries correspond to exactly the same circle S^1 on the sphere; but yet $\boldsymbol{A}_I(\boldsymbol{x})$ cannot be equal to $\boldsymbol{A}_{II}(\boldsymbol{x})$. Otherwise the above integral would vanish, which clearly contradicts the result (38.11). Our only recourse is to introduce a gauge transformation between $\boldsymbol{A}_I$ and $\boldsymbol{A}_{II}$ along the boundary S^1 (the overlap between regions I and II). According to (38.2), this transformation of the vector potential entails a corresponding gauge transformation of the wave function on the region of overlap also. We will write

$$\boldsymbol{A}_I(\boldsymbol{x}) = \boldsymbol{A}_{II}(\boldsymbol{x}) + \frac{\hbar c}{e}\,\nabla\lambda(\boldsymbol{x}) , \quad \psi_I(\boldsymbol{x}) = \psi_{II}(\boldsymbol{x})\exp\{i\lambda(\boldsymbol{x})\} , \tag{38.13}$$

where $\lambda(\boldsymbol{x})$ is a function (not necessarily single-valued) defined on the region of overlap between the regions I and II, namely, the circle S^1. It is crucially important to note at this point that, while $\lambda(\boldsymbol{x})$ is not necessarily single-valued, the so-called **transition function** $\exp\{i\lambda(\boldsymbol{x})\}$, governing the gauge transformation of the quantum mechanical wave function, must be single-valed on the overlap S^1. Equations (38.11) and (38.12) then imply

$$4\pi M = \frac{\hbar c}{e}\oint \nabla\lambda\cdot d\boldsymbol{l}_I = \frac{\hbar c}{e}\Delta\lambda = \frac{\hbar c}{e}\,2\pi n , \tag{38.14}$$

where $\Delta\lambda$ is the change in the function $\lambda(\boldsymbol{x})$ on going once around S^1, and n is an integer: $n = 0, \pm 1, \pm 2, \cdots \in \mathbb{Z}$. This integral provides another example of a nonintegrable geometric phase. The inescapable conclusion is that

$$\boxed{M = n\left(\frac{\hbar c}{2e}\right) , \qquad n \in \mathbb{Z}} \, . \tag{38.15}$$

This fascinating result states that the magnetic charge, if it exists at all, is related to the electron charge in a fundamental way, and furthermore, *it must be quantized.* If we invert the above equation to write

$$e = n\left(\frac{\hbar c}{2M}\right) , \tag{38.16}$$

we have a possible rationale for the *quantization of the electron charge.* Dirac already noticed this in 1931 (Dirac 1931), and concluded presciently: "The presence of even just one magnetic monopole *anywhere* in the universe would explain the quantization of the electron charge". Equation (38.14) also allows us to identify

$$\phi_0 \equiv \frac{hc}{e} \tag{38.17}$$

as the fundamental unit of magnetic flux.

The integer n in (38.15) is an example of a so-called **topological quantum number** (**Chern number** in mathematical terminology): it arises from the topological properties of the manifold S^2, on which the vector potential (gauge field) $\boldsymbol{A}(\boldsymbol{x})$ and the wave function $\psi(\boldsymbol{x})$ are defined. As we have seen above, these properties force the necessity of gauge transformations, which in turn lead to the quantization of physical quantities. Since $\exp\{i\lambda(\boldsymbol{x})\} \in U(1)$ (the unitary group of dimension one) for each $\boldsymbol{x}$ in the overlap region, the vector potential specified by (38.12) is said to be that of a $U(1)$ magnetic monopole. In subsequent chapters, after more physical examples have been introduced, we will formulate more precisely a unifying group of differential geometric notions that form the mathematical basis of a geometrical theory of diverse physical phenomena involving geometric phases.

The above quantization result had been obtained without specifying the functional form of the vector potentials $\boldsymbol{A}_I(\boldsymbol{x})$ and $\boldsymbol{A}_{II}(\boldsymbol{x})$, or the function $\lambda(\boldsymbol{x})$. We will now give a specific solution for $\boldsymbol{A}(\boldsymbol{x})$, known as the **Wu-Yang potential** for the magnetic monopole:

$$\boldsymbol{A}_I(\boldsymbol{x}) = \frac{M(1-\cos\theta)}{r\sin\theta}\,\hat{e}_\varphi\,, \quad \boldsymbol{A}_{II}(\boldsymbol{x}) = -\frac{M(1+\cos\theta)}{r\sin\theta}\,\hat{e}_\varphi\,, \tag{38.18}$$

where (r,θ,φ) are the usual spherical coordinates. We note that the above $\boldsymbol{A}_I$ is actually valid over the entire 2-sphere S^2 except at $\theta=\pi$, while $\boldsymbol{A}_{II}$ is valid everywhere on S^2 except at $\theta = 0$. From (38.18),

$$\boldsymbol{A}_I(\boldsymbol{x}) - \boldsymbol{A}_{II}(\boldsymbol{x}) = \frac{2M}{r\sin\theta}\,\hat{e}_\varphi = \nabla(2M\varphi)\,, \quad (\theta\neq 0\,,\,\theta\neq\pi)\,, \tag{38.19}$$

which is of the form given by (38.13). Thus the line integral of $\boldsymbol{A}_I(\boldsymbol{x}) - \boldsymbol{A}_{II}(\boldsymbol{x})$ over the overlap circle S^1 can be evaluated as follows:

$$\begin{aligned}&\oint(\boldsymbol{A}_I - \boldsymbol{A}_{II})\cdot d\boldsymbol{l}_I = \oint\nabla(2M\varphi)\cdot d\boldsymbol{l}_I \\ &= 2M\oint\frac{1}{r\sin\theta}\,\hat{e}_\varphi\cdot d\boldsymbol{l}_I = 2M\int_0^{2\pi}\frac{\hat{e}_\varphi\cdot(r\sin\theta\,d\varphi)\,\hat{e}_\varphi}{r\sin\theta} = 4\pi M\,.\end{aligned} \tag{38.20}$$

This verifies (38.11).

Problem 38.2 Show that the Wu-Yang potential can be expressed in Cartesian coordinates as

$$A_I^x = -\frac{My}{r(r+z)}\,, \qquad A_I^y = \frac{Mx}{r(r+z)}\,, \qquad A_I^z = 0\,,$$
$$A_{II}^x = \frac{My}{r(r-z)}\,, \qquad A_{II}^y = -\frac{Mx}{r(r-z)}\,, \qquad A_{II}^z = 0\,.$$

With these expressions verify directly that

$$\nabla\times\boldsymbol{A}_I = \frac{M\hat{r}}{r^2} + 4\pi M\delta(x)\delta(y)\theta(-z)\,, \quad \nabla\times\boldsymbol{A}_{II} = \frac{M\hat{r}}{r^2} + 4\pi M\delta(x)\delta(y)\theta(z)\,,$$

where $\theta(z)$ is the step function.

Chapter 39

The Berry Phase in Molecular Dynamics

The Berry phase, which is a kind of geometric phase, comes into play in quantum mechanics when the Hamiltonian depends parametrically on points in a smooth manifold. For example, in molecular dynamics, electronic motion is much faster than nuclear motion. The electronic Hamiltonian $H(\boldsymbol{R}, \boldsymbol{r})$, where $\boldsymbol{R}$ stands for the collective nuclear coordinates and $\boldsymbol{r}$ for the collective electronic coordinates, can then be viewed as an operator in the Hilbert space of electronic wave functions which depend parametrically on $\boldsymbol{R}$. If we further adopt the **semiclassical approach** where the nuclear degrees of freedom are determined by classical trajectories $\boldsymbol{R}(t)$, where t is the time, then the Hamiltonian $H(\boldsymbol{r}, \boldsymbol{R}(t))$ can be regarded as an operator depending parametrically on the time t. In this chapter we will study how the semiclassical approach leads to a geometrical phase, called the Berry phase (Berry 1984, reprinted in Shapere and Wilczek 1989).

The time-dependent Schrödinger equation in the present context is

$$H(\boldsymbol{R}(t))\,|\,\psi(t)\,\rangle = i\hbar\,\frac{\partial|\,\psi(t)\,\rangle}{\partial t}\,. \tag{39.1}$$

We assume that there exists a complete set of instantaneous orthonormalized eigenstates $|\,n, \boldsymbol{R}(t)\,\rangle$ of $H(\boldsymbol{R}(t))$ satisfying

$$H(\boldsymbol{R}(t))\,|\,n, \boldsymbol{R}(t)\,\rangle = \epsilon_n(\boldsymbol{R}(t))\,|\,n, \boldsymbol{R}(t)\,\rangle\,. \tag{39.2}$$

Suppose that the system is in the n-th eigenstate of the electronic Hamiltonian $H(\boldsymbol{R}(0))$ at $t = 0$, then

$$H(\boldsymbol{R}(0))\,|\,n, \boldsymbol{R}(0)\,\rangle = \epsilon_n(\boldsymbol{R}(0))\,|\,n, \boldsymbol{R}(0)\,\rangle\,. \tag{39.3}$$

We will also assume the so-called **adiabatic approximation**, which states that if the $\boldsymbol{R}(t)$ are slow-varying functions of t (slow in comparison to electronic motion), then the system always stays in the same electronic eigenstate, which itself is a slowly varying state vector in the Hilbert space of electronic states that satisfy (39.2). Naively we might then expect the formal solution to (39.1) to be given by

$$|\,\psi(t)\,\rangle = \exp\left(-\frac{i}{\hbar}\int_0^t dt'\,\epsilon_n(\boldsymbol{R}(t'))\right)|\,n,\boldsymbol{R}(t)\,\rangle\,.$$

This, however, is not correct, as can be easily checked. On assuming its validity we have

$$\begin{aligned} i\hbar\,\frac{\partial|\,\psi\,\rangle}{\partial t} &= (i\hbar)\left(-\frac{i}{\hbar}\right)\epsilon_n(t)\exp\left(-\frac{i}{\hbar}\int_0^t \epsilon_n(t')dt'\right)|\,n,\boldsymbol{R}(t)\,\rangle \\ &+ (i\hbar)\exp\left(-\frac{i}{\hbar}\int_0^t \epsilon_n(t')dt'\right)\frac{d}{dt}|\,n,\boldsymbol{R}(t)\,\rangle \\ &= H(\boldsymbol{R}(t))\,|\,\psi(t)\,\rangle + (i\hbar)\exp\left(-\frac{i}{\hbar}\int_0^t \epsilon_n(t')dt'\right)\frac{d}{dt}|\,n,\boldsymbol{R}(t)\,\rangle\,, \end{aligned} \tag{39.4}$$

which clearly contradicts (39.1).

We will express the general solution as follows:

$$\boxed{|\,\psi(t)\,\rangle = \exp\left[\frac{i}{\hbar}\left(\gamma_n(t) - \textstyle\int_0^t \epsilon_n(t')dt'\right)\right]|\,n,\boldsymbol{R}(t)\,\rangle}\,. \tag{39.5}$$

The quantity $\gamma_n(t)$ is yet another example of a geometric phase (introduced in the last chapter), as opposed to the **dynamic phase** $\int_0^t \epsilon_n(t')dt'$. It can be determined by the following procedure. Inserting the above into the Schrödinger equation (39.1) we find that, on using (39.2),

$$\begin{aligned} &\exp\left[\frac{i}{\hbar}\left(\gamma_n - \int_0^t dt'\,\epsilon_n(t')\right)\right]\epsilon_n(\boldsymbol{R}(t))\,|\,n,\boldsymbol{R}(t)\,\rangle \\ &= (i\hbar)\frac{i}{\hbar}\left(\frac{d\gamma_n}{dt} - \epsilon_n(t)\right)\exp\left[\frac{i}{\hbar}\left(\gamma_n - \int_0^t dt'\,\epsilon_n(t')\right)\right]|\,n,\boldsymbol{R}(t)\,\rangle \\ &+ (i\hbar)\exp\left[\frac{i}{\hbar}\left(\gamma_n - \int_0^t dt'\,\epsilon_n(t')\right)\right]\frac{d}{dt}|\,n,\boldsymbol{R}(t)\,\rangle\,. \end{aligned} \tag{39.6}$$

On taking the inner product of this equation with $\langle\, n,\boldsymbol{R}(t)\,|$, we find

$$\frac{d\gamma_n}{dt} = i\hbar\,\langle\, n,\boldsymbol{R}(t)\,|\,\frac{d}{dt}\,|\,n,\boldsymbol{R}(t)\,\rangle\,. \tag{39.7}$$

Thus

$$\gamma_n(t) = i\hbar\int_0^t dt'\,\langle\, n,\boldsymbol{R}(t')\,|\,\frac{d}{dt'}\,|\,n,\boldsymbol{R}(t')\,\rangle = i\hbar\int_{\boldsymbol{R}(0)}^{\boldsymbol{R}(t)}\langle\, n,\boldsymbol{R}\,|\,\nabla_{\boldsymbol{R}}\,|\,n,\boldsymbol{R}\,\rangle\cdot d\boldsymbol{R}\,. \tag{39.8}$$

Note that even in the case $\boldsymbol{R}(0) = \boldsymbol{R}(t)$, so that the above integral is over a closed loop in the smooth manifold of the nuclear coordinates $\boldsymbol{R}$, $\gamma_n(t)$ is not necessarily zero, since the integrand is not necessarily an **exact differential**. The geometric phase $\gamma_n(t)$ accumulated by the wave function over a closed loop in parameter space is called a **Berry phase**. We can verify that $\gamma_n(t)$ is necessarily a real quantity, since

$$\begin{aligned}
&2Re\,\langle\, n, \boldsymbol{R}(t)\,|\,\frac{d}{dt}\,|\, n, \boldsymbol{R}(t)\,\rangle \\
&= \langle\, n, \boldsymbol{R}(t)\,|\,\frac{d}{dt}\,|\, n, \boldsymbol{R}(t)\,\rangle + \left(\frac{d}{dt}\langle\, n, \boldsymbol{R}(t)\,|\right)\,|\, n, \boldsymbol{R}(t)\,\rangle \\
&= \frac{d}{dt}\langle\, n, \boldsymbol{R}(t)\,|\, n, \boldsymbol{R}(t)\,\rangle = 0\,.
\end{aligned} \tag{39.9}$$

The last equality follows from the fact that $|\, n, \boldsymbol{R}(t)\,\rangle$ is chosen to be normalized, that is, $\langle\, n, \boldsymbol{R}(t)\,|\, n, \boldsymbol{R}(t)\,\rangle = 1$.

The geometric (Berry) phase $\gamma_n(t)$ is a deeply interesting theoretical construct. A comparison of the second equality of (39.8) with (38.4) on the one hand, and (39.5) with (38.5) on the other, shows that the quantity $\langle\, n, \boldsymbol{R}\,|\,\nabla_{\boldsymbol{R}}\,|\, n, \boldsymbol{R}\,\rangle$ plays the same role mathematically as the vector potential $\boldsymbol{A}$ in electrodynamics (which is an **abelian gauge field theory**). Thus the same mathematical description seems to underlay the very different physical phenomena of electrodynamics and molecular collisions. Indeed, it is precisely the differential geometric notion of connections on fiber bundles that provides a unified mathematical basis for the description of disparate physical phenomena involving geometric phases (and their generalizations) (not only electrodynamics and molecular collisions, but many others as well, both classical and quantal). In fact the quantity $\langle\, n, \boldsymbol{R}\,|\,\nabla_{\boldsymbol{R}}\,|\, n, \boldsymbol{R}\,\rangle$ and its non-adiabatic generalization is very often called the **Berry-Simon connection** (Simon 1983, reprinted in Shapere and Wilczek 1989). In the remainder of this chapter we will use the example of molecular collision systems to demonstrate more carefully the close formal relationship between the electrodynamic vector potential and the Berry-Simon connection for the case of non-adiabatic collisions. As we will see, the condition of non-adiabaticity leads to the appearance of **non-abelian gauge potentials (Yang-Mills fields)**.

For a general molecular system (involving any number of atoms and/or molecules in interaction) the total Hamiltonian can be written as

$$H(\boldsymbol{R}, \boldsymbol{r}) = -\hbar^2 \sum_I \frac{1}{2M_I}\nabla^2_{R_I} + H_e(\boldsymbol{R}; \boldsymbol{r})\,, \tag{39.10}$$

where $\boldsymbol{R}$ and $\boldsymbol{r}$ again denote the collective nuclear and electronic coordinates of the system, respectively; and the sum is over all the nuclei. The electronic Hamiltonian H_e, parametrized by the nuclear coordinates $\boldsymbol{R}$, is given by

$$H_e(\boldsymbol{R}; \boldsymbol{r}) = -\frac{\hbar^2}{2m}\sum \nabla^2_r + V_N(\boldsymbol{R}) + V_e(\boldsymbol{R}, \boldsymbol{r})\,. \tag{39.11}$$

The first term on the right-hand side of the above equation represents the electronic kinetic energy, the second term represents the interaction potential energy between the nuclei, and the last term represents the sum of the electron-electron and electron-nucleus interaction potential energies. The separation of the Hamiltonian according to (39.10) is again based on the physical understanding that the electronic degrees of freedom are "fast", while the nuclear degrees of freedom are "slow". The electronic Hamiltonian $H_e(\boldsymbol{R}, \boldsymbol{r})$ can then be viewed as a "snapshot" Hamiltonian with the nuclear degrees of freedom $\boldsymbol{R}$ frozen, and corresponds to the time-dependent Hamiltonian introduced in (39.1). Analogous to (39.2) we assume that the time-independent Schrödinger equation for $H_e(\boldsymbol{R}; \boldsymbol{r})$ can be solved for fixed $\boldsymbol{R}$, with a discrete set of eigenstates $|\,\phi_n(\boldsymbol{R})\,\rangle$ and corresponding energy eigenvalues $\epsilon_n(\boldsymbol{R})$ (both parametrized by $\boldsymbol{R}$):

$$H_e\,|\,\phi_n(\boldsymbol{R})\,\rangle = \epsilon_n(\boldsymbol{R})\,|\,\phi_n(\boldsymbol{R})\,\rangle\,. \tag{39.12}$$

Working with the coordinate representation ($\boldsymbol{r}$-representation) of $|\,\phi_n(\boldsymbol{R})\,\rangle$, that is, the electronic wave functions given by

$$\phi_n(\boldsymbol{r}, \boldsymbol{R}) \equiv \langle\,\boldsymbol{r}\,|\,\phi_n(\boldsymbol{R})\,\rangle \tag{39.13}$$

[recall (9.20) and the discussion in Chapter 10], Eq. (39.12) can alternatively be written as

$$H_e\,\phi_n(\boldsymbol{r}, \boldsymbol{R}) = \epsilon_n(\boldsymbol{R})\,\phi_n(\boldsymbol{r}, \boldsymbol{R})\,. \tag{39.14}$$

One can further assume that the set of electronic eigenstates $\{|\,\phi_n(\boldsymbol{R})\,\rangle\}$ are orthonormal and complete for each $\boldsymbol{R}$:

$$\langle\,\phi_n\,|\,\phi_m\,\rangle = \int dr\,\phi_n^*(\boldsymbol{r}, \boldsymbol{R})\phi_m(\boldsymbol{r}, \boldsymbol{R}) = \delta_{nm} \qquad \text{(orthonormality)}\,, \tag{39.15}$$

$$\sum_n |\,\phi_n(\boldsymbol{R})\,\rangle\langle\,\phi_n(\boldsymbol{R})\,| = 1 \qquad \text{(completeness)}\,. \tag{39.16}$$

Note that in (39.15) the integration for the inner product is over the electronic coordinates only. The set of electronic energy eigenvalues $\epsilon_n(\boldsymbol{R})$ obtained from the solution of (39.14), as functions of the nuclear coordinates $\boldsymbol{R}$, are usually referred to as potential energy "surfaces" for nuclear motion.

The time-independent Schrödinger equation for the total Hamiltonian H [(39.10)] can be written

$$H\,|\,\Psi\,\rangle = E\,|\,\Psi\,\rangle\,, \tag{39.17}$$

or

$$H\,\Psi(\boldsymbol{R}, \boldsymbol{r}) = E\,\Psi(\boldsymbol{R}, \boldsymbol{r})\,, \tag{39.18}$$

where

$$\Psi(\boldsymbol{R}, \boldsymbol{r}) \equiv \langle\,\boldsymbol{R}\,|\langle\,\boldsymbol{r}\,|\,\Psi\,\rangle\,. \tag{39.19}$$

Since $\{\phi_n(\boldsymbol{R}, \boldsymbol{r})\}$ is a complete set for each $\boldsymbol{R}$, we can expand $\Psi(\boldsymbol{R}, \boldsymbol{r})$ as

$$\Psi(\boldsymbol{R}, \boldsymbol{r}) = \sum_n \Phi_n(\boldsymbol{R})\,\phi_n(\boldsymbol{r}, \boldsymbol{R})\,. \tag{39.20}$$

Using (39.10) the Schrödinger equation (39.18) can then be written as

$$\sum_n \left\{ \sum_I \left(-\frac{\hbar^2}{2M_I} \nabla^2_{R_I} \right) + H_e \right\} \Phi_n(\boldsymbol{R})\phi_n(\boldsymbol{r}, \boldsymbol{R}) = E \sum_n \Phi_n(\boldsymbol{R})\phi_n(\boldsymbol{r}, \boldsymbol{R}) . \tag{39.21}$$

Using (39.14) and taking the inner product on both sides of the above equation with $\phi_m(\boldsymbol{r}, \boldsymbol{R})$ [that is, applying $\int dr\, \phi^*_m(\boldsymbol{r}, \boldsymbol{R})$ to both sides], we obtain

$$\sum_n \int dr\, \phi^*_m(\boldsymbol{r}, \boldsymbol{R}) \left(\sum_I -\frac{\hbar^2}{2M_I} \nabla^2_{R_I} \right) \Phi_n(\boldsymbol{R})\phi_n(\boldsymbol{r}, \boldsymbol{R}) + \sum_n \Phi_n(\boldsymbol{R})\, \delta_{nm}\, \epsilon_n(\boldsymbol{R}) = E\Phi_m(\boldsymbol{R}) , \tag{39.22}$$

where the orthonormality of $\{\phi_n\}$ [(39.15)] has also been used. The first term on the left-hand side can be manipulated as follows:

$$\begin{aligned} &\int dr\, \phi^*_m(\boldsymbol{r}, \boldsymbol{R}) \nabla_{R_I} \cdot \{\nabla_{R_I}(\Phi_n(\boldsymbol{R})\phi_n(\boldsymbol{r}, \boldsymbol{R}))\} \\ &= \int dr\, \phi^*_m\, \nabla_{R_I} \cdot \{\Phi_n \nabla_{R_I}\phi_n + (\nabla_{R_I}\Phi_n)\phi_n\} \\ &= \int dr\, \phi^*_m \{\Phi_n \nabla^2_{R_I}\phi_n + 2(\nabla_{R_I}\Phi_n)\cdot(\nabla_{R_I}\phi_n) + (\nabla^2_{R_I}\Phi_n)\phi_n\} \\ &= \Phi_n \langle\, \phi_m \,|\, \nabla^2_{R_I}\phi_n \,\rangle + 2\langle\, \phi_m \,|\, \nabla_{R_I}\phi_n \,\rangle \cdot (\nabla_{R_I}\Phi_n) + \delta_{nm}\nabla^2_{R_I}\Phi_n , \end{aligned} \tag{39.23}$$

where on the right-hand side of the last equality, the Dirac bracket of (39.15) has been used. Also note that

$$\begin{aligned} &\langle\, \phi_m \,|\, \nabla^2_{R_I}\phi_n \,\rangle = \nabla_{R_I} \langle\, \phi_m \,|\, \nabla_{R_I}\phi_n \,\rangle - \langle\, \nabla_{R_I}\phi_m \,|\, \nabla_{R_I}\phi_n \,\rangle \\ &= \nabla_{R_I} \langle\, \phi_m \,|\, \nabla_{R_I}\phi_n \,\rangle - \sum_k \langle\, \nabla_{R_I}\phi_m \,|\, \phi_k \,\rangle\langle\, \phi_k \,|\, \nabla_{R_I}\phi_n \,\rangle \\ &= \nabla_{R_I} \langle\, \phi_m \,|\, \nabla_{R_I}\phi_n \,\rangle + \sum_k \langle\, \phi_m \,|\, \nabla_{R_I}\phi_k \,\rangle\langle\, \phi_k \,|\, \nabla_{R_I}\phi_n \,\rangle , \end{aligned} \tag{39.24}$$

where in the second equality we have used the completeness condition for $\{|\, \phi_k \,\rangle\}$ [(39.16)] and in the third we have used the fact that

$$\nabla_{R_I} \langle\, \phi_m \,|\, \phi_k \,\rangle = \nabla_{R_I}\, \delta_{nm} = 0 = \langle\, \nabla_{R_I}\phi_m \,|\, \phi_k \,\rangle + \langle\, \phi_m \,|\, \nabla_{R_I}\phi_k \,\rangle . \tag{39.25}$$

We now define the vector-valued matrix elements

$$\boxed{(\boldsymbol{A}^{(I)})^m_n \equiv i\hbar \langle\, \phi_m \,|\, \nabla_{R_I}\phi_n \,\rangle = i\hbar \int dr\, \phi^*_m(\boldsymbol{r}, \boldsymbol{R})\nabla_{R_I}\phi_n(\boldsymbol{r}, \boldsymbol{R})} \;, \tag{39.26}$$

where m is the row index and n is the column index. The Schrödinger equation for nuclear motion (39.21) can then be written in the matrix form

$$\sum_n H^m_n\, \Phi_n(\boldsymbol{R}) = E\, \Phi_m(\boldsymbol{R}) , \tag{39.27}$$

where the matrix of operators (H^m_n) are given by

$$H^m_n(\boldsymbol{R}) = \sum_I \left(-\frac{\hbar^2}{2M_I}\right)(D^{(I)})^m_k \cdot (D^{(I)})^k_n + \delta^m_n\, \epsilon(\boldsymbol{R})\,. \tag{39.28}$$

In the above equation, the sum over I is over the nuclei and the matrices of vector operators $(D^{(I)})^m_k$ are given by

$$\boxed{(D^{(I)})^m_k \equiv \delta^m_k\, \nabla_{R_I} - \frac{i}{\hbar}(\boldsymbol{A}^{(I)})^m_k(\boldsymbol{R})}\,. \tag{39.29}$$

The above two equations reveal that the nuclear motion is described by a Schrödinger equation [(39.27)] that formally resembles that for a charged particle moving in an electromagnetic field, with the operator $(D^{(I)})^m_k$ in (39.29) appearing as a covariant derivative describing a minimal coupling [cf. (34.4)], and $(\boldsymbol{A}^{(I)})^m_k(\boldsymbol{R})$ as non-abelian "vector potentials", or **non-abelian gauge potentials** (so-called because the matrix-valed vector potentials at different $\boldsymbol{R}$ do not commute). Equation (39.27) also shows explicitly how the different electronic states are coupled by the nuclear motion.

The gauge potentials $(\boldsymbol{A}^{(I)})^m_n(\boldsymbol{R})$ can be related to the electronic energy difference $\epsilon_n(\boldsymbol{R}) - \epsilon_m(\boldsymbol{R})$, the matrix elements $\langle\, \phi_m \,|\, (\nabla_{R_I} H_e) \,|\, \phi_n \,\rangle$, and the gradients of the nuclear potential energies $\nabla_{R_I}\epsilon_n$. We have

$$\begin{aligned}\langle\, \phi_m \,|\, \nabla_{R_I}(H_e \,|\, \phi_n\,\rangle) &= \langle\, \phi_m \,|\, \nabla_{R_I}(\epsilon(\boldsymbol{R}) \,|\, \phi_n\,\rangle) \\ &= \langle\, \phi_m \,|\, (\nabla_{R_I} H_e) \,|\, \phi_n\,\rangle + \langle\, \phi_m \,|\, H_e \,|\, \nabla_{R_I}\phi_n\,\rangle\,.\end{aligned} \tag{39.30}$$

Hence

$$\begin{aligned}&\epsilon_n(\boldsymbol{R})\,\langle\, \phi_m \,|\, \nabla_{R_I}\phi_n\,\rangle + \nabla_{R_I}\,\epsilon_n(\boldsymbol{R})\,\delta_{mn} \\ &\qquad = \langle\, \phi_m \,|\, (\nabla_{R_I} H_e) \,|\, \phi_n\,\rangle + \epsilon_m(\boldsymbol{R})\,\langle\, \phi_m \,|\, \nabla_{R_I}\phi_n\,\rangle\,,\end{aligned} \tag{39.31}$$

or

$$\begin{aligned}&(\epsilon_n(\boldsymbol{R}) - \epsilon_m(\boldsymbol{R}))\,\langle\, \phi_m \,|\, \nabla_{R_I}\phi_n\,\rangle \\ &\qquad = \langle\, \phi_m \,|\, (\nabla_{R_I} H_e) \,|\, \phi_n\,\rangle - \delta_{mn}\,\nabla_{R_I}\epsilon_n(\boldsymbol{R})\,.\end{aligned} \tag{39.32}$$

Thus, for non-degenerate states $(\epsilon_m \neq \epsilon_n)$,

$$(\boldsymbol{A}^{(I)})^m_n(\boldsymbol{R}) = i\hbar\,\frac{\langle\, \phi_m \,|\, (\nabla_{R_I} H_e) \,|\, \phi_n\,\rangle}{\epsilon_n(\boldsymbol{R}) - \epsilon_m(\boldsymbol{R})}\,, \quad (\epsilon_n \neq \epsilon_m)\,. \tag{39.33}$$

We will now illustrate the above formal development with a specific example in molecular physics: the lambda (Λ) doublets of a homonuclear diatomic molecule (such as O_2), where Λ denotes the projection $\boldsymbol{J}\cdot\hat{n}$ of the total electronic angular momentum $\boldsymbol{J}$ (including spin) of the molecule along the internuclear axis $\hat{n}$ (see Fig. 39.1). We will show that the gauge potential $\boldsymbol{A}^m_n(\boldsymbol{R})$ for this

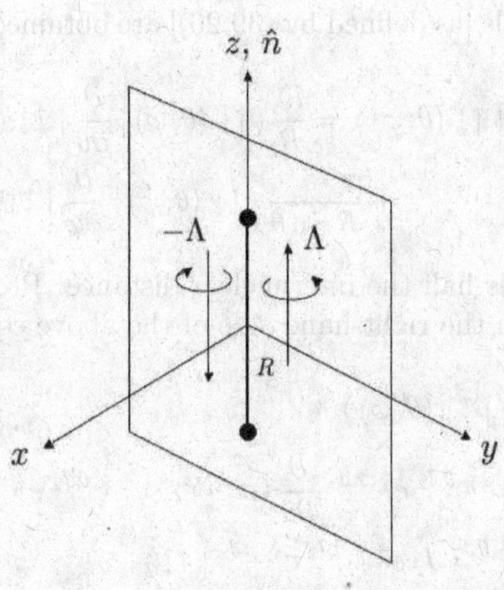

Fig. 39.1

system will emerge as that for a **non-abelian magnetic monopole**, and has the same mathematical form as that of the Wu-Yang potential for an abelian magnetic monopole [cf. (38.18)]!

The system obviously does not have spherical symmetry, so $J^2 = \boldsymbol{J}\cdot\boldsymbol{J}$ is not conserved (since it does not commute with the electronic Hamiltonian H_e), and J_z is not a good quantum number. However $\boldsymbol{J}\cdot\hat{n}$ is conserved and its eigenvalues Λ can be used to label the electronic ststes $|\,\phi_n\,\rangle$. Moreover, since the system has a mirror symmetry about any plane through the internuclear axis, and $\boldsymbol{J}\cdot\hat{n}$ changes sign under such a parity operation, states with $\boldsymbol{J}\cdot\hat{n} = \Lambda\hbar$ are degenerate with those with $\boldsymbol{J}\cdot\hat{n} = -\Lambda\hbar$. Such degenerate pairs of states are called **lambda doublets**.

For a particular value of Λ, label the doublet electronic states $|\,\phi_\Lambda(\hat{n}=\hat{z})\,\rangle$ and $|\,\phi_{-\Lambda}(\hat{n}=\hat{z})\,\rangle$ with the internuclear axis along the z-direction (Fig. 39.1) by $|\uparrow\rangle$ and $|\downarrow\rangle$, respectively. Supposing the molecule to be modeled by a rigid dumbbell (vibrations frozen) and working in the nuclear center-of-mass (CM) frame, the nuclear coordinates $\boldsymbol{R}$ are just the polar angles (θ,φ) marking points on the surface of a two-sphere S^2. The doublet electronic states with $\hat{n}$ pointed along arbitrary directions [specified by (θ,φ)] are those obtained from $|\uparrow\rangle$ and $|\downarrow\rangle$ by rotations (in nuclear coordinate space). We will label these by $|\uparrow(\theta,\varphi)\,\rangle$ and $|\downarrow(\theta,\varphi)\,\rangle$. In fact, recalling Euler's theorem [Eq. (19.29) in Theorem 19.1] and (17.60), we have (on using the symbol $|\uparrow\downarrow\rangle$ to represent either $|\uparrow\rangle$ or $|\downarrow\rangle$),

$$|\uparrow\downarrow(\theta,\varphi)\,\rangle = e^{-\frac{i}{\hbar}\varphi J_3}\, e^{-\frac{i}{\hbar}\theta J_2}\, e^{\frac{i}{\hbar}\varphi J_3}\,|\uparrow\downarrow\rangle\,, \tag{39.34}$$

where the operator $\exp(\frac{i}{\hbar}\varphi J_3)$ produces no effect on $|\uparrow\downarrow\rangle$ [since a rotation by an angle φ about $\hat{z}$ does not move the north (or south) pole], and has been inserted as a computational device.

The gauge potentials [as defined by (39.26)] are obtained from the quantities

$$i\hbar\,\langle\uparrow\downarrow(\theta,\varphi)\,|\,\nabla_{(\theta,\varphi)}\,|\uparrow\downarrow(\theta,\varphi)\,\rangle = \frac{i\hbar}{R}\,\langle\uparrow\downarrow(\theta,\varphi)\,|\,\frac{\partial}{\partial\theta}\,|\uparrow\downarrow(\theta,\varphi)\,\rangle\,e_\theta + \frac{i\hbar}{R\sin\theta}\,\langle\uparrow\downarrow(\theta,\varphi)\,|\,\frac{\partial}{\partial\varphi}\,|\uparrow\downarrow(\theta,\varphi)\,\rangle\,e_\varphi\,, \tag{39.35}$$

where the constant R is half the internuclear distance. Proceeding to calculate the matrix elements on the right-hand side of the above equation, we have

$$\begin{aligned} &i\hbar\,\langle\uparrow\downarrow(\theta,\varphi)\,|\,\frac{\partial}{\partial\theta}\,|\uparrow\downarrow(\theta,\varphi)\,\rangle \\ &= i\hbar\,\langle\uparrow\downarrow\,|e^{-\frac{i}{\hbar}\varphi J_3}\,e^{\frac{i}{\hbar}\theta J_2}\,e^{\frac{i}{\hbar}\varphi J_3}\,\frac{\partial}{\partial\theta}\,(e^{-\frac{i}{\hbar}\varphi J_3}\,e^{-\frac{i}{\hbar}\theta J_2}\,e^{\frac{i}{\hbar}\varphi J_3})\,|\uparrow\downarrow\rangle \\ &= \langle\uparrow\downarrow\,|e^{-\frac{i}{\hbar}\varphi J_3}\,e^{\frac{i}{\hbar}\theta J_2}\,J_2\,e^{-\frac{i}{\hbar}\theta J_2}\,e^{\frac{i}{\hbar}\varphi J_3}\,|\uparrow\downarrow\rangle \\ &= \langle\uparrow\downarrow\,|e^{-\frac{i}{\hbar}\varphi J_3}\,J_2\,e^{\frac{i}{\hbar}\varphi J_3}\,|\uparrow\downarrow\rangle \\ &= \langle\uparrow\downarrow\,|\cos\varphi\,J_2 - \sin\varphi\,J_1\,|\uparrow\downarrow\rangle\,, \end{aligned} \tag{39.36}$$

where, in the last equality, we have used the operator identity

$$e^{-\frac{i}{\hbar}\alpha J_i}\,J_j\,e^{\frac{i}{\hbar}\alpha J_i} = (\cos\alpha)J_j + (\sin\alpha)\varepsilon_{ij}{}^k J_k\,,\quad \alpha\in\mathbb{R}\,, \tag{39.37}$$

with $\varepsilon_{ij}{}^k$ being the completely antisymmetric (Levi-Civita) tensor.

Problem 39.1 Show that the identity (39.37) results from the general operator identity

$$e^{\xi A}\,B\,e^{-\xi A} = B + \xi\,[A,B] + \frac{\xi^2}{2!}\,[A,[A,B]] + \frac{\xi^3}{3!}\,[A,[A,[A,B]]] + \dots\,,\quad \xi\in\mathbb{C}\,,$$

and the Lie algebra structure of the angular momentum operators as given by (30.35a):

$$[J_i, J_j] = i\hbar\,\varepsilon_{ij}{}^k\,J_k\,.$$

Similarly, by repeated application of (39.37), we obtain

$$i\hbar\,\langle\uparrow\downarrow(\theta,\varphi)\,|\,\frac{\partial}{\partial\varphi}\,|\uparrow\downarrow(\theta,\varphi)\,\rangle = \langle\uparrow\downarrow\,|(\cos\theta-1)J_3 - \sin\theta\cos\varphi\,J_1 - \sin\theta\sin\varphi\,J_2\,|\uparrow\downarrow\rangle\,. \tag{39.38}$$

Since the states $|\uparrow\downarrow\rangle$ have internuclear axis along $e_3 = \hat{z}$, we have

$$J_3 \,|\uparrow\rangle = \Lambda\hbar\,|\uparrow\rangle\,, \quad J_3\,|\downarrow\rangle = -\Lambda\hbar\,|\downarrow\rangle\,. \tag{39.39}$$

To calculate the matrix elements of J_1 and J_2, we introduce the hermitian conjugate pairs

$$J_\pm \equiv \frac{J_1 \pm iJ_2}{\sqrt{2}}\,, \tag{39.40}$$

which imply

$$J_1 = \frac{1}{\sqrt{2}}\,(J_+ + J_-)\,, \quad J_2 = \frac{1}{i\sqrt{2}}\,(J_+ - J_-)\,. \tag{39.41}$$

Temporarily writing the states $|\uparrow\rangle$ and $|\downarrow\rangle$ as $|\Lambda\rangle$ and $|-\Lambda\rangle$, respectively ($\Lambda \geq 0$), we have

$$J_+\,|\pm\Lambda\rangle = \alpha_+(\Lambda)\,|\pm\Lambda + 1\rangle\,, \tag{39.42}$$

$$J_-\,|\pm\Lambda\rangle = \alpha_-(\Lambda)\,|\pm\Lambda - 1\rangle\,, \tag{39.43}$$

where $\alpha_+(\Lambda)$ and $\alpha_-(\Lambda)$ are constants depending on Λ.

Now from our discussion of general theory of the representations of $SO(3)$ (Chapters 19, 20 and 22), Λ can assume the values $n/2$; $n = 0, 1, 2, 3, \ldots$. Equations (39.36) and (39.38) clearly show that the gauge potential $\boldsymbol{A}_n^m$ vanishes for $|\phi_n\rangle$ and $|\phi_m\rangle$ corresponding to different values of $|\Lambda|$. Thus only the Λ-doublet states (for fixed Λ) are possibly coupled by the gauge potential. For $\Lambda \neq 1/2$, the doublet states are actually decoupled (the 2×2 gauge potential matrices are diagonal). In fact, for $\Lambda \neq 1/2$,

$$(A_\theta)_n^m = 0\,, \tag{39.44}$$

$$(A_\varphi)_n^m = \begin{pmatrix} -\dfrac{\Lambda\hbar(1-\cos\theta)}{R\sin\theta} & 0 \\ 0 & \dfrac{\Lambda\hbar(1-\cos\theta)}{R\sin\theta} \end{pmatrix}\,, \tag{39.45}$$

where the row and column indices are m and n, respectively, with the values 1 for $\uparrow$ and 2 for $\downarrow$. On comparison with (38.18) for the Wu-Yang potential of a magnetic monopole, we see that each of the nuclear wave functions $\Phi_\uparrow(\theta,\varphi)$ and $\Phi_\downarrow(\theta,\varphi)$ of (39.27) satisfies a Schrödinger equation which formally resembles that describing the motion of a charged particle in a field due to a $U(1)$-magnetic monopole [(38.1)], with $\Lambda\hbar$ playing the role of the magnetic charge for the state $\Phi_\downarrow$, and $-\Lambda\hbar$ playing the same role for $\Phi_\uparrow$.

For $\Lambda = 1/2$, the states $|\uparrow\rangle$ and $|\downarrow\rangle$ are no longer decoupled. In fact, it is straightforward, using (39.39) to (39.43), to show that, for $\Lambda = 1/2$,

$$A_\theta = \hbar \begin{pmatrix} 0 & \lambda e^{-i\varphi}/R \\ \lambda^* e^{i\varphi}/R & 0 \end{pmatrix}, \tag{39.46}$$

$$A_\varphi = \hbar \begin{pmatrix} -(1-\cos\theta)/(2R\sin\theta) & i\lambda e^{-i\varphi}/R \\ -i\lambda^* e^{i\varphi}/R & (1-\cos\theta)/(2R\sin\theta) \end{pmatrix}, \tag{39.47}$$

where

$$\hbar\lambda \equiv \langle \uparrow \mid J_2 \mid \downarrow \rangle\,. \tag{39.48}$$

The above matrix-valued vector potentials [given by (39.44) to (39.47)] are those of a so-called $U(2)$ magnetic monopole.

Problem 39.2 Verify (39.46) and (39.47).

Chapter 40

The Dynamic Phase: Riemann Surfaces in the Semiclassical Theory of Non-Adiabatic Collisions; Homotopy and Homology

Within the semiclassical approach to molecular collisions, the collective nuclear coordinates $\boldsymbol{R}(t)$ are described by classical trajectories (as functions of the time t). The time-dependent Schrödinger equation describing electronic motion can be written [cf. (39.1)]

$$H(t)\Psi(t) = i\hbar\,\frac{\partial\Psi}{\partial t}\,, \tag{40.1}$$

where the Hamiltonian $H(t)$ depends on time implicitly through the nuclear trajectories $\boldsymbol{R}(t)$. The instantaneous eigenfunctions $\phi(t)$ of $H(t)$ satisfying

$$H(t)\phi_n(t) = \epsilon_n(t)\phi_n(t) \tag{40.2}$$

are assumed to form a complete orthonormal set, where $\epsilon_n(t)$, for real t, is the energy curve [corresponding to the potential energy surface $\epsilon_n(\boldsymbol{R})$ of (39.12)] governing nuclear motion on the n-th electronic level in the adiabatic approximation [cf. (39.2) and the discussion following]. We will assume that the energy curves are non-degenerate: $\epsilon_i(t) \neq \epsilon_j(t)$ for all real t and for $i \neq j$. The complete set $\{\phi_n\}$ is said to form an **adiabatic representation** of the Hilbert space $\mathcal{H}$ that it generates. For molecular collision problems, one usually works with a finite-dimensional subspace of $\mathcal{H}$ invariant under H (when collisions may cause transitions only among a finite number of electronic levels). For an N-level problem, and in an arbitrary representation $\{\theta_n(t)\}$ in which the $N \times N$ hermitian matrix representing $H(t)$ is not diagonal (**diabatic representation**), the

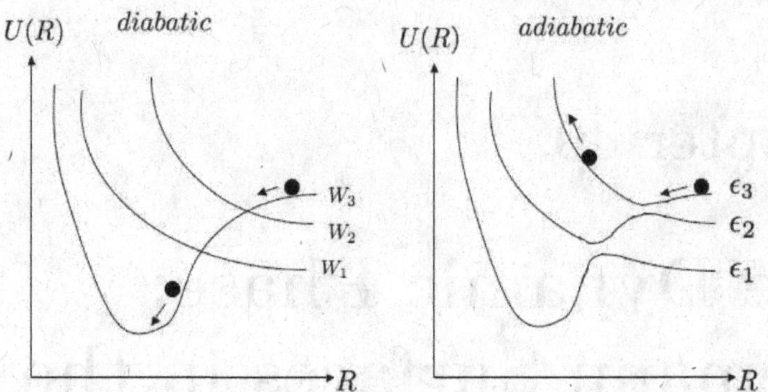

Fig. 40.1

$\epsilon_n(t)$ are obtained by diagonalization, that is, from solutions of the characteristic equation

$$det \,|\, H_{ij}(t) - \epsilon \,| = 0 \,, \tag{40.3}$$

where

$$H_{ij}(t) \equiv \langle\, \theta_i \,|\, H \,|\, \theta_j \,\rangle \,. \tag{40.4}$$

Equation (40.3) is in general an algebraic polynomial equation of the form

$$\epsilon^N + a_1(t)\epsilon^{N-1} + \cdots + a_N(t) = 0 \,, \tag{40.5}$$

where the coefficients $a_i(t)$ are functions of $H_{ij}(t)$. We note that in general the diabatic representation is suitable for problems involving "fast" nuclear motion, in the sense that the couplings $\langle \theta_i \,|\, \dfrac{\partial}{\partial t} \,|\, \theta_j \,\rangle \longrightarrow 0$, for $i \neq j$. Figure 40.1 depicts schematically nuclear motion with respect to the diabatic and adiabatic potential energy surfaces. In this figure $W_i = H_{ii}$ [(40.4)] and $U(R)$ denotes the potential energy for nuclear motion. Note the so-called "avoided crossings" in the potential energy surfaces in the adiabatic representation.

For fixed real t, non-degeneracy and the **fundamental theorem of algebra** imply that there are N distinct real roots for (40.5) – the eigenvalues $\epsilon_i(t)$, $i = 1, \ldots, N$. However, and this is the crucial point in the use of (40.5) for the treatment of non-adiabatic collisions involving transitions between different electronic levels, there always exist complex values of t for which (40.5) has repeated roots. These complex t values are the **branch points** of the multi-valued function $\epsilon(t)$ determined by (40.5), considered as a function on $\mathbb{C}$ (the complex t-plane). Alternatively (and mathematically more correctly), $\epsilon(t)$ is

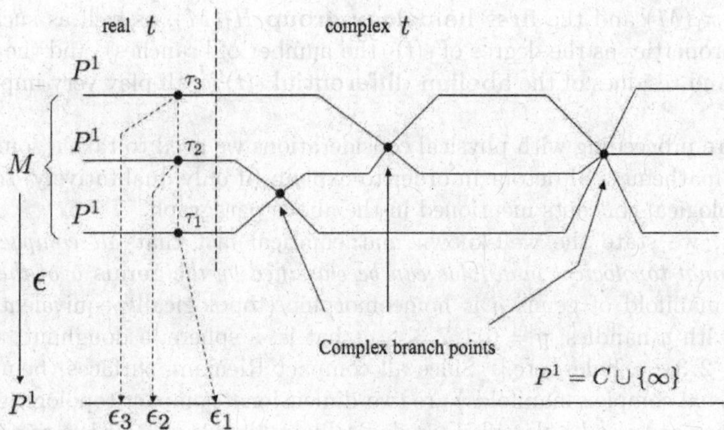

Fig. 40.2

made single-valued by considering its domain as an N-sheeted **Riemann surface** (or a one-dimensional connected **complex manifold**) M whose sheets are joined together at the branch points and along the cuts. When the coefficients $a_i(t)$ are rational functions of t, Eq. (40.5) is said to define an **algebraic curve** in the two-dimensional complex space $\mathbb{C}^2$, and M is said to be the corresponding Riemann surface of the curve. To handle the analytic properties of $\epsilon(t)$ more easily (as will be demonstrated below), it is advantageous to add "the point at infinity" $\{\infty\}$ (or **complex projective point** $\mathbb{P}^0$) to the (non-compact) complex plane $\mathbb{C}$ to obtain the so-called **Riemann sphere** (or **complex projective line** $\mathbb{P}^1$), which is a compact space. This process is called the one-point compactification of the complex plane. The algebraic curve $\epsilon(t)$ can then be considered as the function $\epsilon : M \longrightarrow \mathbb{P}^1$, where the N-sheeted Riemann surface M is made from N copies of $\mathbb{P}^1$ "glued" together along the various cuts terminating at various branch points. We represent this situation schematically in Fig. 40.2. Note that in general one can perform the following sequence of compactifications to obtain the so-called n-dimensional **complex projective spaces** $\mathbb{P}^n$, which are all compact topological spaces:

$$\mathbb{P}^1 \sim \mathbb{C} \cup \mathbb{P}^0 \, , \, \mathbb{P}^2 \sim \mathbb{C}^2 \cup \mathbb{P}^1 \, , \, \dots \, , \mathbb{P}^n \sim \mathbb{C}^n \cup \mathbb{P}^{n-1} \, . \tag{40.6}$$

In the language of the theory of algebraic curves, $\mathbb{P}^0 = \{\infty\}$ is also called the point at infinity, $\mathbb{P}^1$ the line at infinity, and in general, with respect to $\mathbb{P}^n$, $\mathbb{P}^{n-1}$ a hyperplane at infinity.

Thus the dynamics of a multi-(electronic) level system can conveniently be viewed as dynamics on a Riemann surface whose topological and complex structures determine the nature of quantum transitions. We will see that such topological properties as the **genus**, the **fundamental group** (**first homotopy group**) $\pi_1(M)$ and the **first homology group** $H_1(M)$, as well as such analytical properties as the degree of $\epsilon(t)$ (the number of branches), and the singularities and residues of the **abelian differential** $\epsilon(t)dt$, all play very important roles.

Before proceeding with physical considerations we need to take a somewhat lengthy mathematical detour in order to explain (if only qualitatively) some of the topological concepts mentioned in the above paragraph.

First, we state the well-known mathematical fact that *all compact two-dimensional topological manifolds can be classified by the* **genus** g *of the manifold* [a manifold of genus g is homeomorphic (topologically equivalent) to a sphere with g handles, $g = 0, 1, 2, 3, \ldots$, that is, a sphere, a doughnut, a pretzel with $2, 3, \ldots$ holes, etc.]. Since all compact Riemann surfaces (being one-dimensional complex manifolds) are two-dimensional compact topological manifolds, they can also be classified topologically by the genus g. It is a remarkable theorem in classical function theory (see, for example, Section 2.1 of Siegel 1969, or Section 4.14 of Jones and Singerman 1987), as mentioned before, that if the coefficients $a_i(t)$ in (40.5) are all rational functions of t, then the Riemann surface of $\epsilon(t)$ is compact, and the meromorphic function $\epsilon(t)$ is an example of an algebraic function. The converse of this fact is also true: any compact Riemann surface is **conformally equivalent** to the Riemann surface of some algebraic function.

The genus g of the Riemann surface of the algebraic function $\epsilon(t)$ can be easily determined by another remarkable theorem for compact Riemann surfaces – the **Riemann-Hurwitz formula** (see Siegel 1969), which relates g to the number of branch points (B), and the degree (N), of $\epsilon(t)$. Specifically

$$g = \frac{B}{2} - N + 1\,. \tag{40.7}$$

The number of branch points is counted with multiplicities, is necessarily even, and is defined by

$$B = \sum_{p \in M} b_\epsilon(p)\,, \tag{40.8}$$

where the sum is over all points p in the Riemann surface M, and $b_\epsilon(p)$ is the number of sheets meeting at p minus one. (For a point p that is not a branch point, $b_\epsilon(p) = 0$.) An example of the use of the Riemann-Hurwitz formula is shown in Fig. 40.3. It turns out that the topological properties of the Riemann surface corresponding to an algebraic function $\epsilon(t)$ is also related closely to the analytic properties of the **meromorphic** (abelian) **differential** $\epsilon(t)dt$ by the **Poincaré-Hopf Index Formula**:

$$Z - P = 2g - 2 = -\chi\,, \tag{40.9}$$

where Z and P are, respectively, the number of zeroes (counting multiplicities) and the number of poles (counting multiplicities) of the differential $\epsilon(t)dt$; and

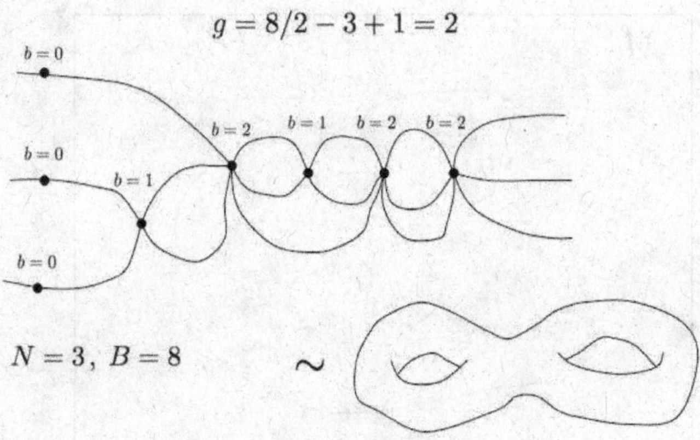

Fig. 40.3

χ is the so-called **Euler characteristic** of the Riemann surface corresponding to $\epsilon(t)$. An example of this formula will be given below. A deep theorem relating the topological and analytic properties of compact Riemann surfaces is the **Riemann-Roch Theorem**. Roughly speaking, it answers the question: What is the dimension of a linear vector space consisting of certain "cooked-to-order" abelian differentials with singularities (zeroes and poles) of specified types? Even the statement of this remarkable theorem is beyond the scope of the present text. (For a relatively non-technical discussion, the reader may consult Siegel 1971.)

Next we turn to the topological notion of homotopy. This has to do with the classification of loops on a topological manifold M that can be continuously deformed into each other. More technically, we first define the homotopy classes of mappings $f : S^1 \to M$ as classes of *homotopic continuous maps* from the circle to M, where two maps f_1 and f_2 are said to be homotopic (written $f_1 \sim f_2$) if there exists a continuous map $F : I \times I \to M$ (I being the closed interval $[0, 1]$) such that

$$F(s, 0) = f_1(s)\ , F(s, 1) = f_2(s)\ , \qquad s \in I\ , \tag{40.10a}$$

$$F(0, t) = F(1, t) = x_0 \in M\ , \qquad t \in I\ , \tag{40.10b}$$

where the argument $s \in [0, 1]$ in $f(s)$ is a parameter specifying a point on S^1 (see Fig. 40.4). Clearly, two homotopic maps can be thought of as two loops in M that can be continuously deformed into each other. The set of homotopy classes of maps with an element denoted by $[f]$, where f is a representative, can

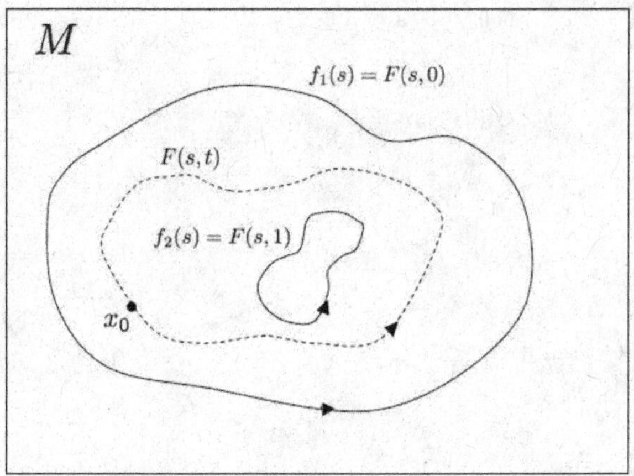

Fig. 40.4

be given a group structure, if we define the following group multiplication:

$$[f_1] \cdot [f_2] \equiv [f_1 \cdot f_2], \tag{40.11}$$

and inverse operation

$$[f]^{-1} \equiv [f^{-1}], \tag{40.12}$$

where on the right-hand side of (40.11) the product of loops with a common point x_0 is defined as a loop consisting of traversing f_1 from x_0 to x_0 first, and then traversing f_2 from x_0 to x_0. If two loops f_1 and f_2 do not have a common point, we can replace one of them on the left-hand side, say f_2, by a homotopic map $f_2' (\sim f_2)$ which does have a common point with f_1. In (40.12) f^{-1} simply means a loop traversed in the opposite sense as f.

Definition 40.1. *Let M be a topological space. The set of homotopy classes of loops, or maps $f : S^1 \to M$, such that $f(0) = f(1) = x_0 \in M$ (where the argument parametrizes points on the circle S^1), with the product rule and inverse given by (40.11) and (40.12), is called the* **fundamental group** *(or* **first homotopy group***) of M at x_0, and is denoted by $\pi_1(M, x_0)$.*

Problem 40.1

(a) Show that the identity of the fundamental group of M at x_0 is the class of loops that can be continuously contracted to the point x_0. Show that $\pi_1(M, x_0)$ is indeed a group, that is, it satisfies the group axioms (of Def. 18.1).

(b) Show that the fundamental groups $\pi_1(M, x_0)$ at different base points $x_0 \in M$ are isomorphic to each other. Thus one can just refer to $\pi_1(M)$.

An important example of fundamental groups is

$$\pi_1(S^1) = \mathbb{Z} \quad \text{(the integers)} . \tag{40.13}$$

This result can be understood intuitively. Each integer (positive, negative, or zero) represents the number of times (winding number) of looping a band completely around a circle (with orientation). Two loops with different winding numbers cannot be continuously deformed into each other. Hence each integer represents a unique element of the fundamental group of S^1. The mathematical result (40.13) has a direct bearing on the physics of the quantization of the magnetic charge of the $U(1)$ magnetic monopole [cf. (38.15)]. Indeed, since $U(1) \sim S^1$ (topologically), the transition functions $\exp(i\lambda(\boldsymbol{x}))$ [cf. (38.13)], which are mappings from S^1 to $U(1)$, can be considered equivalently as mappings from S^1 to S^1, and hence can be classified by $\pi_1(S^1)$, or the integers. Our result indicates that $\pi_1(S^1)$ happens to be abelian. In general, however, fundamental groups are non-abelian, and can be quite difficult to calculate. Other useful examples of fundamental groups are:

$$\pi_1(SU(2)) = \{e\} , \tag{40.14}$$

$$\pi_1(SO(3)) = \mathbb{Z}/2\mathbb{Z} \equiv \mathbb{Z}_2 , \tag{40.15}$$

$$\pi_1(S^n) = \{e\} , \quad n \geq 2 , \tag{40.16}$$

$$\pi_1(T^n) = \underbrace{\mathbb{Z} \oplus \mathbb{Z} \oplus \cdots \oplus \mathbb{Z}}_{n \text{ times}} , \tag{40.17}$$

where T^n is the n-torus (genus $g = n/2$). Note that the result (40.15) is due to the double-connectedness of $SO(3) \sim S^3$ as a topological space, while (40.14) is due to the fact that $SU(2)$ is simply connected, as discussed in Chapter 19. One can generalize to higher order homotopy groups $\pi_n(M)$ by considering homotopy classes of maps $f : S^n \to M$. All the homotopy groups $\pi_n(M)$ are **topological invariants** of the topological space M.

Closely related to the homotopy groups of a topological manifold M are the **homology groups** of M. Two closed oriented loops c_1 and c_2 in M are said to be **homologous** if there exists a 2-dimensional region A in M such that $c_1 - c_2 = \partial A$, where ∂A denotes the boundary of the region A. The set of equivalence classes of homologous loops in a manifold M has a linear as well as an abelian group structure, and is called the **first homology group** of M with integer coefficients, denoted $H_1(M, \mathbb{Z})$, when the loops are allowed to be multiplied by integers. Figure 40.5 shows two homologous loops on the 2-torus T^2 that are not boundaries. Thus $H_1(M, \mathbb{Z})$ is both a vector space over the integers and an abelian group, with the identity of the group being the equivalence class of all closed loops that are also boundaries. We will call all

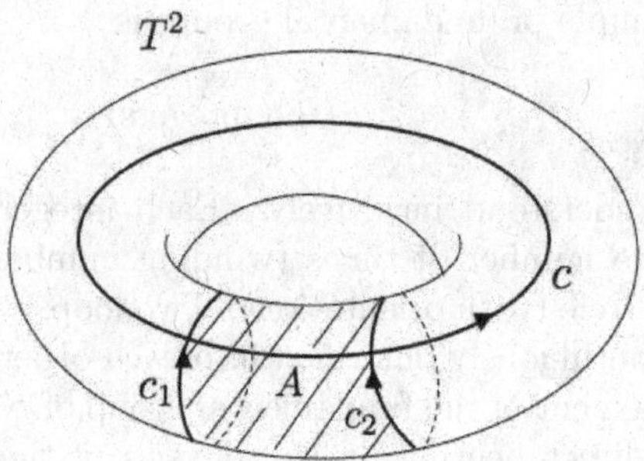

Fig. 40.5

closed loops in M 1-cycles in M (one-dimensional "simplexes" on M which have no boundaries). $H_1(M,\mathbb{Z})$ can be expressed as a quotient group as follows:

$$H_1(M,\mathbb{Z}) = \frac{Z_1(M)}{B_1(M)}\,, \tag{40.18}$$

where $Z_1(M)$ is the group of all 1-cycles in M and $B_1(M)$ is the subgroup of $Z_1(M)$ consisting of all 1-cycles in M that are also boundaries.

The notion of simplexes originates from the intuitive notion of the triangulation of a polygon (which is a bounded 2-dimensional region) in $\mathbb{R}^2$, that is, the division of the polygon into triangles. In $\mathbb{R}^2$ the oriented triangles are called 2-simplexes. Without going into details, we have in general r-simplexes S_r, bounded and oriented regions in the Euclidean space $\mathbb{R}^n$, with $0 \le r \le n$, each bounded by $(r-1)$-simplexes. Picture a line (a 1-simplex) bounded by 2 points (0-simplexes), a triangle (a 2-simplex) bounded by 3 lines (1-simplexes), and a pyramid (a 3-simplex) bounded by 4 triangles (2-simplexes). Now for an arbitrary n-dimensional topological manifold M, consider a smooth map $h : S_r \to M$, which is not necessarily one-to-one. The image of S_r under h in M, denoted by s_r, is called a **singular** r**-simplex** in M. Such simplexes are called singular because h may not be invertible, and hence the s_r do not necessarily provide a **triangulation** of M (a carving M up into the topological equivalents of triangles on a plane). Nevertheless, the r-simplexes in M can furnish vital topological information on M, in the form of the r-th homology groups $H_r(M,\mathbb{Z})$ on M. We will build up to their definitions (Def. 40.4) as follows.

Definition 40.2. *Denote a set of singular r-simplexes in M by $\{s_{r,i}\}$. An r-***chain** *in M is a formal sum of the form*

$$c = \sum_i a_i s_{r,i}\,, \quad a_i \in \mathbb{Z}\,. \tag{40.19}$$

The set of r-chains in M forms an abelian group $C_r(M)$ (under addition), called the r-**chain group** on M. The notion of the boundary of an r-simplex in $\mathbb{R}^n$ also generalizes to that of the boundary ∂s_r of a singular r-simplex, with an induced orientation:

$$\partial s_r \equiv h(\partial S_r)\,, \tag{40.20}$$

where S_r is any r-simplex in $\mathbb{R}^n$ such that $h(S_r) = s_r$. By linearity, the domain of ∂ extends to $C_r(M)$. We thus have the boundary map

$$\partial : C_r(M) \longrightarrow C_{r-1}(M)\,.$$

As in the case of r-simplexes in $\mathbb{R}^n$, the boundary operator acting on C_r is also nilpotent: $\partial^2 = 0$ (the boundary of a boundary is zero).

Definition 40.3. *An r-chain c_r such that $\partial c_r = 0$ is called an r-***cycle***; an r-chain c_r such that $c_r = \partial c_{r+1}$, where c_{r+1} is some $(r+1)$-chain is called an r-***boundary**.

Due to the fact that $\partial^2 = 0$, an r-boundary is necessarily an r-cycle, but an r-cycle need not be an r-boundary.

Definition 40.4. *Let $Z_r(M)$ and $B_r(M)$ be the groups of r-cycles and r-boundaries in the manifold M, respectively. The quotient group $H_r(M,\mathbb{Z})$ defined by*

$$\boxed{H_r(M,\mathbb{Z}) \equiv Z_r(M)/B_r(M)} \tag{40.21}$$

is called the r-th **singular homology group** *on M with integer coefficients.*

Two elements $c_1, c_2 \in Z_r$ belong to the same class: $[c_1] = [c_2] \in H_r(M,\mathbb{Z})$, if there exists some $(r+1)$-chain A such that $c_1 - c_2 = \partial A$. The group operation of $H_r(M,\mathbb{Z})$ (an addition) is defined by

$$[c_1] + [c_2] = [c_1 + c_2]\,. \tag{40.22}$$

For example, consider the two 1-cycles c_1 and c_2 in the 2-dimensional torus T_2 (see Fig. 40.5). It is clear that $c_1 - c_2 = \partial A$, where A is the shaded region, and so $[c_1] = [c_2]$. On the other hand, c and c_1 belong to different classes. In fact $H_1(T^2,\mathbb{Z})$ is generated by those loops which are not boundaries; and there are only two such classes, $[c_1]$ and $[c]$. Thus

$$H_1(T^2,\mathbb{Z}) \sim \mathbb{Z} \oplus \mathbb{Z}\,. \tag{40.23}$$

In general, for a $2g$-torus (g being the genus or the number of holes),

$$H_1(T^{2g}, \mathbb{Z}) \sim \underbrace{\mathbb{Z} \oplus \mathbb{Z} \oplus \cdots \oplus \mathbb{Z}}_{2g} . \tag{40.24}$$

Problem 40.2 Prove the following fact:

$$H_0(T^{2g}, \mathbb{Z}) = H_2(T^{2g}, \mathbb{Z}) = \mathbb{Z} ,$$

where g is the genus of the $2g$-torus T^{2g}.

An important relationship between the first homotopy group (fundamental group) and the first homology group is given by the following theorem (stated without proof):

Theorem 40.1. *Let M be a connected topological space. Then $\pi_1(M)$ is isomorphic to $H_1(M, \mathbb{Z})$ if and only if $\pi_1(M)$ is commutative.*

As examples of the above theorem, we have the following results [cf. (40.13), (40.16), (40.17) and (40.24)]:

$$\pi_1(S^1) = H_1(S^1, \mathbb{Z}) = \mathbb{Z} , \tag{40.25}$$

$$\pi_1(S^n) = H_1(S^n, \mathbb{Z}) = \{e\} , \quad n \geq 2 , \tag{40.26}$$

$$\pi_1(T^{2g}) = H_1(T^{2g}, \mathbb{Z}) = \underbrace{\mathbb{Z} \oplus \mathbb{Z} \oplus \cdots \oplus \mathbb{Z}}_{2g} . \tag{40.27}$$

We will now return to the physics of the problem. The **adiabatic theorem** on the real time (t) axis states that for a slowly-varying, time-dependent Hamiltonian $H(t)$ with non-degenerate (time-dependent) eigenvalues $\epsilon_n(t)$, the solution $\Psi(t)$ [cf. (39.5)] of the Schrödinger (40.1) with the initial condition $\Psi(0) = \phi_n(0)$ [cf. (40.2)] is

$$\Psi(t) = \phi_n(t) \exp\left\{ \frac{i}{\hbar} \left(\gamma_n(t) - \int_0^t \epsilon_n(t')dt' \right) \right\} \tag{40.28}$$

in the semiclassical limit $\hbar \to 0$. The quantity

$$\gamma_n(t) \equiv i\hbar \int_0^t \langle \phi_n \,|\, \dot{\phi_n} \rangle \, dt' \tag{40.29}$$

is the Berry phase introduced in (39.8), and $\int_0^t \epsilon_n(t')dt'$ is called the dynamic phase. The adiabatic theorem can be loosely paraphrased as: "once in an eigenstate, always in an eigenstate".

The condition of adiabaticity [slow time variation of $H(t)$] amounts to the limit $T \to \infty$, where T is the time interval over which $H(t)$ changes significantly.

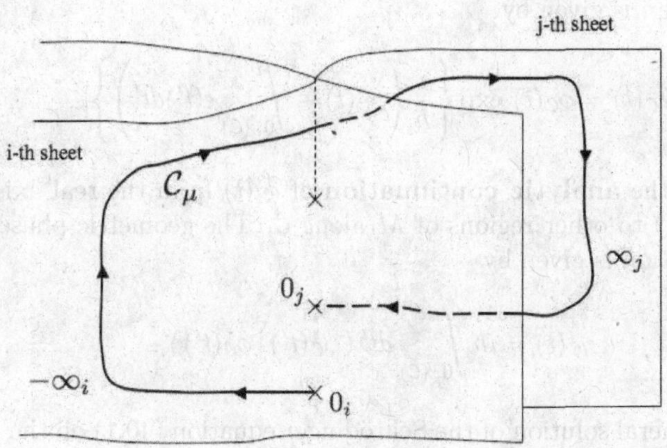

Fig. 40.6

There is a close connection between the adiabatic limit and the classical limit $\hbar \to 0$, as revealed by the time-scaling $\tau = t/T$, which leads to the time-scaled Schrödinger equation

$$i\hbar' \frac{\partial \Psi'}{\partial \tau} = H'(\tau)\Psi'(\tau) , \tag{40.30}$$

where $\hbar' \equiv \hbar/T$, $H'(\tau) \equiv H(T\tau)$, and $\Psi'(\tau) \equiv \Psi(T\tau)$. However, the two limits are not in general equivalent, since $H'(\tau)$ in the above equation may have a concealed $\hbar$-dependence. Thus the classical limit $\hbar \to 0$ (on both sides of the Schrödinger equation) with $T =$ constant may not be equivalent to the limit $T \to \infty$ while keeping $\hbar$ constant. To restore the equivalence, we make a distinction between the classical limit [$\hbar \to 0$ on both sides of the Schrödinger equation (40.1)] and the semiclassical limit [$\hbar \to 0$ on only the right-hand side of (40.1)]. This distinction is justified in the present context since $H(t)$ is presumably determined from a quantum mechanical treatment of the electronic motion, where $\hbar \neq 0$ strictly.

The adiabatic theorem can be generalized from the real t axis to the complex t-plane to incorporate quantum transitions from state i to state j. This result can be stated in terms of the t-Riemann surface M: for the energy function $\epsilon(t) : M \to \mathbb{C} \cup \{\infty\}$, there exists a contour $\mathcal{C}$ in M leading from the i-th sheet to the j-th sheet of M such that (40.28) holds along $\mathcal{C}$ in the semiclassical limit $\hbar \to 0$. Figure 40.6 illustrates one such contour $\mathcal{C}_\mu$, where μ is a homotopy index, and the contour begins at the point $t = 0$ on the real axis of the i-th Riemann sheet (0_i) and ends at the point $t = 0$ on the real axis of the j-th Riemann sheet (0_j). In this figure the points at infinity on the two sheets [denoted by

$-\infty_i (= \infty_i)$ and ∞_j] are also shown. The adiabatic theorem thus implies that, along $\mathcal{C}$, the solution to the Schrödinger equation (40.1) describing a system initially in state i is given by

$$\Psi_{\mathcal{C}}(t) = \phi_{\mathcal{C}}(t)\, \exp\left\{ \frac{i}{\hbar} \left(\gamma_{\mathcal{C}}(t) - \int_{0_i\,(\mathcal{C})}^{t} \epsilon(t')dt' \right) \right\} , \tag{40.31}$$

where $\phi_{\mathcal{C}}(t)$ is the **analytic continuation** of $\phi_i(t)$ from the real axis (on the i-th sheet of M) to other regions of M along $\mathcal{C}$. The geometric phase $\gamma_{\mathcal{C}}(t)$ in the above equation is given by

$$\gamma_{\mathcal{C}}(t) = i\hbar \int_{0_i\,(\mathcal{C})}^{t} dt' \,\langle\, \phi_{\mathcal{C}}(t') \,|\, \dot{\phi}_{\mathcal{C}}(t') \,\rangle \,. \tag{40.32}$$

Now the general solution of the Schrödinger equation (40.1) on the real time axis is

$$\Psi(t) = \sum_{l=1}^{N} C_l(t)\phi_l(t)\exp\left\{ -\frac{i}{\hbar} \int_0^t \epsilon_l(t')dt' \right\} . \tag{40.33}$$

The solution given by (40.31) thus satisfies the boundary conditions

$$\Psi_{\mathcal{C}}(-\infty_i) = C_i(-\infty)\phi_i(-\infty)\exp\left\{ -\frac{i}{\hbar} \int_0^{-\infty} \epsilon_i(t')dt' \right\} \quad \text{and} \tag{40.34a}$$

$$\Psi_{\mathcal{C}}(\infty_j) = C_j(\infty)\phi_j(\infty)\exp\left\{ -\frac{i}{\hbar} \int_0^{\infty} \epsilon_j(t')dt' \right\} , \tag{40.34b}$$

where in the above two equations all time values are real and

$$|\, C_i(-\infty)\,|^2 = 1 \,. \tag{40.35}$$

The initial condition (40.34a) is equivalent to the following initial condition on the geometric phase

$$\exp\left(\frac{i}{\hbar}\gamma_{\mathcal{C}}(-\infty_i) \right) = C_i(-\infty) = e^{i\theta} \quad (\theta \text{ real}) \,. \tag{40.36}$$

Comparing (40.31) with (40.34b), we have

$$\exp\left\{ \frac{i}{\hbar}\gamma_{\mathcal{C}}(\infty_j) \right\} \exp\left\{ -\frac{i}{\hbar} \int_{0_i\,(\mathcal{C})}^{\infty_j} \epsilon(t)dt \right\} = C_j(\infty_j)\exp\left\{ -\frac{i}{\hbar} \int_{0_j}^{\infty_j} \epsilon(t)dt \right\} . \tag{40.37}$$

Thus for a path $\mathcal{C}$ from 0_i to 0_j on which the adiabatic theorem holds, the transition amplitude from state i to state j is given by

$$C_j(\infty_j) = \exp\left\{ \frac{i}{\hbar}\gamma_{\mathcal{C}}(\infty_j) \right\} \exp\left\{ -\frac{i}{\hbar} \int_{\mathcal{C}} \epsilon(t)dt \right\} . \tag{40.38}$$

The adiabatic theorem also necessarily implies that the path $\mathcal{C}(0_i \to 0_j)$ is one where the Abelian integral

$$\int_{\mathcal{C}} \epsilon(t)dt$$

has a negative imaginary part, to ensure that as $\hbar \to 0$ the transition probability does not increase exponentially. This requirement, however, applies only to the entire path and not necessarily to a localized segment. Indeed, any valid path $\mathcal{C}$ can be written as

$$\mathcal{C} = \mathcal{C} + \Delta\mathcal{C} - \Delta\mathcal{C} \ ,$$

where $\Delta\mathcal{C}$ is a localized segment on which the above Abelian integral has a positive imaginary part.

In general there may be several non-homotopic paths $\mathcal{C}_\lambda$ in M from 0_i to 0_j satisfying the adiabatic theorem, where λ is a homotopy index. Consider $\lambda_1 \neq \lambda_2$ such that $\mathcal{C}_{\lambda_1}$ and $\mathcal{C}_{\lambda_2}$ correspond to the same path. (This situation is possible since $\mathcal{C}_{\lambda_1}$ and $\mathcal{C}_{\lambda_2}$ may consist of non-commuting loops traversed in different orders.) One still has

$$\int_{\mathcal{C}_{\lambda_1}} \epsilon(t)dt = \int_{\mathcal{C}_{\lambda_2}} \epsilon(t)dt \qquad \text{or} \qquad \int_{\mathcal{C}_{\lambda_1}\mathcal{C}_{\lambda_2}^{-1}} \epsilon(t)dt = 0 \ . \tag{40.39}$$

Now define a map $f_\epsilon : \pi_1(M) \longrightarrow \mathbb{C}$ by

$$f_\epsilon(\mathcal{C}_\lambda) \equiv \int_{\mathcal{C}_\lambda} \epsilon(t)dt \ . \tag{40.40}$$

This is a group homomorphism provided two different paths $\mathcal{C}_{\lambda k_1}$ and $\mathcal{C}_{\lambda k_2}$ ($k_1 \neq k_2$) within the same homotopy class do not enclose singularities of the Abelian differential $\epsilon(t)dt$ with non-zero residues. The fact that $\mathbb{C}$ is abelian then implies that the map f_ϵ defined by (40.40) is more appropriately considered as an injective map from $H_1(M)$ to $\mathbb{C}$, since, by a well-known theorem, $H_1(M)$ is the abelianization of $\pi_1(M)$. We will not go into the specific details regarding the term "abelianization" here, except to recall the important result of Theorem 40.1. The paths leading to distinct values of the Abelian integral $\int \epsilon(t)dt$ will then be labeled by $\mathcal{C}_\mu$, where μ is a homology index, rather than a homotopy index.

Finally, to take into account the possible existence of singularities of $\epsilon(t)dt$ with non-zero residues, we recognize that, within each homology class μ, there may be distinct paths $\mathcal{C}_{\mu k_1}$ and $\mathcal{C}_{\mu k_2}$ such that

$$\int_{\mathcal{C}_{\mu k_1}} \epsilon(t)dt \neq \int_{\mathcal{C}_{\mu k_2}} \epsilon(t)dt$$

if the loop $\mathcal{C}_{\mu k_1}\mathcal{C}_{\mu k_2}^{-1}$ encloses such singularities. The total transition amplitude is thus given by

$$C_j(\infty_j) = \sum_\mu \exp\left\{i\gamma_{\mathcal{C}_\mu}(\infty_j)\right\} \sum_k \exp\left\{-\frac{i}{\hbar}\int_{\mathcal{C}_{\mu k}} \epsilon(t)dt\right\} , \tag{40.41}$$

where the sum over μ is over the homology classes and that over k is over the paths in each homology class leading to distinct values, with negative imaginary parts, for the Abelian integral $\int \epsilon(t)dt$. We see that since $\gamma_{C_\mu}(\infty_j)$ is real for all μ, the geometric phase factor $\exp\{i\gamma_{C_\mu}(\infty_j)\}$ gives a $U(1)$ representation of $H_1(M)$. Equation (40.41) displays clearly the importance of the topology of M and the analytic properties of the Abelian integral $\int \epsilon(t)dt$ in the calculation of semiclassical transition amplitudes.

To conclude this chapter we will present a very simple and yet interesting example on the use of (40.41). This example is based on the so-called **Landau-Zener model** for a two-state problem. As we will see, it involves a Riemann surface M with $g = 0$ (topologically a 2-sphere), for which $\pi_1(M)$ is abelian and so $\pi_1(M) = H_1(M)$. [For a $g = 1$ example, where it is also true that $\pi_1(M) = H_1(M)$, see Problem 40.4 below, and Lam 1996.]

In the Landau-Zener model the electronic Hamiltonian $H(t)$ is given by

$$H(t) = \begin{pmatrix} at & b \\ b & -at \end{pmatrix}, \tag{40.42}$$

where a and b are positive real constants. Physically the diagonal elements represent two intersecting potential curves. Thus the quantity a can be interpreted as the product of the instantaneous speed of nuclear motion and the gradient of the potential energy (as a function of internuclear coordinates) at the crossing point, while b is the nonadiabatic coupling. The solutions of Eq. (40.4) [or Eq. (40.5)] for the adiabatic potential energy surfaces are immediately obtained by the solutions of a quadratic algebraic equation and given by

$$\epsilon(t) = \mp\sqrt{\left(t + i\frac{b}{a}\right)\left(t - i\frac{b}{a}\right)} \ . \tag{40.43}$$

The algebraic function $\epsilon(t)$ thus has two branch points (at $t = \pm ib/a$) and two branches. Setting $B = N = 2$ in (40.7) (the Riemann-Hurwitz formula), the Riemann surface M of $\epsilon(t)$ is seen to be of genus zero ($g = 0$), or topologically a 2-sphere. This fact can also be seen very clearly by the topological illustrations of Fig. 40.7, which shows one way of constructing M from two distinct (cut) copies of the complex t-plane $\mathbb{C}$. (Other ways result from alternate ways of cutting the t-plane beginning and ending at the branch points $t = \pm ib/a$. All lead to a 2-sphere topologically for M.) In Fig. 40.7 one proceeds from (a) distinct, cut, sheets of $\mathbb{C}$ [labeled 1 and 2, corresponding to the $-$ and $+$ signs in (40.43), respectively] with branch points and cuts as shown ($t_B = ib/a$), through 1-point compactifications to (b) the corresponding Riemann spheres with cuts, through homeomorphic deformations of the cut-spheres to (c) the corresponding hemispheres, and finally, by joining the hemispheres with matching boundaries, to (d) the 2-sphere.

The Abelian differential $\epsilon(t)dt$, with $\epsilon(t)$ given by (40.43), clearly has singularities only at the two (distinct) points at infinity: $t = \infty_1, \infty_2$ (see Fig.

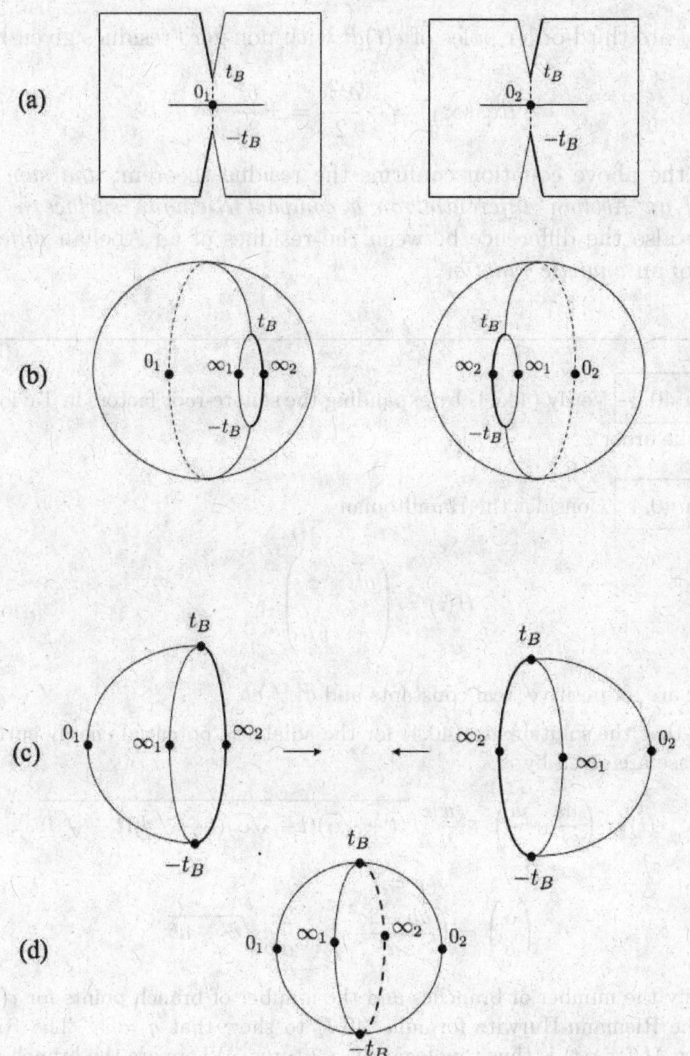

(a)
t_B
0_1
$-t_B$
t_B
0_2
$-t_B$
(b)
t_B
0_1
∞_1
∞_2
$-t_B$
t_B
∞_2
∞_1
0_2
$-t_B$
(c)
t_B
0_1
∞_1
∞_2
$-t_B$
t_B
∞_2
0_2
∞_1
$-t_B$
(d)
t_B
0_1
∞_1
∞_2
0_2
$-t_B$

Fig. 40.7

40.7). To determine the nature of these singularities, we let $z \equiv 1/t$ and expand around $z = 0$. Thus, letting $t_B \equiv ib/a$, we have

$$\epsilon(t)dt = \pm\frac{a}{z^3}(1+t_B z)^{1/2}(1-t_B z)^{1/2}dz = \pm\left(\frac{a}{z^3} - \frac{at_B^2}{2z} + \ldots\right)dz\,. \quad (40.44)$$

Hence $\infty_{1,2}$ are third-order poles of $\epsilon(t)dt$ with non-zero residues given by

$$Res(\infty_{\frac{1}{2}}) = \mp\frac{at_B^2}{2} = \pm\frac{b^2}{2a}\,. \quad (40.45)$$

Note that the above equation confirms the residue theorem: *the sum of the residues of an Abelian differential on a compact Riemann surface is always zero.* Note also the difference between the residues of an Abelian *differential* and those of an *analytic function.*

Problem 40.3 Verify (40.44) by expanding the square-root factors in Taylor series to at least the order z^2.

Problem 40.4 Consider the Hamiltonian

$$H(t) = \begin{pmatrix} at & c \\ c & b/t \end{pmatrix},$$

where a, b, c are all positive, real constants and $c^2 \neq ab$.

(a) Show that the solutions to (40.4) for the adiabatic potential energy surfaces in this case are given by

$$\epsilon(t) = \left(\frac{at}{2} + \frac{b}{t}\right) \mp \frac{a}{2t}\sqrt{(t+\sqrt{\alpha})(t-\sqrt{\alpha})(t+\sqrt{\beta})(t-\sqrt{\beta})}\,,$$

where

$$\begin{pmatrix} \alpha \\ \beta \end{pmatrix} = \left(\frac{ab-2c^2}{a^2}\right) \pm \frac{2c}{a^2}\sqrt{c^2-ab}\,.$$

(b) Identify the number of branches and the number of branch points for $\epsilon(t)$, and use the Riemann-Hurwitz formula (40.7) to show that $g = 1$. The Riemann surface M for $\epsilon(t)$ is thus topologically a 2-torus. Where are the branch points?

(c) Verify the following Laurent expansions for $\epsilon(t)dt$:

$$\epsilon(t)dt = \left[\frac{b \mp a\sqrt{\alpha\beta}}{2t} + \frac{at}{2}\left\{1 \pm \frac{\sqrt{\alpha\beta}}{2}\left(\frac{1}{\alpha}+\frac{1}{\beta}\right)\right\} + O(t^3)\right]dt$$

around $t = 0$; and

$$\epsilon(t)dt = \left[-\frac{a}{2}(1\mp 1)\frac{1}{z^3} + \left\{-\frac{b}{2} \mp \frac{a(\alpha+\beta)}{4}\right\}\frac{1}{z} \mp \frac{a}{16}(\alpha-\beta)^2 z + O(z^3)\right]dz$$

around $t = \infty$ (equivalently around $z = 0$, for $z \equiv 1/t$). Using the branch indices 1 and 2 to denote the branches corresponding to the $-$ and $+$ signs, respectively, in the above expression for $\epsilon(t)$, check that the upper (lower) sign in the above equations for $\epsilon(t)dt$ corresponds to branch 1(2).

(d) Show that $0_1, \pm\sqrt{\alpha}, \pm\sqrt{\beta}$ are all *simple zeros*, 0_2 and ∞_1 are both *simple poles*, and ∞_2 is a *third-order pole* of $\epsilon(t)dt$; and verify that the three poles have residues given as follows

$$Res(0_2) = \frac{b + a\sqrt{\alpha\beta}}{2} = b\,,$$

$$Res(\infty_1) = -\frac{b}{2} - \frac{a\sqrt{\alpha+\beta}}{4}\,,$$

$$Res(\infty_2) = -\frac{b}{2} + \frac{a\sqrt{\alpha+\beta}}{4}\,.$$

Verify that the sum of the residues vanishes, in agreement with the residue theorem on compact Riemann surfaces.

(e) Verify that the above information on the zeros and poles of $\epsilon(t)dt$ satisfies the Poincare-Hopf index formula (40.9).

From (40.26) we see that the fundamental group $\pi_1(M)$ for a 2-sphere [which is equal to $H_1(M)$] is trivial (consisting only of the identity element). Hence in (40.41) for the transition amplitude there is no summation over μ (the homology classes of M); and the geometric phase factor $\exp\{i\gamma_{C_\mu}(\infty_j)\}$ [being a $U(1)$ representation of $H_1(M)$] can be set equal to unity. For the transition $1 \to 2$, (40.41) then reads

$$C_2(\infty_2) = \sum_k \exp\left\{-\frac{i}{\hbar}\int_{\mathcal{C}_k} \epsilon(t)dt\right\}\quad, \tag{40.46}$$

where $\mathcal{C}_k$ is a path on M from 0_1 to 0_2 [cf. Figs. 40.6 and 40.7], and the sum over k is over paths leading to distinct values (with negative imaginary parts) for the Abelian integral $\int \epsilon(t)dt$. Keep in mind that distinct values arise because of the presence of singularities of $\epsilon(t)dt$ with non-zero residues.

There are two direct paths on M, designated I and II in Fig. 40.8(a), that give rise to distinct values of $\epsilon(t)dt$. These can also be represented on the Riemann sheets (cut-planes) of $\epsilon(t)$ as in Fig. 40.8(b), in which the contours run entirely along the imaginary axes of the sheets. Since both branches of $\epsilon(t)$ are real on the real axis [cf. (40.43)], the *Schwarz Reflection Principle* on $\mathbb{C}$ (cf. Theorem 36.1) implies

$$\int_I \epsilon(t)dt = \left(\int_{II} \epsilon(t)dt\right)^*\quad. \tag{40.47}$$

Furthermore, we can also conclude from (40.43) that both $\int_I \epsilon(t)dt$ and $\int_{II} \epsilon(t)dt$ are purely imaginary. Thus on setting $\int_I \epsilon(t)dt = iX$, X real, the residue theorem implies [see Fig. 40.8(a)]

$$\int_{II} \epsilon(t)dt - \int_I \epsilon(t)dt = -2iX = 2\pi i Res(\infty_1) = (2\pi i)\left(\frac{b^2}{2a}\right)\,, \tag{40.48}$$

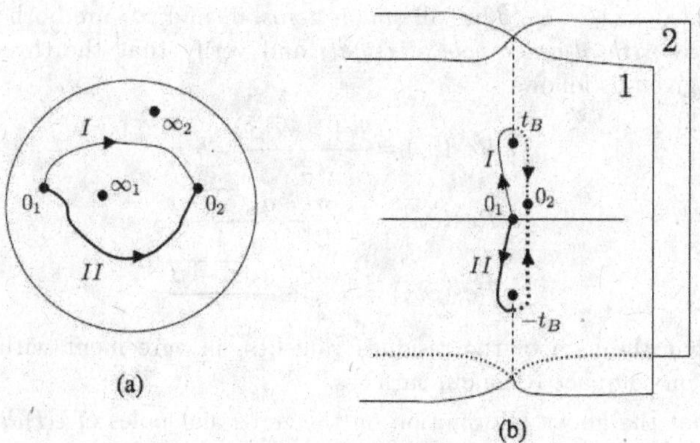

Fig. 40.8

where the last equality follows from (40.45). Now, according to (40.46), we have to sum over *all* paths $\mathcal{C}_k$ such that each $\int_{\mathcal{C}_k} \epsilon(t)dt$ has a negative imaginary part. Fig. 40.8(a) and the value of $Res(\infty_1)$ [from (40.45)] show that, for a positive integer k, $\mathcal{C}_k$ is of the form

$$\mathcal{C}_k = I + k \text{ clockwise loops around } \infty_1 \,, \tag{40.49}$$

since the anti-clockwise loops yield positive imaginary parts for the integral. The transition amplitude can finally be written as

$$C_2(\infty) = \exp\left(-\frac{\pi b^2}{2a\hbar}\right)\sum_{k=0}^{\infty}\exp\left(-\frac{\pi k b^2}{a\hbar}\right) = \frac{\exp(-\pi b^2/(2a\hbar))}{1-\exp(-\pi b^2/(a\hbar))} \,. \tag{40.50}$$

The leading term of the transition probability is thus

$$\boxed{P_{1\to 2} \approx \exp\left(-\frac{\pi b^2}{a\hbar}\right)} \,, \tag{40.51}$$

which is the well-known **Landau-Zener formula** for non-adiabatic transitions. We note that $C_2(\infty)$ as given by (40.50) can only be interpreted as a transition amplitude if it is less than one, which is the case only when $\pi b^2/(2a\hbar)$ is larger than ~ 0.5 as $\hbar \to 0$. The Landau-Zener formula [(40.51)], however, suffers from no such limitation.

Chapter 41

"The Connection is the Gauge Field and the Curvature is the Force": Some Differential Geometry

This chapter is almost entirely a mathematical detour. It attempts to demonstrate how the differential geometric notions of the *connection* and *curvature* correspond to the physical ones of the gauge potential and field strength, respectively. We have already studied in some detail two examples of gauge potentials: the (abelian) electromagnetic vector potential (Chapters 34 and 38), and the (in general non-abelian) Berry-Simon gauge potentials in molecular dynamics (Chapter 39). In fact the entire edifice of the *standard model* of elementary particles is built upon a quantum field theory of gauge fields. (For more detailed and rigorous mathematical expositions on the differential geometric notions introduced throughout this and later chapters, the physics-oriented reader may consult, for example, Nakahara 2003; Lam 2003; and Chern, Chen and Lam 1999.)

There is, in fact, a direct path leading from the classical mechanical concept of the angular velocity of rigid bodies to the field-theoretic concept of gauge fields. This may not be so surprising since gauge field theory is essentially geometric in nature. [This theme is discussed from a mathematical perspective in, for example, Bleecker 1981; and from a physical perspective in, for example, Bertlmann (2000), and Moriyasu (1983).] The word **gauge** refers to the choice of a particular frame of reference at a particular point in space to carry out geometric measurements, and the motion of rigid bodies in space is essentially described by the geometry of the motion of body-fixed frames (moving frames) in space (see, for example, Appendix B in Chern, Chen and Lam 1999), which, by definition, vary from point to point. Whereas in classical mechanics, the space that the rigid body moves in (called the **base manifold** in geometric jargon) is (usually) understood to be 3-dimensional Euclidean space and the moving frames are orthonormal frames (xyz-unit vectors) attached to different points of the manifold, the base manifold in gauge field theory is (usually) 4-dimensional Minkowski spacetime, and the moving frames are frames with "axes" describing different "internal degrees of freedom" (be it neutron or proton, neutrino or electron, the different colors of quarks, etc.). Just as in classical mechanics, where the physics of the motion of a rigid body is independent of the choice of body-fixed frames at different points of space, the physics of the interaction of elementary particles is independent of the choice of frames describing the internal degrees of freedom at different points of spacetime. This seemingly inocuous observation underlies the very fundamental **principle of gauge invariance** (see, for example, Yang 1981).

We will begin our study of differential geometry by considering rotations in rigid body dynamics. This may seem an odd topic to discuss in a book on quantum theory, but we hope the reader will soon be convinced of its relevance here.

We will denote the body-fixed frames (**moving frames**) by $\{e_i\}$, $i = 1, 2, 3$, as distinguished from space-fixed frames $\{\delta_i\}$. With respect to a body-fixed frame, the position vector $\boldsymbol{r}(t)$ of a *fixed* point in the rigid body is given by

$$\boldsymbol{r}(t) = X^i\, \boldsymbol{e}_i \ , \tag{41.1}$$

where the Einstein summation convention has been used. The components X^i naturally remain constant in time as the rigid body rotates. We therefore have, for the velocity,

$$\boldsymbol{v}(t) = \frac{d\boldsymbol{r}(t)}{dt} = X^i\, \frac{d\boldsymbol{e}_i}{dt} \ . \tag{41.2}$$

The geometry (but not the dynamics) of the rotational motion is apparently completely described by the information contained in the quantities $d\boldsymbol{e}_i$. How is one to compute these "differentials" of basis vectors? In fact, what do these "differentials" mean? Naively, they are expected to tell us something about how the basis vectors of the moving frame $\boldsymbol{e}_i$ at one point are compared to those at another. But here is the mathematical difficulty: Vectors constituting the axes of a frame erected at one point in a **differentiable manifold** live in a linear vector space associated with only that point (known as the **tangent space** at the point), and there is no way a priori to compare vectors in tangent spaces at different points. It might be asked: Why not translate a vector at one point "parallelly" in space along some curve until it sits at the other point, and then we can add or subtract (and thus compare) the vectors as we please? The answer is that this cannot be done until a prescription is given on how to do **parallel translation** of vectors. This difficulty can readily be appreciated if we imagine a vector tangent at some point on the surface of a 2-dimensional sphere and ask how we can parallelly translate the vector from that point to some neighboring point. The procedure seems to be obvious only if we view the translation from the vantage point of the 3-dimensional Euclidean space in which the 2-dimensional sphere is embedded. But then we are doing parallel translation in 3-dimensional Euclidean space, not in the 2-dimensional sphere. From the vantage point of an ant living on the surface of the sphere, who does not know that the sphere is embeddable in a higher dimensional space, it is not obvious at all how to translate the vector. Indeed, the notion of **parallelism** of vectors is not defined in an arbitrary differentiable manifold (of which the 3-dimensional Euclidean space in which a rigid body moves is only a very special example) until a structure known as an **affine connection** is imposed on the manifold. A given affine connection is encoded in the prescription for parallel translation in terms of the so-called **covariant derivative** in the following way (note the Einstein summation convention again):

$$\boxed{De_i = \omega_i^j e_j} \quad . \tag{41.3}$$

A **tangent vector** $\boldsymbol{v}$ is said to be **parallelly translated** (transported, or displaced) along a curve on the base manifold if $Dv/dt = 0$ along the curve.

Equation (41.3) needs to be decoded a bit. First we assume that a smooth assignment of a frame $\{e_i\}$ at each point of the base manifold has been made. This assignment is called a **frame field**. The big "D" has replaced the small "d" [in the right-hand side of the last equality in (41.2)] to denote the covariant derivative, as opposed to the so-called **exterior derivative** (which is usually denoted by d). In general, the covariant derivative is used to differentiate vector fields on a manifold, and only makes sense when a connection is given, whereas the exterior derivative is used to differentiate **differential forms**, and does not depend on the connection. The exterior derivative generalizes the various differential operators in vector calculus (div, grad, curl and all that). The connection is given locally by the **connection matrix** of 1-forms (ω_i^j). (The subscript i will be the row index and the superscript j the column index, if the e_i's are interpreted as forming a column vector. In this chapter, unless otherwise specified, we will adopt this convention for matrix indices. The reader should bear in mind that, elsewhere in this book, the other convention – superscript for row index and subscript for column index of a matrix – is adopted instead; but equations with indices displayed explicitly are always unambiguous.) If the base manifold is n-dimensional, then the matrix is $n \times n$. It is important to realize that a connection matrix is always given with respect to a particular frame field.

Before proceeding with further discussion on the covariant derivative, it will be useful at this juncture to provide a brief introduction to the notion of differential forms and the associated operations of exterior multiplication and exterior differentiation.

Consider an n-dimensional differentiable manifold with **local coordinates** $(x^1, \dots, x^n)$. Loosely speaking, an r-differential form (r-form for short) consists a sum of terms each of which is some function on the manifold (that is, a function of the variables $x^1, \dots, x^n$) multiplied by an *antisymmetric product* (called the **exterior product** and denoted by $\wedge$) of r differentials $dx^{i_1}, \dots, dx^{i_r}$. A 0-form is simply a function on the manifold. (The exterior product generalizes the vector cross product in vector algebra.) Using the generic symbol ω to denote an r-form, we can write,

$$\omega = a_{i_1 \cdots i_r}(x^1, \ldots, x^n)\, dx^{i_1} \wedge \cdots \wedge dx^{i_r} \quad , \tag{41.4}$$

where each of the $a_{i_1 \cdots i_r}$ is a smooth function of the local coordinates. In the above equation, each of the indices i_1 to i_r runs from 1 to n. If the local coordinates of a 3-dimensional manifold are x, y and z, examples of a 0-form, a 1-form and a 2-form are x^2+y^2-2xz, $(x^2+y^2)dx+2ydy$, and $(x^2+y^2)\,dx\wedge dy + 2xy\,dy\wedge dz$, respectively. Because of the antisymmetry of the exterior product, an r-form on an n-dimensional manifold is automatically zero if $r > n$.

The exterior product applied to differential forms satisfies the normal distributive and associative rules. In addition, it satisfies the following anti-commutative rule: If ω_1 is a k-form and ω_2 is an l-form, then

$$\omega_1 \wedge \omega_2 = (-1)^{kl}\, \omega_2 \wedge \omega_1 \quad . \tag{41.5}$$

The computational formula for the exterior derivative (denoted d) applied to a differential form [given in general by (41.4)] can be written as

$$\boxed{d\omega = \frac{\partial a_{i_1 \cdots i_r}}{\partial x^j}\, dx^j \wedge dx^{i_1} \wedge \cdots \wedge dx^{i_r}} \quad . \tag{41.6}$$

Note again that the index j is summed over. If ω_1 is a k-form and ω_2 is a form of any order, then the above formula implies the following "product rule" for exterior differentiation:

$$d(\omega_1 \wedge \omega_2) = d\omega_1 \wedge \omega_2 + (-1)^k\, \omega_1 \wedge d\omega_2 \quad . \tag{41.7}$$

Also, by (41.6), the exterior derivative applied twice to any differential form always "kills" it:

$$d^2\omega = 0 \quad , \quad \text{for any form } \omega \quad . \tag{41.8}$$

This is an extremely useful fact about exterior differentiation. Finally, we note that exterior differentiation turns a k-form into a $(k+1)$-form.

Problem 41.1 Use the rule for exterior differentiation given by (41.6) to verify directly the results (41.7) and (41.8).

To better appreciate the usefulness of the exterior product and the exterior derivative, let us apply the above rules in a familiar context, that is, to differential forms on 3-dimensional Euclidean space $\mathbb{R}^3$, with local coordinates x, y, z. Let f be a smooth function (a 0-form) on $\mathbb{R}^3$. Then (41.6) implies

$$df = \frac{\partial f}{\partial x}dx + \frac{\partial f}{\partial y}dy + \frac{\partial f}{\partial z}dz \quad . \tag{41.9}$$

The coefficients of this 1-form are just the components of the vector field ∇f. Next consider the 1-form

$$\omega_1 = Adx + Bdy + Cdz \quad , \tag{41.10}$$

where A, B, C are smooth functions on $\mathbb{R}^3$. Application of (41.6) gives

$$\begin{aligned} d\omega_1 &= \frac{\partial A}{\partial y}\, dy \wedge dx + \frac{\partial A}{\partial z}\, dz \wedge dx + \frac{\partial B}{\partial x}\, dx \wedge dy \\ &= \frac{\partial B}{\partial z}\, dz \wedge dy + \frac{\partial C}{\partial x}\, dx \wedge dz + \frac{\partial C}{\partial y}\, dy \wedge dz \\ &= \left(\frac{\partial C}{\partial y} - \frac{\partial B}{\partial z}\right) dy \wedge dz + \left(\frac{\partial A}{\partial z} - \frac{\partial C}{\partial x}\right) dz \wedge dx + \left(\frac{\partial B}{\partial x} - \frac{\partial A}{\partial y}\right) dx \wedge dy \quad . \end{aligned} \tag{41.11}$$

We recognize the coefficients of this 2-form to be the components of the vector field $\nabla \times \boldsymbol{X}$, where $\boldsymbol{X}$ is the vector field with components A, B and C. Finally, consider the 2-form

$$\omega_2 = A\, dy \wedge dz + B\, dz \wedge dx + C\, dx \wedge dy \quad , \tag{41.12}$$

where, again, A, B and C are smooth functions on $\mathbb{R}^3$. On applying (41.6) we obtain

$$d\omega_2 = \left(\frac{\partial A}{\partial x} + \frac{\partial B}{\partial y} + \frac{\partial C}{\partial z}\right) dx \wedge dy \wedge dz \quad . \tag{41.13}$$

The quantity inside the parentheses is seen to be the divergence of the vector field $\boldsymbol{X} = (A, B, C)$. Thus *the exterior derivative completely subsumes the vector calculus operators of div, grad and curl.* Many vector calculus identities are easily proved by using exterior differentiation.

We will now return to the covariant derivative, and immediately derive an important property of the connection matrix with respect to an orthonormal frame. Suppose $\{\boldsymbol{e}_i\}$ is an orthonormal frame. Then

$$\boldsymbol{e}_i \cdot \boldsymbol{e}_j = \delta_{ij} \quad , \tag{41.14}$$

where δ_{ij} is the Kronecker delta symbol. On covariantly differentiating (using a generalization of the product rule for ordinary differentiation), we have

$$D\boldsymbol{e}_i \cdot \boldsymbol{e}_j + \boldsymbol{e}_i \cdot D\boldsymbol{e}_j = 0 \quad . \tag{41.15}$$

It follows from (41.3) that

$$\omega_i^k\, \boldsymbol{e}_k \cdot \boldsymbol{e}_j + \boldsymbol{e}_i \cdot \omega_j^l\, \boldsymbol{e}_l = 0 \quad , \tag{41.16}$$

or, on using (41.14),

$$\omega_i^k\, \delta_{kj} + \omega_j^l\, \delta_{il} = 0 \quad . \tag{41.17}$$

Thus, if the frame $\{\boldsymbol{e}_i\}$ is orthonormal, the connection matrix ω with respect to it is antisymmetric:

$$\omega_i^j + \omega_j^i = 0 \quad . \tag{41.18}$$

Returning to (41.2) and using (41.3), the velocity of a fixed point in the rigid body is then given by

$$v(t) = X^i \varphi_i^j \, e_j \quad , \tag{41.19}$$

where φ_i^j is a matrix of ordinary derivatives (with respect to time) defined by

$$\varphi_i^j \equiv \frac{\omega_i^j}{dt} \quad . \tag{41.20}$$

The notation is a little strange, but remember that each matrix element ω_i^j, beng a 1-form, has the general form

$$\omega_i^j = f_1(x^1, \ldots, x^n)dx^1 + \cdots + f_n(x^1, \ldots, x^n)dx^n \quad , \tag{41.21}$$

where $(x^1, \ldots, x^n)$ are the local coordinates of the manifold, so each term on the right-hand side of (41.20) in fact involves an ordinary derivative. In view of the antisymmetry property of ω [Eq. (41.18)] and the definition (41.20) we can define a "vector" $\boldsymbol{\omega}$ with 3 components ω^i with respect to an orthonormal frame $\{e_i\}$ in 3-dimensional Euclidean space $\mathbb{R}^3$ (and only in a manifold of dimension 3) as follows:

$$\varphi_i^j \equiv \varepsilon_i{}^j{}_k \, \omega^k \quad , \tag{41.22}$$

where $\varepsilon_i{}^j{}_k$ is the completely antisymmetric **Levi-Civita tensor**, whose indices can be raised or lowered "with impunity". On recalling the properties of $\varepsilon_i{}^j{}_k$ the above equation can be displayed more clearly in matrix form as follows:

$$(\varphi_i^j) = \begin{pmatrix} 0 & \omega^3 & -\omega^2 \\ -\omega^3 & 0 & \omega^1 \\ \omega^2 & -\omega^1 & 0 \end{pmatrix} \quad , \tag{41.23}$$

where i is the row index and j the column index. Then, from (41.19),

$$v(t) = X^i \varepsilon_i{}^j{}_k \omega^k \, e_j = \varepsilon^j{}_{ki} \omega^k X^i \, e_j \quad . \tag{41.24}$$

By the definition of the cross product this is recognized to be the familiar equation for the angular velocity "vector" $\boldsymbol{\omega}$ with components ω^k:

$$\boxed{v = \boldsymbol{\omega} \times r} \quad . \tag{41.25}$$

The fact that an antisymmetric 3×3 matrix has three independent components is an accident of the dimensionality. Were we to deal with an N-dimensional manifold, the antisymmetric matrix φ_i^j would have $N(N-1)/2$ independent components. This quantity is only equal to N when $N = 3$.

We have put quotes around the word "vector" referring to $\boldsymbol{\omega}$ for good reason: $\boldsymbol{\omega}$ is not a vector, or even an antisymmetric tensor [as might have been concluded from (41.23)], because it does not transform like a vector or tensor under a local change of frames! This can actually be demonstrated rather easily. Suppose the frame $\{e_i\}$ is related to another $\{e'_i\}$ at a particular point x in the manifold by a rotation:

$$e'_i = g_i^j(x)\, e_j \quad , \tag{41.26}$$

where x denotes collectively the local coordinates (x_1, x_2, x_3) of a point in the manifold and $g_i^j(x)$ is an element in the rotation group $SO(3)$. It is important to realize that the matrix elements are in general functions of the position in space since one is free to choose different body-fixed frames at different locations of, say, the center of mass of the rigid body without changing the physics (gauge invariance). We will rewrite the above equation in matrix form as

$$e' = g(x)\, e \quad , \tag{41.27}$$

where e and e' are column matrices of basis vectors and g is a 3×3 matrix in $SO(3)$. We can similarly rewrite (41.3) for the covariant derivative in matrix form as

$$De = \omega\, e \quad . \tag{41.28}$$

Under a different frame field e' the connection matrix ω is transformed to ω', one which satisfies

$$De' = \omega'\, e' = \omega'\, ge \quad , \tag{41.29}$$

where the last equality follows from (41.27). Now we will use another generalization of the product rule of ordinary differentiation for the covariant derivative [one similar to that used for (41.15)] to obtain

$$De' = D(g\, e) = (dg)\, e + g\, De = (dg)\, e + (g\omega)\, e \quad , \tag{41.30}$$

where the last equality follows from (41.28), and the d's denote ordinary differentials, which, in this case, when the quantities being acted on are functions on the manifold (0-forms), are actually the same as exterior derivatives. Note that d acts on functions (0-forms) and D acts on vector fields. Comparing the two rightmost right-hand sides of (41.29) and (41.30) we have

$$\omega' g = dg + g\omega \quad , \tag{41.31}$$

which leads directly, upon multiplication by g^{-1} (the inverse of the matrix g) on the right on both sides of the equation, to the transformation rule for a connection matrix under a **gauge transformation** (local change of frames):

$$\boxed{\omega' = (dg)g^{-1} + g\omega g^{-1}} \quad . \tag{41.32}$$

This is one of the central results in this chapter, as well as in gauge field theory. As will be demonstrated later, the gauge fields corresponding to gauge bosons, which are the particles mediating the fundamental forces of nature, transform exactly according to this rule under a local gauge transformation of frames describing the internal degrees of freedom. Thus the affine connection (and the angular velocity that it gives rise to) is in fact formally a gauge field. Since the vector fields on which the covariant derivative acts transform according to the group $SO(3)$, we say that the **gauge group** of our affine connection is $SO(3)$. One can envision other gauge groups such as $SO(N)$, $SU(N)$ (the special unitary group of $N \times N$ complex matrices) etc., the latter applicable to gauge field theories where the internal degrees of freedom are described by vectors living in unitary spaces attached to different points in spacetime, or other base manifolds.

Equation (41.32) shows that the connection matrix ω and hence the angular velocity (with components φ_i^j) are not tensorial. If they were, the transformation rule would be $\omega' = g\omega g^{-1}$ instead (similarity transformation for matrices under a change of frames). The extra term $(dg)g^{-1}$ on the right-hand side of the above equation renders the connection matrix a non-tensorial object; this term does not vanish in general precisely because, for local gauge transformations, $g(x)$ is a function of the coordinates x^1, x^2 and x^3. Equation (41.32) is actually valid for connections on a manifold of arbitrary dimension.

The geometrical concepts of the affine connection, parallel displacement, and the covariant derivative are thus seen to be equivalent mathematically. The proper mathematical setting for these concepts is the **tangent bundle** $T(M)$ of a **Riemannian manifold** M, which, loosely speaking, is the collection of all tangent spaces at all points of the manifold. A **Riemannian manifold** is a differentiable manifold that comes equipped with a quadratic symmetric, nondegenerate **metric tensor** $g_{ij}(x)$, which gives a prescription on how to measure arclengths and angles between tangent vectors at different points $x \in M$ of the manifold. The Riemannian metric tensor [with respect to the coordinate (natural) frame field (see below)] is defined by

$$ds^2 = g_{ij}(x)\, dx^i dx^j \quad , \tag{41.33}$$

where ds^2 is the square of an infinitesimal arclength. There is a special relationship between a given Riemannian metric and a *unique* affine connection associated with that metric, the so-called **Levi-Civita connection**. To explain this fact, we need to introduce another important geometrical notion, that of the **coframe field** dual to a given frame field on a tangent bundle. This notion derives from the more basic linear algebraic notion of the dual space to a certain vector space, which, as we have seen in Chapter 7, underlies the Dirac bracket notation in quantum theory. Since the dual space to a vector space is simply the space of all linear functions on the vector space and is a vector space in its own right (recall Definition 7.1), we can pair an element in the dual space with an element in the original vector space to produce a number. Denoting the pairing by the Dirac notation $\langle\,|\,\rangle$, with the left slot reserved for an element in the dual space and the right slot for an element in the original vector space, we then define the coframe field $\{\boldsymbol{\omega}^i\}$ dual to the frame field $\boldsymbol{e}_i$ by (note the positions of the indices)

$$\langle\, \boldsymbol{\omega}^j \,|\, \boldsymbol{e}_i \,\rangle = \delta_i^j \quad . \tag{41.34}$$

In general, with local coordinates $(x^1, \dots, x^n)$ of a Riemannian manifold, we have the local (non-orthonormal) coordinate frame field (also called **natural frame field**) $\{\partial/\partial x^i\}$ and the dual coframe fields $\{dx^i\}$ satisfying

$$\left\langle dx^j \,\middle|\, \frac{\partial}{\partial x^i} \right\rangle = \delta^j_i \quad . \tag{41.35}$$

It is natural to express tangent vector fields in terms of partial differential operators since a tangent vector along a certain direction produces a directional derivative of functions $f(x)$. The general rule is

$$\left\langle df \,\middle|\, \frac{\partial}{\partial x^i} \right\rangle = \frac{\partial f}{\partial x^i} \quad . \tag{41.36}$$

Note that the coframe fields are in general 1-forms.

Problem 41.2 Show that for a 2-sphere of radius r with the usual Riemannian metric specified by

$$ds^2 = r^2(d\theta)^2 + r^2 \sin^2\theta (d\phi)^2 \,,$$

where θ and ϕ are the polar and azimuthal angles, respectively, the orthonormal frame field $\{\boldsymbol{e}_1, \boldsymbol{e}_2\}$ and the corresponding coframe field field satisfying (41.34) and expressed in terms of the local coordinates $x^1 = \theta$ and $x^2 = \phi$ are given by

$$\boldsymbol{e}_1 = \frac{1}{r}\frac{\partial}{\partial\theta} \,, \quad \boldsymbol{e}_2 = \frac{1}{r\sin\theta}\frac{\partial}{\partial\phi} \,; \qquad \boldsymbol{\omega}^1 = r\,d\theta \,, \quad \boldsymbol{\omega}^2 = r\sin\theta\, d\phi \,.$$

Among all the possible affine connections on a Riemannian manifold, there are two classes which are particularly important: the **torsion-free connections** and the **metric-compatible connections**. The analytical conditions of these are given as follows. A connection ω^j_i is said to be torsion-free if [compare with (41.3)]

$$d\boldsymbol{\omega}^i = \boldsymbol{\omega}^j \wedge \omega^i_j \qquad \text{(torsion-freeness)} \,. \tag{41.37}$$

A connection ω^j_i is said to be metric-compatible (with a metric g_{ij}) if

$$dg_{ij} = \omega^k_i g_{kj} + \omega^k_j g_{ik} \qquad \text{(metric-compatibility)} \,. \tag{41.38}$$

In the above two equations, the coframe field $\{\boldsymbol{\omega}^i\}$ is dual to an arbitrary frame field $\{\boldsymbol{e}_i\}$ (not necessary orthonormal), and the connection matrix ω^j_i and the metric tensor g_{ij} are with respect to $\{\boldsymbol{e}_i\}$. The geometrical (and more intuitive) meanings of these two equations are illustrated in Fig. 41.1. In Fig. 41.1(a), two infinitesimal tangent vectors at the same point in the Riemannian manifold are parallelly translated, each along the direction of the other (shown by double arrows). If the resulting figure is a closed parallellogram, then the connection

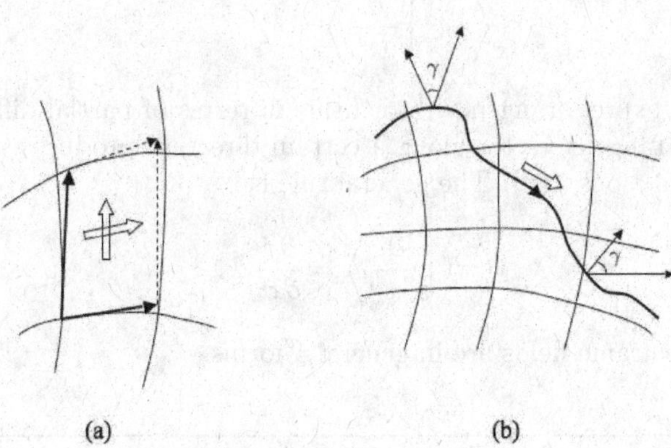

Fig. 41.1

(which determines the parallel translations) is torsion free. In Fig. 41.1(b), tangent vectors X and Y at the same point in the manifold are parallelly translated along a curve in the manifold. If the lengths of the vectors and the angles between the vectors (both are determined by the metric tensor) remain invariant under the parallel translations, then the connection is metric-compatible.

Problem 41.3 Show that the geometrical meanings for torsion-freeness and metric-compatibility of an affine connection as described in the above paragraph are in agreement with the analytical conditions for these properties as given by (41.37) and (41.38), respectively.

Recalling (41.21), the connection matrix elements can be written in terms of local coordinates $\{x^i\}$ as

$$\omega_i^j = \Gamma_{ik}^j(x^1, \dots, x^n)\, dx^k \quad . \tag{41.39}$$

The quantities Γ_{ik}^j are called **Christoffel symbols**, and are functions of the local coordinates x^i. (This notation is usually used in the physics literature on general relativity.) A fundamental theorem of Riemannian geometry is the following. *Given a Riemannian metric g_{ij}, there exists a unique affine connection ω, called the Levi-Civita connection, that is both torsion-free and metric* *compatible.* The Levi-Civita connection with respect to the natural frame field $\{\partial/\partial x^i\}$ is given in terms of the Riemannian metric by

$$\boxed{\Gamma^j_{ik} = \frac{1}{2} g^{jl} \left(\frac{\partial g_{il}}{\partial x^k} + \frac{\partial g_{kl}}{\partial x^i} - \frac{\partial g_{ik}}{\partial x^l} \right)} , \tag{41.40}$$

where g^{ij} is the inverse matrix of g_{ij}. (For a derivation of this basic formula in Riemannian geometry see, for example, Chern, Chen, and Lam 1999.) In terms of the Christoffel symbols the condition of torsion-freeness (41.37) is easily seen to be

$$\Gamma^j_{ik} = \Gamma^j_{ki} \qquad \text{(torsion freeness)} \quad . \tag{41.41}$$

With respect to the orthonormal frame field $\{e_i\}$ the Riemannian metric is by definition $g_{ij} = \delta_{ij}$, and the metric-compatibility condition (41.38) becomes simply $\omega^j_i + \omega^i_j = 0$. This is just the antisymmetry condition obtained previously as (41.18).

Problem 41.4 Prove the torsion-freeness condition (41.41) by using (41.39) and setting $\boldsymbol{\omega}^i = dx^i$ in (41.37), and the antisymmetry property of exterior products of differential one-forms.

Problem 41.5 For a 2-sphere of radius r with $x^1 = \theta$ and $x^2 = \phi$, we have the following expression for the metric tensor [cf. (41.33) and the first equation in Problem 41.2]:

$$g_{ij} = \begin{pmatrix} r^2 & 0 \\ 0 & r^2 \sin^2\theta \end{pmatrix} .$$

Use (41.39) and (41.40) to show that, with respect to the natural frame field $\{\partial/\partial\theta,\, \partial/\partial\phi\}$, the Levi-Civita connection is given by the following matrix of one-forms:

$$(\omega^j_i)_{natural} = \begin{pmatrix} 0 & \cot\theta d\phi \\ -\sin\theta\cos\theta d\phi & \cot\theta d\theta \end{pmatrix} .$$

By using the gauge transformation rule (41.32) and

$$g = \begin{pmatrix} \dfrac{1}{r} & 0 \\ 0 & \dfrac{1}{r\sin\theta} \end{pmatrix}$$

in that equation, show that with respect to the orthonormal frame field $\{e_1,\, e_2\}$ the *same* Levi-Civita connection is given by the following (different) matrix of one-forms:

$$(\omega^j_i)_{orthonormal} = \begin{pmatrix} 0 & \cos\theta d\phi \\ -\cos\theta d\phi & 0 \end{pmatrix} .$$

Problem 41.6 For a 2-sphere identify all the Christoffel symbols Γ^j_{ik} of the Levi-Civita connection using the result for $(\omega_i^j)_{natural}$ in the previous problem and check that they satisfy the torsion-freeness condition (41.41). Also check the metric-compatibi condition (41.38) for $(\omega_i^j)_{natural}$.

We finally come to the notion of curvature, which is defined in terms of the connection. The **curvature matrix** of 2-forms Ω is given by

$$\boxed{\Omega \equiv d\omega - \omega \wedge \omega} \quad , \tag{41.42}$$

where the wedge $\wedge$ indicates matrix multiplication as well as an exterior product. [In the matrix multiplication the subscript and the superscript in (ω_i^j) correspond to the row and column indices, respectively (see Problem 41.7 below).] Using (41.39) we can write the curvature matrix elements in terms of the Christoffel symbols as follows.

$$\begin{aligned}\Omega_i^j &= d\omega_i^j - \omega_i^k \wedge \omega_k^j \\ &= \frac{\partial \Gamma^j_{ik}}{\partial x^l}\, dx^l \wedge dx^k - \Gamma^h_{il}\Gamma^j_{hk}\, dx^l \wedge dx^k \\ &= \frac{1}{2}\left(\frac{\partial \Gamma^j_{il}}{\partial x^k} - \frac{\partial \Gamma^j_{ik}}{\partial x^l} + \Gamma^h_{il}\Gamma^j_{hk} - \Gamma^h_{ik}\Gamma^j_{hl} \right) dx^k \wedge dx^l \\ &\equiv \frac{1}{2}\, R^j_{ikl}\, dx^k \wedge dx^l \quad .\end{aligned} \tag{41.43}$$

The quantity within the parentheses on the right-hand side of the last equality in the above equation is known as the **Riemann curvature tensor**:

$$\boxed{R^j_{ikl} = \frac{\partial \Gamma^j_{il}}{\partial x^k} - \frac{\partial \Gamma^j_{ik}}{\partial x^l} + \Gamma^h_{il}\Gamma^j_{hk} - \Gamma^h_{ik}\Gamma^j_{hl}} \quad . \tag{41.44}$$

The many indices present in the curvature tensor shows the power and elegance of expressing quantities as differential forms, as done in (41.42). With respect to the $\{\mathbf{e}_1, \mathbf{e}_2\}$ frame, it is easily seen that the curvature matrix (of 2-forms) corresponding to $(\omega_i^j)_{orthonormal}$ (see Problem 41.5) is given by

$$(\Omega_i^j) = \begin{pmatrix} 0 & -\sin\theta d\theta \wedge d\phi \\ \sin\theta d\theta \wedge d\phi & 0 \end{pmatrix} \quad . \tag{41.45}$$

Unlike the connection matrix ω, which transforms according to (41.32) under a gauge transformation [recall (41.27)], the curvature matrix is a tensor. Under the gauge transformation (41.27), Ω transforms according to

$$\Omega' = g\Omega g^{-1} \quad . \tag{41.46}$$

This follows simply from exteriorly differentiating (41.31). The fact that the connection is not a tensor while the curvature is one has given rise to a lot of confusion about the physical importance of the gauge field versus that of the field tensor, for example, in electrodynamics. The modern viewpoint, and the more correct one, is that *in quantum mechanics and quantum field theory the gauge potential is of primary importance, while the field strength is a derived quantity.*

Problem 41.7 Show that, if in Eqs. (41.3) and (41.26), that is,

$$De_i = \omega_i^j \, e_j \quad \text{and} \quad e'_i = g_i^j(x) \, e_j \ ,$$

the superscript j and the subscript i [in the quantities (ω_i^j) and $g_i^j(x)$] are interpreted as the row and column indices, respectively, of matrix elements, then the gauge transformation rule (41.32) would appear as

$$\omega' = g^{-1} dg + g^{-1} \omega g \ ,$$

the definition for curvature (41.42) would appear as

$$\Omega \equiv d\omega + \omega \wedge \omega \ ,$$

and the tensorial transformation rule (41.48) for the curvature would appear as

$$\Omega' = g^{-1} \Omega g \ .$$

We finally consider the application of the geometrical concepts introduced in this chapter to gauge field theories. In this case we no longer deal with the Levi-Civita connection and the Riemann curvature tensor on Riemannian manifolds. Instead the more general concepts of connections and curvatures on vector bundles are more appropriate. *A* ***vector bundle*** *is, loosely speaking, a collection of (isomorphic) vector spaces associated with different points in the base manifold, with one vector space, called a* ***fiber****, associated with each point.* It is usually denoted by $\pi : E \to M$, where E is called the **total space** and M the **base manifold** of the bundle, and π is a projection map that projects a fiber onto the point in the base manifold with which the fiber is associated. An arbitrary connection (satisfying the general rules of covariant differentiation of vector fields given below) can always be imposed on a vector bundle. A tangent bundle is then a special example of a vector bundle, and an affine connection is then defined as a connection on a tangent bundle. The formal definitions of the connection [(41.3)] and the curvature [(41.42)] still apply. Basically, a connection on a vector bundle is a specific way to differentiate vector fields on M. A vector field on M is also known as a **section** of the vector bundle E. Since there is no canonical relationship between points on different fibers, one needs to impose a structure on E, beyond its differentiable structure, so that points on "neighboring" fibers can be "connected". The two general rules of covariant differentiation that any connection must satisfy are given as follows: Suppose $D\boldsymbol{s}$ is the covariant derivative of the vector field (section) $\boldsymbol{s}$, then,

(1) for any two vector fields $\boldsymbol{s}$ and $\boldsymbol{t}$ on M,

$$D(\boldsymbol{s}+\boldsymbol{t}) = D\boldsymbol{s} + D\boldsymbol{t} \; ; \tag{41.47}$$

(2) for any vector field $\boldsymbol{s}$ and any smooth function α on M,

$$D(\alpha\boldsymbol{s}) = (d\alpha)\boldsymbol{s} + \alpha D\boldsymbol{s} \; . \tag{41.48}$$

The above rules guarantee that the covariant derivative D is a linear map on the space of smooth vector fields on M (sections of E) to the space of smooth vector fields on $T^*(M)\otimes E$, where $T^*(M)$ is the **cotangent bundle** of M (the bundle of cotangent spaces of M, a **cotangent space** being the dual space of a tangent space). In symbols we write $D:\Gamma(E)\to\Gamma(T^*(M)\otimes E)$, where $\Gamma(E)$ is the space of smooth vector fields (sections) of E. If $\boldsymbol{X}$ is a smooth tangent field on M [a vector field on $T(M)$] and $\boldsymbol{s}$ is a vector field on E [$\boldsymbol{s}\in\Gamma(E)$], one can define the **directional covariant derivative** of $\boldsymbol{s}$ along $\boldsymbol{X}$, $D_{\boldsymbol{X}}\boldsymbol{s}$, as being the vector field (section) on E given by [cf. (41.34) to (41.36)]

$$D_{\boldsymbol{X}}\boldsymbol{s} \equiv \langle\, D\boldsymbol{s} \,|\, \boldsymbol{X}\,\rangle \in \Gamma(E) \; , \tag{41.49}$$

where the Dirac bracket denotes the pairing between a vector field in the tangent bundle $T(M)$ (right slot) and one in the cotangent bundle $T^*(M)$ (left slot). It is not difficult to see that the directional covariant derivative satisfies the following properties: If $\boldsymbol{X}$ and $\boldsymbol{Y}$ are any two smooth tangent fields on M, $\boldsymbol{s}$, $\boldsymbol{s}_1$ and $\boldsymbol{s}_2$ are arbitrary vector fields of E, and α is a smooth function on M, then

$$D_{\boldsymbol{X}+\boldsymbol{Y}}\boldsymbol{s} = D_{\boldsymbol{X}}\boldsymbol{s} + D_{\boldsymbol{Y}}\boldsymbol{s} \; , \tag{41.50a}$$

$$D_{\alpha\boldsymbol{X}} = \alpha D_{\boldsymbol{X}}\boldsymbol{s} \; , \tag{41.50b}$$

$$D_{\boldsymbol{X}}(\boldsymbol{s}_1+\boldsymbol{s}_2) = D_{\boldsymbol{X}}\boldsymbol{s}_1 + D_{\boldsymbol{X}}\boldsymbol{s}_2 \; , \tag{41.50c}$$

$$D_{\boldsymbol{X}}(\alpha\boldsymbol{s}) = \langle\, d\alpha \,|\, \boldsymbol{X}\,\rangle\boldsymbol{s} + \alpha D_{\boldsymbol{X}}\boldsymbol{s} \; , \tag{41.50d}$$

where the meaning of the Dirac bracket in the last equation is given by (41.36) and the linearity of the bracket.

Problem 41.8 Verify the above four properties [Eqs. (41.50)] of the directional covariant derivative.

In the context of the Yang-Mills gauge field theories of elementary particles one usually uses Greek indices $(\mu, \nu, \dots)$ to label spacetime local coordinates of the base manifold and Latin indices $(i, j, \dots)$ to label **fiber coordinates** (referring to internal degrees of freedom). Equations (41.39) and (41.43) are then rewritten as, respectively,

$$\omega_i^j = \mathcal{A}_{i\mu}^j \, dx^\mu = (\mathcal{A}_\mu)_i^j \, dx^\mu \quad , \tag{41.51}$$

and

$$\Omega = \frac{1}{2} \, \mathcal{F}_{\mu\nu} \, dx^\mu \wedge dx^\nu \quad . \tag{41.52}$$

Assuming the gauge group to be the Lie group $SU(N)$, $\mathcal{A}_\mu$ and $\mathcal{F}_{\mu\nu}$ (for particular values of μ and ν) are *Lie algebra-valued*, and thus may be represented by $N \times N$ anti-hermitian matrices in the so-called defining representation, whose matrix elements are all functions of the spacetime coordinates x^μ. Writing ∂_μ for $\partial/\partial x^\mu$ the matrix equation analogous to (41.44) is seen to be

$$\mathcal{F}_{\mu\nu} = \partial_\mu \mathcal{A}_\nu - \partial_\nu \mathcal{A}_\mu - [\, \mathcal{A}_\mu, \, \mathcal{A}_\nu \,] \quad , \tag{41.53}$$

where the commutator $[A, B] \equiv AB - BA$. The physical **gauge fields** (potentials) A_μ, required to be hermitian, are given by

$$\mathcal{A}_\mu \equiv -iA_\mu \quad . \tag{41.54}$$

Defining the physical field strength tensor $F_{\mu\nu}$ by

$$\mathcal{F}_{\mu\nu} \equiv -iF_{\mu\nu} \quad , \tag{41.55}$$

Eq. (41.53) implies the following well-known (matrix) equation relating field strengths and gauge potentials in non-abelian gauge field theories:

$$\boxed{F_{\mu\nu} = \partial_\mu A_\nu - \partial_\nu A_\mu + i\, [\, A_\mu, \, A_\nu \,]} \quad . \tag{41.56}$$

The two pairs of equations, (41.51) and (41.54) on the one hand and (41.53) and (41.55) on the other, show explicitly that *the physical gauge field arises from a connection on a vector bundle and the physical field strength arises from the curvature corresponding to the connection.*

Clearly, $F_{\mu\nu}$ is antisymmetric with respect to μ and ν. For the special case of Maxwell's electrodynamics, the gauge group is $U(1)$ and $A_\mu, F_{\mu\nu}$ (for a particular μ and ν) are just single functions of x^μ. The commutator in the above equation vanishes and we have the familiar relationship between the antisymmetric **electromagnetic field tensor** $F_{\mu\nu}$ (with tensorial indices μ and ν) and the **vector potentials** A_μ: $F_{\mu\nu} = \partial_\mu A_\nu - \partial_\nu A_\mu$. An element in the group $U(1)$ can be written as $g(x^\mu) = \exp(-i\lambda(x^\mu))$, where $\lambda(x^\mu)$ is an arbitrary smooth function of x^μ, and so $(dg)g^{-1} = -i\partial_\mu \lambda dx^\mu$. Setting $\omega = -iA_\mu dx^\mu$ the gauge transformation rule (41.32) then reduces to the following familiar rule for the local gauge transformation of vector potentials in electrodynamics:

$$A'_\mu(x^\nu) = A_\mu(x^\nu) + \partial_\mu \lambda(x^\nu) \quad . \tag{41.57}$$

We have already seen this as one of the equations in Eq. (34.3).

We will complete our discussion on gauge fields in non-relativistic quantum theory by considering how the covariant derivative acts on wave functions. A multi-component wave function $\boldsymbol{\psi}$ is essentially a vector field (section) on a vector bundle. We will write it as

$$\boldsymbol{\psi} = \psi^i \, \boldsymbol{e}_i \quad , \tag{41.58}$$

where, in analogy to (41.1), $\boldsymbol{e}_i$ stand for a certain choice of basis vectors of a fiber space. Rule (2) of covariant differentiation [(41.48)] and (41.3) then give

$$D\boldsymbol{\psi} = (d\psi^j + \psi^i \omega_i^j)\, \boldsymbol{e}_j = \left(\partial_\nu \psi^j dx^\nu - i\psi^i (A_\nu)_i^j dx^\nu \right) \boldsymbol{e}_j \quad . \tag{41.59}$$

It follows from (41.49) and (41.36) that the directional covariant derivative D_μ (along the μ-th direction in the spacetime base manifold) is given by

$$D_\mu \boldsymbol{\psi} \equiv \left\langle D\boldsymbol{\psi} \,\middle|\, \frac{\partial}{\partial x^\mu} \right\rangle = \left(\partial_\mu \psi^j - i(A_\mu)_i^j \psi^i \right) \boldsymbol{e}_j \quad . \tag{41.60}$$

The quantity within parentheses is often called a **minimal coupling** describing the interaction of the wave function $\boldsymbol{\psi}$ with the gauge potential A_μ. Note that in general the components of the vector potential $(A_\mu)_i^j$ describing the gauge field is an $m \times m$ hermitian matrix if the fiber (internal) space is an m-dimensional representation space of the **gauge group**. Recall that, as in the case of the tangent bundle, the gauge group of a general vector bundle is a certain group of transformations of the fiber space of the vector bundle. In the Standard Model of the interactions between particles and gauge fields, the gauge groups are often the special unitary groups $SU(N)$. If the gauge group is non-abelian, we say that the gauge field is non-abelian. Different representations of the gauge group may describe different multiplets of elementary particles. For example, in quantum chromodynamics, different representations of $SU(3)$ describe different flavors of quarks or colors of gluons. The above formalism constitutes the mathematical basis of the so-called **Yang-Mills** non-abelian **gauge field theories** (see Yang 1981, Yang and Mills 1954), with the exception of the second-quantization of the gauge fields and the wave functions of the particles required in (relativistic) quantum field theory. A specific abelian example of this in the context of the $U(1)$ electromagnetic coupling between the photon gauge field and the electron wave function has been discussed at the beginning of Chapter 34.

How does the covariant derivative of a Schrödinger wave function transform under a gauge transformation given by (41.27), in which $g(x) \in G$, where G is the gauge group of the vector bundle in which the wave function lives? First note that (41.32) and (41.27) imply, respectively, the following matrix equations:

$$A_\mu \to (A')_\mu = i(\partial_\mu g)g^{-1} + gA_\mu g^{-1} \quad , \tag{41.61}$$

and

$$\boldsymbol{\psi} \to \boldsymbol{\psi}' = \boldsymbol{\psi} g^{-1} \quad . \tag{41.62}$$

In the last equation ψ and ψ' stand for the row vectors of their components, so that the equation is equivalent to: $\psi^i \longrightarrow (\psi')^i = \psi^j (g^{-1})^i_j$. We then have

$$\begin{aligned} D_\mu \psi \to (D_\mu \psi)' &= \partial_\mu \psi' - i\psi'(A')_\mu \\ &= \partial_\mu(\psi g^{-1}) - i\psi g^{-1}\left(i(\partial_\mu g)g^{-1} + gA_\mu g^{-1}\right) \\ &= (\partial_\mu \psi)g^{-1} + \psi\partial_\mu(g^{-1}) + \psi g^{-1}(\partial_\mu g)g^{-1} - i\psi A_\mu g^{-1} \,. \end{aligned} \tag{41.63}$$

Since $gg^{-1} = 1$, it follows that $(\partial_\mu g)g^{-1} = -g\partial_\mu(g^{-1})$, which implies

$$\psi g^{-1}(\partial_\mu g)g^{-1} = -\psi g^{-1} \cdot g\partial_\mu(g^{-1}) = -\psi\partial_\mu(g^{-1}) \,.$$

Thus

$$(D_\mu \psi)' = (\partial_\mu \psi - i\psi A_\mu) \cdot g^{-1} = (D_\mu \psi) \cdot g^{-1} \,, \tag{41.64}$$

which shows that $D_\mu \psi$ transforms in the same way as ψ, that is, as the components of a vector field under a gauge transformation. Hence the directional covariant derivative of a vector field is also a vector field, as expected. This fact is responsible for the necessity to replace ∂_μ by D_μ (the directional covariant derivative) in order for the Lagrangian density in quantum field theory or the Schrödinger equation in non-relativistic quantum mechanics to be invariant under a gauge transformation, so as to satisfy the principle of gauge invariance. In the case of the Schrödinger equation we have seen examples for abelian cases manifested by the Aharonov-Bohm effect and the Maxwell magnetic monopole (cf. Chapter 38), and for a non-abelian case manifested by the Berry-Simon connection in molecular dynamics (cf. Chapter 39). The geometrical developments in this chapter demonstrate that the descriptions of various physical effects involving gauge potentials and the accompanying geometric phases can be very effectively unified by the mathematical language of connections on vector bundles. In fact, geometrical phases occur in much broader physical contexts than the quantum mechanical ones we have studied, and appear in many classical phenomena as well. For exhaustive surveys of these we refer the reader to Shapere and Wilczek 1989; Bohm, Mostafazadeh, Koizumi, Niu, and Zwanziger 2003; and Chruściński and Jamiołkowski 2004.

Problem 41.9 Consider an (abelian) electromagnetic gauge field $\boldsymbol{A}(\boldsymbol{x})$ for which the covariant derivative of a wave function $\psi(\boldsymbol{x})$ is given by

$$D_i\psi(\boldsymbol{x}) = \left(\partial_i - \frac{ie}{\hbar c}\, A_i(\boldsymbol{x})\right)\psi(\boldsymbol{x}) \,.$$

Show that **parallel translation** (displacement) of $\psi(\boldsymbol{x})$ along a curve $C(t) = \{x^i(t)\}$ (parametrized by t) is described by the differential equation

$$\frac{d\psi}{dt} - \frac{ie}{\hbar c}\, A_i\, \frac{dx^i}{dt}\, \psi = 0 \,,$$

where ψ and $\boldsymbol{A}$ depend implicitly on t through $\boldsymbol{x}(t)$. Show that the formal solution to this equation is

$$\psi(\boldsymbol{x}) = \psi(\boldsymbol{x}_0) \exp\left(\frac{ie}{\hbar c} \int_{\boldsymbol{x}_0\,(C)}^{\boldsymbol{x}} \boldsymbol{A} \cdot d\boldsymbol{x} \right) ,$$

where the line integral is along the path $C(t)$ from $\boldsymbol{x}_0$ to $\boldsymbol{x}$. Discuss the relationship of this solution to the Aharonov-Bohm result (38.5).

Chapter 42

Topological Quantum (Chern) Numbers: The Integer Quantum Hall Effect

A topological quantum number arises in quantum theory as a result of topological properties of the (spacetime) manifold on which the wave function of a quantum system is defined (see Thouless 1998). We have already seen an example in the quantization of the magnetic charge (cf. Chapter 38). Mathematically, it is known as a **Chern number**, which is obtained from integrals of geometric objects called **Chern classes** related to the curvature of a connection on vector bundles. In this chapter we will explain these mathematical concepts as far as they are applicable to quantum theoretical applications, and illustrate them with one of the most fascinating topics in condensed matter physics – the **integer quantum Hall effect**. As usual, we will begin with the physics first.

Before describing the quantum Hall effect, it is useful to recall the basic physics behind the classical Hall effect. We consider a number of electrons occupying an area A on a planar surface (the xy plane) and subjected to a longitudinal electric field $\boldsymbol{E}$ (along some direction on the surface) and a transverse magnetic field $\boldsymbol{B}$ (say, along the positive z-direction). For simplicity, the electrons are considered to be otherwise free. At equilibrium, the net force on each electron is zero, and so

$$\boldsymbol{E} + \frac{1}{c}\,\boldsymbol{v} \times \boldsymbol{B} = 0\,, \tag{42.1}$$

where c is the speed of light and $\boldsymbol{v}$ is the velocity of an electron. Introducing the current density $\boldsymbol{j}$ (measured in charge per unit time per unit length transverse to the flow of charges) given by

$$\boldsymbol{j} = -n|e|\boldsymbol{v} , \tag{42.2}$$

where n is the electron density on the plane (or the number of electrons per unit area) and $|e|$ is the absolute value of the electron charge, we have

$$\boldsymbol{E} = \frac{1}{nc|e|}\,\boldsymbol{j}\times\boldsymbol{B} = \frac{1}{nc|e|}\begin{vmatrix} \hat{\boldsymbol{x}} & \hat{\boldsymbol{y}} & \hat{\boldsymbol{z}} \\ j_x & j_y & 0 \\ 0 & 0 & B \end{vmatrix} = \frac{B}{nc|e|}\,(j_y\hat{\boldsymbol{x}} - j_x\hat{\boldsymbol{y}}) . \tag{42.3}$$

In matrix form, we have

$$\begin{pmatrix} E_x \\ E_y \end{pmatrix} = \frac{B}{nc|e|}\begin{pmatrix} 0 & 1 \\ -1 & 0 \end{pmatrix}\begin{pmatrix} j_x \\ j_y \end{pmatrix} . \tag{42.4}$$

On inversion one obtains

$$\begin{pmatrix} j_x \\ j_y \end{pmatrix} = \frac{nc|e|}{B}\begin{pmatrix} 0 & -1 \\ 1 & 0 \end{pmatrix}\begin{pmatrix} E_x \\ E_y \end{pmatrix} . \tag{42.5}$$

One can then identify the classical **Hall conductivity** matrix σ_H by:

$$\sigma_H = \frac{nc|e|}{B}\begin{pmatrix} 0 & -1 \\ 1 & 0 \end{pmatrix} . \tag{42.6}$$

In particular, if the electric field $\boldsymbol{E}$ is along the x-direction, then the Hall current $\boldsymbol{j}$ is along the y-direction, as shown in Fig. 42.1.

Initial contact with quantum theory can be made at this point by recalling that [cf. (38.17)] the fundamental magnetic flux quantum is $\phi_0 = hc/|e|$. We can then write

$$|\,\sigma_{xy}\,| = \frac{nc|e|}{B} = \frac{n}{B/\left(\dfrac{hc}{|e|}\right)}\,\frac{e^2}{h} = \nu\,\frac{e^2}{h} , \tag{42.7}$$

where ν, the so-called **filling factor** is defined by.

$$\nu \equiv \frac{n}{B/\left(\dfrac{hc}{|e|}\right)} = \frac{nA}{BA/\left(\dfrac{hc}{|e|}\right)} = \frac{\text{number of electrons}}{\text{number of flux quata}} . \tag{42.8}$$

From (42.5) we see that the **Hall resistivity** (which is also the **Hall resistance** for dimensional reasons) is given by

$$\rho_{xy} = \frac{B}{nc|e|} = \frac{1}{\nu}\left(\frac{h}{e^2}\right) . \tag{42.9}$$

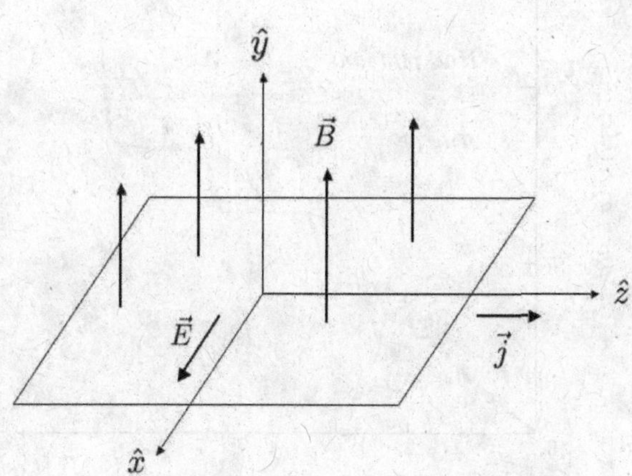

Fig. 42.1

Classically, then, one should expect that the Hall resistance should vary smoothly with $\boldsymbol{B}$. Experimental observations, however, for large magnetic fields ($B \to$ 10 teslas) and low temperatures ($T \to 0$), reveal that, for a particular temperature, there are regions of $|\boldsymbol{B}|$ over which the Hall resistance remains remarkably constant, corresponding to integer values for the filling factor ν (with an accuracy of 1 in 10^8). This discovery, made by von Klitzing et al., is known as the **integer quantum Hall effect** (see von Klitzing, Doda and Pepper 1980). The high accuracy of the experiments in fact allowed yet another highly accurate determination of the combination of fundamental constants:

$$\frac{e^2}{h} = \frac{1}{25,812.807572\ldots}\,\Omega^{-1} = 1\,(\text{Klitzing})^{-1}\,. \tag{42.10}$$

The regions of constant ρ are known as **Hall plateaus**. A schematic representation of the experimental results are shown in Fig. 42.2. (Subsequent to the discovery of the integer quantum Hall effect, it was actually discovered that for certain *fractional* values of ν, there are Hall plateaus also. This so-called **fractional quantum Hall effect**, however, involves very different physics from the integer effect, and will be discussed in the final chapter of this text.)

To approach the quantum theory we first consider a system of two-dimensional non-interacting electron gas in a uniform magnetic field $\boldsymbol{B}$. The time-independent Schrödinger equation for an electron in the system is [cf. (34.1)]

$$\frac{1}{2m}\left(\boldsymbol{p} + \frac{|e|}{c}\boldsymbol{A}\right)^2 \psi(x,y) = E\psi(x,y)\,, \tag{42.11}$$

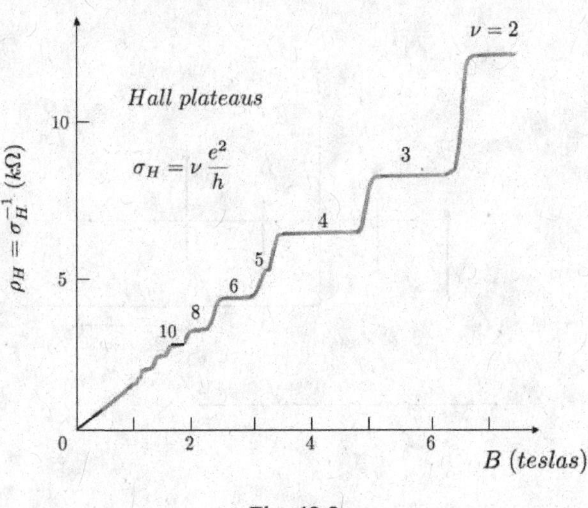

Fig. 42.2

where $\boldsymbol{A}$, the vector potential corresponding to $\boldsymbol{B}$, is determined only up to a gauge transformation. Suppose $\boldsymbol{B} = B\hat{\boldsymbol{z}}$, where B is a constant. We will use the **Landau gauge**, in which $A_x = 0$, $A_y = Bx$. (The reader should verify that this indeed gives $\nabla \times \boldsymbol{A} = B\hat{\boldsymbol{z}}$.) The Schrödinger equation then becomes

$$\frac{1}{2m}\left[\left(-i\hbar\frac{\partial}{\partial x}\right)^2 + \left(-i\hbar\frac{\partial}{\partial y} + \frac{|e|}{c}Bx\right)^2\right]\psi = E\psi\,. \tag{42.12}$$

This equation clearly possesses translational symmetry in the y-direction, so we can write

$$\psi(x,y) = e^{iqy}\,\varphi(x)\,. \tag{42.13}$$

Since

$$\left(-i\hbar\frac{\partial}{\partial x}\right)^2 (e^{iqy}\,\varphi(x)) = e^{iqy}\left(-i\hbar\frac{\partial}{\partial x}\right)^2\varphi(x)\,, \tag{42.14}$$

$$\left(-i\hbar\frac{\partial}{\partial y} + \frac{|e|}{c}Bx\right)^2 (e^{iqy}\,\varphi(x)) = \left(\hbar q + \frac{|e|}{c}Bx\right)^2 (e^{iqy}\,\varphi(x))\,, \tag{42.15}$$

the Schrödinger equation can be written as the following one-dimensional one:

$$\frac{1}{2m}\left[\left(-i\hbar\frac{\partial}{\partial x}\right)^2 + \left(\hbar q + \frac{|e|}{c}Bx\right)^2\right]\varphi(x) = E\varphi(x)\,. \tag{42.16}$$

This is none other than the Schrödinger equation for the one-dimensional simple harmonic oscillator with the minimum of the potential at $x = -\hbar qc/(|e|B)$ [cf.

(14.36) and Problem 14.1]. The displacement $q\hbar c/(|e|B)$ can be written as ql_B^2, where l_B, the **magnetic length** corresponding to a magnetic field B, is defined by

$$l_B^2 \equiv \frac{\hbar c}{|e|B}\,. \tag{42.17}$$

According to (14.57), the solution to the Schrödinger equation (42.16) is then given by

$$\varphi(x) = \varphi_n(x + l_B^2 q) \propto \exp\left(-\frac{(x + l_B^2 q)^2}{2l_B^2}\right) H_n(x + l_B^2 q)\,, \tag{42.18}$$

where φ_n is the n-th eigenfunction of the simple harmonic oscillator and H_n is the n-th Hermite polynomial [cf. (14.54)]. Furthermore the energy eigenvalues, the so-called **Landau levels**, are given by

$$E_n = \left(n + \frac{1}{2}\right)\hbar\omega_c\,, \quad n = 0, 1, 2, \dots, \tag{42.19}$$

where

$$\omega_c \equiv \frac{|e|B}{mc} \tag{42.20}$$

is the so-called **cyclotron frequency**. Each Landau level is highly degenerate. Suppose the area of the two-dimensioanl electron system is given by $A = L_x L_y$. The degeneracy can be obtained by imposing the periodic boundary condition

$$\psi(x, y) = \psi(x, y + L_y)\,. \tag{42.21}$$

From (42.13) this requires that $\exp(iqL_y) = 1$, and so the possible values of q are quantized:

$$q = l\left(\frac{2\pi}{L_y}\right), \quad l = 0, \pm 1, \pm 2, \dots\,. \tag{42.22}$$

Hence, by (42.18), the wave functions φ_n are centered at

$$x = -l\left(\frac{2\pi}{L_y}\right)\left(\frac{\hbar c}{|e|B}\right), \quad l = 0, \pm 1, \pm 2, \dots\,. \tag{42.23}$$

Thus the allowed number of l values within the length L_x of the specimen (or the degeneracy of each Landau level) is given by

$$\begin{aligned} \text{degeneracy} &= \frac{L_x}{\left(\frac{2\pi}{L_y}\right)\left(\frac{\hbar c}{|e|B}\right)} = (L_x L_y)\frac{|e|B}{hc} = \frac{BA}{(hc/|e|)} \\ &= \#\text{ of flux quanta in area } A\,. \end{aligned} \tag{42.24}$$

Problem 42.1 Verify Eq. (42.15).

Problem 42.2 Fill in the details in the estimation of the Landau energy levels and the degeneracy of each level by the following elementary arguments. For an electron moving in a magnetic field the Lamor radius r of its circular orbit is given by

$$\frac{mv^2}{r} = \frac{|e|vB}{c} .$$

It follows that the cyclotron frequency ω_c can be estimated to be

$$\omega_c \approx \frac{v}{r} = \frac{|e|B}{mc} .$$

Setting the angular momentum mvr equal to $\hbar$ by the Bohr-Sommerfeld quantization rule, one gets

$$B(\pi r^2) = \frac{\pi \hbar c}{|e|} \approx \frac{hc}{|e|} ,$$

which is the fundamental unit of magnetic flux. The area πr^2 of each Lamor circle is then approximately given by $hc/(|e|B)$. Since this circle can be anywhere in the area A of the two-dimensional electron gas system, translational invariance implies that the degeneracy is roughly given by the number of Lamor circles that can be fitted into the area A, that is,

$$\text{degeneracy} \approx \frac{A}{\pi r^2} \approx \frac{BA}{(hc)/|e|} ,$$

in agreement with (42.24).

In addition to the uniform magnetic field along the z-direction we will now introduce a uniform longitudinal electric field $\boldsymbol{E} = (0, E_y, 0)$, with E_y being constant. Recalling that

$$\boldsymbol{E} = -\frac{1}{c}\frac{\partial \boldsymbol{A}}{\partial t} - \nabla\phi , \tag{42.25}$$

where ϕ is the scalar electrostatic potential, we can set (in the Landau gauge again),

$$\boldsymbol{A} = (0, Bx - cE_y t, 0) . \tag{42.26}$$

The Schrödinger equation with both the electric and magnetic fields then reads

$$\frac{1}{2m}\left[\left(-i\hbar\frac{\partial}{\partial x}\right)^2 + \left(-i\hbar\frac{\partial}{\partial y} + \frac{|e|}{c}(Bx - cE_y t)\right)^2\right]\psi = E\psi . \tag{42.27}$$

Note that in this equation the time t is treated as a parameter, and the time-variation of the Hamiltonian is considered as slow enough so that the adiabatic condition applies [see discussion following (40.27)]. Translational invariance along the y-direction is still retained, and (42.13) is still valid. Analogous to (42.16) we obtain

$$\frac{1}{2m}\left[\left(-i\hbar\frac{\partial}{\partial x}\right)^2 + \left(\frac{|e|B}{c}\right)^2\left\{\frac{q\hbar c}{|e|B} + x - \frac{cE_y t}{B}\right\}^2\right]\varphi(x) = E\varphi(x) . \tag{42.28}$$

This equation is of the same form as (42.16) and so the energy levels are the same Landau levels given by (42.19). But now the eigenfunctions are centered at $x + \dfrac{q\hbar c}{|e|B} - \dfrac{cE_y t}{B}$. We can interpret this physically to mean that the electrons are moving with a velocity along the x-direction, given by

$$\boldsymbol{v} = \frac{cE_y}{B}\,\hat{\boldsymbol{x}}\,. \tag{42.29}$$

Thus the current density is

$$\boldsymbol{j} = -n|e|\boldsymbol{v} = -\frac{n|e|cE_y}{B}\,\hat{\boldsymbol{x}}\,. \tag{42.30}$$

This is in agreement with (42.5).

At this point a very qualitative and crude explanation of the integer quantum Hall effect can be given as follows. In the presence of impurities the discrete Landau levels are broadened into Landau bands, with non-localized states having energies near the band centers and localized states having energies away from the band centers. With constant magnetic field B one can picture the energy levels filling up as the electron number density n increases, according to the Pauli exclusion principle. Equivalently, according to (42.8), as the filling fraction ν increases, the Fermi level climbs up the ladder of the Landau bands. When the Fermi level goes through a band center, non-localized states prevail and the Hall conductivity σ_H goes through a steep jump. As ν continues to increase, the Fermi level traverses (in the energy scale) a region of localized states until it reaches the next band center, and the conductivity remains relatively constant. This leads to the step-like behavior of σ_H (recall Fig. 42.2). The remarkable fact is that this state of affairs is independent of the specific nature of the impurities.

More interesting details emerge when a two-dimensional periodic potential $U(x,y)$ is added to the Hamiltonian, so that the Schrödinger equation becomes

$$\left[\frac{1}{2m}\left(\boldsymbol{p} + \frac{|e|}{c}\boldsymbol{A}\right)^2 + U(x,y)\right]\psi = E\psi\,. \tag{42.31}$$

In our presentation we will follow the very instructive treatment by Kohmoto (see Kohmoto 1985, reprinted in Thouless 1998). Suppose the periods of this potential along the x and y axes are a and b, respectively, so that

$$U(x+a,y) = U(x,y+b) = U(x,y)\,. \tag{42.32}$$

It will be useful to introduce the **Bravais lattice vectors**

$$\boldsymbol{R} \equiv n\boldsymbol{a} + m\boldsymbol{b}\,, \quad n, m \text{ integers}\,, \tag{42.33}$$

and the translational operators (for each $\boldsymbol{R}$) [recall (17.53)]

$$T_{\boldsymbol{R}} = \exp\left(\frac{i}{\hbar}\boldsymbol{R}\cdot\boldsymbol{p}\right)\,, \quad \boldsymbol{p} = -i\hbar(\partial/\partial x,\, \partial/\partial y)\,, \tag{42.34}$$

which satisfy

$$T_{\boldsymbol{R}} f(\boldsymbol{r}) = f(\boldsymbol{r} + \boldsymbol{R}) \,, \tag{42.35}$$

where $f(\boldsymbol{r})$ is a smooth function on the xy-plane. The Hamiltonian H in (42.31), however, is not invariant under $T_{\boldsymbol{R}}$, since $\boldsymbol{A}$ is not. Thus

$$[\, T_{\boldsymbol{R}},\, H\,] \neq 0 \,, \quad \text{or} \quad (T_{\boldsymbol{R}})^{\dagger} H T_{\boldsymbol{R}} \neq H \,. \tag{42.36}$$

With this in mind we introduce the following **magnetic translation operators:**

$$\hat{T}_{\boldsymbol{R}} \equiv \exp\left\{ \frac{i}{\hbar} \boldsymbol{R} \cdot \left(\boldsymbol{p} + \frac{|e|}{2c} (\boldsymbol{r} \times \boldsymbol{B}) \right) \right\} = \exp\left\{ \frac{i}{\hbar} \boldsymbol{R} \cdot \boldsymbol{p} + \frac{i|e|}{\hbar c} \frac{\boldsymbol{r} \cdot (\boldsymbol{B} \times \boldsymbol{R})}{2} \right\} , \tag{42.37}$$

where we have used the **symmetric** (isotropic) **gauge** for the vector potential $\boldsymbol{A}$ given by

$$\boldsymbol{A} = \frac{1}{2} (\boldsymbol{B} \times \boldsymbol{r}) = \frac{1}{2} (-By\, \hat{\boldsymbol{x}} + Bx\, \hat{\boldsymbol{y}}) \,. \tag{42.38}$$

(Again, as for the Landau gauge, the reader should verify that the symmetric gauge also leads to $\nabla \times \boldsymbol{A} = B\hat{\boldsymbol{z}}$.) Unlike the ordinary translation operators $T_{\boldsymbol{R}}$, the magnetic translation operators $\hat{T}_{\boldsymbol{R}}$ do commute with the Hamiltonian in (42.31), that is,

$$[\, \hat{T}_{\boldsymbol{R}},\, H\,] = 0 \,, \quad H = \frac{1}{2m} \left[\left(-i\hbar \partial_i + \frac{|e|}{c} A_i \right)^2 + U(x, y) \right] , \tag{42.39}$$

where, in the symmetric gauge [(42.38)], the components of the vector potential can be written as

$$A_i = -\frac{1}{2} B \epsilon_{ij} x^j \,, \quad i, j = 1, 2 \,, \tag{42.40}$$

with ϵ_{ij} being the antisymmetric tensor (in two dimensions). Analogous to (42.35) one can also show that

$$\hat{T}_{\boldsymbol{R}}\, \psi(\boldsymbol{r}) = \exp\left(\frac{i|e|B}{2\hbar c} R^i \epsilon_{ij} x^j \right) \psi(\boldsymbol{r} + \boldsymbol{R}) \tag{42.41}$$

for any wave function $\psi(\boldsymbol{r})$ defined on the xy-plane. Note that the exponential factor in the above equation is precisely the geometric phase factor in the Aharonov-Bohm result given by (38.5).

Problem 42.3 Show that the Landau and symmetric gauges are related by an electrodynamic gauge transformation of the vector potential of the form given by the first equation in Eqs. (34.3), that is,

$$(A_{symmetric})_i = (A_{Landau})_i + \partial_i \lambda(x, y) \,, \quad i = x, y \,.$$

Give an explicit expression for $\lambda(x, y)$.

Problem 42.4 Prove (42.39), that is, that the magnetic translation operators commute with the Hamiltonian. It will be convenient to write

$$\hat{T}_{\boldsymbol{R}} = e^{\frac{i}{\hbar}\boldsymbol{\kappa}\cdot\boldsymbol{R}} ,$$

where

$$\kappa_i \equiv -i\hbar\partial_i + \frac{|e|B}{2c}\epsilon_{ij}x^j .$$

To prove (42.39), it is then sufficient to prove that

$$[P_i, \kappa_j] = 0 ,$$

where

$$P_i \equiv -i\hbar\partial_i - \frac{|e|B}{2c}\epsilon_{ij}x^j .$$

Problem 42.5 Verify (42.41). (Hint: Recall that $[p_i, \epsilon_{ij}x^j] = 0$.)

Even though the magnetic translation operators $\hat{T}_{\boldsymbol{R}}$ all commute with the Hamiltonian, they do not in general commute with each other. We will demonstrate this by calculating the actions of $\hat{T}_{\boldsymbol{a}}\hat{T}_{\boldsymbol{b}}$ and $\hat{T}_{\boldsymbol{b}}\hat{T}_{\boldsymbol{a}}$ on an arbitrary wave function $\psi(x, y)$, where $\boldsymbol{a}$ and $\boldsymbol{b}$ are directed along the x and y axes, respectively [recall (42.33)]. Using the expressions for $\kappa_{1,2}$ in Problem 42.4 and the result (42.41), we have

$$\begin{aligned} \hat{T}_{\boldsymbol{a}}\hat{T}_{\boldsymbol{b}}\,\psi(x,y) &= e^{\frac{i}{\hbar}\kappa_1 a}e^{\frac{i}{\hbar}\kappa_2 b}\,\psi(x,y) = e^{\frac{i}{\hbar}\kappa_1 a}\left(e^{-\frac{i}{\hbar}\frac{|e|B}{2c}bx}\,\psi(x,y+b)\right) \\ &= e^{\frac{i}{\hbar}\frac{|e|B}{2c}ay}\,e^{-\frac{i}{\hbar}\frac{|e|B}{2c}b(x+a)}\,\psi(x+a,y+b) = e^{-\frac{i|e|Bab}{2\hbar c}}\,\Psi(x,y) , \end{aligned} \tag{42.42}$$

where

$$\Psi(x,y) \equiv e^{\frac{i|e|B}{2\hbar c}(ay-bx)}\,\psi(x+a,y+b) . \tag{42.43}$$

Similarly it can be shown that

$$\hat{T}_{\boldsymbol{b}}\hat{T}_{\boldsymbol{a}}\,\psi(x,y) = e^{\frac{i|e|Bab}{2\hbar c}}\,\Psi(x,y) . \tag{42.44}$$

Comparing (42.42) and (42.44) we see that

$$\hat{T}_{\boldsymbol{a}}\hat{T}_{\boldsymbol{b}} = \exp\left(-2\pi i\,\frac{Bab}{(hc)/|e|}\right)\hat{T}_{\boldsymbol{b}}\hat{T}_{\boldsymbol{a}} . \tag{42.45}$$

In general, for $\boldsymbol{a}$ and $\boldsymbol{b}$ running along arbitrary directions, we have

$$\hat{T}_{\boldsymbol{a}}\hat{T}_{\boldsymbol{b}} = e^{-2\pi i N_\phi}\,\hat{T}_{\boldsymbol{b}}\hat{T}_{\boldsymbol{a}} , \tag{42.46}$$

where

$$N_\phi \equiv \frac{B(\boldsymbol{a}\times\boldsymbol{b})\cdot\hat{\boldsymbol{z}}}{(hc)/|e|} = \#\text{ of flux quanta in the unit cell with area }(\boldsymbol{a}\times\boldsymbol{b})\cdot\hat{\boldsymbol{z}}\,. \tag{42.47}$$

If $N_\phi = p/q$, where p and q are integers that are relatively prime, we can define a **magnetic unit cell** spanned by $q\boldsymbol{a}$ and $\boldsymbol{b}$ such that the magnetic translation operators $\hat{T}_{\boldsymbol{R}'}$, where $\boldsymbol{R}' = m(q\boldsymbol{a}) + n\boldsymbol{b}$, commute with each other. We then have a complete set of mutually commuting observables: $\{\hat{T}_{\boldsymbol{R}'}\,(\boldsymbol{R}' = m(q\boldsymbol{a}+n\boldsymbol{b})), H\}$, with a corresponding set of complete simultaneous eigenfunctions $\psi_{\boldsymbol{k}}(\boldsymbol{r})$ [cf. Chapter 15], known as the **generalized Bloch functions**, which satisfy:

$$\hat{T}_{\boldsymbol{R}'}\,\psi_{\boldsymbol{k}}(\boldsymbol{r}) = e^{i\boldsymbol{k}\cdot\boldsymbol{R}'}\,\psi_{\boldsymbol{k}}(\boldsymbol{r}) = \exp\left(\frac{i|e|B}{2\hbar c}\,(R')^i\epsilon_{ij}x^j\right)\psi_{\boldsymbol{k}}(\boldsymbol{r}+\boldsymbol{R}')\,, \tag{42.48}$$

$$H\psi_{\boldsymbol{k}}(\boldsymbol{r}) = \epsilon_{\boldsymbol{k}}\,\psi_{\boldsymbol{k}}(\boldsymbol{r})\,, \tag{42.49}$$

where the last equality in (42.48) follows from (42.41). It is useful to rewrite N_ϕ [given by (42.47)] as (when $\boldsymbol{a}$ and $\boldsymbol{b}$ are orthogonal)

$$N_\phi = \frac{ab}{\left(\dfrac{hc}{|e|}\right)/B} = \frac{\text{area of lattice unit cell}}{\text{area occupied by one flex quantum for a given } B}\,. \tag{42.50}$$

Without going into details we mention the following qualitative features of the energy spectrum (for a more complete discussion see, for example, Chapter 13 of Bohm, Mostafazadeh, Koizumi, Niu and Zwanziger 2003). When the two area measures noted in the above equation are commensurate ($N_\phi = p/q$, p and q relatively prime), each magnetic unit cell has a whole number p of flux quanta going through it. In this case each parent Landau band is split up into p subbands (in the limit of strong magnetic field and weak periodic modulation). If, however, the two area measures are incommensurate, an intricate fractal structure of the energy spectrum, known as the **Hofstadter's butterfly**, will emerge (see Hofstadter 1976).

To simplify our calculations in what follows we will again assume that $\boldsymbol{a} = a\hat{\boldsymbol{x}}$, $\boldsymbol{b} = b\hat{\boldsymbol{y}}$. The generalized Bloch functions then satisfy the following generalized Bloch conditions:

$$\hat{T}_{qa}\,\psi_{\boldsymbol{k}}(\boldsymbol{r}) = e^{ik_1qa}\,\psi_{\boldsymbol{k}}(x,y) = \exp\left(i\frac{|e|B}{2\hbar c}\,qay\right)\psi_{\boldsymbol{k}}(x+qa,y)\,, \tag{42.51a}$$

$$\hat{T}_{\boldsymbol{b}}\,\psi_{\boldsymbol{k}}(\boldsymbol{r}) = e^{ik_2b}\,\psi_{\boldsymbol{k}}(x,y) = \exp\left(-i\frac{|e|B}{2\hbar c}\,bx\right)\psi_{\boldsymbol{k}}(x,y+b)\,, \tag{42.51b}$$

with

$$0 \le k_1 \le \frac{2\pi}{qa}\,,\quad 0 \le k_2 \le \frac{2\pi}{b}\,. \tag{42.52}$$

The region in $\boldsymbol{k}$-space given by the above inequalities is called the first **magnetic Brillouin zone**. With the inclusion of a band index α we now define a gauge transformation of the generalized Bloch functions given by [cf. the second equation of (34.3)]:

$$\psi_{\boldsymbol{k}}^{(\alpha)}(\boldsymbol{r}) \longrightarrow u_{\boldsymbol{k}}^{(\alpha)}(\boldsymbol{r}) \equiv e^{-i\boldsymbol{k}\cdot\boldsymbol{r}}\psi_{\boldsymbol{k}}^{(\alpha)}(\boldsymbol{r})\,. \tag{42.53}$$

Using the condition that $N_\phi = p/q$, one can show quite straightforwardly that the generalized Bloch conditions (42.51) for $\psi_{\boldsymbol{k}}^{(\alpha)}(\boldsymbol{r})$ are equivalent to the following conditions for $u_{\boldsymbol{k}}^{(\alpha)}(\boldsymbol{r})$:

$$u_{\boldsymbol{k}}^{(\alpha)}(x+qa,y) = e^{-\frac{i\pi py}{b}}\,u_{\boldsymbol{k}}^{(\alpha)}(x,y)\,, \tag{42.54a}$$

$$u_{\boldsymbol{k}}^{(\alpha)}(x,y+b) = e^{\frac{i\pi px}{qa}}\,u_{\boldsymbol{k}}^{(\alpha)}(x,y)\,. \tag{42.54b}$$

Problem 42.6 Verify the Bloch conditions for the wave functions $u_{\boldsymbol{k}}^{(\alpha)}(x,y)$ given by (42.54).

According to the first equation of (34.3) [on setting $-i\boldsymbol{k}\cdot\boldsymbol{r} = -i\frac{|e|}{\hbar c}\lambda(\boldsymbol{r})$], the gauge transformation (42.53) on the wave function $\psi_{\boldsymbol{k}}^{(\alpha)}(\boldsymbol{r})$ entails the following gauge transformation on the vector potential:

$$\boldsymbol{A}(\boldsymbol{r}) \longrightarrow \boldsymbol{A}'(\boldsymbol{r}) = \boldsymbol{A}(\boldsymbol{r}) + \frac{\hbar c}{|e|}\nabla(\boldsymbol{k}\cdot\boldsymbol{r}) = \boldsymbol{A}(\boldsymbol{r}) + \frac{\hbar c}{|e|}\boldsymbol{k}\,. \tag{42.55}$$

Hence the Schrödinger equation for $u_{\boldsymbol{k}}^{(\alpha)}(\boldsymbol{r})$ is the following gauge-transformed equation from (42.31):

$$\left\{\frac{1}{2m}\left(\boldsymbol{p}+\hbar\boldsymbol{k}+\frac{|e|}{c}\boldsymbol{A}\right)^2 + U(\boldsymbol{r})\right\} u_{\boldsymbol{k}}^{(\alpha)}(\boldsymbol{r}) = E^{(\alpha)}\,u_{\boldsymbol{k}}^{(\alpha)}(\boldsymbol{r})\,. \tag{42.56}$$

Using the Bloch conditions (42.54), it can be checked directly that on traversing a rectangular magnetic unit cell completely once in $\boldsymbol{r}$-space in the counter-clockwise direction:

$$(x,y) \longrightarrow (x+qa,y) \longrightarrow (x+qa,y+b) \longrightarrow (x,y+b) \longrightarrow (x,y)\,, \tag{42.57}$$

the wave functions $u_{\boldsymbol{k}}^{(\alpha)}(\boldsymbol{r})$ gain a geometric phase in each of the respective steps as follows:

$$\begin{aligned} &u(x,y) \longrightarrow e^{-\frac{i\pi py}{b}}\,u(x,y) \longrightarrow e^{\frac{i\pi p(x+qa)}{qa}}\,e^{-\frac{i\pi py}{b}}\,u(x,y) \\ &e^{2i\pi p}\,e^{\frac{i\pi px}{qa}}\,u(x,y) \longrightarrow e^{2i\pi p}\,u(x,y)\,. \end{aligned} \tag{42.58}$$

Suppose, in addition to the constant magnetic field, a *slowly varying* electric field $\boldsymbol{E}(t)$ is also present, where "slow" in the present context means that the part of the vector potential giving rise to the electric field is given by $\delta \boldsymbol{A}(t) = -c\boldsymbol{E}(t)t$ [cf. (42.25) and (42.26)], and that the following adiabatic condition is satisfied [cf. (40.2)]:

$$H(t)\, u_{\boldsymbol{k}}^{(\alpha)}(t) = \epsilon_{\boldsymbol{k}}^{(\alpha)}(t)\, u_{\boldsymbol{k}}^{(\alpha)}(t)\,, \tag{42.59}$$

with the explicitly time-dependent Hamiltonian given by

$$H(t) = \frac{1}{2m}\left(\boldsymbol{p} + \hbar\boldsymbol{k} + \frac{|e|}{c}\{\boldsymbol{A} - c\boldsymbol{E}(t)t\}\right)^2 + U(\boldsymbol{r})\,, \tag{42.60}$$

in which $\boldsymbol{A}$ is the vector potential giving rise to the magnetic field only. Introducing the eigenkets $|\,\boldsymbol{k},\alpha;\,t\,\rangle$ by

$$u_{\boldsymbol{k}}^{(\alpha)}(t) = \langle\,\boldsymbol{r}\,|\,\boldsymbol{k},\alpha;\,t\,\rangle\,, \tag{42.61}$$

Eq. (42.59) can be rewritten as

$$H(t)\,|\,\boldsymbol{k},\alpha;\,t\,\rangle = \epsilon_{\boldsymbol{k}}^{(\alpha)}(t)\,|\,\boldsymbol{k},\alpha;\,t\,\rangle\,. \tag{42.62}$$

We will assume the orthonormality condition

$$\langle\,\boldsymbol{k},\beta;\,t\,|\,\boldsymbol{k},\alpha;\,t\,\rangle = \delta_{\alpha\beta}\,, \tag{42.63}$$

and the completeness condition

$$\sum_{\alpha} |\,\boldsymbol{k},\alpha;\,t\,\rangle\langle\,\boldsymbol{k},\alpha;\,t\,| = 1\,. \tag{42.64}$$

Our objective is to solve the time-dependent Schrödinger equation

$$H(t)\,|\,\Psi\,\rangle = i\hbar\,\frac{\partial|\,\Psi\,\rangle}{\partial t}\,, \tag{42.65}$$

by the methods of time-dependent perturbation theory introduced in Chapter 33. Recalling (33.5) we write

$$|\,\Psi_{\boldsymbol{k}}(t)\,\rangle = \sum_{\alpha} c_\alpha(t)\,\exp\left(-\frac{i}{\hbar}\int_0^t \epsilon_{\boldsymbol{k}}^{(\alpha)}(t')dt'\right)|\,\boldsymbol{k},\alpha;\,t\,\rangle\,. \tag{42.66}$$

Substituting this expression directly into (42.65) and using (42.62), we obtain

$$\sum_{\alpha}\frac{\partial c_\alpha}{\partial t}\,\exp\left(-\frac{i}{\hbar}\int_0^t \epsilon_{\boldsymbol{k}}^{(\alpha)}(t')dt'\right)|\,\boldsymbol{k},\alpha;t\,\rangle$$
$$= -\sum_{\alpha} c_\alpha(t)\,\exp\left(-\frac{i}{\hbar}\int_0^t \epsilon_{\boldsymbol{k}}^{(\alpha)}(t')dt'\right)\frac{\partial}{\partial t}|\,\boldsymbol{k},\alpha;\,t\,\rangle\,. \tag{42.67}$$

Taking the inner product with $\langle\, \boldsymbol{k}, \beta;\, t\,|$ from the left and using the orthonormality condition (42.63) we obtain [analogous to (33.22)] the following system of differential equations for the c_α's:

$$\frac{\partial c_\beta}{\partial t} = -\sum_\alpha c_\alpha(t)\, \exp\left(-\frac{i}{\hbar}\int_0^t \{\epsilon_{\boldsymbol{k}}^{(\alpha)}(t') - \epsilon_{\boldsymbol{k}}^{(\beta)}(t')\}dt'\right) \langle\, \boldsymbol{k}, \beta;\, t\,|\frac{\partial}{\partial t}|\, \boldsymbol{k}, \alpha;\, t\,\rangle\,. \tag{42.68}$$

Assume that the initial condition is $|\, \Psi_{\boldsymbol{k}}(0)\,\rangle = |\, \boldsymbol{k}, \alpha;\, 0\,\rangle$. To first order in the perturbation we can then put, on the right hand side of the above equation,

$$c_\alpha(t) \approx c_\alpha(0) \approx 1\,; \quad c_\beta(t) \approx 0 \quad \text{for } \beta \neq \alpha\,. \tag{42.69}$$

Thus

$$\frac{\partial c_\beta}{\partial t} \approx -\exp\left(-\frac{i}{\hbar}\int_0^t \{\epsilon_{\boldsymbol{k}}^{(\alpha)}(t') - \epsilon_{\boldsymbol{k}}^{(\beta)}(t')\}dt'\right) \langle\, \boldsymbol{k}, \beta;\, t\,|\frac{\partial}{\partial t}|\, \boldsymbol{k}, \alpha;\, t\,\rangle \;(\beta \neq \alpha)\,. \tag{42.70}$$

If, as is assumed to be the case, the electric field $\boldsymbol{E}(t)$ is slowly time-varying, then the matrix element involving the partial derivative in the above equation is also slowly time-varying, and we have

$$c_{\beta\neq\alpha}(t) \approx -\left\{\int_0^t dt'\, \exp\left(-\frac{i}{\hbar}\int_0^{t'} [\epsilon_{\boldsymbol{k}}^{(\alpha)}(t'') - \epsilon_{\boldsymbol{k}}^{(\beta)}(t'')]\, dt''\right)\right\} \langle\, \boldsymbol{k}, \beta;\, t\,|\frac{\partial}{\partial t}|\, \boldsymbol{k}, \alpha;\, t\,\rangle\,, \tag{42.71}$$

which yields

$$c_{\beta\neq\alpha}(t) \approx (i\hbar)\, \frac{\exp\left(-\frac{i}{\hbar}\int_0^t \{\epsilon_{\boldsymbol{k}}^{(\alpha)}(t') - \epsilon_{\boldsymbol{k}}^{(\beta)}(t')\}dt'\right)}{\epsilon_{\boldsymbol{k}}^{(\beta)}(t) - \epsilon_{\boldsymbol{k}}^{(\alpha)}(t)} \langle\, \boldsymbol{k}, \beta;\, t\,|\frac{\partial}{\partial t}|\, \boldsymbol{k}, \alpha;\, t\,\rangle\,. \tag{42.72}$$

Finally, to first order in the perturbation,

$$\begin{aligned} |\, \Psi_{\boldsymbol{k}}^{(\alpha)}(t)\,\rangle \approx{}& \exp\left(-\frac{i}{\hbar}\int_0^t \epsilon_{\boldsymbol{k}}^{(\alpha)}(t')dt'\right) \\ &\times \left\{ |\, \boldsymbol{k}, \alpha;\, t\,\rangle + i\hbar \sum_{\beta\neq\alpha} \frac{|\, \boldsymbol{k}, \beta;\, t\,\rangle\langle\, \boldsymbol{k}, \beta;\, t\,|\, \frac{\partial}{\partial t}\,|\, \boldsymbol{k}, \alpha;\, t\,\rangle}{\epsilon_{\boldsymbol{k}}^{(\beta)}(t) - \epsilon_{\boldsymbol{k}}^{(\alpha)}(t)} \right\}. \end{aligned} \tag{42.73}$$

Next we rewrite the Hamiltonian (42.60) as

$$H(t) = \frac{1}{2m}\left(\boldsymbol{p} + \hbar\boldsymbol{k}(t) + \frac{|e|}{c}\boldsymbol{A}\right)^2 + U(\boldsymbol{r})\,, \tag{42.74}$$

where

$$\boldsymbol{k}(t) \equiv \boldsymbol{k} - \frac{|e|}{\hbar}\boldsymbol{E}(t)t\,, \tag{42.75}$$

and make the further approximation that $\boldsymbol{E}(t) \approx \boldsymbol{E}$, which is constant in time. Then

$$\frac{\partial}{\partial t} = \frac{\partial}{\partial k^i}\frac{\partial k^i}{\partial t} \approx -\frac{|e|E^i}{\hbar}\frac{\partial}{\partial k^i}\,, \tag{42.76}$$

and

$$|\,\Psi_{\boldsymbol{k}}^{(\alpha)}(t)\,\rangle \approx \exp\left(-\frac{i}{\hbar}\int_0^t \epsilon_{\boldsymbol{k}}^{(\alpha)}(t')dt'\right) \times \left\{ |\,\boldsymbol{k},\alpha;\,t\,\rangle - i|e|E^i \sum_{\beta\neq\alpha} \frac{|\,\boldsymbol{k},\beta;\,t\,\rangle\langle\,\boldsymbol{k},\beta;\,t\,|\,\frac{\partial}{\partial k^i}\,|\,\boldsymbol{k},\alpha;\,t\,\rangle}{\epsilon_{\boldsymbol{k}}^{(\beta)}(t)-\epsilon_{\boldsymbol{k}}^{(\alpha)}(t)} \right\}. \tag{42.77}$$

To calculate the Hall conductivity we set [recall (42.5)]

$$j_x = \sigma_{xy}\,E_y\,, \tag{42.78}$$

with [recall (42.2)]

$$j_x = -n|e|\,\langle\,v_x\,\rangle\,, \tag{42.79}$$

where

$$\langle\,v_x\,\rangle = \langle\,\Psi_{\boldsymbol{k}}^{(\alpha)}(t)\,|\,v_x\,|\,\Psi_{\boldsymbol{k}}^{(\alpha)}(t)\,\rangle\,. \tag{42.80}$$

Using (42.77) for $|\,\Psi_{\boldsymbol{k}}^{(\alpha)}(t)\,\rangle$, we have, after some straightforward steps,

$$\langle\,v_x\,\rangle \approx \langle\,\boldsymbol{k},\alpha;\,t\,|\,v_x\,|\,\boldsymbol{k},\alpha;t\,\rangle + i|e|E_y \sum_{\beta\neq\alpha} \frac{\left\langle \frac{\partial u_{\boldsymbol{k}}^{(\alpha)}}{\partial k_y} \middle|\, u_{\boldsymbol{k}}^{(\beta)} \right\rangle \langle\,\boldsymbol{k},\beta;\,t\,|\,v_x\,|\,\boldsymbol{k},\alpha;\,t\,\rangle}{\epsilon_{\boldsymbol{k}}^{(\beta)}(t)-\epsilon_{\boldsymbol{k}}^{(\alpha)}(t)} - i|e|E_y \sum_{\beta\neq\alpha} \frac{\left\langle u_{\boldsymbol{k}}^{(\beta)} \middle|\, \frac{\partial u_{\boldsymbol{k}}^{(\alpha)}}{\partial k_y} \right\rangle \langle\,\boldsymbol{k},\alpha;\,t\,|\,v_x\,|\,\boldsymbol{k},\beta;\,t\,\rangle}{\epsilon_{\boldsymbol{k}}^{(\beta)}(t)-\epsilon_{\boldsymbol{k}}^{(\alpha)}(t)} + \dots\,, \tag{42.81}$$

where

$$\left\langle u_{\boldsymbol{k}}^{(\alpha)} \middle|\, \frac{\partial u_{\boldsymbol{k}}^{(\beta)}}{\partial k_y} \right\rangle \equiv \langle\,\boldsymbol{k},\alpha;\,t\,|\,\frac{\partial}{\partial k_y}\,|\,\boldsymbol{k},\beta;\,t\,\rangle\,. \tag{42.82}$$

We will show that this quantity is given by the following useful result:

$$\left\langle u_{\boldsymbol{k}}^{(\alpha)} \middle|\, \frac{\partial u_{\boldsymbol{k}}^{(\beta)}}{\partial k_y} \right\rangle = \frac{\left\langle \boldsymbol{k},\alpha;\,t \,\middle|\, \frac{\partial H}{\partial k_y} \,\middle|\, \boldsymbol{k},\beta; \right\rangle}{\epsilon_{\boldsymbol{k}}^{(\beta)}(t)-\epsilon_{\boldsymbol{k}}^{(\alpha)}(t)}\,. \tag{42.83}$$

It is clear that

$$\frac{\partial}{\partial k_y}\langle\,\boldsymbol{k},\alpha;\,t\,|\,H(t)\,|\,\boldsymbol{k},\beta;\,t\,\rangle = \langle\,\frac{\partial u_{\boldsymbol{k}}^{(\alpha)}(t)}{\partial k_y}\,|\,H(t)\,|\,\boldsymbol{k},\beta;\,t\,\rangle + \langle\,\boldsymbol{k},\alpha;\,t\,|\,\frac{\partial H}{\partial k_y}\,|\,\boldsymbol{k},\beta;t\,\rangle + \langle\,\boldsymbol{k},\alpha;\,t\,|\,H(t)\,|\,\frac{\partial u_{\boldsymbol{k}}^{(\beta)}(t)}{\partial k_y}\,\rangle\,. \tag{42.84}$$

The left-hand side vanishes for $\alpha \neq \beta$ since, by (42.62) and (42.63),

$$\langle \boldsymbol{k}, \alpha;\, t \,|\, H(t) \,|\, \boldsymbol{k}, \beta;\, t \rangle = \epsilon_{\boldsymbol{k}}^{(\beta)}(t) \langle \boldsymbol{k}, \alpha;\, t \,|\, \boldsymbol{k}, \beta;\, t \rangle = 0 \,. \tag{42.85}$$

Using this fact and (42.62) on the right-hand side, and rearranging terms, one obtains (42.83). To calculate the matrix elements $\langle \boldsymbol{k}, \alpha; t \,|\, v_x \,|\, \boldsymbol{k}, \beta; t \rangle = \langle u_{\boldsymbol{k}}^{(\alpha)} \,|\, v_x \,|\, u_{\boldsymbol{k}}^{(\beta)} \rangle$ appearing in (42.81), we exploit the slow time-variation of $H(t)$ and use (the classical) Hamilton's equations of motion to write

$$\frac{\partial H}{\partial k_i} = \frac{\partial H}{\partial \pi_i}\frac{\partial \pi_i}{\partial k_i} = \hbar\, \frac{\partial H}{\partial \pi_i} = \hbar v^i \,, \tag{42.86}$$

where

$$\boldsymbol{\pi} \equiv \boldsymbol{p} + \hbar \boldsymbol{k} + \frac{|e|}{c}\{\boldsymbol{A} - c\boldsymbol{E}(t)t\} \tag{42.87}$$

is the canonical momentum in the Hamiltonian given by (42.60). Thus we have

$$\begin{aligned} &\langle u_{\boldsymbol{k}}^{(\alpha)} \,|\, v^i \,|\, u_{\boldsymbol{k}}^{(\beta)} \rangle = \frac{1}{\hbar} \left\langle u_{\boldsymbol{k}}^{(\alpha)} \,\middle|\, \frac{\partial H}{\partial k_i} \,\middle|\, u_{\boldsymbol{k}}^{(\beta)} \right\rangle \\ &= \frac{1}{\hbar}\left(\epsilon_{\boldsymbol{k}}^{(\beta)}(t) - \epsilon_{\boldsymbol{k}}^{(\alpha)}(t)\right) \left\langle u_{\boldsymbol{k}}^{(\alpha)} \,\middle|\, \frac{\partial u_{\boldsymbol{k}}^{(\beta)}}{\partial k_i} \right\rangle = -\frac{1}{\hbar}\left(\epsilon_{\boldsymbol{k}}^{(\beta)}(t) - \epsilon_{\boldsymbol{k}}^{(\alpha)}(t)\right) \left\langle \frac{\partial u_{\boldsymbol{k}}^{(\alpha)}}{\partial k_i} \,\middle|\, u_{\boldsymbol{k}}^{(\beta)} \right\rangle \,, \\ &(\text{for } \alpha \neq \beta) \,, \end{aligned} \tag{42.88}$$

where the second equality follows from (42.83). Substituting this result in (42.81) we obtain

$$\begin{aligned} &\langle v_x \rangle = \langle u_{\boldsymbol{k}}^{(\alpha)} \,|\, v_x \,|\, u_{\boldsymbol{k}}^{(\alpha)} \rangle - \frac{i|e|E_y}{\hbar} \quad \times \\ &\sum_{\beta \neq \alpha} \left(\left\langle \frac{\partial u_{\boldsymbol{k}}^{(\alpha)}}{\partial k_y} \,\middle|\, u_{\boldsymbol{k}}^{(\beta)} \right\rangle \left\langle u_{\boldsymbol{k}}^{(\beta)} \,\middle|\, \frac{\partial u_{\boldsymbol{k}}^{(\alpha)}}{\partial k_x} \right\rangle - \left\langle \frac{\partial u_{\boldsymbol{k}}^{(\alpha)}}{\partial k_x} \,\middle|\, u_{\boldsymbol{k}}^{(\beta)} \right\rangle \left\langle u_{\boldsymbol{k}}^{(\beta)} \,\middle|\, \frac{\partial u_{\boldsymbol{k}}^{(\alpha)}}{\partial k_y} \right\rangle \right) . \end{aligned} \tag{42.89}$$

This expression can then be further simplified by using the completeness relation (42.64) in the form

$$\sum_{\beta \neq \alpha} |\, u_{\boldsymbol{k}}^{(\alpha)} \rangle \langle \, u_{\boldsymbol{k}}^{(\beta)} \,| = 1 - |\, u_{\boldsymbol{k}}^{(\alpha)} \rangle \langle \, u_{\boldsymbol{k}}^{(\alpha)} \,| \,. \tag{42.90}$$

The second term on the right-hand side does not contribute in (42.89) since

$$\begin{aligned} &\left\langle \frac{\partial u_{\boldsymbol{k}}^{(\alpha)}}{\partial k_y} \,\middle|\, u_{\boldsymbol{k}}^{(\alpha)} \right\rangle \left\langle u_{\boldsymbol{k}}^{(\alpha)} \,\middle|\, \frac{\partial u_{\boldsymbol{k}}^{(\alpha)}}{\partial k_x} \right\rangle - \left\langle \frac{\partial u_{\boldsymbol{k}}^{(\alpha)}}{\partial k_x} \,\middle|\, u_{\boldsymbol{k}}^{(\alpha)} \right\rangle \left\langle u_{\boldsymbol{k}}^{(\alpha)} \,\middle|\, \frac{\partial u_{\boldsymbol{k}}^{(\alpha)}}{\partial k_y} \right\rangle \\ &= \left(- \left\langle u_{\boldsymbol{k}}^{(\alpha)} \,\middle|\, \frac{\partial u_{\boldsymbol{k}}^{(\alpha)}}{\partial k_y} \right\rangle \right) \left(- \left\langle \frac{\partial u_{\boldsymbol{k}}^{(\alpha)}}{\partial k_x} \,\middle|\, u_{\boldsymbol{k}}^{(\alpha)} \right\rangle \right) - \left\langle \frac{\partial u_{\boldsymbol{k}}^{(\alpha)}}{\partial k_x} \,\middle|\, u_{\boldsymbol{k}}^{(\alpha)} \right\rangle \left\langle u_{\boldsymbol{k}}^{(\alpha)} \,\middle|\, \frac{\partial u_{\boldsymbol{k}}^{(\alpha)}}{\partial k_y} \right\rangle \\ &= 0 \,. \end{aligned} \tag{42.91}$$

Also, by the first equality of (42.88),

$$\begin{aligned}\langle u_{\boldsymbol{k}}^{(\alpha)} \,|\, v_x \,|\, u_{\boldsymbol{k}}^{(\alpha)} \rangle &= \frac{1}{\hbar} \left\langle u_{\boldsymbol{k}}^{(\alpha)} \left| \frac{\partial H}{\partial k_x} \right| u_{\boldsymbol{k}}^{(\alpha)} \right\rangle \\ &= \frac{1}{\hbar} \left(\frac{\partial}{\partial k_x} \langle u_{\boldsymbol{k}}^{(\alpha)} \,|\, H \,|\, u_{\boldsymbol{k}}^{(\alpha)} \rangle - \langle \frac{\partial u_{\boldsymbol{k}}^{(\alpha)}}{\partial k_x} \,|\, H \,|\, u_{\boldsymbol{k}}^{(\alpha)} \rangle - \langle u_{\boldsymbol{k}}^{(\alpha)} \,|\, H \,|\, \frac{\partial u_{(\boldsymbol{k}}^{(\alpha)}}{\partial k_x} \rangle \right) \\ &= \frac{1}{\hbar} \left(\frac{\partial \epsilon_{\boldsymbol{k}}^{(\alpha)}}{\partial k_x} - \epsilon_{\boldsymbol{k}}^{(\alpha)} \left\langle \frac{\partial u_{\boldsymbol{k}}^{(\alpha)}}{\partial k_x} \right| \left. u_{\boldsymbol{k}}^{(\alpha)} \right\rangle - \epsilon_{\boldsymbol{k}}^{(\alpha)} \left\langle u_{\boldsymbol{k}}^{(\alpha)} \left| \frac{\partial u_{\boldsymbol{k}}^{(\alpha)}}{\partial k_x} \right. \right\rangle \right) \\ &= \frac{1}{\hbar} \frac{\partial \epsilon_{\boldsymbol{k}}^{(\alpha)}}{\partial k_x} \,. \end{aligned} \tag{42.92}$$

Equation (42.89) then yields

$$\langle v_x \rangle = \frac{1}{\hbar} \frac{\partial \epsilon_{\boldsymbol{k}}^{(\alpha)}}{\partial k_x} - \frac{i|e|E_y}{\hbar} \left(\left\langle \frac{\partial u_{\boldsymbol{k}}^{(\alpha)}}{\partial k_y} \right| \left. \frac{\partial u_{\boldsymbol{k}}^{(\alpha)}}{\partial k_x} \right\rangle - \left\langle \frac{\partial u_{\boldsymbol{k}}^{(\alpha)}}{\partial k_x} \right| \left. \frac{\partial u_{\boldsymbol{k}}^{(\alpha)}}{\partial k_y} \right\rangle \right) . \tag{42.93}$$

At this point we observe that in (42.79), the quantity n (electrons per unit area) is to be replaced by the following integral operator, integrated over values of $\boldsymbol{k}$ in a magnetic Brillouin zone:

$$n \longrightarrow \frac{1}{(2\pi)^2} \int_0^{\frac{2\pi}{qa}} dk_x \int_0^{\frac{2\pi}{b}} dk_y \,, \tag{42.94}$$

and the expression for $\langle v_x \rangle$ as given by (42.93) is ultimately to be integrated over $\boldsymbol{k}$. But the first term does not contribute since

$$\int_0^{\frac{2\pi}{b}} dk_y \int_0^{\frac{2\pi}{qa}} dk_x \frac{\partial \epsilon_{\boldsymbol{k}}^{(\alpha)}}{\partial k_x} = \int_0^{\frac{2\pi}{b}} dk_y \left[\epsilon_{\boldsymbol{k}}^{(\alpha)} \left(k_x = \frac{2\pi}{qa} \right) - \epsilon_{\boldsymbol{k}}^{(\alpha)} (k_x = 0) \right] = 0 \,, \tag{42.95}$$

as the integrand vanishes because of periodicity. It finally follows from (42.78) and (42.79) that the contribution to the Hall conductivity from the α-th filled band is given by

$$\boxed{\sigma_{xy}^{(\alpha)} = \frac{e^2}{h} \frac{i}{2\pi} \int dk_x \int dk_y \left(\left\langle \frac{\partial u_{\boldsymbol{k}}^{(\alpha)}}{\partial k_y} \right| \left. \frac{\partial u_{\boldsymbol{k}}^{(\alpha)}}{\partial k_x} \right\rangle - \left\langle \frac{\partial u_{\boldsymbol{k}}^{(\alpha)}}{\partial k_x} \right| \left. \frac{\partial u_{\boldsymbol{k}}^{(\alpha)}}{\partial k_y} \right\rangle \right)} \,, \tag{42.96}$$

where the region of integration is a magnetic Brillouin zone.

The remarkable fact is that $\sigma_{xy}^{(\alpha)}/(e^2/h)$ is a topological invariant, in fact always an integer, referred to as a **Chern number** or **topological quantum number** [cf. the discussion of the quantization of the magnetic charge in

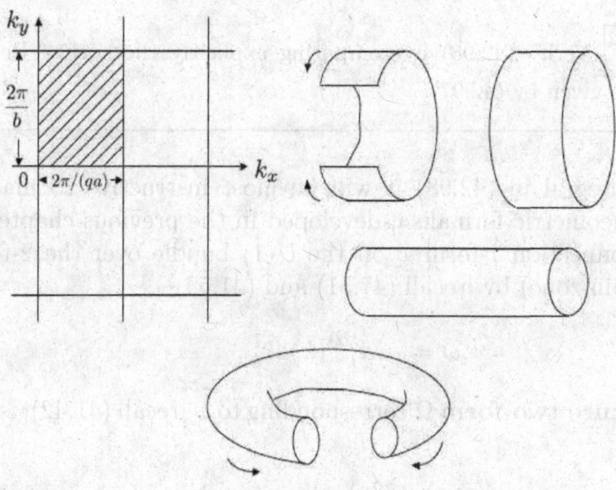

Fig. 42.3

Chapter 38]! To see this we define a Berry-Simon gauge potential (connection) $\hat{\boldsymbol{A}}^{(\alpha)}(\boldsymbol{k})$ in $\boldsymbol{k}$-space by [cf. (39.26)]

$$\hat{\boldsymbol{A}}^{(\alpha)}(\boldsymbol{k}) \equiv i\langle\, u_{\boldsymbol{k}}^{(\alpha)} \,|\, \nabla_{\boldsymbol{k}} \,|\, u_{\boldsymbol{k}}^{(\alpha)} \rangle = i \int d^2 r\, (u_{\boldsymbol{k}}^{(\alpha)})^* \nabla_{\boldsymbol{k}} u_{(\alpha)} \,. \tag{42.97}$$

Then it is straightforward to show that

$$\sigma_{xy}^{(\alpha)} = -\frac{e^2}{h}\frac{1}{2\pi} \int d^2 k\, [\nabla_{\boldsymbol{k}} \times \hat{\boldsymbol{A}}^{(\alpha)}(\boldsymbol{k})]_z \,. \tag{42.98}$$

Note that this integral has the form of a "magnetic" flux through a surface area in $\boldsymbol{k}$-space [compare with (38.6)]. If we divide the infinite 2-dimensional $\boldsymbol{k}$-space under consideration into periodic lattices of magnetic Brillouin zones [as given by (42.52), for example], the opposite sides of a zone can be identified, and each zone is then topologically equivalent (homeomorphic) to a 2-torus (see Fig. 42.3), which, of course, is closed and without boundary. The above integral over an individual Brillouin zone is then [as that in (38.6) for the actual magnetic flux through a spherical surface in coordinate space enclosing a magnetic monopole], also a flux integral through a closed surface, this time a toroidal surface in momentum space. The point is that formally it resembles a magnetic flux through a closed surface, and so may be expected to depend crucially on the topological properties of the closed surface, regardless of the specific details of the gauge potential leading to the "magnetic" field, as we have demonstrated for the case of an actual magnetic monopole in Chapter 38.

Problem 42.7 Verify (42.98) by computing explicitly the curl of Berry-Simon gauge potential as given by (42.97).

To do the integral in (42.98) it will be most instructive to make use of the differential geometric formalism developed in the previous chapter. Let us introduce the connection 1-form ω on the $U(1)$ bundle over the 2-torus (the magnetic Brillouin zone) by [recall (41.51) and (41.54)]

$$\omega = -i\hat{A}_i^{(\alpha)}(\boldsymbol{k})dk^i \,. \tag{42.99}$$

Then the **curvature two-form** Ω corresponding to ω [recall (41.42)] is obtained as follows:

$$\begin{aligned}\Omega = d\omega - \omega\wedge\omega = d\omega &= -i\,\frac{\partial \hat{A}_j^{(\alpha)}}{\partial k^i}\,dk^i\wedge dk^j \\ &= -i\left(\frac{\partial \hat{A}_j^{(\alpha)}}{\partial k^i} - \frac{\partial \hat{A}_i^{(\alpha)}}{\partial k^j}\right) dk^i\wedge dk^j\,, \quad (i<j\,;\, i,j=1,2) \\ &= -i\left(\frac{\partial \hat{A}_2^{(\alpha)}}{\partial k^1} - \frac{\partial \hat{A}_1^{(\alpha)}}{\partial k^2}\right) dk^1\wedge dk^2\,.\end{aligned} \tag{42.100}$$

The Hall conductivity given by (42.98) can then be expressed geometrically, and most succinctly, as

$$\sigma_{xy} = -\frac{e^2}{h}\left(\frac{i}{2\pi}\int\Omega\right) = -\frac{e^2}{h}\int C_1(\Omega) = -\frac{e^2}{h}\,c_1\,, \tag{42.101}$$

where

$$C_1(\Omega) = \frac{i}{2\pi}\,\Omega \qquad \text{and} \qquad c_1 = \int C_1(\Omega) \tag{42.102}$$

are called the **first Chern class** of the curvature Ω and the **first Chern number**, respectively. The first Chern number occurring in the expression for the Hall conductivity is another example, together with the magnetic charge discussed in Chapter 38, of topological quantum numbers. *It is a known mathematical fact that the Chern numbers are always integers if the base manifold is closed and without boundary.* (This fact will be discussed more fully in the next chapter.) Equation (42.101) thus explains the integer quantum Hall effect.

To conclude this chapter we will show explicitly using elementary methods that c_1 for the present example, with the connection one-form given by (42.97) and the domain of integration in $\boldsymbol{k}$ space (the base manifold of our $U(1)$ bundle) topologically a 2-torus, has to be an integer. The procedure will be similar to that used in the derivation of the magnetic charge leading to (38.15). Suppose we (arbitrarily) divide the domain of integration – the magnetic Brillouin zone

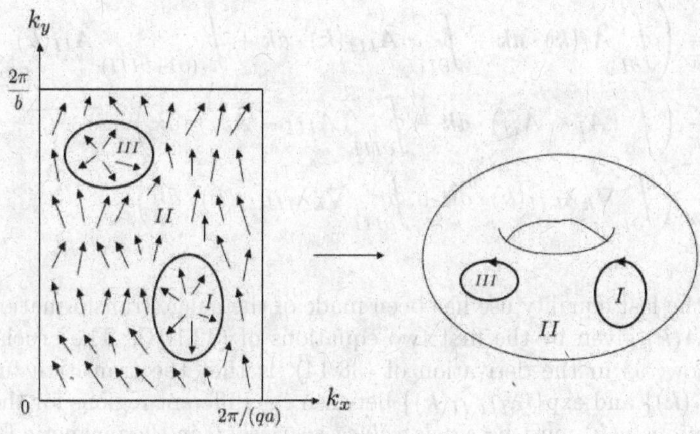

Fig. 42.4

(with opposite sides identified) – into three regions I, II and III, with region II overlapping both regions I and III, which are both entirely in the interior of the torus (see Fig. 42.4). Take the overlap regions to be the boundaries of regions I and III, and introduce gauge transformations for the gauge potential $\hat{\boldsymbol{A}}$ and the Bloch wave functions $u(\boldsymbol{k})$ in the overlap regions as follows:

$$\begin{aligned}\hat{\boldsymbol{A}}_I(\boldsymbol{k}) = \hat{\boldsymbol{A}}_{II}(\boldsymbol{k}) + \nabla\lambda_{I,II}(\boldsymbol{k}) , \quad \hat{\boldsymbol{A}}_{III}(\boldsymbol{k}) = \hat{\boldsymbol{A}}_{II}(\boldsymbol{k} + \nabla\lambda_{III,II}(\boldsymbol{k}) ;\\ u_I(\boldsymbol{k}) = \exp\{i\lambda_{I,II}(\boldsymbol{k})\}\, u_{II}(\boldsymbol{k}) , \quad u_{III}(\boldsymbol{k}) = \exp\{i\lambda_{III,II}(\boldsymbol{k})\}\, u_{II}(\boldsymbol{k}) ,\end{aligned} \tag{42.103}$$

where in the above equations we have dispensed with the band index α in order not to clutter the notation, and emphasized the $\boldsymbol{k}$-dependence of both the gauge potential and the wave functions. Then

$$c_1 = \frac{i}{2\pi}(-i)\left(\int_I + \int_{II} + \int_{III}\right) \nabla_{\boldsymbol{k}} \times \hat{\boldsymbol{A}}(\boldsymbol{k})\, d^2k \, . \tag{42.104}$$

By the Stokes Theorem of vector calculus it follows that

$$c_1 = \frac{1}{2\pi}\left(\oint_{\partial I} \hat{\boldsymbol{A}}_I(\boldsymbol{k}) \cdot d\boldsymbol{k} + \oint_{\partial III} \hat{\boldsymbol{A}}_{III}(\boldsymbol{k}) \cdot d\boldsymbol{k} + \oint_{\partial II} \hat{\boldsymbol{A}}_{II}(\boldsymbol{k}) \cdot d\boldsymbol{k}\right) , \tag{42.105}$$

where $\partial I, \partial II$ and ∂III are the boundaries of the regions I, II and III, respectively. Since the entire region of integration, the magnetic Brillouin zone M (topologically a 2-torus), is without boundary, we see (Fig. 42.2) that

$\partial II = -\partial I - \partial III$. Thus

$$\begin{aligned} c_1 &= \frac{1}{2\pi}\left(\oint_{\partial I} \hat{\boldsymbol{A}}_I(\boldsymbol{k})\cdot d\boldsymbol{k} + \oint_{\partial III} \hat{\boldsymbol{A}}_{III}(\boldsymbol{k})\cdot d\boldsymbol{k} + \oint_{-(\partial I + \partial III)} \hat{\boldsymbol{A}}_{II}(\boldsymbol{k})\cdot d\boldsymbol{k}\right) \\ &= \frac{1}{2\pi}\left(\oint_{\partial I} (\hat{\boldsymbol{A}}_I - \hat{\boldsymbol{A}}_{II})\cdot d\boldsymbol{k} + \oint_{\partial III} (\hat{\boldsymbol{A}}_{III} - \hat{\boldsymbol{A}}_{II})\cdot d\boldsymbol{k}\right) \\ &= \frac{1}{2\pi}\left(\oint_{\partial I} \nabla_{\boldsymbol{k}}\lambda_{I,II}(\boldsymbol{k})\cdot d\boldsymbol{k} + \oint_{\partial III} \nabla_{\boldsymbol{k}}\lambda_{III,II}(\boldsymbol{k})\cdot d\boldsymbol{k}\right), \end{aligned} \tag{42.106}$$

where in the last equality use has been made of the gauge transformation properties of $\hat{\boldsymbol{A}}(\boldsymbol{k})$ given by the first two equations of (42.103). The crucial point to note now, as in the derivation of (38.14), is that the transition functions $\exp\{i\lambda_{I,II}(\boldsymbol{k})\}$ and $\exp\{i\lambda_{III,II}(\boldsymbol{k})\}$ between the different regions for the Bloch wave functions $u(\boldsymbol{k})$ must be *single-valued everywhere* in the magnetic Brillouin zone M. So the total change of both $\lambda_{I,II}(\boldsymbol{k})$ and $\lambda_{III,II}(\boldsymbol{k})$ over closed loops in M must be integer multiples of 2π. We finally have

$$c_1 = \frac{1}{2\pi}(2n\pi + 2m\pi) = n + m\,, \tag{42.107}$$

where n and m are both integers. It can be readily seen that this result does not depend on the specific partition of M into different regions.

Chapter 43

de Rham Cohomology and Chern Classes: Some More Differential Geometry

In this chapter we will consolidate the geometrical development presented in Chapter 41, and lay the mathematical groundwork for the next one, where we will examine the use of the so-called Chern-Simons form in the description of the fractional quantum Hall effect. The Chern-Simons class is a secondary characteristic class derived from the Chern characteristic classes, which, as we will see, are (global) topological constructs derived from the curvature of a fiber bundle. We have already seen an example, the first Chern class, in the last chapter [recall (42.101) and (42.102) for the Hall conductivity], and actually in Chapter 38 also, where in the example of the magnetic monopole, the first Chern class is proportional to the integrand for the magnetic flux in (38.11). As in Chapter 41, we will here attempt to give an overview of these central and useful mathematical notions in differential geometry, providing just enough details for their roles in physical applications to be appreciated, without an exhaustive discussion of underlying mathematical issues. (For these we again refer the reader to Nakahara 2003; Lam 2003; or Chern, Chen and Lam 1999.) To understand Chern classes, we first have to familiarize ourselves with the so-called de-Rham cohomology classes and groups, which in turn are related intimately to the homology groups introduced in Chapter 40.

An n-dimensional multiple integral of the form, for instance,

$$\int \cdots \int f(x_1, \dots, x_n)\, dx_1 \dots dx_n$$

over some n-dimensional region D is more properly expressed as an integral over D of the differential n-form $\omega = f(x_1, \dots, x_n)\, dx_1 \wedge \cdots \wedge dx_n$:

$$\int_D \omega = \int_D f(x_1, \dots, x_n)\, dx_1 \wedge \cdots \wedge dx_n \,,$$

where the antisymmetry property of the wedge products (generalizations of the vector cross products discussed in Chapter 41) does the book-keeping for the orientation of the region of integration. A most important result on integrals of differential forms is the Stokes Theorem, stated (loosely) below.

Theorem 43.1 (Stokes). *Suppose D is a region in an n-dimensional oriented manifold M, and ω is a differential $(n-1)$-form on M, then*

$$\int_D d\omega = \int_{\partial D} \omega \,, \tag{43.1}$$

where ∂D is the boundary of D with the induced orientation. If $\partial D = \emptyset$ (the empty set), then the integral on the right-hand side is zero.

This theorem subsumes the Fundamental Theorem of calculus (for one-dimensional integrals), Green's Theorem (for two-dimensional integrals), and the Gauss Theorem and Stokes Theorem (for three-dimensional integrals) in vector calculus.

Next we define the notions of closed and exact differential forms.

Definition 43.1. *If a differential form ω (of arbitrary order) on a manifold M is such that $d\omega = 0$, then it is said to be a* ***closed form****. If an n-form ω on M is such that $\omega = d\lambda$, where λ is an $(n-1)$-form on M, then ω is said to be an* ***exact form****.*

These are generalizations of the curl-free (vector) fields and conservative (vector) fields, respectively, in vector calculus. A very common example in physics is the notion of a force field $\boldsymbol{F}(\boldsymbol{x})$, corresponding to which we have the "work" one-form $\boldsymbol{F}(\boldsymbol{x}) \cdot d\boldsymbol{x}$. Indeed, from (41.11), a closed 1-form corresponds precisely to a curl-free vector field; while an exact 1-form ω entails the existence of a (single-valued) scalar (potential) function λ such that $\omega = d\lambda$, and consequently the integral of ω over any closed loop vanishes:

$$\oint \omega = \oint d\lambda = 0 \,. \tag{43.2}$$

Equation (41.8) implies that *an exact form is always closed.* Applied to the case of a 1-form, this result is equivalent to the fact that *a conservative vector field is always curl-free.* This raises the interesting question of whether the converse is also true, that is, whether a curl-free field is always conservative. When the vector field is defined on the 3-dimensional Euclidean manifold, the answer is in the affirmative, as stated in most classical mechanics textbooks. However, the following simple example will show that, for manifolds with different topologies from that of 3-d Euclidean space, this need not be true. Consider the 1-form

$$\omega = \sin^2\theta \, d\theta \,, \tag{43.3}$$

defined on the unit circle in $\mathbb{R}^2$, that is, on the 1-sphere S^1, with the local coordinate chosen to be the polar angle θ ($0 \le \theta \le 2\pi$). Since the top form on this manifold is a 1-form (the manifold being 1-dimensional), all 1-forms on the manifold are necessarily closed. So ω is closed. But on choosing a closed loop C to be S^1 itself, we have

$$\int_C \omega = \int_0^{2\pi} \sin^2\theta\, d\theta = \pi \neq 0 . \tag{43.4}$$

Thus ω is not exact. It may be objected that since we can write $\sin^2\theta\, d\theta = dU$, where $U(\theta) = \theta/2 - \sin^2\theta/4$, ω should be exact. But this expression for U is valid only *locally*, and fails to be globally valid because $U(0) \neq U(2\pi)$. Thus U is not a 0-form over all of S^1. On the other hand, the 1-form $\omega = \cos\theta\, d\theta$ is both closed and exact, as can be verified easily. In this case, $\omega = dU$, where $U(\theta) = \sin\theta$, a single-valued function on S^1. As we shall see, the fact that there are closed forms on S^1 that are not exact is dictated by the topological property that there are closed paths in S^1 which are *not* boundaries. Indeed, the closed path $C = S^1$ is not the boundary of anything in S^1.

Problem 43.1 Consider the 1-form on the two-dimensional manifold $M = \mathbb{R}^2 - \{0\}$ (the two-dimensional Euclidean plane with the origin removed) given by

$$\omega = -\frac{\alpha y}{2\pi r^2}\, dx + \frac{\alpha x}{2\pi r^2}\, dy ,$$

where $r^2 = x^2 + y^2$ and α is a positive constant. Show that in polar coordinates the vector field corresponding to this 1-form is

$$\boldsymbol{F}(\boldsymbol{r}) = \frac{\alpha}{2\pi r}\, \hat{\boldsymbol{\theta}} ,$$

and hence that

$$\oint_C \boldsymbol{F}(\boldsymbol{r}) \cdot d\boldsymbol{r} \neq 0 ,$$

where C is the circle $x^2 + y^2 = R^2$, $R \neq 0$ (say, in the counter-clockwise direction). On the other hand, show by direct calculation of the exterior derivative of ω that $d\omega = 0$. Thus $\boldsymbol{F}(\boldsymbol{r})$ is curl-free but not conservative (ω is closed but not exact). What are the topological properties of M that account for this fact?

Problem 43.2 Consider the 1-form on a 2-torus T^2 given by

$$\omega = A(\theta_1) d\theta_1 + B(\theta_2) d\theta_2 ,$$

where the local coordinates θ_1 and θ_2, with $0 \leq \theta_1, \theta_2 \leq 2\pi$, are the angles parametrizing the position of a point on a loop around the "inner tube" of the torus and a loop around the cross section of the tube, respectively. The functions $A(\theta_1)$ and $B(\theta_2)$ are both periodic functions with period 2π. Show that ω is closed, that is, $d\omega = 0$. Consider the loops C_1 and C_2 characterized above, and the integrals

$$\oint_{C_1} \omega = \int_0^{2\pi} A(\theta_1) d\theta_1 , \quad \oint_{C_2} \omega = \int_0^{2\pi} B(\theta_2) d\theta_2 .$$

State the conditions under which ω is not exact. What are the topological properties of the 2-torus that lead to the possible non-exactness of ω?

Since an exact form is always closed but a closed form is not necessarily exact, we can introduce the notion of the de Rham cohomology group as follows.

Definition 43.2. *Let $Z^r(M)$ and $B^r(M)$ be the vector spaces (over the integers) of closed and exact differential r-forms on a manifold M, respectively. The quotient space*

$$\boxed{H^r(M,\mathbb{Z}) \equiv \frac{Z^r(M)}{B^r(M)}} \tag{43.5}$$

is called the r-th **de Rham cohomology group** *with integer coefficients of the manifold M.*

Note that one can similarly define cohomology groups $H^r(M,\mathbb{R})$ with real coefficients.

If $\alpha \in Z^r(M)$, we will denote the correspond cohomology class by $[\alpha] \in H^r(M,\mathbb{Z})$. Two elements $\alpha, \alpha' \in Z^r(M)$ belong to the same class (are cohomologous to each other) if $\alpha - \alpha' \in B^r(M)$, that is, if $\alpha - \alpha' = d\beta$, for some $(r-1)$-form β. The group operation of $H^r(M,\mathbb{Z})$ (an addition) is defined by

$$[\alpha] + [\beta] \equiv [\alpha + \beta] \,, \tag{43.6}$$

which makes $H^r(M,\mathbb{Z})$ an abelian group. If $[\alpha] \in H^r(M,\mathbb{Z})$ and $[\beta] \in H^r(M,\mathbb{Z})$, one can also define an exterior product by

$$[\alpha] \wedge [\beta] \equiv [\alpha \wedge \beta] \,, \tag{43.7}$$

which is an element in $H^{r+s}(M,\mathbb{Z})$. This is indeed an exterior product since it follows the same rule for ordinary exterior products [cf. (41.5)]:

$$[\alpha] \wedge [\beta] = (-1)^{rs} [\beta] \wedge [\alpha] \,. \tag{43.8}$$

This exterior product then extends to the direct sum

$$H^*(M,\mathbb{Z}) = \sum_{\oplus r} H^r(M,\mathbb{Z}) \,, \tag{43.9}$$

called the **cohomology ring** of M.

Problem 43.3 Show that the product rule (43.7) makes sense, that is, show that

1) $d(\alpha \wedge \beta) = 0$;
2) If $[\alpha] = [\alpha']$ and $[\beta] = [\beta']$, then $[\alpha \wedge \beta] = [\alpha' \wedge \beta']$.

Problem 43.4 Verify (43.8).

The de Rham cohomology group $H^r(M,\mathbb{Z})$ is so named because of the so-called **de Rham's Theorem** (for a proof of this deep theorem, see, for example, Singer and Thorpe 1976):

Theorem 43.2 (de Rham). *Suppose M is a compact manifold. Then the r-th homology group $H_r(M,\mathbb{Z})$ [recall (40.21)] and the r-th de Rham cohomology group $H^r(M,\mathbb{Z})$ are dual vector spaces to each other, with the pairing bewteen them given by*

$$\langle\, [c] \,|\, [\omega] \,\rangle \equiv \langle\, c \,|\, \omega \,\rangle \equiv \int_c \omega \,, \tag{43.10}$$

for $c \in Z_r(M)$ (the group of r-cycles in M) and $\omega \in Z^r(M)$.

The soundness of the above definition for the pairing (denoted by a Dirac bracket) is guaranteed by the Stokes Theorem (Theorem 43.1). Indeed, suppose $[c] = [c']$ and $c - c' = \partial c''$. Then, for any $\omega \in H^r(M,\mathbb{Z})$,

$$\begin{aligned}\langle\, [c'] \,|\, [\omega] \,\rangle &= \langle\, c' \,|\, \omega \,\rangle = \langle\, c - \partial c'' \,|\, \omega \,\rangle = \langle\, c \,|\, \omega \,\rangle - \langle\, \partial c'' \,|\, \omega \,\rangle \\ &= \langle\, c \,|\, \omega \,\rangle - \langle\, c'' \,|\, d\omega \,\rangle = \langle\, c \,|\, \omega \,\rangle = \langle\, [c] \,|\, [\omega] \,\rangle \,,\end{aligned} \tag{43.11}$$

where the fourth equality follows from Stokes Theorem and the fifth from the fact that $d\omega = 0$. Likewise, suppose $[\omega] = [\omega']$, so that $\omega - \omega' = d\omega''$. Then, for any $[c] \in H_r(M,\mathbb{Z})$,

$$\begin{aligned}\langle\, [c] \,|\, [\omega'] \,\rangle &= \langle\, c \,|\, \omega' \,\rangle = \langle\, c \,|\, \omega - d\omega'' \,\rangle = \langle\, c \,|\, \omega \,\rangle - \langle\, c \,|\, d\omega'' \,\rangle \\ &= \langle\, c \,|\, \omega \,\rangle - \langle\, \partial c \,|\, \omega'' \,\rangle = \langle\, c \,|\, \omega \,\rangle = \langle\, [c] \,|\, [\omega] \,\rangle \,.\end{aligned} \tag{43.12}$$

Note that using the Dirac bracket notation, the Stokes Theorem can be written as

$$\langle\, c \,|\, d\omega \,\rangle = \int_c d\omega = \int_{\partial c} \omega = \langle\, \partial c \,|\, \omega \,\rangle \,. \tag{43.13}$$

Thus *the exterior derivative d and the boundary operator ∂ can be regarded as adjoint operators of each other.* Because of de Rham's Theorem, the results (40.23) to (40.26) for the first homology groups apply to the respective first cohomology groups also. In addition, we can list the following results, stated in terms of first cohomology groups but valid for the respective homology groups also:

$$H^1(\mathbb{R}^3,\mathbb{Z}) = H^1(\mathbb{R}^2,\mathbb{Z}) = H^1(S^2,\mathbb{Z}) = \{0\} \,, \tag{43.14}$$

$$H^1(S^1,\mathbb{Z}) = H^1(\mathbb{R}^2 - \{0\},\mathbb{Z}) = \mathbb{Z} \,. \tag{43.15}$$

The de Rham cohomology groups derive from the differentiable structure of a manifold, while the homology groups are purely topological in character. Thus the de Rham Theorem establishes a deep connection between the local and global properties of a manifold. This is a recurrent and central theme in modern differential geometry and its applications to physics. We can now appreciate better the assertion made earlier in this chapter, that *the existence of closed but non-exact forms on a differentiable manifold relates directly to the topological property that there are cycles that are non-boundaries in the manifold.*

Problem 43.5 Verify (43.14) and (43.15) by means of de Rham's Theorem, that is, by considering the corresponding first homology groups, or the topology of closed loops, on the different manifolds.

We now recall the general definition of the curvature matrix of 2-forms Ω on a general complex vector bundle $\pi : E \to M$ (corresponding to a connection matrix of 1-forms ω) given by (41.42):

$$\Omega = d\omega - \omega \wedge \omega \,, \tag{43.16}$$

where the wedge product indicates both matrix and exterior multiplication, and, as discussed in Chapter 41, the minus sign above has to be used with the convention that in a particular matrix element ω_i^j, the lower(upper) index is the row(column) index of the matrix ω. If the dimension of the fiber space (a complex vector space) is q, then the matrices ω and Ω are both $q \times q$. We also recall the important facts that under a gauge (frame field) transformation [cf. (41.26)]

$$\mathbf{e}'_i = g_i^j(x)\mathbf{e}_j \,, \quad (i = 1, \dots, q) \,, \tag{43.17}$$

where $\{\mathbf{e}_i\}$ is a local frame field and x represents the local coordinates of M, the connection transforms non-tensorially as [cf. (41.32)]

$$\omega' = (dg)g^{-1} + g\omega g^{-1} \,, \tag{43.18}$$

while the curvature transforms tensorially as [cf. (41.46)]

$$\Omega' = g\Omega g^{-1} \,. \tag{43.19}$$

The transformation property (43.19), sometimes called an **adjoint transformation**, led Chern to consider the expansion of the form

$$\boxed{\det\left(1 + \frac{i}{2\pi}\Omega\right) = 1 + \cdots + C_j(\Omega) + \cdots + C_q(\Omega)} \quad , \tag{43.20}$$

where $C_j(\Omega)$ is a differential $2j$-form globally defined on M. The expansion has to terminate with $C_q(\Omega)$ at most since Ω itself is a $q \times q$ matrix of 2-forms. In addition, if $\dim(M) = n$, $C_j(\Omega)$ vanishes for $2j > n$, since the top form on M is an n-form. Obviously, for any $g \in GL(q;\mathbb{C})$,

$$1 + \frac{i}{2\pi}\, g\Omega g^{-1} = g\left(1 + \frac{i}{2\pi}\,\Omega\right) g^{-1}\,; \tag{43.21}$$

and so

$$\det\left(1 + \frac{i}{2\pi}\, g\Omega g^{-1}\right) = \det\left(1 + \frac{i}{2\pi}\,\Omega\right)\,. \tag{43.22}$$

It follows that the $C_j(\Omega)$ are **adjoint-invariant** (or **ad-invariant**):

$$C_j(g\Omega g^{-1}) = C_j(\Omega)\,. \tag{43.23}$$

It can also be shown (although the proof will not be given here) that $C_j(\Omega)$ are real differential forms:

$$(C_j(\Omega))^* = C_j(\Omega)\,. \tag{43.24}$$

This is guaranteed by the factor i in the left-hand side of (43.20). By direct expansion of the left-hand side of (43.20) (using the general definition of the determinant of a matrix) one can show that

$$C_j(\Omega) = \frac{1}{j!}\left(\frac{i}{2\pi}\right)^j \sum_{1\le \alpha_r,\beta_r\le q} \delta^{\alpha_1\dots\alpha_j}_{\beta_1\dots\beta_j}\, \Omega^{\beta_1}_{\alpha_1}\wedge\cdots\wedge\Omega^{\beta_j}_{\alpha_j}\,. \tag{43.25}$$

The following explicit expressions follow from the above equation:

$$C_0(\Omega) = 1\,, \tag{43.26}$$

$$C_1(\Omega) = \left(\frac{i}{2\pi}\right) Tr\,\Omega\,, \tag{43.27}$$

$$C_2(\Omega) = \frac{1}{2}\left(\frac{i}{2\pi}\right)^2 [Tr\,\Omega\wedge Tr\,\Omega - Tr\,(\Omega\wedge\Omega)]\,, \tag{43.28}$$

$$\vdots \tag{43.29}$$

$$C_q(\Omega) = \left(\frac{i}{2\pi}\right)^q \det\,\Omega\,, \tag{43.30}$$

where $Tr\,A$ is the trace of the matrix A.

Problem 43.6 Verify (43.25) by expanding the left-hand side of (43.20) and using the general definition of the determinant of a matrix:

$$\det(A) = \sum_{\sigma\in S_n} (sgn\,\sigma)\, a_1^{\sigma(1)}\dots a_n^{\sigma(n)}\,,$$

where S_n is the symmetric group of the permutations of n objects [cf. Chapter 29].

Problem 43.7 Use (43.25) to verify (43.26) to (43.30).

The importance of the quantities $C_j(\Omega)$ rests on the following central result (stated without proof) in differential geometry, which is a special case of the so-called **Chern-Weil Theorem**:

Theorem 43.3 (**Chern-Weil**). *Suppose* $\pi : E \to M$ *is a* q*-dimensional complex vector bundle on a smooth manifold* M*, and* Ω *is the curvature associated with a given connection* ω *on* E*. Then*

i) $C_j(\Omega)$ *is closed, that is,*

$$dC_j(\Omega) = 0\,, \quad (0 \leq j \leq q)\,. \tag{43.31}$$

ii) Suppose $\tilde{\omega}$ *is another connection on* E *with corresponding curvature* $\tilde{\Omega}$*, then* $C_j(\tilde{\Omega}) - C_j(\Omega)$ *is exact, that is, there exists a differential* $(2j-1)$*-form* Q *on* M *such that*

$$C_j(\tilde{\Omega}) - C_j(\Omega) = dQ\,. \tag{43.32}$$

This is a most remarkable result. It implies that $C_j(\Omega)$ *determines a de Rham cohomology class (an element in the de Rham cohomology group* $H^{2j}(M)$*), independent of the choice of the connection.* The $2j$-form $C_j(\Omega)$ is called the j-th **Chern characteristic class** (or **Chern class**) of the complex vector bundle $\pi : E \to M$. In fact, the factor 2π on the left-hand side of (43.20) is a normalization constant that makes the Chern classes cohomology classes with integer coefficients (see, for example, Kobayashi and Nomizu 1969, Vol. II, Chapter XII):

$$C_j(\Omega) \in H^{2j}(M, \mathbb{Z})\,. \tag{43.33}$$

In other words, we have another remarkable result, that *the integrals of Chern classes over integral cycles in* M *are always integers* (a result known as the **integrality condition** of Chern classes). These integers are known as **Chern numbers**. When they appear in physics applications, as we have seen in the examples of the magnetic charge and the Hall conductivity [cf. (38.15) and (42.102)], they are called **topological quantum numbers**. The integrality condition also implies the following useful result (stated here without proof):

$$\int_M C_{i_1}(\Omega) \wedge \cdots \wedge C_{i_p}(\Omega) = m \in \mathbb{Z}\,, \tag{43.34}$$

for even-dimensional manifolds M with $dim\, M = n$ and $i_1 + \cdots + i_p = n/2$.

Since the Chern classes are defined in terms of the curvature Ω, they are calculated by means of local data, but yet they determine de Rham cohomology classes, which, by the de Rham Theorem discussed above, are **topological invariants** revealing purely global (topological) information on M (and E). This is a marvellous example of the deep relationship between the local (differential) and global (topological) properties of a differentiable manifold, and this is what makes the Chern classes useful in physics applications where topology plays a fundamental role.

Chapter 44

Chern-Simons Forms: The Fractional Quantum Hall Effect, Anyons and Knots

The Chern-Simons form can be used as the Lagrangian in an effective field theory to describe the physics of fractional quantum Hall systems. It turns out that such a theory does not depend on the metric (rulers and clocks) of the space-time on which it is formulated, and is hence a good example of what is called a **topological quantum field theory** (Nash 1991). In this chapter we will provide introductory accounts of the physics of the fractional quantum Hall effect, the mathematical origin of the Chern-Simons forms (which arise from the Chern classes descussed in the last chapter), and how the Chern-Simons forms achieve the stated goal. In the process, we will encounter the physical concepts of **fractional statistics** and **anyons**, and get a glimpse of the relevance of **knot theory** in the fractional quantum Hall effect. We will also present briefly the astounding results of the Chern-Simons-Witten Theory, which display the deep connections between Chern-Simons topological quantum field theories and knot invariants in 3-dimensional topology in the form of the **Jones polynomials**.

Recall from (42.7) and (42.8) that the Hall conductivity σ_H can be written as

$$\sigma_H = \nu \, \frac{e^2}{h} \, , \tag{44.1}$$

where the filling factor ν is given by

$$\nu = \frac{N}{d} \, , \tag{44.2}$$

with N being the total number of electrons in an area A and d the degeneracy of a Landau level. The degeneracy is in turn given by the number of flux quanta in A:

$$d = \frac{\phi}{\phi_0} \, , \tag{44.3}$$

where $\phi = BA$ is the total magnetic flux through the area A and $\phi_0 = hc/|e|$ is the fundamental unit of flux. We thus have the charge per state e^* given by

$$e^* = \frac{eN}{d} = \nu e \,. \tag{44.4}$$

The fractional quantum Hall effect arises when the filling factor assumes fractional values of the form $\nu = m/(2n+1)$, where m and n are non-zero positive integers such that the numerator and denominator are relatively prime (for example $\nu = 1/3, 2/5, 3/7, 4/9, 5/9$, etc.), leading to corresponding Hall plateaus (cf. Fig. 42.2) observed in the Hall resistance as a function of the magnetic field B (Tsui, Stömer, and Gossard 1982).

Phenomenologically, a fractional quantum Hall system can then be thought of as a system of charged quasiparticles with magnetic flux, or charged-particle flux-tube composites (also called **charged vortices**), each carrying a fractional charge of $e^* = e/(2n+1)$ and a unit magnetic flux of ϕ_0. We will now demonstrate that a Chern-Simons gauge field A_μ in an appropriate Chern-Simons Lagrangian $\mathcal{L}_{CS}$ will endow charged quasiparticles with magnetic flux.

The theory will be formulated in a $(2+1)$-dimensional space-time manifold, where $\mu = 0$ will denote the time dimension and $\mu = 1, 2$ the planar space dimensions. We introduce the following Chern-Simons Lagrangian:

$$\mathcal{L}_{CS} = \left(\frac{e\theta}{2\phi_0}\right) \epsilon^{\mu\nu\lambda} A_\mu \partial_\nu A_\lambda + A_\mu j^\mu + \mathcal{L}_0 \,, \tag{44.5}$$

where θ is a free parameter, $\epsilon^{\mu\nu\lambda}$ is the completely antisymmetric (Levi-Civita) tensor, A_μ is the gauge potential giving rise to the magnetic field B, j^μ is the "current density" vector, and $\mathcal{L}_0$ is a Lagrangian describing other interactions not involving A_μ. The standard field theory procedure of varying the Lagrangian with respect to the gauge potential (which is beyond the scope of this book) leads to the following field equation:

$$j^\mu = -\left(\frac{e\theta}{\phi_0}\right) \epsilon^{\mu\nu\lambda} \partial_\nu A_\lambda \,. \tag{44.6}$$

Since the "time" component j^0 is the (two-dimensional) charge density, it is straightforward to calculate the total charge q within a certain area A of the system:

$$q = \int_A j^0 \, d^2x = \left(\frac{e\theta}{\phi_0}\right) \int_A (\partial_2 A_1 - \partial_1 A_2) \, d^2x = \theta e \left(\frac{\phi}{\phi_0}\right) \,. \tag{44.7}$$

We thus see that the charge associated with the unit flux (when $\phi = \phi_0$) is $q = \theta e$. The Chern-Simons Lagrangian $\mathcal{L}_{CS}$ can then be used in an effective (Ginsburg-Landau) field theory to describe the physics of fractional quantum Hall systems when $\theta = 1/(2n+1)$ $(n = 1, 2, 3, \ldots)$. A charged vortex in this theory can be pictured as a quasiparticle carrying a fractional charge of θe and a flux-tube with flux ϕ_0.

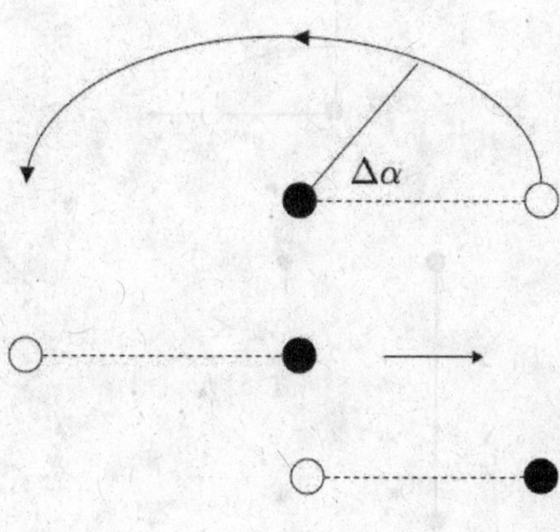

Fig. 44.1

Let us now imagine one such charged vortex looping once around another identical one on a plane. According to our discussions in Chapter 38 [cf. (38.8)], the wave function $\psi(\boldsymbol{x})$ of the quasiparticle would acquire an Aharonov-Bohm (geometrical) phase:

$$\begin{aligned}\psi(\boldsymbol{x}) \longrightarrow \psi(\boldsymbol{x}) \exp\left(\frac{2\pi i}{hc}(\theta e) \oint \boldsymbol{A}\cdot d\boldsymbol{x}\right) &= \psi(\boldsymbol{x}) \exp\left(\frac{2\pi i}{hc}(\theta e)\phi_0\right) \\ = \psi(\boldsymbol{x}) \exp(2\pi i\theta)\,. \end{aligned} \tag{44.8}$$

When θ is a fraction (for example, $1/(2n+1)$, as seen above), we have the possibility of **fractional statistics** – a quantum statistics other than either bosonic or fermionic. This can be appreciated as follows. An exchange of two identical charged vortices can be achieved geometrically on the two-dimensional (spatial) plane of a quantum Hall system by moving one of them around the other half way around and then spatially translating the two-particle system so that the final position of the looping particle (represented by the white circle in Fig. 44.1) occupies the original position of the particle that is being looped around (represented by the black circle). In doing so the wave function of the system acquires an Aharonov-Bohm phase given by

$$\psi \longrightarrow \psi\, e^{i\lambda\Delta\alpha}\,, \tag{44.9}$$

where $\Delta\alpha = \pi$ for the above described process (looping half way around) and λ is the so-called **statistical factor**. (There is no geometrical phase accumulation for the translation part of the process.) Thus, when $\lambda = 0$, we have bosons;

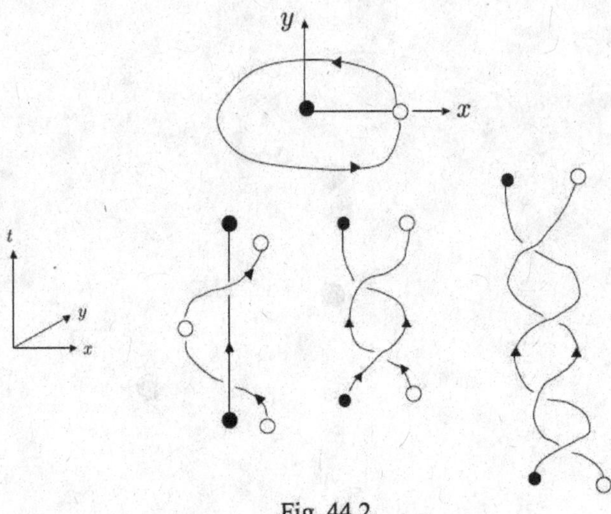

Fig. 44.2

and when $\lambda = 1$, we have fermions. When $\Delta\alpha = 2\pi$ (looping once all the way around), comparison with (44.8) shows that we must have $\lambda = \theta$. In other words, the value of θ controls the quantum statistics of the charged vortex. When $\theta = 1/(2n+1)$, as in fractional quantum Hall systems, the charged vortices must obey fractional statistics (neither bosons nor fermions). These charged vortices are then called **anyons**, a term coined by F. Wilczek (see Wilczek 1982).

It turns out that *anyons can be classified according to different representations of the so-called **braid groups** B_N, just as bosons and fermions are particles belonging to different representations of the symmetric groups S_N* (as discussed in Chapter 29). Braids arise naturally from the worldlines of anyons moving in a 2-dimensional spatial region, or a $(2+1)$-dimensional spacetime region. This can be seen by considering the simple motion of one anyon circling another identical one on the xy-plane. The worldlines of the two quasiparticles can then be topologically represented by the two strands of a braid in the $(2+1)$-dimensional (xyt)-space, as shown in Fig. 44.2.

Without going into details we mention that the braid group B_N of N strands forming braids is generated by the N elements σ_I, $I = 1, \dots, N$, as shown in Fig. 44.3. These generators satisfy the commutation relation

$$\sigma_I \sigma_J = \sigma_J \sigma_I \,, \quad \text{for} \quad |I - J| \geq 2 \,, \tag{44.10}$$

and the so-called **Artin (Yang-Baxter) relation**

$$\sigma_I \, \sigma_{I+1} \, \sigma_I = \sigma_{I+1} \, \sigma_I \, \sigma_{I+1} \,, \quad (I = 1, 2, \dots, N-2) \,. \tag{44.11}$$

The above two relations are also represented graphically in Fig. 44.3. The braid groups B_N also arise formally as the fundamental (first homotopy) groups of the configuration spaces of N identical anyons living in $(2+1)$-dimensional spacetime. Suppose X is the d-dimensional spatial manifold in which the anyons move. Then the restricted configuration space of N identical anyons (in which configurations corresponding to all N anyons occupying the same position are not allowed) is given by

$$R(X^N) \equiv \frac{X^N}{S_N} - \{(x,\ldots,x)\}\,, \tag{44.12}$$

where $(x_1,\ldots,x_N) \in X^N$ is a general point in X^N, and X^N/S_N is a quotient space with the symmetric group S_N being the permutation group of N objects. For example, if $X = \mathbb{R}^2$, then we have the following topological equivalence:

$$R(X^2) \sim \mathbb{R}^2 \times \frac{\mathbb{R}^2 - \{0\}}{I}\,, \tag{44.13}$$

where I represents the inversion of a point in $\mathbb{R}^2 - \{0\}$, considered as a vector space. The relevant mathematical result here (stated without proof) for the fundamental groups is

$$\pi_1(R(X^N)) = \begin{cases} B_N\,; & d = 2\,, \\ S_N\,; & d \geq 3\,. \end{cases} \tag{44.14}$$

This result indicates the special role that topology plays in two-dimensional physical systems.

Problem 44.1 Verify (44.13) by geometrical considerations. (Hints: A point in $\mathbb{R}^2 \times \mathbb{R}^2$ can be represented by $(\boldsymbol{r}_1, \boldsymbol{r}_2)$, where $\boldsymbol{r}_1$ is the position of anyon number 1 and $\boldsymbol{r}_2$ that of anyon number 2. This can equally well be specified by $(\boldsymbol{r}_1, \boldsymbol{r}_2 - \boldsymbol{r}_1)$. One can then identify $(\boldsymbol{r}_1, \boldsymbol{r}_2 - \boldsymbol{r}_1)$ and $(\boldsymbol{r}_1, \boldsymbol{r}_1 - \boldsymbol{r}_2)$ as a unique point in $R((\mathbb{R}^2)^2)$, for $\boldsymbol{r}_1 \neq \boldsymbol{r}_2$.)

Problem 44.2 Verify (44.14) geometrically for the case $d = 2$ by using the result (44.13) and the pictorial representation of the braid group B_2 as shown in Fig. 44.3. (Hints: Verify that, topologically, the space $(\mathbb{R}^2 - \{0\})/I$ is equivalent to the surface of an infinite cone punctured at the tip.)

There is a one-to-one mapping between each braid pictured in Figs. 44.2 and 44.3 and a **link** made up of **knots**. This equivalence is illustrated in Fig. 44.4 for a braid made up of three strands, corresponding to a link made up of two knots. The equivalence is not accidental, as the topological invariants of

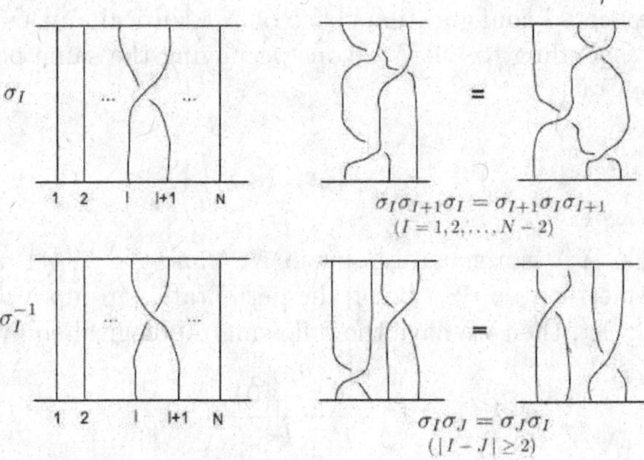

Fig. 44.3

oriented knots and links in three dimensional space, such as linking numbers and writhes, can be "measured" by the Chern-Simons form defined on 3-dimensional space. Before we illustrate this remarkable fact concretely, we need to define precisely the concepts of the linking number between two knots, and the writhes of a knot and a link.

When oriented 3-dimensional knots and links are projected onto a 2-dimensional plane as link or knot diagrams (Fig. 44.5), each crossing is marked by one strand overlaying another. Depending on the orientations of the overlaying strands, we can assign a sign [denoted by $sgn(c_i)$] of either $+1$ or -1 to the i-th crossing c_i (Fig. 44.5). The **linking number** $L(K, K')$ of an oriented link made up of two oriented knots K and K' is then defined by

$$L(K, K') \equiv \frac{1}{2} \sum_i sgn(c_i) , \tag{44.15}$$

where the sum is over all the crossings c_i arising from the link diagram. Figure 44.5 illustrates two simple and intuitive examples. The **writhe** $W(K)$ of an oriented knot K is defined by

$$W(K) \equiv N_+ - N_- , \tag{44.16}$$

where $N_+(N_-)$ is the total number of positive (negative) crossings of the knot diagram of K. Figure 44.5 again shows an example. The **writhe** $W(L)$ of an oriented link L made up of several oriented knots K_i is defined by

$$W(L) \equiv \sum_{i \neq j} L(K_i, K_j) + \sum_i W(K_i) . \tag{44.17}$$

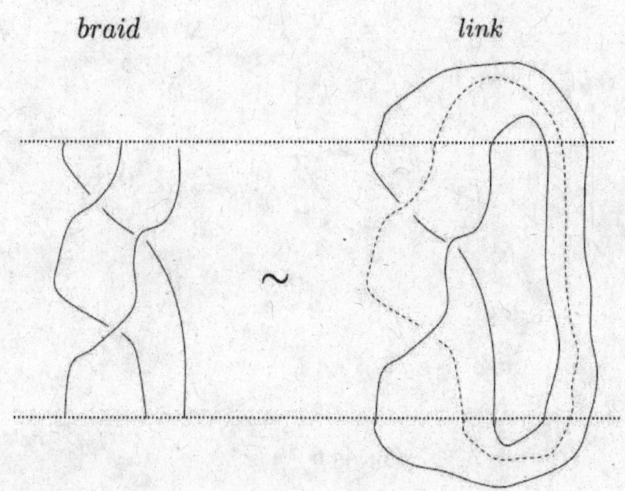

Fig. 44.4

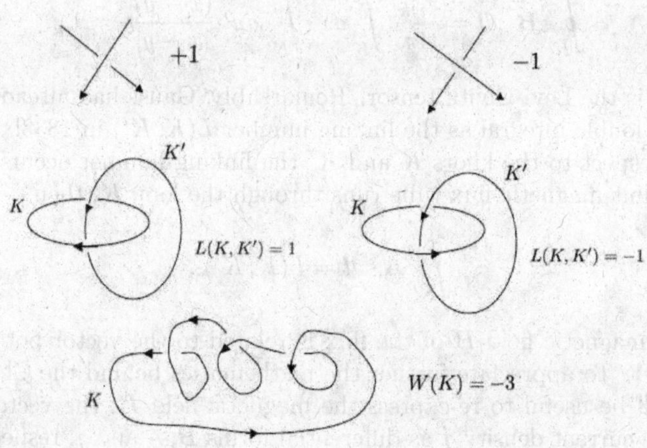

Fig. 44.5

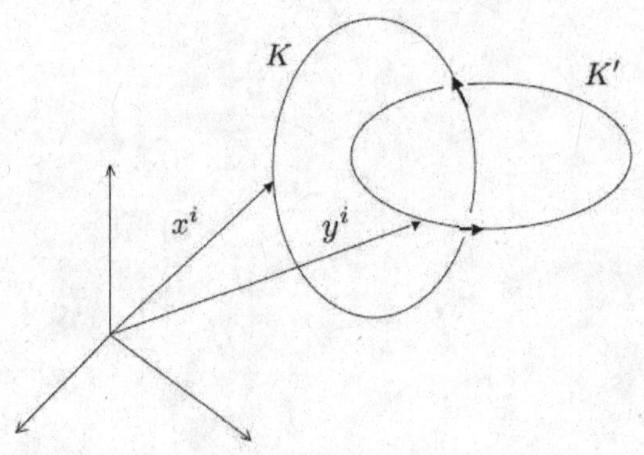

Fig. 44.6

In fact, the relevance of knots and links in physics is already evident in Maxwell's electrodynamics, in particular, in Ampere's law. Consider a unit current flowing through a closed loop K (Fig. 44.6). Then the Biot-Savart law (equivalent to Ampere's law) states that the line integral of the magnetic field $\boldsymbol{B}$ over a loop K' linking with K as shown in Fig. 44.6 is given by

$$\oint_{K'} \boldsymbol{B} \cdot d\boldsymbol{l} = \frac{\epsilon_{ijk}}{4\pi} \int_K dx^i \int_{K'} dy^j \, \frac{(x-y)^k}{|x-y|^3} = 1\,, \tag{44.18}$$

where ϵ_{ijk} is the Levi-Civita tensor. Remarkably, Gauss had already identified the above double integral as the linking number $L(K, K')$ in 1833!

With respect to the knots K and K' the linking number occurs in another way: If a unit magnetic flux tube runs through the loop K, then

$$\int_{K'} \boldsymbol{A} \cdot d\boldsymbol{l} = L(K, K')\,, \tag{44.19}$$

where the magnetic field $\boldsymbol{B}$ of the flux is related to the vector potential $\boldsymbol{A}$ by $\boldsymbol{B} = \nabla \times \boldsymbol{A}$. To appreciate further the mathematics behind the Chern-Simons form it will be useful to re-express the magnetic field $\boldsymbol{B}$, the vector potential $\boldsymbol{A}$, and the current density $\boldsymbol{j}$ as differential forms B, A and j, respectively:

$$B = B_x\, dy \wedge dz + B_y\, dz \wedge dx + B_z\, dx \wedge dy \quad \text{(a 2-form)}\,, \tag{44.20}$$

$$A = A_x\, dx + A_y\, dy + A_z\, dz \qquad \text{(a 1-form)}\,, \tag{44.21}$$

$$j = j_x\, dx + j_y\, dy + j_z\, dz \qquad \text{(a 1-form)}\,. \tag{44.22}$$

It will also be convenient to introduce the following so-called **Hodge-star** forms:

$$\star B \equiv B_x\, dx + B_y\, dy + B_z\, dz \qquad \text{(a 1-form)}\,, \tag{44.23}$$

$$\star j \equiv j_x\, dy \wedge dz + j_y\, dz \wedge dx + j_z\, dx \wedge dy \qquad \text{(a 2-form)}\,. \tag{44.24}$$

Note that the Hodge-star operator takes a 2-form to a 1-form and vice-versa, for 3-dimensional vector fields. Equations (44.18) and (44.19) can then be rewritten as

$$\int_{K'} \star B = L(K, K') \qquad \text{(unit current flowing through } K)\,, \tag{44.25}$$

$$\int_{K'} A = L(K, K') \qquad \text{(unit flux tube running through } K)\,. \tag{44.26}$$

Indeed, we have the following re-expressions of two of Maxwell's equations in terms of differential forms:

$$d(\star B) = \star j \qquad (\nabla \times \boldsymbol{B} = \boldsymbol{j})\,, \tag{44.27}$$

$$dB = 0 \qquad (\nabla \cdot \boldsymbol{B} = 0)\,, \tag{44.28}$$

where d denotes the exterior derivative [cf. (41.4) and (41.6)]. In a general situation with unit magnetic flux tubes running through each oriented knot in a link L situated in 3-dimensional space $\mathbb{R}^3$, Maxwell's equations imply that, since $B = dA$,

$$\int_{\mathbb{R}^3} A \wedge B = \int_{\mathbb{R}^3} A \wedge dA = W(L)\,, \tag{44.29}$$

where $W(L)$ is the writhe of the link L defined in (44.17). We note that $A \wedge dA$ corresponds exactly to the first term in the Chern-Simons Lagrangian introduced in (44.5), in the following sense:

$$A \wedge dA = \epsilon^{ijk}\, A_i \partial_j A_k\, dx^1 \wedge dx^2 \wedge dx^3 \qquad (i, j, k = 1, 2, 3)\,. \tag{44.30}$$

The term $A \wedge dA$ is a special example of a so-called Chern-Simons form.

Problem 44.3 For this problem refer to Fig. 44.7.

(a) If a unit current flows through knot K show that $\int_{K'} \star B = 1$.

(b) If a unit magnetic flux tube runs through knot K show that $\int_{K'} A = 1$.

(c) If unit magnetic flux tubes run through both knots K and K' show that $\int_{\mathbb{R}^3} A \wedge dA = -7$.

It is now time to trace the mathematical origin of the general Chern-Simons form in terms of the Chern classes introduced in the last chapter. Chern-Simons forms in general arise from the Chern classes by a mathematical procedure called

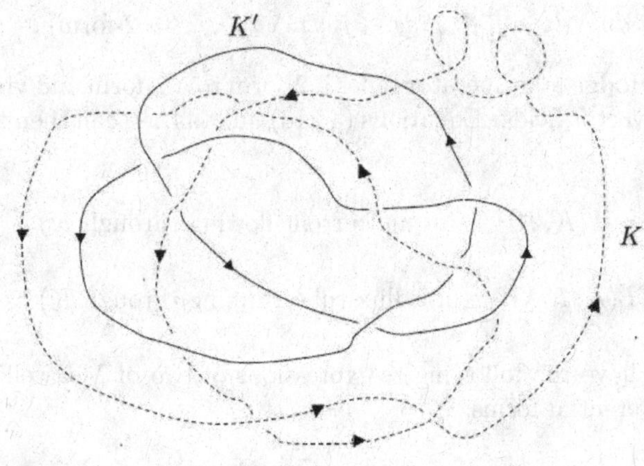

Fig. 44.7

transgression. This involves "pulling" a characteristic form, for example, the Chern classes $C_j(\Omega)$, originally defined on the base manifold M of a vector bundle $\pi : E \to M$, back to the total space P of the frame bundle $\pi' : P \to M$ associated with the vector bundle. This pullback, done by the **pullback operator** $(\pi')^*$, makes the pullback form an exact form on P:

$$(\pi')^* C_j = d(TC_j) , \tag{44.31}$$

where TC_j, denoting the transgression of the j-th Chern class, is referred to as a **secondary characteristic class** or a **Chern-Simons class**. Roughly speaking, a **frame bundle** arises from a vector bundle $\pi : E \to M$ in the following manner. For a particular point $x \in M$, the fiber space $\pi^{-1}(x)$ is by definition a vector space. A frame in the fiber space is just a particular choice of basis in that space. The set of all possible frames at all points of M is then, loosely speaking, the frame bundle $\pi' : P \to M$. The fiber space $(\pi')^{-1}(x)$ is isomorphic to some **gauge group** of invertible linear transformations on $\pi^{-1}(x)$, such as $SO(n)$, and is not a vector space. The frame bundle is thus not a vector bundle, and is an example of what is called a **principal bundle**. As it turns out the Chern classes play a very important role in the problem of the classification of principal bundles.

We will not define rigorously the notion of a pullback operator here, but will just illustrate it by means of the following procedure. Suppose in a local coordinate neighborhood of M we have a local expression ω for a given connection on the vector bundle $\pi : E \to M$ with respect to a local frame field $\{e_i\}$ on E. Under a local change of gauge, or equivalently, a local change of frame field given by [cf. (41.26)]

$$e'_i = g_i^j\, e_j \;, \quad (g_i^j) = g \in G \;, \tag{44.32}$$

where G is the gauge group referred to above, we have [cf. (41.32)] the following expression for the connection matrix A with respect to $\{e'_i\}$:

$$A = (dg)g^{-1} + g\omega g^{-1} \;. \tag{44.33}$$

The matrix elements g_i^j of the matrix g can be considered as local fiber coordinates of the frame bundle $\pi' : P \to M$. Then A becomes the pullback of ω by $(\pi')^*$, that is,

$$A = (\pi')^*(\omega) \;, \tag{44.34}$$

and is a well-defined connection matrix of one-forms on P. The corresponding curvature matrix of 2-forms [cf. (41.42) and (41.46)]

$$F = dA - A \wedge A = g\Omega g^{-1} \;, \tag{44.35}$$

where $\Omega = d\omega - \omega \wedge \omega$, is also well-defined on P.

Now consider the first Chern class [cf. (43.27)]

$$C_1(F) = \frac{i}{2\pi}\, Tr\, F \;. \tag{44.36}$$

Since

$$Tr\,(A \wedge A) = 0 \;, \tag{44.37}$$

we have

$$Tr\, F = Tr\, dA = d(Tr\, A) \;. \tag{44.38}$$

Thus

$$C_1(F) = \frac{i}{2\pi}\, d(Tr\, A) \;. \tag{44.39}$$

In other words, the first Chern class can be written as an exact form on P. Next consider the second Chern class [cf. (43.28)]

$$C_2(F) = \frac{1}{2}\left(\frac{i}{2\pi}\right)^2 [Tr\, F \wedge Tr\, F - Tr\,(F \wedge F)] \;. \tag{44.40}$$

From (44.38) we have

$$Tr\, F \wedge Tr\, F = d(Tr\, A) \wedge d(Tr\, A) = d(Tr\, A \wedge d(Tr\, A)) \;. \tag{44.41}$$

Using the rules for exterior differentiation and exterior products, and the so-called **Bianchi identity** for the curvature form F given by

$$dF = A \wedge F - F \wedge A \;, \tag{44.42}$$

we can also establish that

$$Tr\,(F \wedge F) = d(Tr\,(A \wedge F)) + Tr\,(A \wedge A \wedge F) \;, \tag{44.43}$$

$$Tr\,(A \wedge A \wedge F) = \frac{1}{3}\, d\,[Tr\,(A \wedge A \wedge A)] \;. \tag{44.44}$$

It follows that

$$C_2(F) = \frac{1}{2}\left(\frac{i}{2\pi}\right)^2 d\left[Tr\, A \wedge d(Tr\, A) - CS(A)\right], \tag{44.45}$$

where $CS(A)$, known as the **Chern-Simons 3-form**, is given by

$$CS(A) \equiv Tr\,(A \wedge F) + \frac{1}{3} Tr\,(A \wedge A \wedge A)\,. \tag{44.46}$$

Thus, as for the first Chern class $C_1(F)$, the second Chern class $C_2(F)$ is also exact on P. Using (44.35), the Chern-Simons 3-form can also be written as

$$\boxed{CS(A) = Tr\left(A \wedge dA - \frac{2}{3} A \wedge A \wedge A\right)}\,. \tag{44.47}$$

We will not give here the general expressions for the Chern-Simons form TC_j [cf. (44.31)] corresponding to the j-th Chern class. The interested reader is referred to Lam 2003.

When the fiber space of the vector bundle is a 1-dimensional vector space, the connection matrix of one-forms is a 1×1 matrix, or simply a 1-form, and the corresponding gauge field, such as the Maxwell vector potential, is an abelian field. In this case $A \wedge A \wedge A$ automatically vanishes, and the Chern-Simons 3-form reduces to our earlier expression in (44.30) [consisting of only the first term in (44.47)]. When the dimension of the fiber space is larger than one, the gauge field is non-abelian, and (44.47) gives the general expression for the Chern-Simons 3-form.

Problem 44.4 Verify (44.37).

Problem 44.5 Verify the Bianchi identity (44.42).

Problem 44.6 Verify (44.43) by calculating $d(Tr\,(A \wedge F))$ and using the Bianchi identity.

Problem 44.7 Verify the following results:

$$Tr\,(F \wedge A \wedge A) = Tr\,(A \wedge A \wedge F)\,,$$
$$Tr\,(A \wedge A \wedge A \wedge A) = 0\,.$$

Problem 44.8 Verify (44.44) by calculating $d\,[Tr\,(A \wedge A \wedge A)]$. Use the Bianchi identity and the results in the last problem.

Beyond the application of an abelian Chern-Simons theory to anyon dynamics as discussed earlier, we will conclude this chapter with a very cursory description of the mysteriously deep relationship between a topological quantum field theory based on the non-abelian Chern-Simons 3-form and 3-dimensional topology as manifested by knot invariants. This remarkable relationship was discovered by E. Witten (Witten 1989), and provided yet another beautiful example of the synergy between mathematics and physics at a fundamental level.

We have already seen some examples of knot invariants: the linking numbers and writhes. Another very useful and important invariant is known as the **Jones polynomial**, which is a topological link invariant that distinguishes chirality. It originally arose from a (purely mathematical) study of finite-dimensional von Neumann algebras, and so its connection to physics through topological quantum field theories was all the more astounding. It is quite beyond the scope of this book to engage in any systemic study of these polynomials. We will simply give two examples, that for the **trefoil knot** (clover leaf) L and its mirror image $\tilde{L}$ (see Fig. 44.8). Denoting these Jones polynomials by $V_L(t)$ and $V_{\tilde{L}}(t)$, respectively, we have

$$V_L(t) = t + t^3 - t^4\,, \qquad V_{\tilde{L}}(t) = \frac{1}{t} + \frac{1}{t^3} - \frac{1}{t^4}\,. \tag{44.48}$$

All Jones polynomials (as well as other knot invariants, for example the **Alexander polynomial**) satisfy a very important equation called the **skein relation**. Let L_+, L_- and L_0 be three links which are identical except for the interior of a small disc where they differ as shown in Fig. 44.9. Then the skein relation reads

$$tV_{L_+}(t) - \frac{1}{t}\,V_{L_-}(t) + \left(t^{1/2} - \frac{1}{t^{1/2}}\right) V_{L_0}(t) = 0\,. \tag{44.49}$$

Iteration of this relation expresses any $V_L(t)$ in terms of the Jones polynomial for a finite number of unlinked, unknotted circles, which is given by

$$V_L(t) = \left\{ -\frac{(t - t^{-1})}{(t^{1/2} - t^{-1/2})} \right\}^{p-1}\,, \tag{44.50}$$

where p is the number of unlinked, unknotted circles in the link L.

Problem 44.9 Use the skein relation (44.49) and the result (44.50) to verify the results for the Jones polynomials for the trefoil knots given by (44.48).

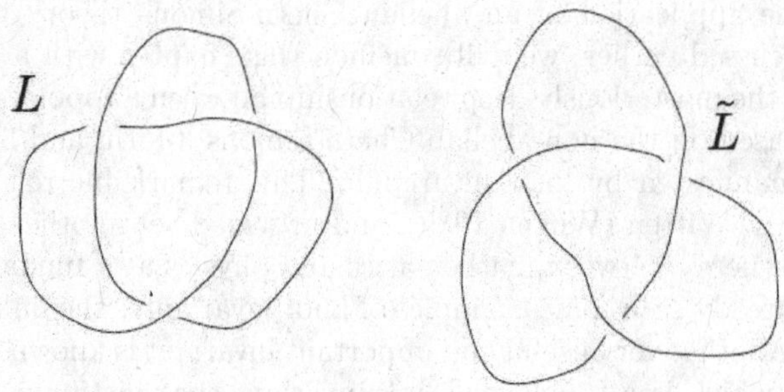

Fig. 44.8

The setup of a Chern-Simons 3-dimensional topological quantum field theory is as follows. Suppose the configuration space of a physical system is some 3-dimensional topological manifold M [such as the $(2+1)$-dimensional spacetime of a fractional quantum Hall system introduced earlier in this chapter]. Then one can introduce a connection (gauge field) A on a principal G-bundle $\pi : P \to M$ with gauge group G. Most frequently the gauge groups considered are the special unitary groups $SU(n)$ of physical interest, such as $U(1)$, $SU(2)$, etc. Let R be a certain irreducible representation of G and K be an oriented knot in M. Then one can define a non-abelian generalization of the Aharonov-Bohm geometric phase [cf. (38.5)], called a **Wilson loop** as follows:

$$W(R,K) \equiv Tr\,\mathcal{P}\left\{\exp\left(\int_K A\right)\right\}, \tag{44.51}$$

where $\mathcal{P}$ denotes a **path-ordered product**, the necessity of which is dictated by the fact that, at different points along the oriented knot K, the $m \times m$ matrices of the connection matrix of one-forms A do not necessarily commute, m being the dimension of the irreducible representation R. In the Chern-Simons quantum field theory, the quantities of physical relevance are the so-called **vacuum expectation values** (or **correlation functions**) of Wilson loops. For a link of p component knots $K_1, \ldots, K_p$, and p irreducible representations $R_1, \ldots, R_p$

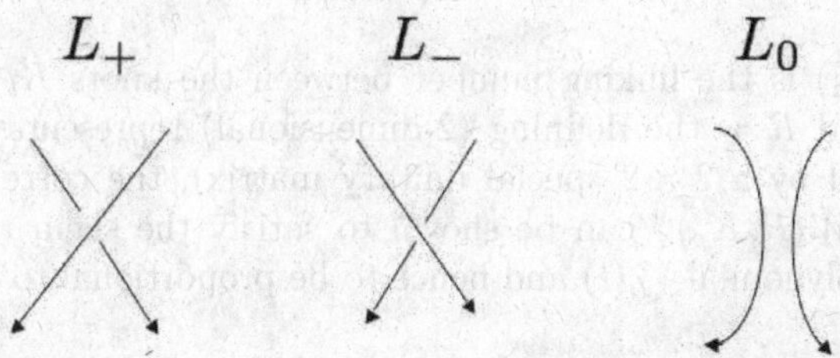

Fig. 44.9

of the gauge group G, these are defined by

$$\begin{aligned}&\langle W(R_1,K_1)\dots W(R_p,K_p)\rangle\\&=\frac{1}{Z(M)}\int_{A/G} DA\,\{W(R_1,K_1)\dots W(R_p,K_p)\}\exp\left\{\frac{ik}{4\pi}\int_M CS(A)\right\}\,,\end{aligned} \tag{44.52}$$

where k is a free parameter and $CS(A)$ is the Chern-Simons form given in (44.47). The quantity $Z(M)$ in this equation is called the **partition function** (a term derived from statistical mechanics), defined by

$$Z(M)\equiv\int_{A/G} DA\exp\left\{\frac{ik}{4\pi}\int_M CS(A)\right\}\,. \tag{44.53}$$

The integration DA in both the above equations is with respect to a **path measure** in the so-called **moduli space** of connections (gauge fields) modulo gauge transformations given by the group G [cf. (41.32)], [reminiscent of the **Feynman path integration** used in standard quantum field theory (see, for example, Zee 2003)]. There is as yet no general rigorous mathematical definition for this measure; but mathematical physicists have proceeded with heuristic manipulations quite successfully for many specific cases. A case in point is the so-called **Chern-Simons-Witten theory** (Atiyah 1990, Baez and Muniain 1994, Hu 2001), whose results we will now mention briefly.

For $G = U(1)$ (an abelian group), every irreducible representations is a map of the form $\theta \mapsto \exp(in\theta)$ $(0 \le \theta < 2\pi,\, n \in \mathbb{Z})$, so the trace and the path-ordering operations in (44.51) are irrelevant. Equation (44.52) yields

$$\langle\, W(n_1, K_1) \dots W(n_p, K_p) \,\rangle = \exp \left\{ i \sum_{l,m=1}^{p} n_l n_m \, L(K_l, K_m) \right\} , \qquad (44.54)$$

where $L(K_l, K_m)$ is the linking number between the knots K_l and K_m. For $G = SU(2)$, and $R =$ the defining (2-dimensional) representation (each $g \in G$ is represented by a 2×2 special unitary matrix), the correlation function $\langle\, W(R, K_1, \dots, W(R, K_p) \,\rangle$ can be shown to satisfy the skein relation (44.49) for the Jones polynomial $V_L(t)$ and hence to be proportional to $V_L(t)$ provided

$$t = \exp \left(-\frac{2\pi i}{k+2} \right) . \qquad (44.55)$$

In fact, the following powerful general result holds: *For any 3-dimensional topological manifold M, the topological quantum field theory based on the Chern-Simons 3-form [(44.47)] yields, through the correlation function (44.52), a different topological link invariant of M for each finite-dimensional representation of each semisimple Lie group.* Such remarkable and somewhat unexpected results beg the question: why does quantum field theory, a fundamentally physics construct that does not at present even have a rigorous mathematical foundation, seem to connect at such a deep level with purely mathematical (in particular topological) constructs, such as the topological invariants of knots and links? The attempt to answer this question satisfactorily will surely fuel much research on the part of both physicists and mathematicians in the near future, and bring about advances in both disciplines.

References

Y. Aharonov and D. Bohm, Phys. Rev., Vol. 115, 485 (1959).

H. Araki, Y. Munakata, and M. Kawaguchi, Prog. Theor. Phys., Vol. 17, 419 (1957).

M. Atiyah, *The Geometry of Physics and Knots*, Cambridge University Press (1990).

J. B. Baez and J. P. Muniain, *Gauge Fields, Knots and Gravity*, World Scientific (1994).

M. V. Berry, Proc. Roy. Soc. London, Vol. A392, 45 (1984).

R. A. Bertlmann, *Anomalies in Quantum Field Theory*, Oxford University Press (2000).

D. Bleecker, *Gauge Theory and Variational Principles*, Addison-Wesley (1981).

A. Bohm, A. Mostafazadeh, H. Koizumi, Q. Niu and J. Zwanziger, *The Geometric Phase in Quantal Systems*, Springer (2003).

M. Born, W. Heisenberg and P. Jordan, Z. Phys., Vol. 35, 557 (1925).

M. Born and P. Jordan, Z. Phys., Vol. 34, 858 (1925).

S. S. Chern, W. H. Chen, and K. S. Lam, *Lectures on Differential Geometry*, World Scientific (1999).

D. Chruściński and A. Jamiołkowski, *Geometric Phases in Classical and Quantum Mechanics*, Birkhäuser (2004).

E. U. Condon and G. H. Shortley, *The Theory of Atomic Spectra*, Cambridge University Press (1970).

V. De Alfaro and T. Regge, *Potential Scattering*, North Holland (1965).

P. A. M. Dirac, Proc. Roy. Soc. A, Vol. 109, 642 (1925).

P. A. M. Dirac, Proc. Roy. Soc. London, Vol. 133, 60 (1931).

P. A. M. Dirac, *The Principles of Quantum Mechanics*, 4 th ed., Oxford University Press (1967).

I. M. Gel'fand and N. Ya Vilenkin (translated by A. Feinstein), *Generalized Functions, Vol. 4: Applications of Harmonic Analysis*, Academic Press (1964).

V. Glaser and G. Källén, Nucl. Phys., Vol. 2, 706 (1956).

M. L. Goldberger and K. M. Watson, *Collision Theory*, Krieger (1975).

W. Heisenberg, Z. Phys., Vol. 33, 879 (1925).

D. R. Hofstadter, Phys. Rev. B, Vol. 14, 2239 (1976).

S. Hu, *Chern-Simons-Witten Theory*, World Scientific (2001).

J. D. Jackson, *Classical Electrodynamics*, John Wiley and Sons, Inc. (1966).

G. A. Jones and D. Singerman, *Complex Functions: an Algebraic and Geometric Viewpoint*, Cambridge University Press (1987).

S. Kobayashi and K. Nomizu, *Foundations of Differential Geometry*, Interscience Publishers, Vol. I (1963), Vol. II (1969).

M. Kohmoto, Annals of Physics, Vol. 160, No. 2, 343 (1985).

K. S. Lam and T. F. George, Phys. Rev. A, Vol. 33, 2491 (1986).

K. S. Lam, J. Phys. A: Math. Gen., Vol. 29, 1055 (1996).

K. S. Lam, *Topics in Contemporary Mathematical Physics*, World Scientific (2003).

T. D. Lee, Phys. Rev., Vol. 95, 1329 (1954)

W. Magnus, Communications Pure and Applied Math., Vol. 7, 649 (1954).

K. Moriyasu, *An Elementary Primer to Gauge Theory*, World Scientific (1983).

M. Nakahara, *Geometry, Topology and Physics*, 2 nd edition, Institute of Physics Publishing (2003).

C. Nash, *Differential Topology and Quantum Field Theory*, Academic Press (1991).

W. Pauli, Z. Phys., Vol. 36, 336 (1926).

M. Reed and B. Simon, *Methods of Modern Mathematical Physics, Vol. 1: Functional Analysis*, Academic Press, New York (1972); *Vol. III: Scattering Theory*, Academic Press, New York (1979).

M. Ross, ed., *Quantum Scattering Theory*, Indiana University Press (1963).

J. J. Sakurai, *Modern Quantum Mechanics*, Addison-Wesley (1985).

E. Schrödinger, Annalen der Physik (4), Vol. 79, (1926).

E. Schrödinger, *Collected Papers on Wave Mechanics by E. Schrödinger, together with His Four Lectures on Wave Mechanics*, Chelsea Publishing Co. (1982).

A. Shapere and F. Wilczek, ed., *Geometric Phases in Physics*, World Scientific (1989).

C. L. Siegel, *Topics in Complex Function Theory, Vol. 1: Elliptic Functions and Uniformization Theory*, Wiley (1969).

C. L. Siegel, *Topics in Complex Function Theory, Vol. II: Automorphic Functions and Abelian Integrals*, Wiley (1971).

B. Simon, Phys. Rev. Lett., Vol. 51, 2167 (1983).

I. M. Singer and J. A. Thorpe, *Lecture Notes in Elementary Topology and Geometry*, Undergrtaduate Texts in Mathematics, Springer-Verlag (1976).

S. Sternberg, *Group Theory in Physics*, Cambridge University Press (1994).

J. R. Taylor, *Scattering Theory: The Quantum Theory of Nonrelativistic Collisions*, Krieger (1983).

D. J. Thouless, *Topological Quantum Numbers in Nonrelativistic Physics*, World Scientific (1998).

D. C. Tsui, H. L. Stömer, and A. C. Gossard, Phys. Rev. Lett., Vol. 48, 1559 (1982).

B. L. Van der Waerden, ed., *Sources of Quantum Mechanics*, Dover Publications (1968).

K. von Klitzing, G. Doda, and M. Pepper, Phys. Rev. Lett., Vol. 45, 494 (1980).

J. von Neumann, *Mathematical Foundations of Quantum Mechanics*, Princeton University Press (1955).

G. H. Weiss and A. A. Maradudin, J. Math. Phys., Vol. 3, 771 (1962).

F. Wilczek, Phys. Rev. Lett., Vol. 49, 957 (1982).

E. Witten, Commun. Math. Phys., Vol. 121, 351 (1989).

C. N. Yang and R. Mills, Phys. Rev., Vol. 96, 191 (1954).

C. N. Yang, *Geometry and Physics*, in *To Fulfill a Vision: Jerusalem Einstein Centennial Symposium in Gauge Theories and Unification of Physical Forces*, edited by Yuval Ne'eman, Addison-Wesley (1981).

A. Zee, *Quantum Field Theory in a Nutshell*, Princeton University Press (2003).

Index